12th International Conference on Applied Physics and Mathematics (ICAPM 2022)

Journal of Physics: Conference Series Volume 2287

Singapore
18-20 February 2022

ISBN: 978-1-7138-5849-2
ISSN: 1742-6588

Printed from e-media with permission by:

Curran Associates, Inc.
57 Morehouse Lane
Red Hook, NY 12571

Some format issues inherent in the e-media version may also appear in this print version.

Institute of Physics
Dirac House, Temple Back
Bristol BS1 6BE UK

Phone: 44 1 17 929 7481
Fax: 44 1 17 920 0979

techtracking@iop.org

Additional copies of this publication are available from:

Curran Associates, Inc.
57 Morehouse Lane
Red Hook, NY 12571 USA
Phone: 845-758-0400
Fax: 845-758-2633
Email: curran@proceedings.com
Web: www.proceedings.com

TABLE OF CONTENTS

ASTROPHYSICS AND SPACE PHYSICS

APPLIED MECHANICS AND ENGINEERING

Preface

Welcome to 2022 12th International Conference on Applied Physics and Mathematics (ICAPM 2022), which should be held in Singapore during February 18-20, 2022. Due to global travel restrictions because of the COVID-19 pandemic, the conference was organized as an online meeting. Despite this, despite the different time zones of participants, despite some unavoidable technical hazards the conference matched by far the expectation of researchers and specialists to be a place of discussions, a vitrine of the new achievements in the field of applied Physics and Mathematics.

It has been successfully held in 11 consecutive years since 2011, respectively. In the past five years, ICAPM 2017, ICAPM 2018, ICAPM 2019, ICAPM 2020, and ICAPM 2021 Conference Proceedings have been archived into IOP Journal of Physics: Conference Series, indexed by ISI Web of Science (Conference Proceedings Citation Index (CPCI), Scopus, Engineering Information (EI), etc.

ICAPM 2022 consists of participants from all over the world, such as Albania, Canada, China, Egypt, France, Hungary, India, Iraq, Japan, Malaysia, Peru, South Korea, Thailand, UK, USA, etc. Thanks to all authors, the Technical Program Committee members and reviewers the accepted paper for presentation in conference and in these proceedings are selected after a tight and rigorous peer review process assuring promotion of outstanding research works from universities, research institutes, industry from all the world.

The questions during the conference session that followed each presentation and discussions that some of them has initiated constitutes a step forward in dissemination of this achievements and the embryonies of new research and new collaboration of this scientific community. We could regret the impossibility for personal discussion that an in-person conference offers, but nevertheless we could be sure that at least some of these discussions initiated during the on-line conference shall continue in private or in other occasions.

Turning towards the future would like to just thank all the public, and the organizing committee for the fantastic job during these days and looking forward the next conference that we hope to be in person.

Conference Chair
Prof. Po-Liang Liu
National Chung-Hsing University, Taiwan

Journal of Physics: Conference Series **2287** (2022) 011001 doi:10.1088/1742-6596/2287/1/011001

Committees

Conference Chairs

Naoyuki Ishimura, Chuo University, Japan

Po-Liang Liu, National Chung-Hsing University, Taiwan

Program Chairs

Kuo-Chih Chuang, Zhejiang University, China

Jen Chuan Tung, Chang Gung University, Taiwan

Publicity Chairs

Xiuhong Li, Beijing Normal University, China

Remedios de Dios Bulos, De La Salle University, Philippines

TPC Members

Abdul Ghaffar Jaffar, Balochistan University of Information Technology, Pakistan

Abdur Rashid, Gomal University, Pakistan

Agus Maman Abadi, Yogyakarta State University, Indonesia

Ahmed Elsaid, Egypt-Japan University of Science and Technology, Egypt

Alexandra B. Santos-Putungan, University of the Philippines Los Banos, Philippines

Badam Singh Kushvah, Indian Institute of Technology (ISM), India

Balqees Aldabbagh, University of Technology, Iraq

Chaiya Luengviriya, Kasetsart University, Thailand

Cheng-Kang Le, Cheng Shiu University, Taiwan

Chontita Rattanakul, Mahidol University, Thailand

Chukiat Chaiboonsri, Chiang Mai University, Thailand

Darmesah Gabda, Universiti Malaysia Sabah, Malaysia

Darwin Barayang Putungan, University of the Philippines Los Banos, Philippines

Ernest Macalalad, Mapua University, Philippines

Farhad Ali, Kohat University of Science and Technology Kohat, Pakistan

Fouad Abolaban, King Abdulaziz University, Saudi Arabia

Fuaada Mohd Siam, Universiti Teknologi Malaysia, Malaysia

Ibrahim Rose Irnawaty, Unversity Sains Islam Malaysia, Malaysia

Ilona Tregub, Financial University under the Government of the Russian Federation, Russia

Jai Prakash Verma, Netaji Subhash Chandra Bose Girls Government Degree College aliganj, India

Joginder Dhiman, Himachal Pradesh University, India

John Irish Lira, National University Manila Philippine, Philippines

Khushboo Verma, University of Lucknow, India
Kwangwon Ahn, Yonsei University, South Korea
M Sharif Uddin, Prince Sattam bin Abdulaziz University Alkharj, Saudi Arabia
Madhusudhan Mishra, North Eastern Regional Institute of Science and Technology, India
Majdi Alnowaimi, King Abdulaziz University, Saudi Arabia
Maria Cecilia Galvez, De La Salle University, Philippines
Mazhar Hussain, Government College University, Pakistan
Mohammad Ekramol Islam, Sonargaon University, Bangladesh
Mohanad Alkhasawneh, Qatar University, Qatar
Natita Wangsoh, Pathumwan Institute of Technology, Thailand
Norihan Md Arifin, Universiti Putra Malaysia, Malaysia
Noufe H. Aljahdaly, King Abdulaziz University, Saudi Arabia
Pankaj Biswas, Shri Mata Vaishno Devi University, India
Parmjit Singh Aperapar Singh, Universiti Teknologi MARA Malaysia, Malaysia
Priscilla Macansantos, University of the Philippines Baguio, Philippines
Puyang Gao, Chang'an University, China
Qi Li, Chang'an University, China
Raja Mohammad Latif, Prince Mohammad Bin Fahd University Al Khobar, Saudi Arabia
Rene Batac, De La Salle University, Philippines
Reza Mohammadi, University of Neyshabur, Iran
Roberto Montemanni, University of Modena and Reggio Emilia, Italy
Samir Hadid, Ajman University, United Arab Emirates
Sherif Ibrahim Rabia, Egypt-Japan University of Science and Technology, Egypt
Siti Suzilliana Putri Mohamed Isa, Universiti Putra Malaysia, Malaysia
Suntaree Unhapipat, Mahidol University, Thailand
Thana Yeeram, Mahasarakham University, Thailand
Umer Farooq, Jiangsu University, China
Vikas Malik, Jaypee Institute of Information Technology, India
Wachirapond Permpoonsinsup, Pathumwan Institute of Technology, Thailand
Zety Sharizat Hamidi, Universiti Teknologi MARA, Malaysia
Zik Hong Joseph Boon, Universiti Teknologi MARA Malaysia, Malaysia

Peer Review Statement

All papers published in this volume have been reviewed through processes administered by the Editors. Reviews were conducted by expert referees to the professional and scientific standards expected of a proceedings journal published by IOP Publishing.

- **Type of peer review:** Double Anonymous
- **Conference submission management system:** Morressier
- **Number of submissions received:** 100
- **Number of submissions sent for review:** 100
- **Number of submissions accepted:** 51
- **Acceptance Rate (Submissions Accepted / Submissions Received × 100):** 51
- **Average number of reviews per paper:** 2
- **Total number of reviewers involved:** 43
- **Contact person for queries:**
  Name: Po-Liang Liu
  Email: pliu@dragon.nchu.edu.tw
  Affiliation: National Chung-Hsing University

The Application of GAP Software in Constructing the Non-Normal Subgroup Graphs of Alternating Groups

N F Rahin[1, 2], N H Sarmin[2] and S Ilangovan[3]

1 Centre for Defence Foundation Studies, Universiti Pertahanan Nasional Malaysia, Kem Sg Besi, 57000 Kuala Lumpur, Malaysia.
2 Department of Mathematical Sciences, Faculty of Science, Universiti Teknologi Malaysia, 81310 UTM Skudai, Johor Bahru, Johor, Malaysia.
3 Faculty of Science and Engineering, University of Nottingham Malaysia, Jalan Broga, 43500 Semenyih, Selangor, Malaysia.
E-mail: Sheila.Ilangovan@nottingham.edu.my

Abstract. A graph in group theory is constructed by using any elements of a group as a set of vertices. Some of the properties of a group are used to form the edges of the graph. A finite group can be represented in a graph by its subgroup structure. A subgroup H of a group G is a subset of G, where H itself is a group under the same operation as in G, whereas a subgroup H is said to be a normal subgroup if its left and right cosets coincide. The non-normal subgroup graph of a group G is defined as a directed graph with a vertex set G and two distinct elements x and y are adjacent if $xy \in H$. In this paper, the non-normal subgroups graph of alternating groups for order twelve is determined by using GAP software. The graphs are found to be a union of complete digraphs and directed graphs with the same pattern depending on the order of the non-normal subgroups.

1. Introduction

A graph associated with a finite group shows a relationship among the group elements. It is constructed by how we define the adjacent vertices using the properties of the finite group. In this paper, the finite group discussed is the alternating group which is represented geometrically by its subgroup structure. Some of the graphs which are related to its subgroup were cyclic subgroup graphs defined by Devi and John Arul Singh [1]. These graphs were using the cyclic subgroups as vertices and two distinct subgroups are adjacent if one of them is a subset of the other. Then, more graphs were constructed using subgroups which are stable subgroup graph defined by Tolue [2] and normal subgroup-based power graph defined by Bhuniya and Sudip Bera [3]. Most of the research on graphs associated with the finite group were simple undirected graphs. But there were some researches on the directed graph related to group theory. For instance, the order graph was defined by Bilal and Ahmad [4] as a directed graph whose vertices are the elements of the group order classes. Next, the subgroup graph was formally defined in 2015 by Kakeri and Erfanian [5]. The subgroup graph is a simple directed graph associated with subgroups of a finite group and this graph was extended to a non-normal subgroup graph defined by Nabilah et al. [6]. Therefore, non-normal subgroup graphs are constructed for the alternating group of order twelve in this paper.

Computer software is widely used in mathematical research as a tool to compute complex computation. Besides, computation related to group theory is one of the first area in pure mathematics to be computerized too. Groups, Algorithms, and Programming software, GAP is a system for computational discrete algebra, with particular emphasis on Computational Group Theory. This software is commonly

used in research and teaching for studying groups and their representations, rings, vector spaces, algebras, combinatorial structures, codes, Lie algebras, and more. The other famous software which is mostly used by pure mathematicians are MAGMA and Maple to observe the properties of groups, constructing groups and graphs for visualization.
The following section is the preliminary section where some basic definitions in graph theory and group theory related are stated. The third section discusses the main result of this research followed by the conclusion section.

2. Preliminaries

In this section, some definitions related to this research are stated. The definitions for the symmetric and the alternating group are as follows:
Definition 1 [7] (Symmetric group)
Let A be the finite set $\{1, 2, \ldots, n\}$. The group of all permutations of A is the symmetric group on n letters and is denoted by S_n. Note that S_n has $n!$ elements, where $n! = n(n-1)(n-2) \ldots (3)(2)(1)$.
Definition 2 [8] (Alternating group)
The set of all even permutations in S_n forms a subgroup of S_n for each $n \geq 2$. This subgroup is called the alternating group of degree n, denoted by A_n.
Definition 3 [6] (The non-normal subgroup graph)
Let H be a non-normal subgroup of a group G. The non-normal subgroup graph $\Gamma_H^{NN}(G)$ is a graph in which the directed edges are formed by the vertex set G; such that the edge starts from x to y if and only if $x \neq y$ and $xy \in H$.

3. Results and discussion

In this section, the results of the non-normal subgroup graphs are determined for the alternating group of order 12, A_4. The elements of A_4 are

$$A_4 = \{(1), (13)(24), (12)(34), (14)(23), (123), (132), (234), (142), (243), (143), (134), (124)\}$$

Groups, Algorithms and Programming (GAP) is used to identify the non-normal subgroups and the directions of the edges for constructing the non-normal subgroup graph. The GAP coding is shown in the first section, while the second section is on the non-normal subgroup graph obtained for the alternating group of order 12.

3.1. GAP coding

In this subsection, the coding for the non-normal subgroups and the direction of the edges of the graph in this research are presented. The GAP software is used to obtain the direction of the edges instead of doing it manually since there are 12 elements in the alternating group. Based on definition 3, the vertices of the non-normal subgroup graph are all the elements of A_4 and the direction of the edges is determined by the product of the two distinct elements. The following GAP coding is for one of the subgroups of the alternating group of order 12.

```
gap> A4:=AlternatingGroup(4);
Alt( [ 1 .. 4 ] )
gap> e:=Elements(A4);
[ (), (2,3,4), (2,4,3), (1,2)(3,4), (1,2,3), (1,2,4), (1,3,2), (1,3,4),
(1,3)(2,4), (1,4,2), (1,4,3), (1,4)(2,3) ]
gap> AllSubgroups(A4);
[ Group(()), Group([ (1,2)(3,4) ]), Group([ (1,3)(2,4) ]),
Group([ (1,4)(2,3) ]), Group([ (2,4,3) ]),
  Group([ (1,2,3) ]), Group([ (1,4,2) ]), Group([ (1,3,4) ]),
Group([ (1,3)(2,4), (1,2)(3,4) ]), Group([ (1,3)
  (2,4), (1,2)(3,4), (2,4,3) ]) ]
gap> N:=NormalSubgroups(A4);
[ Alt( [ 1 .. 4 ] ), Group([ (1,3)(2,4), (1,2)(3,4) ]), Group(()) ]
gap> H1:=Subgroup(A4,[(1,2)(3,4)]);
```

```
Group([ (1,2)(3,4) ])
gap> Elements(H1);
[ (), (1,2)(3,4) ]
gap> for i in e do
> for j in e do;
> if i<>j then
> c:=i*j;
> d:=c in H1;
> if (d=true) then
> Print ("{",i,",",j,",",c,",",d,"},");
> fi;
> fi;
> od;
> od;
{(),(1,2)(3,4),(1,2)(3,4),true},{(2,3,4),(2,4,3),(),true},{(2,3,4),(1,2
,3),(1,2)(3,4),true},{(2,4,3),
(2,3,4),(),true},{(2,4,3),(1,2,4),(1,2)(3,4),true},{(1,2)(3,4),(),(1,2)
(3,4),true},{(1,2,3),(1,3,2),(),true},{(1,2,3),
(1,4,3),(1,2)(3,4),true},{(1,2,4),(1,3,4),(1,2)(3,4),true},{(1,2,4),(1,
4,2),(),true},{(1,3,2),(2,4,3),(1,2)
(3,4),true},{(1,3,2),(1,2,3),(),true},{(1,3,4),(1,3,2),(1,2)(3,4),true}
,{(1,3,4),(1,4,3),(),true},{(1,3)(2,4),(1,4)
(2,3),(1,2)(3,4),true},{(1,4,2),(2,3,4),(1,2)(3,4),true},{(1,4,2),(1,2,
4),(),true},{(1,4,3),(1,3,4),(),true},{(1,4,3),
(1,4,2),(1,2)(3,4),true},{(1,4)(2,3),(1,3)(2,4),(1,2)(3,4),true},
gap> LogTo();
```

From this GAP coding, there are ten subgroups of the alternating group of order twelve which consists of three normal and seven non-normal subgroups, respectively. The order of the non-normal subgroups are two and three for instance, $H_1 = \{(1),(12)(34)\}$, $H_2 = \{(1),(13)(24)\}$, $H_3 = \{(1),(14)(23)\}$, $H_4 = \{(1),(234),(243)\}$, $H_5 = \{(1),(123),(132)\}$, $H_6 = \{(1),(124),(142)\}$ and $H_7 = \{(1),(134),(143)\}$. The direction of the edges for the subgroup H_1 is given and the similar GAP coding can be applied to the other non-normal subgroups which are H_2 until H_7. Moreover, this coding also can be applied for a higher order of the alternating groups. In the next section, the non-normal subgroup graphs are constructed for each subgroup of the alternating group of order 12.

3.2. Main results

In this subsection, the results obtained from GAP software are used to construct the non-normal subgroup graph for each subgroup of the alternating group of order 12.

Proposition 3.2.1

Let G be the alternating group of order 12, A_4 and H_i be a non-normal subgroup of G with order two. Then the non-normal subgroup graph for H_i, $\Gamma_{H_i}^{NN}(G)$ is a union of two complete digraphs with two vertices and a bipartite graph as shown in figure 1.

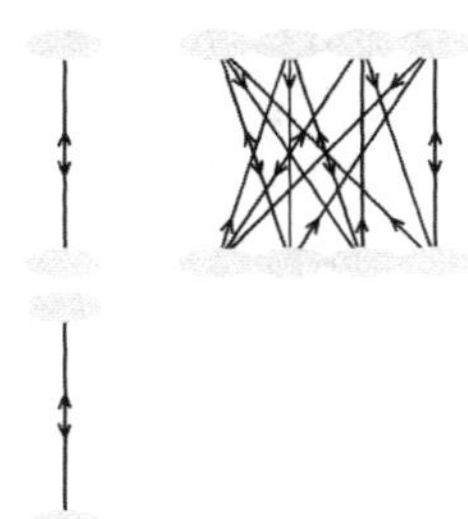

Figure 1. $\Gamma_{H_i}^{NN}(G)$ for H_i of order two.

Proposition 3.2.2
Let G be the alternating group of order 12, A_4 and H_i be a non-normal subgroup of G with order three. Then the non-normal subgroup graph for H_i, $\Gamma_{H_i}^{NN}(G)$ is shown in figure 2.

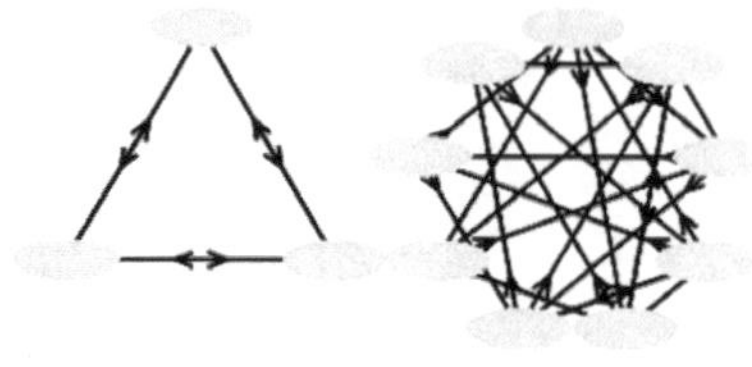

Figure 2. $\Gamma_{H_i}^{NN}(G)$ for H_i of order three.

4. Conclusion
The non-normal subgroup graph for subgroups of the alternating groups, A_4 are determined by using GAP software and the coding are presented in this paper. The non-normal subgroup graphs of A_4 have two different graphs depending on the order of the subgroups. The subgroups of A_4 with order two is a union of two complete digraphs with two vertices and a bipartite graph. Meanwhile, the subgroups of A_4 with order three is a union of a complete digraph with three vertices and a directed graph with nine vertices and 24 directed edges.

References
[1] Devi S and John Arul Singh J 2017 A Note on Cyclic Subgroup Graph of a Finite Groups *Communications in Applied Analysis* **21** No. 3 425-430
[2] Tolue B 2018 The Stable Subgroup Graph *Boletim da Sociedade Paranaense de Matematica* **36** 3 129–139
[3] Bhuniya A K and Bera S 2016 Normal subgroup Based Power Graphs of a Finite Group *Communication in Algebra* DOI: 10.1080/00927872.2016.1236122
[4] Al-Hasanat B N and Al-Hasanat A S 2019 Order Graph: A New Representation of Finite Groups *International Journal of Mathematics and Computer Science* **14** 809-819
[5] Kakeri F and Erfanian A 2015 The Complement of Subgroup Graph of a Group *Journal of Prime Research in Mathematics* **11** 55-60
[6] Rahin N F, Sarmin N H and Ilangovan S 2020 The Non-Normal Subgroup Graph of Some Dihedral Groups *ASM Science Journal* **13** 1-7
[7] Fraleigh J B 2003 *A First Course in Abstract Algebra 7th Edition* Addison-Wesley Boston 78
[8] Durbin J R 2009 *Modern Algebra: An Introduction 6th Edition* John Wiley & Sons 35

Acknowledgments
The first author would like to acknowledge Universiti Teknologi Malaysia (UTM) for the funding of this research through UTM Fundamental Research Grant (UTMFR) Vote Number 20H70 and Universiti Pertahanan Nasional Malaysia (UPNM) for the support. The authors would also like to thank the Ministry of Higher Education Malaysia (MoHE) for the financial support through Fundamental Research Grant Scheme (FRGS/1/2020/STG06/UTM/01/2) and University of Nottingham Malaysia.

A Markov Decision Process Framework for Optimal Cancer Chemotherapy Dose Selection

A O Adeyiola[1], S I Rabia[1], A Elsaid[1], S Fadel[2] and A Zakaria[3]

1 Institute of Basic and Applied Sciences, Egypt-Japan University of Science and Technology, P O Box 179, New Borg El-Arab, 21934, Alexandria, Egypt
2 Department of Clinical Oncology and Nuclear Medicine, Alexandria University, 22, El-Guish Road, El-Shatby, 21526, Alexandria, Egypt
3 Alexandria Pediatric Oncology Center, Borg El-Arab University Hospital, New Borg El-Arab, 21934, Alexandria, Egypt
abdulramon.olawale@ejust.edu.eg

Abstract. Chemotherapy is arguably the most effective treatment for treating cancer. However, it triggers toxic effects on patients due to its interaction with normal cells. In this paper, we propose a finite-horizon Markov decision process framework for optimal chemotherapy dose selection during cancer treatment. In contrast to many other research works, our model accommodates all possible patient's clinical health states that could be considered during treatment. We study the use of our model by applying it on Wilms tumor. We explore three different cases of varying transition probabilities and rewards. Our numerical results agree with clinical intuition and demonstrate the potential applicability of our model.

1. Introduction

Chemotherapy is arguably the most effective and extensively used modality for most cancers [1]. Although it is highly useful for treating cancer, it triggers toxic effects due to the interaction of the drug with normal cells. This prevents sufficient doses from being administered to obtain a complete cure [2]. Thus, it is important to design treatment strategies for cancer that will ensure a desired rate of tumor cells are damaged without overdosing the host.

Several studies have used optimal theory to model chemotherapy treatment problems. Nanda et al. [3] proposed a mathematical model that seeks to find the treatment regimens that minimize cancer cell count and deleterious effects of drug therapy on patients. They represented the disease dynamics by a system of ordinary differential equations and solve them using the fourth-order Runge-Kutta method. Pillis et al. [4] considered the dynamics of the interaction of the triplet: tumor, immune system, and anti-cancer drug. They represented the dynamics with a coupled differential equation and evaluated two optimal control strategies: a quadratic control, and a linear control. Ledzewicz and Schaettler [5] proposed a mathematical model that takes into account the heterogeneity of the tumor. They considered the interaction between two cancer cell sub-populations of different sensitivities with respect to chemotherapeutic agent and proved that administration of lower dose chemotherapy is a viable option for population with varying chemotherapeutic sensitivities.

As can be noted, all of these works mainly considered the problem of optimal chemotherapy treatment strategy that minimizes the size of tumor population without quantifying the benefit/disutility of the patient over the treatment period. In addition, the methods used are highly deterministic and do not capture effectively the stochastic dynamics of cancer chemotherapy treatment process. Therefore, we propose a Markov decision process (MDP) framework for the optimum strategy during chemotherapy treatment. MDPs are powerful analytical tools used to solve complex, stochastic and dynamic problems [6].

MDP models have found sizeable usability in modelling common chronic diseases that involve time to intervention and treatment strategy. Shechter et al. [7] considered the problem of optimal initiation of HIV treatment based on patient CD4 count that maximizes the patient's quality-adjusted lifetime. They proposed a stationary, infinite-horizon MDP model and proved the existence of a control-limit policy in terms of the patient's CD4 count. Alagoz et al. [8] presented an MDP model that seeks to find the optimal timing of living-donor liver transplantation based on the patient health state which is described by the Model for End-stage Liver Disease (MELD) score. Akhavan-Tabatabaei et al. [9] proposed a finite-horizon MDP model for cervical cancer screening decision. The optimal policy was formulated in terms of current diagnosis, age, human papillomavirus (HPV) contraction risk, and screening test result. They further evaluated their model by analyzing the trade-off between cost of screening and early treatment procedures, against the risk of letting the disease advance.

Furthermore, several studies have designed MDP models for optimal treatment strategy for cancer. Kim et al. [10] developed a finite-horizon MDP model for optimal fraction of radiotherapy based on number of organs at risk, and targeted tumor volume. They built the model mainly on medical intuition by considering some key features, as well as state-action transition. Maass et al. [11] proposed a finite-horizon MDP model for multi-modality cancer treatment based on tumor progression and normal tissue side effect. They categorized the treatment options into three types in terms of tumor reduction, risk to normal tissue, and repeatability. More recently, Bazrafshan et al. [12] used MDP framework to propose a model for optimizing cancer chemotherapy in clinical trials which captured the treatment dynamics. However, the authors only considered the toxicity level of the patient. They did not consider the utility of chemotherapy with respect to its primary purpose. Moreover, the treatment regimens are for gastric and gastroesophageal cancers and can not be used for other types of cancers.

In this study, we propose a general MDP-based framework for cancer chemotherapy treatment. Our contributions to the existing literature on cancer treatment are:

1) We propose a novel approach for optimum chemotherapy dose selection that can be used for treatment of any type of cancer.
2) Our proposed model does not only consider the tumor progression, but also considers the clinical health states of the patient.
3) We propose a multi-objective model that solves the trade-off problem between chemotherapy treatment utility and disutility.

The remainder of this paper is organized as follows: In section 2, we present our MDP model and give a detailed description of its components. In section 3, we present numerical results. Finally, we highlight our main conclusions and possible extensions in section 4.

2. MDP model

We construct a discrete-time finite-horizon MDP model with decision epochs $t \in T = \{1,2,\dots,E\}$. At each decision epoch t, the decision maker (i.e., oncologist) observes the patient clinical health states by conducting some clinical tests, such as complete blood count (CBC), temperature, weight, etc. Afterwards, the decision maker considers the tumor progression level. If an absorbing state is attained (e.g., cancer free, death, etc.) treatment is stopped and the decision process terminates. Otherwise, the decision maker decides (based on the patient clinical health state and tumor progression level) on the optimal dose of chemotherapy to be given (if any). This decision continue till an absorbing state is attained or the treatment period ends. For each action the decision maker takes, there is an immediate reward representing the treatment utility (if the tumor progression is expected to improve based on this action) or penalty to represent action disutility (if this action is expected to get the patient health state worse). Our goal is to solve the trade-off problem between maximizing the total chemotherapy treatment utility and minimizing its total disutility. A formal definition of the core components of our MDP model is given as follows:

2.1. States

The patient state is composed of two state variables: clinical health state (vector), and tumor progression state (scalar). The clinical health state is defined by the pre-assessment tests done by the decision maker at each decision epoch t, and is denoted by $\boldsymbol{c}_t \in \mathcal{C} = \{(c_1, c_2, \dots, c_N): c_i \in \{1,2,\dots,l_i\}\}$, where N is the number of pre-assessment tests, and l_i is the number of possible outcomes of the ith pre-assessment test. For any test state c_i, we assume that $c_i = 1$ represents the best clinical health state, while $c_i = l_i$, represents the worst clinical health state.

Let $\tau_t \in \mathcal{T} = \{t_0, t_1, \dots, t_M\}$ denote the tumor progression observed at decision epoch t. $\mathcal{T}$ is ordered such that $\tau_t = t_0$ is the best patient state (e.g., cancer-free) while $\tau_t = t_M$ is the worst patient state (death due to tumor progression). We note that t_0 and t_M are absorbing states.

The patient state at time t is given by $s_t = (\boldsymbol{c}_t, \tau_t) \in S$, where $S = \mathcal{C} \times \mathcal{T}$.

2.2. Actions

We denote the action space by $A = \{a_0, a_1, \dots, a_K\}$, where a_0 represents $wait$, a_K represents $full\ dose$, and $a_1, \dots, a_{K-1}$ are different fractions of chemotherapy. We assume that if a_0 is chosen, the patient's clinical health state can either improve or remain the same, while the tumor progression state can either remain the same or get worse. For any other action $a_k, k = 1,2,\dots,K$, the patient's clinical health state can either get worse or remain the same, while the tumor progression state can either get better or remain the same. We assume that for any action a_k, the clinical health and tumor progression states can only improve or get worse by at most one step.

2.3. Transition probabilities

When the decision maker chooses action $a_t \in A$ at decision epoch t for a patient in state s_t, the patient state moves to s_{t+1} at $t+1$ with probability $P_t(s_{t+1}|s_t, a_t)$. We assume that

$$P_t(s_{t+1}|s_t, a_t) = P_t^{\mathcal{C}}(\boldsymbol{c}_{t+1}|\boldsymbol{c}_t, a_t) \times P_t^{\mathcal{T}}(\tau_{t+1}|\tau_t, a_t), \quad (1)$$

where $P_t^{\mathcal{C}}(c_t, a_t)$ and $P_t^{\mathcal{T}}(\tau_{t+1}|\tau_t, a_t)$ are the transition probabilities for the clinical health state and tumor progression state, respectively. This assumption is consistent with our intuition that clinical health and tumor progression do not depend on each other but depend on the decision maker's action only.

Moreover, we order A in terms of its effect on the patient's clinical health state and tumor progression state. More specifically, we assume that

$$P_t^{\mathcal{C}}(c_{t+1}|\boldsymbol{c}_t, a_1) > P_t^{\mathcal{C}}(c_{t+1}|\boldsymbol{c}_t, a_2) > \cdots > P_t^{\mathcal{C}}(c_{t+1}|\boldsymbol{c}_t, a_K), \quad (2)$$

where $\boldsymbol{c}_{t+1}$ is a better clinical health state than $\boldsymbol{c}_t$. Also,

$$P_t^{\mathcal{T}}(\tau_{t+1}|\tau_t, a_1) < P_t^{\mathcal{T}}(\tau_{t+1}|\tau_t, a_2) < \cdots < P_t^{\mathcal{T}}(\tau_{t+1}|\tau_t, a_K), \quad (3)$$

where τ_{t+1} is a better tumor progression state than τ_t.

We note that a_0 is excluded from the above ordering because its effect on the patient's clinical health state and tumor progression state is reversed compared to any other action.

Since t_0, and t_M are absorbing states, we have

$$P_t^{\mathcal{T}}(t_0|t_0, a_t) = P_t^{\mathcal{T}}(t_M|t_M, a_t) = 1. \quad (4)$$

2.4. Reward functions

Our model includes two reward functions: immediate reward, and terminal reward. A formal description of the two rewards is given as follows:

- The immediate reward function $r_t(s_t, a_t)$ reflects the utility/disutility the decision maker incurs when the patient health state is s_t and action a_t is taken at decision epoch t. It is defined as

$$r_t(s_t, a_t) = \sum_{i=1}^{N} r_{t,i}(c_i, a_t) + r_t(\tau_t, a_t), \quad (5)$$

where $r_{t,i}(c_i, a_t)$ refers to the immediate reward regarding clinical health state c_i, and $r_t(\tau_t, a_t)$ is the the immediate reward regarding tumor progression state τ_t.

- The terminal reward function quantifies patient's utility/disutility at the end of the treatment period and is denoted by $r_{T+1}(s)$ where s is the terminal state. This function is defined as

$$r_{T+1}(s) = \sum_{i=1}^{N} r_{T+1}(c_i) + r_{T+1}(\tau), \tag{6}$$

where $r_{T+1}(c_i)$ is the terminal reward regarding clinical health state c_i, and $r_{T+1}(\tau)$ is the terminal reward regarding τ.

Recall that the two state variables, c_t and τ_t are ordered according to their benefit to the patient. Let c_i and c_i', denote two different patient clinical health states where $c_i > c_i'$, and τ and τ' denote two different tumor progression states where $\tau > \tau'$. Hence, the corresponding reward functions are assumed to follow this inequality for all t:

$$r_t(c_i, a_t) < r_t(c_i', a_t), r_t(\tau, a_t) < r_t(\tau', a_t). \tag{7}$$

2.5. Value function

The goal of our MDP model is to find the optimal strategy for cancer chemotherapy treatment. That is, we seek a rule for taking action at each state that will maximize the expected total reward of a patient over the treatment period. This can be achieved by solving the following Bellman's recursive equations for all $s_t \in S$ and $t = 1,2, \dots, T$:

$$V_t(s_t) = \max_{a_t \in A} \left\{ r_t(s_t, a_t) + \sum_{s_{t+1} \in S} P_t(s_{t+1} | s_t, a_t) V_{t+1}(s_{t+1}) \right\}, \tag{8}$$

where $V_t(s_t)$ is the patient's maximum expected total reward at the decision epoch t when the patient is in state s_t, with the boundary condition $V_{T+1}(s) = r_{T+1}(s)$.

3. Results and discussion

In this section, we present numerical study to illustrate the applicability of our proposed MDP model. Specifically, we base our numerical examples on Wilms tumor (WT). We chose WT because it is the most common primary renal tumor of all childhood cancer [13]. The incidence of WT in the United States is approximately 7 new cases for every 1 million children with top occurrence between 2 and 3 years [14]. Meanwhile, there have been some recent advances in management of WT with more than 85% overall survival rate [13]. There is need to further improve its treatment over the current risk stratification to better direct therapy beyond its present limitation [13].

For simplicity, we consider complete blood count (CBC) as the only patient's clinical health state. We denote this variable by $c \in \mathcal{C} = \{1,2\}$, where 1 and 2 represent *normal* and *abnormal* CBC, respectively. Furthermore, we represent the tumor progression states by $\mathcal{T} = \{0, 1/4, 1/2, 3/4, 1\}$, and recall that $\mathcal{T}$ is an ordered set with 0 representing the best state (cancer free) and 1 representing the worst state (death). The action space is denoted by $A = \{W, 33\%, 66\%, F\}$, where W denotes $wait$, and $33\%, 66\%$, and F denote $33\%, 66\%$, and $full$ dose, respectively. We consider a treatment period of length 4 weeks, i.e., $E = 4$. Backward induction was used to solve (8) and compute the optimal treatment policy over the treatment period. This was implemented on MATLAB using the MDP toolbox.

The remainder of this section is organized as follows: In section 3.1, we present a base case example for numerical illustration, including the transition probabilities, rewards, as well as some general assumptions made to make them clinically applicable. In section 3.2, we investigate the sensitivity of our model by exploring how changes in transition probabilities and reward function affect our optimal policy. In all our numerical examples, we consider stationary probabilities and rewards, i.e., $P_t(s_{t+1} | s_t, a_t) = P(s_{t+1} | s_t, a_t)$, and $r_t(s_{t+1} | s_t, a_t) = r(s_{t+1} | s_t, a_t)$ for all t.

3.1. Base case

Tables 1-3 contain the state-action transition probabilities for both tumor progression, and CBC. The CBC immediate rewards for all the possible actions are contained in Table 4. We penalize the decision maker for not giving $full$ dose when the patient's CBC is $normal$. Moreover, if $full$ dose is given when the patient's CBC is $abnormal$, the decision maker is penalized. The maximum reward/penalty is taken to $100/-100$, respectively. Since in WT treatment, the tumor progression is usually measured once after the surgery (by the end of treatment period), we assign 0 immediate reward for tumor progression and 0 terminal reward for CBC. We then define a linear terminal reward function for tumor progression which is given by

$$r_{T+1}(\tau) = 100(1-2\tau), \tau = 0, 1/4, 1/2, 3/4, 1. \quad (9)$$

The resulting optimal policy for our base case example is presented in Table 5. The structure of the policy is quite simple and follows the medical intuition. If at any decision epoch the patient's CBC is normal, $full$ dose should be selected. Meanwhile, at any decision epoch in the absorbing states, $wait$ is the only available action. For the intermediate tumor progression states, 33% should be selected if the patient's CBC is $abnormal$ for all decision epochs, except in the first decision epoch when the tumor progression is still considerably good ($\tau = 1/4$).

Table 1. Tumor progression state-action transition matrix for $wait$ and 33% action.

wait						33%					
τ/τ'	0	1/4	1/2	3/4	1	τ/τ'	0	1/4	1/2	3/4	1
0	1	0	0	0	0	0	1	0	0	0	0
1/4	0	0.4	0.6	0	0	1/4	0.6	0.4	0	0	0
1/2	0	0	0.4	0.6	0	1/2	0	0.5	0.5	0	0
3/4	0	0	0	0.4	0.6	3/4	0	0	0.4	0.6	0
1	0	0	0	0	1	1	0	0	0	0	1

Table 2. Tumor progression state-action transition matrix for 66% and $full$ action.

66%						*full*					
τ/τ'	0	1/4	1/2	3/4	1	τ/τ'	0	1/4	1/2	3/4	1
0	1	0	0	0	0	0	1	0	0	0	0
1/4	0	0.7	0.3	0	0	1/4	0.8	0.2	0	0	0
1/2	0	0.6	0.4	0	0	1/2	0	0.7	0.3	0	0
3/4	0	0	0.5	0.5	0	3/4	0	0	0.6	0.4	0
1	0	0	0	0	1	1	0	0	0	0	1

Table 3. CBC state-action transition matrix for all actions.

wait			33%			66%			*full*		
	c'			c'			c'			c'	
c	1	2	c	1	2	c	1	2	c	1	2
1	0.9	0.1	1	0.8	0.2	1	0.7	0.3	1	0.6	0.4
2	0.8	0.2	2	0.5	0.5	2	0.3	0.7	2	0	1

Table 4. CBC immediate rewards for all actions.

Base case					Case 1				
c	*wait*	33%	66%	*full*	c	*wait*	33%	66%	*full*
1	-100	-75	-50	100	1	-100	50	75	100
2	100	75	50	-100	2	100	-50	-75	-100

3.2. Sensitivity analysis

In this section, we investigate how changes in our model parameters affect the structure of the optimal policy. First, in subsection 3.2.1, we demonstrate how change in the immediate rewards for CBC affects the optimal policy (case 1). Then, we consider how slight increment in the effect of chemotherapy on tumor progression could affect the optimal policy (case 2).

3.2.1. Case 1. In contrast to the assumptions made on the immediate reward in section 3.1, we penalize the decision maker for selecting any dose of chemotherapy in an $abnormal$ CBC state. Meanwhile, the decision maker receives a reward for administering chemotherapy when the patient's CBC is $normal$, and penalized for taking action $wait$. The immediate rewards used for this example are contained in Table 6. We maintain all other parameters and assumptions used in our base case example.

As shown in Table 6, the structure of the optimal policy is particularly simple. Irrespective of the patient's tumor progression state, the optimal action is $full$ when the patient's CBC is normal for all decision epochs. In addition, whenever the patient is in an $abnormal$ CBC state, the optimal action is $wait$ for all decision epochs and all tumor progression states. This agrees with the general chemotherapy guidelines of the Renal Tumor Study Group of the International Society of Paediatric Oncology (SIOP-RTSG) 2016 Umbrella Protocol [15].

3.2.2. Case 2. Here, we consider a scenario whereby the effect of chemotherapy on tumor progression states is slightly higher than initially assumed in the base case. We therefore increased the probability that the patient's tumor progression state will increase by 0.1. Thus, the following transition probabilities are changed: $P(0|1/4,33\%) = 0.7$, $P(1/4|1/2,33\%) = 0.6$, $P(1/2|3/4,33\%) = 0.5$, $P(0|1/4,66\%) = 0.8$, $P(1/4|1/2,66\%) = 0.7$, $P(1/2|3/4,66\%) = 0.6$, $P(0|1/4,F) = 0.9$, $P(1/4|1/2,F) = 0.8$, $P(1/2|3/4,F) = 0.7$.

We maintain all other assumptions and parameters used in our base case example, including the state-action transition matrix for $wait$. We note that the structure of the optimal policy (Table 7) is similar to that of our base case example, except that the optimal action is $wait$ in the first and second decision period when patient's CBC state is $abnormal$ and the tumor progression is in a considerably good state.

As can be noted in Tables 5-7, the optimal action is to give $full$ dose whenever the clinical health state of the patient is normal. This is consistent with the clinical intuition that as much as the patient is in a good clinical health state, full dose chemotherapy should be initiated. Furthermore, we note that the proposed approach gives a policy that prioritizes treating the patient for Wilms tumor, but with a careful approach by initiating 33% dose when the patient's clinical health state is abnormal.

4. Conclusion

With cancer being one of the major life threatening diseases worldwide, many patient's characteristics are considered during its treatment. In this study, we demonstrated the potential of using mathematical framework for solving complex sequential decision making problems in cancer chemotherapy treatment by proposing a finite horizon MDP model. Specifically, our model accommodates all patient's clinical factors that could be considered during chemotherapy treatment. This will enhance decision making during cancer chemotherapy treatment and can be personalized for each individual patient based on their profile. As a case study, we applied our model to Wilms tumor. Although we considered only one clinical factor, our results are consistent with clinical intuition and demonstrate the potential applicability of our model. In the future, we could apply our model to other types of cancers, such as leukemia, colorectal cancer, lymphoma, etc. We note that the problem of curse of dimensionality may arise as the patient clinical health state increases. This problem can be solved by approximate dynamic programming techniques [16].

Table 5. Optimal policy for the base case.

	$\tau = 0$				$\tau = 1/4$				$\tau = 1/2$				$\tau = 3/4$				$\tau = 1$			
	Decision period																			
c	1	2	3	4	1	2	3	4	1	2	3	4	1	2	3	4	1	2	3	4
1	F	F	F	F	F	F	F	F	F	F	F	F	F	F	F	F	F	F	F	F
2	W	W	W	W	W	33%	33%	33%	33%	33%	33%	33%	33%	33%	33%	33%	W	W	W	W

Table 6. Optimal policy for case 1.

	$\tau = 0$				$\tau = 1/4$				$\tau = 1/2$				$\tau = 3/4$				$\tau = 1$			
	Decision period																			
c	1	2	3	4	1	2	3	4	1	2	3	4	1	2	3	4	1	2	3	4
1	F	F	F	F	F	F	F	F	F	F	F	F	F	F	F	F	F	F	F	F
2	W	W	W	W	W	W	W	W	W	W	W	W	W	W	W	W	W	W	W	W

Table 7. Optimal policy for case 2.

	$\tau = 0$				$\tau = 1/4$				$\tau = 1/2$				$\tau = 3/4$				$\tau = 1$			
	Decision period																			
c	1	2	3	4	1	2	3	4	1	2	3	4	1	2	3	4	1	2	3	4
1	F	F	F	F	F	F	F	F	F	F	F	F	F	F	F	F	F	F	F	F
2	W	W	W	W	W	W	33%	33%	33%	33%	33%	33%	33%	33%	33%	33%	W	W	W	W

References

[1] Abbas Z and Rehman S 2018 An overview of cancer treatment modalities *Neoplasm* **1** 139–57
[2] Iliadis A and Barbolosi D 2000 Optimizing drug regimens in cancer chemotherapy by an efficacy–toxicity mathematical model *Computers and Biomedical Research* **33** 211–26
[3] Nanda S, Moore H and Lenhart S 2007 Optimal control of treatment in a mathematical model of chronic myelogenous leukemia *Mathematical Biosciences* **210** 143–56
[4] de Pillis L G, Gu W, Fister K R, Head T, Maples K, Murugan A, Neal T and Yoshida K 2007 Chemotherapy for tumors: An analysis of the dynamics and a study of quadratic and linear optimal controls *Mathematical Biosciences* **209** 292–315
[5] Ledzewicz U and Schättler H 2014 On optimal chemotherapy for heterogeneous tumors *Journal of Biological Systems* **22** 177–97
[6] Schaefer A J, Bailey M D, Shechter S M and Roberts M S 2004 Modeling medical treatment using Markov decision processes *Operations Research and Health Care: A Handbook of Methods and Applications* ed M L Brandeau, F Sainfort and W P Pierskalla (Boston, MA: Springer US) pp 593–612
[7] Shechter S M, Bailey M D, Schaefer A J and Roberts M S 2008 The optimal time to initiate HIV therapy under ordered health states *Operations Research* **56** 20–33
[8] Alagoz O, Maillart L M, Schaefer A J and Roberts M S 2007 Choosing among living-donor and cadaveric livers *Management Science* **53** 1702–15
[9] Akhavan-Tabatabaei R, Sánchez D M and Yeung T G 2016 A Markov decision process model for cervical cancer screening policies in Colombia *Medical Decision Making* **37** 196–211
[10] Kim M, Ghate A and Phillips M H 2009 A Markov decision process approach to temporal modulation of dose fractions in radiation therapy planning *Physics in Medicine & Biology* **54** 4455
[11] Maass K and Kim M 2020 A Markov decision process approach to optimizing cancer therapy using multiple modalities *Mathematical Medicine and Biology: A Journal of the IMA* **37** 22–39
[12] Bazrafshan N and Lotfi M M 2020 A finite-horizon Markov decision process model for cancer chemotherapy treatment planning: an application to sequential treatment decision making in clinical trials *Annals of Operations Research* **295** 483–502
[13] Oostveen R M and Pritchard-Jones K 2019 Pharmacotherapeutic management of Wilms tumor: An Update *Pediatric Drugs* **21** 1–13
[14] Lorenzo A and Lopes R I 2017 Recent advances in the management of Wilms' tumor *F1000Research* **6** 670
[15] Vujanić G M, Gessler M, Ooms A H A G, Collini P, Coulomb-l'Hermine A, D'Hooghe E, de Krijger R R, Perotti D, Pritchard-Jones K, Vokuhl C, van den Heuvel-Eibrink M M and Graf N 2019 Publisher Correction: The UMBRELLA SIOP–RTSG 2016 Wilms tumour pathology and molecular biology protocol (Nature Reviews Urology, (2018), 15, 11, (693-701), 10.1038/s41585-018-0100-3) *Nature Reviews Urology* **16** 563
[16] Powell W B 2007 *Approximate Dynamic Programming: Solving the curses of dimensionality (Wiley series in probability and statistics)* (USA: Wiley-Interscience)

Some Generalizations of Random Broken and Pick-up Stick Problems

Yucheng Liu[1], Zhichao Wang*[2]

1 United World College of Changshu China, Jiang Su, China
2 University of California San Diego, La Jolla, CA
E-mail: ycliu20@uwcchina.org; zhw036@ucsd.edu

Abstract. We consider sticks with random lengths, which are further randomly broken into several pieces. The probability that these sticks can form a polygon is computed in this article. This Broken Pick-up Stick Problem was first asked in [1] and we give a general formula to solve this problem. Besides, we study the probability that any three sticks with independent and uniformly distributed lengths can form a triangle. Our results generalize the famous Spaghetti Problem and the Pick-up Stick Problem in probability. We present a way to transfer these continuous probability problems into discrete problems and apply combinatorial methods to address the discrete version of the problems.

1. Introduction

There is an old probability problem called the Spaghetti Problem, which is asking the probability of forming a triangle when a stick was broken into three pieces randomly. It has been already studied and extended by many mathematicians for a long time back to the early 19th century. We refer [2,3,4] for the history and background of this famous probability problem. Here, we repeat the original question as follows.

Problem 1.1 (The classical Spaghetti Problem). Randomly and independently pick two points on a spaghetti stick with length 1, and divide the stick into three parts. Calculate the probability that the resulting three pieces form a triangle.

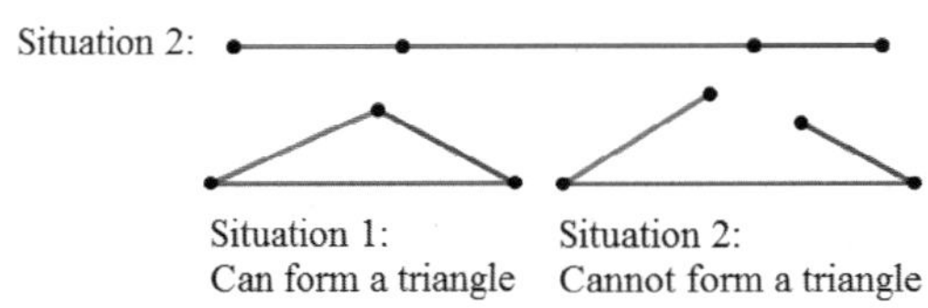

Figure 1. Two examples on the situations of Spaghetti Problem: three sticks can form a triangle in Situation 1, but cannot in Situation 2.

As the examples shown in Figure 1, to calculate the probability for Problem 1.1, it suffices to find out the condition that the lengths of three broken sticks can form a triangle. We can either compute the probability by some integrals or use the geometry to compute the feasible region. In fact, this classical problem has many different methods to solve and the answer should be $\frac{1}{4}$. Moreover, this problem has plenty of generalizations in probability theory and has a close relation with random geometry (see [5]).

Illinois Geometry Lab has been studying a sequence of generalizations of the broken sticks problem for several years in [4, 6]. Also, there are several problems related to Problem 1.1 are raised in The American Mathematical Monthly, for instance [7, 8, 9]. Even now, there are still several open and interesting questions in this direction. Most of the extensions require new techniques in probability and geometry to address. Our current work tackle some of these open problems using discrete probability and combinatorics.

One extension is asking the probability of forming a k-gon if we randomly pick $k-1$ points in a stick. It was solved by Carlos and Emiliano in [7], which states as follows.

Theorem 1.2 (See [10]). Given any natural number $n \geq 3$. If we randomly and independently select $n-1$ points in a stick to break the stick into n pieces, then the probability that such n pieces form a n-gon is

$$1 - \frac{n}{2^{n-1}} \tag{1}$$

There are other forms of related problems, for instance, asking the possibility that any three broken sticks can form a triangle when we randomly divide a stick into more than three pieces. In our work, we will consider a related but more general problem, which is called Arbitrariness Problem. Besides, instead of these broken stick problems, there is another kind of problem, called Pick-up Stick Problem. The Pick-up Stick Problem discusses the probability of forming a polygon with multiple random length sticks. In the simplest form of this question, it discusses the situation of three sticks with random lengths but without breaking. Here, we can consider these lengths are independently and identically distributed by Uniform(0, 1). The answer to this question is $\frac{1}{2}$, obviously different the simplest Spaghetti Problem. A general version of this question is stated in the following Theorem 1.3. An example of a such Pick-up Problem is shown in Figure 2.

Theorem 1.3 (Pick-up Stick Problem [1]). Randomly and independently select k sticks, which k is a positive integer, with lengths chosen in a uniform distribution Uniform(0, 1). Then, the probability that these k sticks can form a k-gon is

$$1 - \frac{1}{(k-1)!} \tag{2}$$

Following Theorem 1.2 and Theorem 1.3, this article will generalize these results and prove probability formulas for several problems related to broken and pick-up stick problems. We will combine the Spaghetti Problem and the Pick-up Stick Problem together to study a more general problem stated below. In terms of [1], we call this new problem as the Broken Pick-up Sticks Problem. As a result of our, we fully solve the open question raised in the last section of [1].

Problem 1.4 (The Broken Pick-up Sticks Problem). Given any n natural number $a_1, \ldots, a_n$, such that $k := \sum_{i=1}^{n} a_i \geq 3$. We consider n sticks with independent lengths distributed by

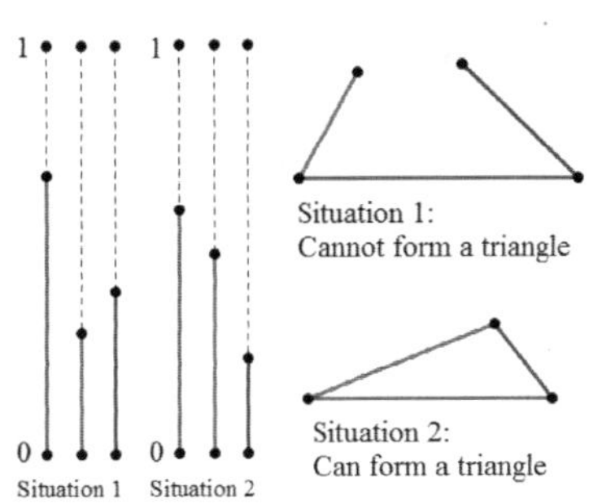

Figure 2. Two examples of the situations of Pick-up Problem stated in Theorem 1.3, the triangle can be formed in situation 1 but cannot in situation 2.

Uniform(0, 1). For $1 \leq i \leq n$, we randomly break the i-th stick into a_i pieces. Denote the collection of all k broken sticks by a set $\mathbf{Stick}(a_1, \ldots, a_n)$. Then, what is the probability that these k parts can form a k-gon?

Furthermore, we are interested in the Arbitrariness Problem in this topic. More precisely, we want to know the probability that any three broken sticks can form a triangle. This arbitrariness problem in the set-up of Theorem 1.2 has been studied in [11] by order statics and exponential random variables. In this article, we focus on the arbitrariness of Pick-up Stick Problem as below.

Problem 1.5 (Arbitrariness of Forming Triangles in Pick-up Problem). Given n sticks with independent lengths distributed by Uniform(0, 1), what is the probability that any three sticks in these n sticks can form a triangle?

1.1. Notation

In this work, we will always use $\mathbb{P}(A)$ to describe the probability that the situation fit the problem (or the probability of success), which is the answer of our problem. Here, A will always be denoted as the successful event corresponding to each problem. Relatively, $\mathbb{P}(A^c)$ as the complement describe the probability that the situation is not fit. We denote Ω as the probability sample space. In the discrete case, the number of all possible cases is $\#\{\Omega\}$. We use $\mathbb{P}_{dis}$ to represent the probability in discrete situations. We denote the rounding function $[x]$ as the integer part of x.

2. Main results

In this article, we will present our main results: the solutions to Broken Pick-up Stick Problem 1.4 and Arbitrariness of forming triangles in Pick-up Problem 1.5. Before introducing the theorem, we first define some notions which we will use in the proofs.

Definition 2.1. Given natural numbers a_i, for $1 \leq i \leq n \in \mathbb{N}$. We randomly and independently pick up n sticks with lengths distributed by Uniform(0, 1). Denote the length of each stick by l_i. Then for each l_i, we randomly break it into a_i pieces. Let $\mathbf{Stick}(a_1, \ldots, a_n)$ be the collection of all broken sticks.

Theorem 2.2 (The Broken Pick-up Sticks Problem). Given positive integers n and $a_1, \ldots, a_n$. Consider the collection of sticks $\mathbf{Stick}(a_1, \ldots, a_n)$ defined above. Let $k = \sum_{i=1}^{n} a_i$. The probability that all k parts in $\mathbf{Stick}(a_1, \ldots, a_n)$ are able to form a k-gon is

$$\mathbb{P}(A) := \mathbb{P}(\mathbf{Stick}(a_1, \ldots, a_n)\ can\ form\ a\ k-gon) = 1 - \frac{1}{n} \cdot \sum_{i=1}^{n} \frac{a_i!}{(a_i+n-2)!} \cdot \frac{1}{2^{a_i-1}} \tag{3}$$

Remark 2.3. We can consider some special cases in Theorem 2.2. If n sticks have the same number of broken parts, $\mathbf{Stick}(a, a, \ldots, a)$, then the successful probability of forming an an-gon is

$$\mathbb{P}(A) = 1 - \frac{a!}{2^{(a-1)}(a+n-2)!} \tag{4}$$

Hence, when $a = 1$, we return to Pick-up Stick Problem in Theorem 1.3. Besides, when $n = 1$, for $\mathbf{Stick}(a)$, the formula becomes

$$\mathbb{P}(A) = 1 - \frac{a}{2^{(a-1)}} \tag{5}$$

which is same as the original Spaghetti Problem for polygons in Theorem 1.2.

For Problem 1.5, we compute the probability that any three sticks in Pick-up Stick Problem form a triangle. The proof is finished in Section 4. The idea of the proof is analogous to the proof of Theorem 2.2 but has more difficulties in compute the limit of the probability for discrete problem.

Theorem 2.4 (Arbitrariness of Forming Triangles in Pick-up Problem). For any integer $n \in \mathbb{N}$, if we independently select n sticks with lengths from the distribution Uniform(0, 1), then the probability that any three sticks among them can form a triangle is

$$\frac{1}{2^{(n-2)}} \tag{6}$$

Based on Definition 2.1, the collection of sticks with random lengths is $\mathbf{Stick}(1, \ldots, 1)$ for n ones. Like the previous question, we find a sufficient and necessary condition for any three sticks form a triangle, namely the sum of shortest two sticks are longer than the longest stick. See Lemma 4.1. Theorem 2 in [4] and Theorem 1 in [11] studied the Arbitrariness of Forming Triangles in Broke Stick

Problem. Consider an n-piece randomly broken stick. The probability that any triple of pieces chosen from the n pieces can form a triangle is

$$\frac{1}{\binom{2n-2}{n}} \tag{7}$$

which is much less than the probability in Theorem 2.4 for sufficiently large n. This indicates that it is more difficult to guarantee any three sticks form triangle in a broken stick with lots of pieces.

3. The proof of Theorem 2.2

Our original Problem 1.4 has too many variables to study. Instead, we first introduce another problem, which is a simple and discrete version of Problem 1.4. In this case, all the lengths become integers.

Problem 3.1 (The broken pick-up sticks problem - Simple Version). Given a positive integer n, for the group of $\mathbf{Stick}(a_1, \ldots, a_n)$ with a_i chosen from the natural numbers, every $\mathbf{Stick}(a_i)$ with $a_i \in \{a_i > 1 | a_i \in \mathbb{N}^+\}$ has a fixed length 1 and every $\mathbf{Stick}(1)$ has the length randomly chosen from the real number range (0, 1). There are totally $\sum_{i=1}^{n} a_i = k$ parts, where the i-th stick is randomly divided into a_i parts. What is the probability that these h parts are able to form a k-gon?

Proposition 3.2 (Necessary and Sufficient Condition of Not Forming). In Problem 3.1, the k-gon cannot be form if and only if the longest part L in $\mathbf{Stick}(a_1, \ldots, a_n)$ is longer than the half of the total length of all sticks $\sum_{i=1}^{n} l_i$.

In other words,

$$\mathbb{P}(A^c) = \mathbb{P}\left(L \geq \frac{1}{2}\sum_{i=1}^{n} l_i\right) \tag{8}$$

On the basis of this, we will separate the Problem 3.1 into three situations.

Lemma 3.3 (Two a_i more than 1 situation). If the $\mathbf{Stick}(a_1, \ldots, a_n)$ includes two or more parameters $a_i > 1$, for example $\mathbf{Stick}(3, 4, 1, 1)$, must have the probability $\mathbb{P}(A^c) = 0$ and $\mathbb{P}(A) = 1$.

Proof. Because that there are at least two sticks, suggest they are a_p and a_q have parameters more than one, they have the fixed-length 1, according to the problem, we have:

$$\sum_{i=1}^{n} l_i > (l_p + l_q) = 2 \times 1 \tag{9}$$

Because that all the parts in $\mathbf{Stick}(a_i > 1)$ are less than one and l_j with $a_j = 1$ is less than one, it does not exist a part, which is a side of the h-gon, more than one.

$$0 < L < 1 \tag{10}$$

$$2L < (l_p + l_q) < \sum_{i=1}^{n} l_i \tag{11}$$

And, with the Eq.(8),

$$\mathbb{P}(A^c) = 0 \tag{12}$$

$$\mathbb{P}(A) = 1 - \mathbb{P}(A^c) = 1 \tag{13}$$

Secondly, because we exclude the situation with two or more $a_i > 1$, we discuss the situation that there is only one $a_i > 1$ in the group of $\mathbf{Stick}(a_1, \ldots, a_n)$. However, we will firstly discuss a special situation in this second part, which is $\mathbf{Stick}(a, 1)$, with $a > 1$, instead of a $\mathbf{Stick}(a)$ and a lot of $\mathbf{Stick}(1)$.

Lemma 3.4 (Probability of $\mathbf{Stick}(a, 1)$). Given a positive integer $a \geq 1$, for the $\mathbf{Stick}(a, 1)$ with a length-1-stick, which separated into a parts and a stick with random length between 0 and 1, the probability of forming a h-gon is:

$$\mathbb{P}(A) = 1 - \frac{1}{2^{a-1}} \tag{14}$$

Proof. In this part, we will use discrete thought. Suggest the $\mathbf{Stick}(a)$ has a fixed length of m, which is large enough, instead of 1 and $\mathbf{Stick}(1)$ has an integer length randomly chosen in $(0, m)$. We will divide the $\mathbf{Stick}(a)$ into a parts by randomly select $(a-1)$ points on the integer points of this m-

length stick. We suggest the **Stick**(1) has a length of k, which satisfies $k \in (0, m)$ and k is an integer. To solve this situation from a discrete side, we put up a discrete version of this Problem 3.1:

Problem 3.5 (Discrete version problem). In the group of **Stick**$(a, 1)$ with $a \geq 1$ chosen from the natural numbers, containing a stick with length m, a large enough integer, divided into a parts and a stick with length k, which is an integer selected from range$(0, m)$. There are totally $(a + 1)$ parts. What's the probability of these $(a + 1)$ parts are able to form a $(a + 1)$-gon?

As Eq.(8), we can show the probability by Proposition 3.2:

$$\mathbb{P}_{dis}(A^c) = \frac{\#\left\{L \geq \frac{1}{2}\sum_{i=1}^{n} l_i\right\}}{\#\{\Omega\}} \tag{15}$$

According to the problem, we can get the formula of the total length:

$$\sum_{i=1}^{n} l_i = m + k \tag{16}$$

We consider the total length $\sum_{i=1}^{n} l_i$ as an important value in the whole proof, since that we need to determine the range of $\mathbb{P}(A^c)$ depending on the total length. So that, we made the total length fixed by considering k a parameter, instead of a variable, as a condition.

In this assumption, we can get the universal set $\#\{\Omega\}$ easily. We only need to consider the **Stick**(a) with length m, selecting $(a - 1)$ points from $(m - 1)$ points in total.

$$\#\{\textbf{Stick}(1) = k\} = \binom{m-1}{a-1} \tag{17}$$

In order to satisfy the A^c situation, the longest part(L):

$$L \geq \frac{\sum_{i=1}^{n} l_i}{2} = \frac{m+k}{2} \tag{18}$$

Notice that the longest side cannot be the **Stick**(1), since $k < m$ and $k < \frac{k+m}{2}$, L can only be a part of **Stick**(a), which is divided to be very long. Suggest f_i to be the parts divided in **Stick**(a), with $i \in \{1, 2, \dots, a\}$, the longest side L can be any of f_i and being any of them will not affect the final answer. So that, we can assume $L = f_a$, but we need to multiple a to the result of this step. Thus, we get L on the rightest side of **Stick**(a) and no points in the range that L has occupied. So, all the points chosen in **Stick**(a) should be in the range (0, $[\frac{m-k}{2}]$):

$$\mathbb{P}_{dis}\left(L \geq \frac{1}{2}\sum_{i=1}^{n} l_i\right) = \binom{[\frac{m-k}{2}]}{a-1} \tag{19}$$

Start, we use Eq.(17) times a and divides by Eq.(19) to get $\mathbb{P}(A^c)$ under this **Stick**$(1) = k$ condition:

$$\mathbb{P}_{dis}(A^c|\textbf{Stick}(1) = k) = \frac{a\binom{[\frac{m-k}{2}]}{a-1}}{\binom{m-1}{a-1}} \tag{20}$$

Because the conditions **Stick**$(1) = k$, $k \in \{1, \dots, n\}$ are independent, we can add up to the sum of them and calculate the mean to eliminate this condition and find $\mathbb{P}_{dis}(A^c)$.

$$\mathbb{P}_{dis}(A^c) = \sum_{i=1}^{m-1} \mathbb{P}_{dis}(A^c|\textbf{Stick}(1) = i) \times \frac{1}{m-1} = \sum_{i=1}^{m-1} \frac{a\binom{\left[\frac{m-i}{2}\right]}{a-1}}{\binom{m-1}{a-1}} \times \frac{1}{m-1}$$

$$= \frac{a}{\binom{m-1}{a-1}} \sum_{i=1}^{m-1} \binom{[\frac{m-i}{2}]}{a-1} \times \frac{1}{m-1} \tag{21}$$

Lemma 3.6 (Approximation of the number of combinations). For s, which is big enough, and a fixed smaller r, the approximation of $\binom{S}{r}$ is equal to $\frac{s^r}{r!}$.

As we have m big enough and a relatively small a, we can approximate the formula above into the following versions:

$$\mathbb{P}_{dis}(A^c) = \frac{a\times(a-1)!}{(m-1)\times(m-1)^{(a-1)}}\sum_{i=1}^{m-1}\frac{1}{(a-1)!}\times\left[\frac{m-i}{2}\right]^{(a-1)} = \frac{a}{2^{(a-1)}\times(m-1)^a}\sum_{i=1}^{m-1}(m-i)^{(a-1)} \quad (22)$$

To this step, we can no longer perform discrete simplifications. For the discrete version, we got the answer above, with m the length of the fixed $\mathbf{Stick}(a>1)$ and a the parameter. We are going to require m to be positive infinite instead of a very large integer and thus separate the summation part by using calculus. We take the limit of $\mathbb{P}_{dis}(A^c)$ to get $\mathbb{P}(A^c)$, requiring m to approach positive infinity.

$$\mathbb{P}(A^c) = \lim_{m\to+\infty}\frac{a}{2^{(a-1)}\times(m-1)^a}\sum_{i=1}^{m-1}(m-i)^{(a-1)} = \frac{a}{2^{(a-1)}}\times\lim_{m\to+\infty}\frac{\sum_{i=1}^{m-1}(m-i)^{(a-1)}}{(m-1)^a}$$
$$= \frac{a}{2^{(a-1)}}\times\lim_{(m-1)\to+\infty}\frac{1}{a} = \frac{1}{2^{a-1}} \quad (23)$$

So that, we can get the probability by solving this formula, keeping a as a parameter.

$$\mathbb{P}(A^c) = \frac{1}{2^{a-1}} \quad (24)$$

$$\mathbb{P}(A) = 1-\frac{1}{2^{a-1}} \quad (25)$$

Finally, we will discuss the remaining situation, which also contain only one parameter $a>1$, but many 1 in $\mathbf{Stick}(a,1,1,\dots,1)$. Notice that, the position of a parameter in $\mathbf{Stick}(a,1,1,\dots,1)$ has no affect to the result probability since they all have a same meaning. Suggest the $\mathbf{Stick}(a,1,1,\dots,1)$ with one a and α 1s can be written in the form of $\mathbf{Stick}(a;\alpha)$.

Lemma 3.7 (Probability of $\mathbf{Stick}(a;\alpha)$). For the $\mathbf{Stick}(a;\alpha)$ with a length-1-stick and α sticks with randomly chosen length:

$$\mathbb{P}(A) = 1-\frac{a!}{(a+\alpha-1)!}\times\frac{1}{2^{a-1}} \quad (26)$$

Proof. As the previous proves, we know that the $\mathbf{Stick}(1)$s in $\mathbf{Stick}(a;\alpha)$ cannot be the longest part to satisfy $L>\frac{1}{2}\sum_{i=1}^{n}l_i$, since:

$$l_i < 1, \text{for } i>1 \quad (27)$$

$$l_i < \frac{l_i+1}{2} \quad (28)$$

So that we only need to discuss the probability in the $\mathbf{Stick}(a)$, which is exactly the same in the proof above. The only limitation is that the total length should be under 2.

$$1 > L > \frac{1}{2}\sum_{i=1}^{n}l_i \quad (29)$$

$$\sum_{i=1}^{n}l_i < 2 \quad (30)$$

So, $\sum_{i=1}^{n}l_i$ for $i>1$, α sticks in total, should be less then 1. Still, we use the discrete thought to solve this problem, suggesting that $\mathbf{Stick}(a)=m$ and the α $\mathbf{Stick}(1)$'s have length $\{k_1,\dots,k_\alpha\}$. With the experience of the previous situation, we discuss every possible $K=\sum_{i=1}^{\alpha}k_i$ for the $\mathbb{P}_{dis}(A^c)$. When $K>m$, $\mathbb{P}(A^c)=0$, so that, the range of possible K is less than m and more than α, since every $\mathbf{Stick}(1)$ should at least have length 1.

$$\mathbb{P}_{dis}(A^c) = \sum_{i=b}^{m}\mathbb{P}_{dis}(A^c|\textstyle\sum_{i=1}^{\alpha}k_i = K)\cdot\mathbb{P}(\sum_{i=1}^{\alpha}k_i = K) \quad (31)$$

To find $\mathbb{P}(\sum_{i=1}^{\alpha}k_i=K)$, we can assume a K-length stick, with is the combination of all $\mathbf{Stick}(1)$'s, and find the way of dividing it into α parts. So that we need to choose $(\alpha-1)$ points among $(K-1)$ points. Then, it is easy to find the universal set of this event, which asks the total ways of forming the α $\mathbf{Stick}(1)$'s. That is m^{α} since each stick has m kinds of lengths to choose.

$$\mathbb{P}(\sum_{i=1}^{\alpha} k_i = K) = \frac{\#\{P(K)\}}{\#\{\Omega\}} = \frac{\binom{K-1}{\alpha-1}}{m^{\alpha}} \tag{32}$$

Put this formula into $\mathbb{P}_{dis}(A^c)$, we can get:

$$\mathbb{P}_{dis}(A^c) = \sum_{K=\alpha}^{m}\left(\frac{a\binom{[\frac{m-k}{2}]}{a-1}}{\binom{m-1}{a-1}} \times \frac{\binom{K-1}{\alpha-1}}{m^{\alpha}}\right) \tag{33}$$

By using Lemma 3.6, we can replace the combination parts into the form of powers

$$\mathbb{P}_{dis}(A^c) = \sum_{K=\alpha}^{m}\left(\frac{a(m-K)^{(a-1)}}{2^{(a-1)}\cdot(m-1)^{(a-1)}} \cdot \frac{(K-1)^{(\alpha-1)}}{(\alpha-1)!\cdot m^{\alpha}}\right) = \frac{a}{2^{(a-1)}\cdot(\alpha-1)!}\sum_{K=\alpha}^{m}\left(\frac{(m-K)^{(a-1)}}{(m-1)^{(a-1)}} \cdot \frac{(K-1)^{(\alpha-1)}}{m^{\alpha}}\right) \tag{34}$$

Noticing that, as the previous situation, we limit the length m to positive infinity to get the continuous probability in the real number range. We can change the lower boundary to 0 since α is very small comparing to m.

$$\mathbb{P}(A^c) = \frac{a}{2^{(a-1)}\cdot(\alpha-1)!}\lim_{m\to+\infty}\sum_{K=\alpha}^{m}\left(\left(\frac{m-K}{m}\right)^{(a-1)}\cdot\left(\frac{K}{m}\right)^{(\alpha-1)}\cdot\frac{1}{m}\right)$$

$$= \frac{a}{2^{(a-1)}\cdot(\alpha-1)!}\int_0^m\left((1-\frac{K}{m})^{(a-1)}\cdot\left(\frac{K}{m}\right)^{(\alpha-1)}\cdot\frac{1}{m}\right)dK \tag{35}$$

Because the defined range for the problem is (0, 1), we scale m to 1 and introduce a new variable $y = \frac{K}{m}$. Thus, $y \in (0,1)$ and $\frac{1}{m} = dy$.

$$\mathbb{P}(A^c) = \frac{a}{2^{(a-1)}\cdot(\alpha-1)!}\int_0^1\left((1-y)^{(a-1)}\cdot y^{(\alpha-1)}\right)dy \tag{36}$$

Suggest $\int_0^1\left((1-y)^{(a-1)}\cdot y^{(\alpha-1)}\right)dy = D$, we can use integration by parts to simplify D.

$$D = \frac{y^{\alpha}(1-y)^{(a-1)}}{b}|_0^1 - \int_0^1 -(a-1)(1-y)^{a-2}\cdot\frac{y^{\alpha}}{\alpha}dy = \int_0^1(a-1)(1-y)^{a-2}\cdot\frac{y^{\alpha}}{\alpha}dy \tag{37}$$

By doing this time by time to eliminate the term $(1-y)$ in the formula above, we need to do $(a-1)$ times and then we get:

$$D = \int_0^1(a-1)!\cdot\frac{y^{\alpha|a-2}}{\frac{(a+\alpha-2)!}{(\alpha-1)!}}dy = \frac{(a-1)!\cdot(\alpha-1)!}{(a+\alpha-2)!}\cdot(a+\alpha-2) \tag{38}$$

Then we can put D into the $\mathbb{P}(A^c)$ formula and get:

$$\mathbb{P}(A^c) = \frac{a!}{(a+\alpha-1)!}\cdot\frac{1}{2^{a-1}} \tag{39}$$

Therefore, the answer of this situation can be shown

$$\mathbb{P}(A) = 1 - \frac{a!}{(a+\alpha-1)!}\cdot\frac{1}{2^{a-1}} \tag{40}$$

We finish discussing all three situations that the Probability for **Stick**$(a_1,\dots,a_n)$ to form a k-gon in Problem 3.1. We can conclude the second situation into the third one, since the second one is situation when $\alpha = 1$ in the third situation. Here is the final result: There are two situations:

- More than one parameter $a_i > 1$: $\mathbb{P}(A) = 1$
- Only one parameter $a_i > 1$: $\mathbb{P}(A) = 1 - \frac{a!}{(a+\alpha-1)!}\cdot\frac{1}{2^{a-1}}$

Proposition 3.8 (Longest side finding). Suggest a_{great} is the longest stick among the group **Stick**$(a_1,\dots,a_n)$, the longest part L must be in the **Stick**(a_{great}), if L is longer than $\frac{\sum_{i=1}^{n} l_i}{2}$.

Proof. Suggest the length of **Stick**(a_{great}) is lgreat, and L is a part of l_j, with a_j is not a_{great}.

$$L < l_j \text{ and } l_j < l_{great} \tag{41}$$

$$L < l_j < \frac{l_j + l_{great}}{2} < \frac{\sum_{i=1}^{n} l_i}{2} \tag{42}$$

Seeing that the L not in the longest stick will lead to the contradiction, L can only be a part of $\textbf{Stick}(a_{great})$.

Proof of Theorem 2.2. If we assume that we know the $\textbf{Stick}(a_{great})$, in other words, the chosen of the longest stick, we can build the relationship between the original problem with the Problem 3.1 that we have already solved. We scale the length of all sticks in equal proportions, and make the length of $\textbf{Stick}(a_{great})$ with $l_{great} = 1$ as a fixed length and make all other sticks have a length taken form the range$(0, 1)$. Suppose that $l_{great} = V$, where V is a fixed known parameter, and all other Sticks have length $U_1, \ldots, U_{n-1}$. After the scaling, $l_{great} = 1$ and others have length

$$\frac{U_1}{V}, \ldots, \frac{U_{n-1}}{V} \sim \text{Uniform}(0, 1) \tag{43}$$

So that, this problem turned to the Situation 2 in the Problem 3.1, the simple version. Notice that the number of sticks is n, including the one $\textbf{Stick}(a)$ and α $\textbf{Stick}(1)$ in the Situation 2.

$$n = \alpha + 1 \tag{44}$$

So that, we can give the answer to the Broken Pick-up Stick Problem 1.4 under the assumption that $\textbf{Stick}(a_{great})$ is known:

$$\mathbb{P}(A | a_{great} = a_i) = 1 - \frac{a_i!}{(a_i + n - 2)!} \times \frac{1}{2^{a_i - 1}} \tag{45}$$

However, each in the group $\textbf{Stick}(a_1, \ldots, a_n)$ may be the longest one, since they are all randomly chosen. Therefore, we must find all the conditions when $a_i = a_{great}$ and take the average value, since they have the same probability of getting the longest. In this way we get the answer of this problem, presented above in Section 2.

4. The proof of Theorem 2.4

With the method in Section 3, we discuss the Prolem 1.5, which asks if any three parts of the separating can form a triangle in $\textbf{Stick}(1, \ldots, 1)$. The arbitrariness and existence of the Spaghetti Problem was first given in [4] but without the prove. We will start with the arbitrariness of Pick-up Problem $\textbf{Stick}(1,1, \ldots, 1)$ in the current section. We will always suppose that there are n sticks in this group as shown in Figure 3.

Before proving Theorem 2.4, there is an important condition to claim, which is essential for our arbitrariness problem. This lemma will help us deduce our original geometric problem to an algebraic inequality problem. The full proof of the following lemma is given in Lemma 2 of [11].

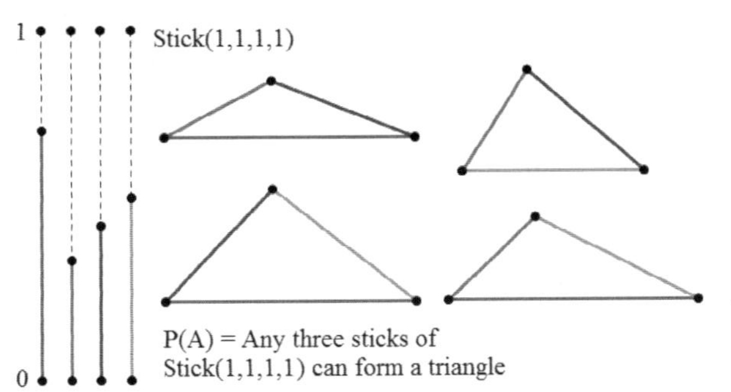

Figure 3. The example of $\textbf{Stick}(1, 1, 1, 1)$ that any three sticks in this collection of sticks can form a triangle.

Lemma 4.1. If any of three parts of a group of sticks can form a triangle, it has a necessary and sufficient condition that the sum of shortest two of them should be longer than the longest one.
The Lemma 4.1 is not hard to prove. Suppose that the sticks have length $\{k_1 < \cdots < k_n\}$ in order, $k_1 + k_2 > k_n$ is necessary for the condition. For any three sticks, $k_a + k_b > k_c$, we know that $k_c \leq k_n$, $k_a + k_b \geq k_1 + k_2$, so that $k_a + k_b > k_c$ must be true. Thus, $k_1 + k_2 > k_n$ is sufficient for the condition.
Proof of the Probability of Arbitrariness of Pick-up Problem. Thinking discretely, we still suggest the longest one among those n sticks have a fixed length M, which is a large enough integer. Thus, all other sticks are selected with the length among the integers in the range $(0, M)$. As every other $(n-1)$ stick have M points to choose, the total number of situations is $M^{(n-1)}$. However, considering the Lemma 4.1, we should select the shortest two sticks. We divide the total number by $(n-1)!$ to get it sorted.

$$\#\Omega = \frac{M^{(n-1)}}{(n-1)!} \tag{46}$$

In order to get the number of situations of the sum of shortest two sticks longer than the longest, we count the situation of those two and multiple the situations of the rest sticks. Suggest the second-shortest stick is k_2 and the shortest one is k_1,

$$\#\{\text{Fit for the problem}\} = \begin{cases} k_1 > M - k_2 \\ k_1 < k_2 < M \\ \text{Others' length are selected between } k_2 \text{ and } M \end{cases} \tag{47}$$

Notice that if $k_1 + k_2 > M$, k_2 should be larger than $\frac{M}{2}$, since k_1 and k_2 are all smaller than M. So that, for all values of k_2 in the range $(\frac{M}{2}, M)$, k_1 can choose from at least $(M - k_2)$ to at most k_2. And the other sticks, $(n-3)$ sticks in total, can choose from the range $(M - k_2)$, since they should be larger than k_2 but less than M.
Thus, we found k_1 has $k_2 - (M - k_2) = 2k_2 - M$ situations to choose and all other sticks have $\binom{M - k_2}{n - 3}$ situations. We can express those situations by using summation of k_2:

$$\#\{\text{Fit for the problem}\} = \sum_{k_2=\frac{M}{2}}^{M} (2k_2 - M) \times \binom{M - k_2}{n - 3} \tag{48}$$

Using the formula to express the probability $\mathbb{P}_{dis}(A)$,

$$\mathbb{P}_{dis}(A) = \frac{\#\{\text{Fit for the problem}\}}{\#\Omega} = \frac{(n-1)! \cdot \sum_{k_2=\frac{M}{2}}^{M} (2k_2 - M) \times \binom{M-k_2}{n-3}}{M^{(n-1)}} \tag{49}$$

Since M is large enough, we can use the Lemma 3.6 to simplify the formula.

$$\mathbb{P}_{dis}(A) = \frac{(n-1)!}{(n-3)!} \cdot \sum_{k_2=\frac{M}{2}}^{M} \frac{(2k_2 - M) \times (M - k_2)^{(n-3)}}{M^{(n-1)}} \tag{50}$$

Similar to the previous problem, we let the M approach the positive infinity to get the $\mathbb{P}(A)$. By taking this limitation, the summation of k, will approach the integral of k.

$$\mathbb{P}(A) = \frac{(n-1)!}{(n-3)!} \cdot \int_{\frac{M}{2}}^{M} \frac{(2k_2 - M) \times (M - k_2)^{(n-3)}}{M^{(n-1)}} dk_2$$

Suggest $x = \frac{k}{M}$,

$$\mathbb{P}(A) = (n-1)(n-2) \cdot \int_{\frac{1}{2}}^{1} (2x - 1) \cdot (1 - x)^{(n-3)} dx \tag{51}$$

Suppose that $y = 1 - x$, then

$$\mathbb{P}(A) = (n-1)(n-2) \cdot \left(\frac{1}{2^{(n-2)}} \left(\frac{1}{n-2} + \frac{1}{n-1} \right) \right) = \frac{1}{2^{(n-2)}} \tag{52}$$

Thus, we get the final arbitrariness probability for n sticks in Pick-up Stick Problem 1.3.

5. Conclusions and Open Problems

In summary, we work out two probability problems related to broken and pick-up sticks in the literature. We first compute the probability of forming a polygon in $\mathbf{Stick}(a_1, \dots, a_n)$. Here, $\mathbf{Stick}(a_1, \dots, a_n)$ is collected by picking up n sticks with independent lengths in Uniform(0, 1), and then breaking each stick into a_i pieces randomly. This result solves the open problem proposed by Petersen and Tenner in [1]. We next compute the probability that arbitrary triple sticks in $\mathbf{Stick}(1, \dots, 1)$ form a triangle. This can be view as a generalization of Theorem 2 in [4]. In the proofs, we first study the discrete version of these two problems, where all lengths of sticks are natural numbers. Then, we obtain the final answer by taking the asymptotic limit of the probability in discrete case.

6. References

[1] T Kyle Petersen and Bridget Eileen Tenner. Broken bricks and the pick-up sticks problem. Mathematics Magazine, 93(3):175–185, 2020.

[2] M. Gardner and D. Richards. The Colossal Book of Short Puzzles and Problems: Combinatorics, Probability, Algebra, Geometry, Topology, Chess, Logic, Cryptarithms, Wordplay, Physics and Other Topics of Recreational Mathematics. Norton, 2006.

[3] G. S. Goodman. The problem of the broken stick reconsidered. The Mathematical Intelligencer, 30:43–49, 2008.

[4] Lingyi Kong, Luvsandondov Lkhamsuren, A Turney, Aananya Uppal, and AJ Hildebrand. Random points, broken sticks, and triangles project report. URL: http://www. math. illinois. edu/~ajh/ugresearch/brokenstick-spring2013report. pdf, 2013.

[5] Art Johnson. Geometric Probability. COMAP, 1995.

[6] Alex Page, Yuliya Semibratova, Yi Xuan, Eva Rui Zhang, M Tip Phaovibul, and AJ Hildebrand. Calculus, geometry, and probability in n dimensions: The broken stick problem in higher dimensions igl project report, spring 2015. 2015.

[7] Carlos D'Andrea and Emiliano Gómez. The broken spaghetti noodle. The American Mathematical Monthly, 113(6):555–557, 2006.

[8] Eugen J. Ionascu. Problem 11443. The American Mathematical Monthly, 116(6):548, 2009.

[9] Eugen J. Ionascu. Problem 11663. The American Mathematical Monthly, 119(8):699–706, 2012.

[10] Eugen J. Ionascu and Gabriel Prajitura. Things to do with a broken stick. arXiv preprint arXiv:1009.0890, 2010.

[11] William Verreault. On the probability of forming polygons from a broken stick. arXiv preprint arXiv:2107.09604, 2021.

Journal of Physics: Conference Series **2287** (2022) 012004 doi:10.1088/1742-6596/2287/1/012004

Application of the Chimera Method to Poisson's Equation with the Homogeneous Dirichlet Boundary Condition

Henry Chien, Yupeng Fan, Ziyun Zeng

No.2 Zongmaotoutiao, Xitiejianghutong, Xicheng District Beijing, China
E-mail: ray.7000@yahoo.com, Williamfyp@163.com, zengziyun@hotmail.com

Abstract. Establishing variational formulation is an effective way to study the existence and uniqueness of the solution of certain elliptic partial differential equation with boundary condition. For the solution of certain elliptic partial differential equation with boundary condition, we know that the numerical solution obtained by the finite element method approximates the solution of this equation. Moreover, to avoid gridding overly complex domains, we can use the Chimera method to decompose the domain into several overlapping sub-domains. In this paper, we study Poisson's equation with the homogeneous Dirichlet boundary condition. By analyzing the existence and uniqueness of the solution of the corresponding variational formulation, we know the existence and uniqueness of the solution of Poisson's equation with the homogeneous Dirichlet boundary condition. We use the Chimera method and the finite element method to deal with Poisson's equation with the homogeneous Dirichlet boundary condition by constructing two iterative sequences and analyzing their properties.

1. Introduction

Partial differential equations (PDE) are equations that contain unknown functions and their partial derivatives. For some specific partial differential equations with boundary conditions, we can analyze many properties of the solutions, such as the existence, uniqueness, boundedness, and regularity of the solutions. In preliminaries, we introduce Sobolev spaces and the variational formulation to conclude the existence and uniqueness of the solution in $H_0^1(\Omega)$ for Poisson's equation with the homogeneous Dirichlet boundary condition. We can transform the original problem into a variational formulation problem using Green's formula, then use Lax-Milgram theorem to analyze the existence and uniqueness of the solution of the variational formulation. Finally, the equivalence between the solution of the variational formulation and the solution of the original equation can be derived by regularity conclusions, which leads to the existence and uniqueness of the solution of the original equation in $H_0^1(\Omega)$ [1].

We are also interested in explicit forms of solutions to partial differential equations, such as Poisson, heat and wave equations under specific boundary conditions [2]. However, we are curious about the explicit form of the solution corresponding to the general boundary conditions. We learned that the finite element method (FEM) can be used to give a numerical solution to Poisson's equation with the homogeneous Dirichlet boundary condition, so that the numerical solution can be used to approximate the solution of the equation. In Subsection 2.5, we introduce the approximate variational formulation. The solution of the approximate variational formulation can be obtained by computing the corresponding matrix. For Poisson's equation with the homogeneous Dirichlet boundary condition (in a polyhedral domain), we have the conclusion that the solution of the variational formulation can be approximated by the solution of the approximate variational formulation in H^1 by the theory related to the finite element method. When the regularity of the solution of the variational form is better, we can also obtain the error estimate [3].

The Chimera method is used in fluid mechanics to avoid meshing overly complicated objects [4][5]. It solves partial differential equations in a domain by decomposing the domain into several overlapping sub-domains. This allows the meshes of each sub-domain to be independent of each other without the need for an overall mesh.

2. Some preliminaires

In this paper, Ω is a domain of $\mathbb{R}^N$ (bounded or unbounded), with boundary denoted $\partial\Omega$. Sometimes we also assume that Ω is a regular and bounded domain. A regular domain is roughly a domain whose boundary is a regular hypersurface (a manifold of dimension $N-1$). This domain is locally located on only one side of its boundary. We define the exterior normal to the boundary $\partial\Omega$ to be the unit vector $n=(n_i)_{1\le i\le N}$ normal to any point on the tangent plane of Ω and pointing to the exterior of Ω. In $\Omega\subset\mathbb{R}^N$ we note dx the volume measure, or the Lebesgue measure of dimension N. In $\partial\Omega$ we denote ds the surface measure, or the Lebesgue measure of dimension $N-1$ on the submanifold $\partial\Omega$.

2.1 Poisson's equation with the homogeneous dirichlet boundary condition and variational formulation

The equal signs in Poisson's equation are regarded as the equal signs in the sense of almost everywhere, without further special explanation.

We consider Poisson's equation with the homogeneous Dirichlet boundary condition:

$$\begin{cases}-\Delta u=f & in\ \Omega\\ u=0 & on\ \partial\Omega,\end{cases} \tag{1}$$

where Ω is a regular and bounded domain in the space $\mathbb{R}^N$, $f\in L^2(\Omega)$. We wonder if (1) has a unique solution in $H_0^1(\Omega)$. We assume that $u\in H^2(\Omega)$ and then we multiply both sides of the first equation in (1) by $v\in H_0^1(\Omega)$ and integrate them to get the following formulation:

$$Find\, u\in H_0^1(\Omega)\ such\ that\ \int_\Omega \nabla u(x)\cdot\nabla v(x)dx=\int_\Omega f(x)v(x)dx\ for\, all\, v\in H_0^1(\Omega). \tag{2}$$

Remark 2.1. *Different elliptic partial differential equations with boundary conditions correspond to different variational formulations. To get the corresponding variational formulation, we have the following rough approach: we first find a Hilbert space V ($H_0^1(\Omega)$in this paper), multiply both sides of the partial differential equation by $v\in V$ and integrate, then use Green's formula to reduce the differential order of the integral equation by one order, and finally obtain the following formulation:*

$$\text{Find } u\in V \text{ such that } a(u,v)=L(v) \text{ for all } v\in V. \tag{3}$$

For (2.2), we have:

$$a(u,v)=\int_\Omega \nabla u(x)\cdot\nabla v(x)dx \tag{4}$$

and

$$L(v)=\int_\Omega f(x)v(x)dx. \tag{5}$$

Theorem 2.2. ***(Lax-Milgram Theorem)*** *Let V be a real Hilbert space, and*
(1) $a(\cdot,\cdot)$ is a continuous bilinear form on V, i.e., there exists $M>0$ such that:

$$|a(w,v)|\le M\parallel w\parallel\parallel v\parallel\ for\ all\ w,v\in V. \tag{6}$$

(2) $a(\cdot,\cdot)$ is coercive(or elliptic), i.e., there exists $\alpha>0$ such that:

$$a(v,v)\ge\alpha\parallel v\parallel^2\ \ for\, all\, v\in V. \tag{7}$$

(3) $L(\cdot)$ is a continuous linear form on V, i.e., there exists $C>0$ such that:

$$|L(v)|\le C\parallel v\parallel\ \ for\, all\, v\in V. \tag{8}$$

then the variational formulation (2.3) has a unique solution in V.

Using the Lax-Milgram theorem, we can analyze the existence and uniqueness of the solution of (2). Since $H_0^1(\Omega)$ is a Hilbert space, we only need to verify (6)-(8) for $a(u,v)$ and $L(v)$ in the form of (4), (5). From Cauchy-Schwarz inequality , we have:

(1) For $u,v \in H_0^1(\Omega)$

$$|a(u,v)| = \left|\int_\Omega \nabla u(x)\cdot\nabla v(x)dx\right| \le \| \nabla u \|_{L^2(\Omega)} \| \nabla v \|_{L^2(\Omega)} \le \| u \|_{H^1(\Omega)} \| v \|_{H^1(\Omega)}. \quad (9)$$

(2) For $v \in H_0^1(\Omega)$

$$a(v,v) = \int_\Omega |\nabla v(x)|^2\, dx \ge \frac{1}{2} \| v \|^2_{H_0^1(\Omega)}. \quad (10)$$

(3) For $v \in H_0^1(\Omega)$

$$|L(v)| = \left|\int_\Omega f(x)v(x)dx\right| \le \| f \|_{L^2(\Omega)} \| v \|_{L^2(\Omega)} \le C \| v \|_{H^1(\Omega)}. \quad (11)$$

Then we can deduce that there exists a unique solution in $H_0^1(\Omega)$ of (2).

Since we have the existence and uniqueness of the solution in $H_0^1(\Omega)$ of (2), We want to show that the solution of (2) solves (1). We can assume that the solution of (2) satisfies: $\Delta u = \nabla\cdot\nabla u$ exists in the weak sense and $\Delta u \in L^2(\Omega)$. Then by the definition of the weak divergence, (2) can be transformed into:

$$\int_\Omega (\Delta u + f)\,\phi dx = 0 \quad \forall \phi \in C_c^\infty(\Omega). \quad (12)$$

Then by the density of $C_c^\infty(\Omega)$ in $L^2(\Omega)$, we can deduce that

$$-\Delta u = f \textit{ almost everywhere in } \Omega. \quad (13)$$

2.2 Finite Element Method

The finite element method is a numerical method for the calculation of solutions of boundary problems. Given a Hilbert space V, a continuous and coercive bilinear form $a(u,v)$, and a continuous linear form $L(v)$, we consider the following variational formulation:

$$\textit{Find } u \in V \textit{ such that } a(u,v) = L(v) \quad \forall v \in V. \quad (14)$$

The internal approximation consists of replacing the Hilbert space V with a finite dimensional subspace V_h, *i.e.*, in search of the solution of:

$$\textit{Find } u_h \in V_h \textit{ such that } a(u_h,v_h) = L(v_h) \quad \forall v \in V. \quad (15)$$

Then we have the following lemma:

Lemma 2.3. *Let V be a real Hilbert space, and V_h a finite dimensional subspace. Let $a(u,v)$ be a continuous and coercive bilinear form on V, and $L(v)$ a continuous linear form on V. Then the internal approximation (15) has a unique solution. Moreover, this solution can be obtained by solving a linear system of positive definite matrix (and symmetric if $a(u,v)$ is symmetric).*

Proof. For finite dimensional space V_h, we have a finite basis $(\phi_j)_{1\le j\le N_h}$. We write $u_h = \sum_{j=1}^{N_h} u_j\,\phi_j$, let $U_h = (u_1,\dots,u_{N_h})$ be the vector in $\mathbb{R}^{N_h}$ of the coordinates of u_h. Then (2.27) is equivalent to

$$\text{Find } U_h \in \mathbb{R}^{N_h} \text{ such that } a\left(\sum_{j=1}^{N_h} u_j\,\phi_j, \phi_i\right) = L(\phi_i) \quad \forall 1 \le i \le N_h.$$

which is written as a linear system:

$$\mathcal{K}_h U_h = b_h \quad (16)$$

Definition 2.4. *Given a mesh $\mathcal{T}_h$ of a connected and open polyhedral Ω, the finite element method $\mathbb{P}_k$, or triangular Lagrangian finite elements of order k, associated with this mesh is defined by the discrete space:*

$$V_h = \{v \in C(\bar{\Omega}) \text{ such that } v|_{K_i} \in \mathbb{P}_k \text{ for all } K_i \in \mathcal{T}_h\}. \quad (17)$$

We also define the subspace V_{0h} by:

$$V_{0h} = \{v \in V_h \text{ such that } v = 0 \text{ on } \partial\Omega\}. \quad (18)$$

Theorem 2.5. *Let $(\mathcal{T}_h)_{h>0}$ be a sequence of regular meshes of Ω. Let $u \in H^1(\Omega)$ be the solution of the Poisson's equation with the homogeneous Dirichlet boundary condition (13), and $u_h \in V_{0h}$, be the solution of its internal approximation (15) by the $\mathbb{P}_k$ finite element method. Then the finite element method $\mathbb{P}_k$ converges, i.e.:*

$$\lim_{h\to 0}\|u-u_h\|_{H^1(\Omega)}=0. \tag{19}$$

Moreover, if $u\in H^{k+1}(\Omega)$ and if $k+1>N/2$, *then we have the error estimate*

$$\|u-u_h\|_{H^1(\Omega)}\le Ch^k\parallel u\parallel_{H^{k+1}(\Omega)}, \tag{20}$$

where C *is a constant independent of* h *and* u.

Proposition 2.6 *Let* $(\mathcal{T}_h)_{h>0}$ *be a sequence of regular meshes of* Ω. *We assume that* $k+1>N/2$. *Then, for any* $v\in H^{k+1}(\Omega)$, *the interpolation* $r_h v=\sum_{i=1}^{n_{dl}} v(\hat{a}_i)\phi_i(x)$ *is well defined, and there exists a constant* C, *independent of* h *and* v, *such that:*

$$\|v-r_h v\|_{H^1(\Omega)}\le Ch^k\parallel v\parallel_{H^{k+1}(\Omega)} \text{ with } r_h v(x)=\sum_{i=1}^{n_{dl}} v(\hat{a}_i)\phi_i(x), \tag{21}$$

where $(\phi_i)_{1\le i\le n_{dl}}$ *is the basis for* V_{0h} *of the finite element method* $\mathbb{P}_k$*(see definition 2.29).*

3. Application of the Chimera Method

Chimera method is used in Fluid Mechanics to avoid meshing overly complicated objects. The Chimera method aims at solving partial differential equations in a domain Ω by decomposition into subdomains Ω_i with overlap. This allows each of the subdomains Ω_i to be meshed independently of each other without the need for a global mesh of $\Omega=\bigcup_i\Omega_i$. In this section, we will use the Chimera method to deal with the Poisson's equation with the homogeneous Dirichlet boundary condition. To simplify the process, we divide the entire domain into two overlapping domains.

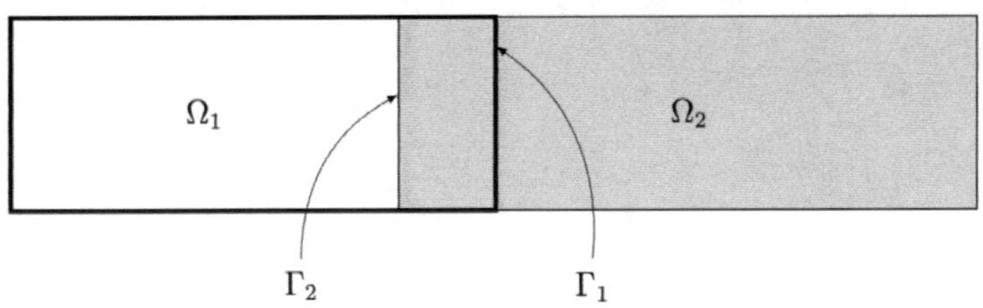

Figure 1. Ω decomposed into two overlapping subdomains Ω_1 and Ω_2.

Let $\Omega\subset\mathbb{R}^2$ be a regular and bounded domain. For a function $f\in L^2(\Omega)$, consider the following problem:

$$\text{Find } u\in H_0^1(\Omega) \text{ such that: } a(u,v)=\int_\Omega f(x)v(x)\mathrm{d}x,\quad \forall v\in H_0^1(\Omega), \tag{22}$$

where

$$a(u,v)=\int_\Omega \nabla u(x)\cdot\nabla v(x)\mathrm{d}x.$$

It is assumed that $\Omega=\Omega_1\cup\Omega_2$ with Ω_1,Ω_2 two open sets such that $\Omega_1\cap\Omega_2\neq\emptyset$. We note $\Gamma_i=\partial\Omega_i\setminus\partial\Omega$ for $i=1,2$, and it is assumed throughout the problem that $\bar{\Gamma}_1\cap\bar{\Gamma}_2=\emptyset$ (see Figure 1 for example).

3.1 Iterative Algorithm Analysis

In this subsection, we will construct two iterative sequences, then prove the weak limit of these sequences relate to the solution of the variational formulation (22). We expand the function $v_i\in H_0^1(\Omega_i)$ with $i=\{1,2\}$. We define $\tilde{v}_i$ by:

$$\tilde{v}_i=\begin{cases} v_i \text{ almost everywhere in } \Omega_i,\\ 0 \text{ almost everywhere in } \Omega\setminus\Omega_i.\end{cases}$$

In the sequel, when there is no ambiguity, we will continue to note v_i the extension $\tilde{v}_i$.

Let $b(\cdot,\cdot)$ be a continuous coercive bilinear form on $L^2(\Omega)$. Now we consider the following iterative algorithm:

- Let $u_1^0\in H_0^1(\Omega_1)\cap H^2(\Omega_1), u_2^0\in H_0^1(\Omega_2)\cap H^2(\Omega_2)$;
- For $n\ge 0$, we assume $u_i^n\in H_0^1(\Omega_i)(i=1,2)$, then we solve the problems :

Find $u_i^{n+1} \in H_0^1(\Omega_i), i = \{1,2\}$, such that :

$$b\left(u_{1,2}^{n+1} - u_{1,2}^{n}, v_{1,2}\right) + a\left(u_{1,2}^{n+1} + u_{2,1}^{n}, v_{1,2}\right) = \int_\Omega f(x) v_{1,2}(x)\mathrm{d}x, \quad \forall v_{1,2} \in H_0^1\left(\Omega_{1,2}\right) \tag{23}$$

And we have the following conclusion for this iterative algorithm:

Proposition 3.1. *The equations (23) define the sequences* $(u_1^n)_{n\geq 0}$ *and* $(u_2^n)_{n\geq 0}$ *uniquely.*

Proof. We can rewrite equation (23) as follows:

Find $u_i^{n+1} \in H_0^1(\Omega_i), i = \{1,2\}$, such that :

$$b\left(u_{1,2}^{n+1}, v_{1,2}\right) + a\left(u_{1,2}^{n+1}, v_{1,2}\right) = \int_\Omega f(x) v_{1,2}(x)\mathrm{d}x + b\left(u_{1,2}^{n}, v_{1,2}\right) - a\left(u_{2,1}^{n}, v_{1,2}\right), \quad \forall v_{1,2} \in H_0^1\left(\Omega_{1,2}\right).$$

By noting:

$$c(u,v) = b(u,v) + a(u,v),$$
$$\ell_{1,2}^n(v) = \int_\Omega f(x)v(x)\mathrm{d}x + b\left(u_{1,2}^n, v\right) - a\left(u_{2,1}^n, v\right)$$

this is the same as solving the problem:

Find $u_i^{n+1} \in H_0^1(\Omega_i), i = \{1,2\}$, such that :

$$c(u_1^{n+1}, v_1) = \ell_1^n(v_1), \quad \forall v_1 \in H_0^1(\Omega_1).$$
$$c(u_2^{n+1}, v_2) = \ell_2^n(v_2), \quad \forall v_2 \in H_0^1(\Omega_2).$$

For $i = \{1,2\}$ we will therefore apply the Lax-Milgram theorem to the Hilbert space $H_0^1(\Omega_i)$ equipped with the $\|\cdot\|_{H_0^1(\Omega_i)}$ norm.

- The bilinear form c is continuous on $H_0^1(\Omega_i)$: if $u_i, v_i \in H_0^1(\Omega_i)$,

$$\begin{aligned} |c(u_i,v_i)| &\leq C_b\|u_i\|_{L^2(\Omega_i)}\|v_i\|_{L^2(\Omega_i)} + \|u_i\|_{H_0^1(\Omega_i)}\|v_i\|_{H_0^1(\Omega_i)} \\ &\leq (C_b C_P^2 + 1)\|u_i\|_{H_0^1(\Omega_i)}\|v_i\|_{H_0^1(\Omega_i)}. \end{aligned}$$

where we used the continuity of the bilinear form b, the Cauchy-Schwarz inequality and Poincaré inequality.

- The bilinear form c is coercive on $H_0^1(\Omega_i)$: if $u_i \in H_0^1(\Omega_i)$,

$$c(u_i,u_i) = b(u_i,u_i) + \|u_i\|_{H_0^1(\Omega_i)}^2 \geq \alpha_b\|u_i\|_{L^2(\Omega_i)}^2 + \|u_i\|_{H_0^1(\Omega_i)}^2 \geq \|u_i\|_{H_0^1(\Omega_i)}^2.$$

where we used the coercivity of the bilinear application b.

- The linear form ℓ_i^n is continuous on $H_0^1(\Omega_i)$: let $v_i \in H_0^1(\Omega_i)$ and $j \in \{1,2\}, i \neq j$

$$\begin{aligned} |\ell_i^n(v_i)| &\leq \left(\| f \|_{L^2(\Omega_i)} + C_b\|u_i^n\|_{L^2(\Omega_i)}\right)\|v_i\|_{L^2(\Omega_i)} + \left\|u_j^n\right\|_{H_0^1(\Omega_i)}\|v_i^n\|_{H_0^1(\Omega_i)} \\ &\leq \left(C_P \| f \|_{L^2(\Omega_i)} + C_p C_b\|u_i^n\|_{L^2(\Omega_i)} + \left\|u_j^n\right\|_{H_0^1(\Omega_i)}\right)\|v_i^n\|_{H_0^1(\Omega_i)}. \end{aligned}$$

Note that we have used the Cauchy-Schwarz inequality, the continuity of the bilinear form b, and Poincaré inequality.

For $i \in \{1,2\}$ fixed, we can then use the Lax-Milgram theorem on the Hilbert space $H_0^1(\Omega_i)$. Note that $u_1^n \in H_0^1(\Omega_1), u_2^n \in H_0^1(\Omega_2)$, there exists a unique $u_i^{n+1} \in H_0^1(\Omega_i)$ such that:

$$c\left(u_i^{n+1}, v_i\right) = l_i^n(v_i), \quad \forall v_i \in H_0^1(\Omega_i).$$

Now we decompose $u \in H_0^1(\Omega)$. We construct $\chi \in C^\infty(\Omega)$ that satisfies the following properties:

- $\chi = 0$ on $\partial\Omega_1 \setminus \partial\Omega = \Gamma_1$ and $\chi = 1$ on $\partial\Omega_2 \setminus \partial\Omega = \Gamma_2$.
- $\chi = 0$ in $\Omega_2 \setminus \Omega_1$ and $\chi = 1$ in $\Omega_1 \setminus \Omega_2$.

Then we have the following decomposition:

$$w_1 = \chi u \in H_0^1(\Omega_1), \quad w_2 = (1-\chi)u \in H_0^1(\Omega_2) \quad \text{and} \quad u = w_1 + w_2. \tag{24}$$

Remark 3.2. *We observe that the function χ which verify the conditions (3.3) is not unique. By example, we can choose $\chi \in C^\infty(\Omega)$ such that:*

$$0 \le \chi \le 1, \quad \chi = 0 \text{ in } \Omega_2 \setminus \Omega_1, \quad \chi = 1 \text{ in } \Omega_1 \setminus \Omega_2.$$

We notice that χ^2 verifies the same conditions and thus the decomposition $u = w_1 + w_2$ is not unique.

Let $u = w_1 + w_2$ be a decomposition of the solution u of problem (3.1) as shown in (24). For $i = 1,2$, we set $w_i^n = u_i^n - w_i$. We want to find a formulation verified by each $w_i^{n+1}, i = \{1,2\}, n \ge 0$ but not u_i^n, so we deal with the original problem as follows:

Since $u = w_1 + w_2$ is a solution of the variational problem (3.1), for all $v \in H_0^1(\Omega)$ we have:

$$a(w_1 + w_2, v) = \int_\Omega f(x)v(x)\mathrm{d}x.$$

Let $i = \{1,2\}$ be fixed and $v_i \in H_0^1(\Omega_i)$. Then $\tilde{v}_i \in H_0^1(\Omega)$ and noting v_i the extension $\tilde{v}_i$, we can write:

$$a(w_1 + w_2, v_i) = \int_\Omega f(x)v_i(x)\mathrm{d}x.$$

By subtracting this equation from the following equation (3.2):

$$b\left(u_i^{n+1} - u_i^n, v_i\right) + a\left(u_i^{n+1} + u_j^n, v_i\right) = \int_\Omega f(x)v_i(x)\mathrm{d}x.$$

we obtain:

$$b\left(u_i^{n+1} - u_i^n, v_i\right) + a\left(w_i^{n+1} + w_j^n, v_i\right) = 0.$$

which can be rewritten as:

$$b\left(u_i^{n+1}, v_i\right) + a\left(w_i^{n+1}, v_i\right) = b(u_i^n, v_i) - a\left(w_j^n, v_i\right).$$

Subtracting the term $b(w_i, v_i)$ on each side, we finally obtain:

$$b\left(w_i^{n+1}, v_i\right) + a\left(w_i^{n+1}, v_i\right) = b(w_i^n, v_i) - a\left(w_j^n, v_i\right),$$

and the variational formulation verified by each w_i^{n+1} is written as follows:

$$\begin{aligned} &\text{Find } w_i^{n+1} \in H_0^1(\Omega_i), i = \{1,2\}, \text{ such that :} \\ &b(w_1^{n+1}, v_1) + a(w_1^{n+1}, v_1) = b(w_1^n, v_1) - a(w_2^n, v_1), \quad \forall v_1 \in H_0^1(\Omega_1). \\ &b(w_2^{n+1}, v_2) + a(w_2^{n+1}, v_2) = b(w_2^n, v_2) - a(w_1^n, v_2), \quad \forall v_2 \in H_0^1(\Omega_2). \end{aligned} \tag{25}$$

By writing the variational formulation verified by each $w_i^{n+1}, i = \{1,2\}, n \ge 0$, we give a formulation that does not show u_i^n.

Now we assumed that the form $b(\cdot,\cdot) = \beta\langle\cdot,\cdot\rangle_{L^2(\Omega)}$, for a constant $\beta > 0$. Then we have the following result:

Proposition 3.3. *By verifying the following formula:*

$$\begin{aligned} &\frac{\beta}{2}\Big(\|w_1^{n+1}\|_{L^2(\Omega)}^2 - \|w_1^n\|_{L^2(\Omega)}^2 + \|w_1^{n+1} - w_1^n\|_{L^2(\Omega)}^2 \\ &+\|w_2^{n+1}\|_{L^2(\Omega)}^2 - \|w_2^n\|_{L^2(\Omega)}^2 + \|w_2^{n+1} - w_2^n\|_{L^2(\Omega)}^2\Big) \\ &+a(w_1^{n+1}, w_1^{n+1}) + a(w_2^n, w_1^{n+1}) + a(w_1^n, w_2^{n+1}) + a(w_2^{n+1}, w_2^{n+1}) = 0, \end{aligned}$$

we can deduce the following inequality:

$$\begin{aligned}\|w_1^{n+1}\|^2_{L^2(\Omega)} & +\|w_2^{n+1}\|^2_{L^2(\Omega)} + \|w_1^{n+1} - w_1^n\|^2_{L^2(\Omega)} + \|w_2^{n+1} - w_2^n\|^2_{L^2(\Omega)} \\ & +\frac{1}{\beta}\left(a(w_1^{n+1}, w_1^{n+1}) + a(w_2^{n+1}, w_2^{n+1})\right) \\ \leq & \ \|w_1^n\|^2_{L^2(\Omega)} + \|w_2^n\|^2_{L^2(\Omega)} + \frac{1}{\beta}\left(a(w_1^n, w_1^n) + a(w_2^n, w_2^n)\right).\end{aligned}$$

Proof. We choose $v_i = w_i^{n+1}$ in the variational formulation obtained in (25), then for $i, j = \{1,2\}, i \neq j$:

$$\beta\langle w_i^{n+1} - w_i^n, w_i^{n+1}\rangle_{L^2(\Omega)} + a(w_i^{n+1}, w_i^{n+1}) + a(w_j^n, w_i^{n+1}) = 0.$$

We observe that $y(y - x) = \frac{1}{2}(y^2 - x^2 + (x - y)^2)$, then we obtain:

$$\frac{\beta}{2}\left(\|w_i^{n+1}\|^2_{L^2(\Omega)} - \|w_i^n\|^2_{L^2(\Omega)} + \|w_i^{n+1} - w_i^n\|^2_{L^2(\Omega)}\right) + a(w_i^{n+1}, w_i^{n+1}) + a(w_j^n, w_i^{n+1}) = 0.$$

By summing over $i = \{1,2\}$, we obtain the expected formula.

To obtain the inequality, we use the Cauchy-Schwartz inequality to get:

$$\begin{aligned}-a(w_2^n, w_1^{n+1}) & = -\int_\Omega \nabla w_2^n(x) \cdot \nabla w_1^{n+1}(x)\mathrm{d}x \leq \|w_2^n\|_{H_0^1(\Omega)}\|w_1^{n+1}\|_{H_0^1(\Omega)} \\ & \leq \frac{1}{2}\|w_2^n\|^2_{H_0^1(\Omega)} + \frac{1}{2}\|w_1^{n+1}\|^2_{H_0^1(\Omega)} = \frac{1}{2}a(w_2^n, w_2^n) + \frac{1}{2}a(w_1^{n+1}, w_1^{n+1}).\end{aligned}$$

The same calculation on the term $a(w_1^n, w_2^{n+1})$ gives the expected inequality.

From the above inequality, we have the following result:

Lemma 3.4. *the sequences $(u_i^n)_{n\geq 0}(i = \{1,2\})$ are bounded in $H_0^1(\Omega_i)$ and furthermore :*

$$\sum_{n=0}^{+\infty}\left(\|u_1^{n+1} - u_1^n\|^2_{L^2(\Omega)} + \|u_2^{n+1} - u_2^n\|^2_{L^2(\Omega)}\right) < \infty. \quad (26)$$

Proof. Let $N \in \mathbb{N}^*$. By summing the inequality obtained in the previous proposition for $n = \{0, \cdots, N\}$, we obtain:

$$\begin{aligned}\|w_1^{N+1}\|^2_{L^2(\Omega)} & +\|w_2^{N+1}\|^2_{L^2(\Omega)} + \sum_{n=0}^{N}\left(\|w_1^{n+1} - w_1^n\|^2_{L^2(\Omega)} + \|w_2^{n+1} - w_2^n\|^2_{L^2(\Omega)}\right) \\ & +\frac{1}{\beta}\left(a(w_1^{N+1}, w_1^{N+1}) + a(w_2^{N+1}, w_2^{N+1})\right) \\ \leq & \ \|w_1^0\|^2_{L^2(\Omega)} + \|w_2^0\|^2_{L^2(\Omega)} + \frac{1}{\beta}\left(a(w_1^0, w_1^0) + a(w_2^0, w_2^0)\right).\end{aligned}$$

Thus, for $i = \{1,2\}$ we have:

$$\begin{aligned}\|w_i^{N+1}\|^2_{H_0^1(\Omega_i)} = & \ a(w_i^{N+1}, w_i^{N+1}) \\ \leq & \ \beta\left(\|w_1^0\|^2_{L^2(\Omega)} + \|w_2^0\|^2_{L^2(\Omega)}\right) + a(w_1^0, w_1^0) + a(w_2^0, w_2^0) \\ \leq & \ 2\beta\left(\|u_1^0\|^2_{L^2(\Omega)} + \|u_2^0\|^2_{L^2(\Omega)} + \|w_1\|^2_{L^2(\Omega)} + \|w_2\|^2_{L^2(\Omega)}\right) \\ & +2\left(\|u_1^0\|^2_{H_0^1(\Omega)} + \|u_2^0\|^2_{H_0^1(\Omega)} + \|w_1\|^2_{H_0^1(\Omega)} + \|w_2\|^2_{H_0^1(\Omega)}\right),\end{aligned}$$

which shows that the sequences $(u_i^n)_{n\geq 0}(i = \{1,2\})$ are well bounded in $H_0^1(\Omega_i)$.

In the same way we have:

$$\sum_{n=0}^{N}\left(\|w_1^{n+1}-w_1^n\|_{L^2(\Omega)}^2+\|w_2^{n+1}-w_2^n\|_{L^2(\Omega)}^2\right)$$
$$\leq \; 2\left(\|u_1^0\|_{L^2(\Omega)}^2+\|u_2^0\|_{L^2(\Omega)}^2+\|w_1\|_{L^2(\Omega)}^2+\|w_2\|_{L^2(\Omega)}^2\right)$$
$$+\frac{2}{\beta}\left(\|u_1^0\|_{H_0^1(\Omega)}^2+\|u_2^0\|_{H_0^1(\Omega)}^2+\|w_1\|_{H_0^1(\Omega)}^2+\|w_2\|_{H_0^1(\Omega)}^2\right).$$

Obeserve that $w_i^{n+1}-w_i^n=(u_i^{n+1}-w_i)-(u_i^n-w_i)=u_i^{n+1}-u_i^n$, *we obtain:*

$$\sum_{n=0}^{N}\left(\|u_1^{n+1}-u_1^n\|_{L^2(\Omega)}^2+\|u_2^{n+1}-u_2^n\|_{L^2(\Omega)}^2\right)\leq C,$$

so the series of partial sums is increased and the series is convergent.

From Lemma 3.4, we can deduce the following result:

Lemma 3.5. *The sequence* $\left(u_i^{n+1}-u_i^n\right)_{n\geq 0}(i=\{1,2\})$ *converges to 0 weakly in* $H_0^1(\Omega_i)$ *and strongly in* $L^2(\Omega_i)$.

Proof. From Lemma 3.4 we know that the sequence $\left(u_i^{n+1}-u_i^n\right)_{n\geq 0}$ is bounded in $H^1(\Omega_i)$. So from Theorem 2.5 and Rellich–Kondrachov theorem, we can extract a sub-sequence that converges weakly in $H^1(\Omega_i)$ and strongly in $L^2(\Omega_i)$ to some limit $u_\infty\in H^1(\Omega_i)$. Again from (3.5), we know that $\lim_{n\to+\infty}\|u_i^{n+1}-u_i^n\|_{L^2(\Omega_i)}=0$ and thus the entire sequence $\left(u_i^{n+1}-u_i^n\right)_{n\geq 0}$ converges strongly to 0 in $L^2(\Omega_i)$. Thus, we have a unique limit point and the whole sequence $\left(u_i^{n+1}-u_i^n\right)_{n\geq 0}$ converges weakly to 0 in $H_0^1(\Omega_i)$.

Now we can introduce the main result here:

Theorem 3.6. *We take* $u_1^0\in H_0^1(\Omega_1)\cap H^2(\Omega_1), u_2^0\in H_0^1(\Omega_2)\cap H^2(\Omega_2)$. *The iterative sequence* $(u_1^n,u_2^n)_{n\geq 0}$ *defined by*

$$b(u_1^{n+1}-u_1^n,v_1)+a(u_1^{n+1}+u_2^n,v_1)=\int_\Omega f(x)v_1(x)\mathrm{d}x,\quad \forall v_1\in H_0^1(\Omega_1).$$
$$b(u_2^{n+1}-u_2^n,v_2)+a(u_2^{n+1}+u_1^n,v_2)=\int_\Omega f(x)v_2(x)\mathrm{d}x,\quad \forall v_2\in H_0^1(\Omega_2).$$

converges weakly to a limit $(u_1^\star,u_2^\star)$ *in* $H_0^1(\Omega_1)\times H_0^1(\Omega_2)$, *where*

$$a(u,v)=\int_\Omega \nabla u(x)\cdot\nabla v(x)\mathrm{d}x,$$
$$b(u,v)=\beta\int_\Omega u(x)v(x)\mathrm{d}x \text{ with } \beta>0.$$

Moreover, the solution in $H_0^1(\Omega)$ *of the variational fomulation:*

$$\int_\Omega \nabla u(x)\cdot\nabla v(x)\mathrm{d}x=\int_\Omega f(x)v(x)\mathrm{d}x,\quad \forall v\in H_0^1(\Omega).$$

satisfies $u=u_1^\star+u_2^\star$.

Proof. From Lemma 3.4, the sequence $(u_1^n,u_2^n)_{n\geq 0}$ is bounded in $H_0^1(\Omega_1)\times H_0^1(\Omega_2)$, then by Theorem 2.5, we can extract a sub-sequence which converges weakly to a limit $(u_1^\star,u_2^\star)$ in $H_0^1(\Omega_1)\times H_0^1(\Omega_2)$. Since the functions $u_1^\star$ and $u_2^\star$ were uniquely defined independently of the decomposition of u into w_1+w_2, we have the uniqueness of the limit $(u_1^\star,u_2^\star)$. We observe that the sequence $(u_1^n,u_2^n)_{n\geq 0}$ admits a unique limit point, so we can conclude the weak convergence of the whole sequence $(u_1^n,u_2^n)_{n\geq 0}$ to $(u_1^\star,u_2^\star)$ in $H_0^1(\Omega_1)\times H_0^1(\Omega_2)$.

We write the variational formulation (35) for the sequence $(u_1^n, u_2^n)_{n\geq 0}$. For $i, j = \{1,2\}, i \neq j$, we consider $v_i \in H_0^1(\Omega_i)$ and we observe that $v_i = 0$ in $\Omega \setminus \Omega_i$, With respect to the first term, knowing that $\lim_{n\to+\infty}\|u_i^{n+1} - u_i^n\|_{L^2(\Omega_i)} = 0$, we have:

$$\int_{\Omega_i} \left(u_i^{n+1}(x) - u_i^n(x)\right) v_i(x)\mathrm{d}x \leq \|u_i^{n+1} - u_i^n\|_{L^2(\Omega_i)}\|v_i\|_{L^2(\Omega_i)} \underset{n\to+\infty}{\to} 0.$$

For the second term, we observe that:

$$\int_{\Omega_i} \left(\nabla u_i^{n+1}(x) + \nabla u_j^n(x)\right) \cdot \nabla v_i(x)\mathrm{d}x$$
$$= \int_{\Omega_i} \nabla \left(u_i^{n+1} - u_i^n\right)(x) \cdot \nabla v_i(x)\mathrm{d}x + \int_{\Omega_i} \left(\nabla u_i^n(x) + \nabla u_j^n(x)\right) \cdot \nabla v_i(x)\mathrm{d}x.$$

By Lemma 3.5, we know that the sequence $\left(u_i^{n+1} - u_i^n\right)_{n\geq 0}$ converges weakly to 0 in $H_0^1(\Omega_i)$, so we can pass to the limit in the first term which tends to 0. The weak convergence of the sequence $(u_1^n, u_2^n)_{n\geq 0}$ to $(u_1^\star, u_2^\star)$ in $H_0^1(\Omega_1) \times H_0^1(\Omega_2)$ gives:

$$\int_{\Omega_i} \left(\nabla u_i^{n+1}(x) + \nabla u_j^n(x)\right) \cdot \nabla v_i(x)\mathrm{d}x \underset{n\to+\infty}{\to} \int_{\Omega_i} \left(\nabla u_1^\star(x) + \nabla u_2^\star(x)\right) \cdot \nabla v_i(x)\mathrm{d}x,$$

so

$$\int_\Omega \nabla\,(u_1^\star + u_2^\star)(x) \cdot \nabla v_i(x)dx = \int_\Omega f\,(x)v_i(x)dx, \quad \forall v_i \in H_0^1(\Omega_i). \tag{27}$$

Let $v \in H_0^1(\Omega)$, then by the decomposition, there exists $(v_1, v_2) \in H_0^1(\Omega_1) \times H_0^1(\Omega_2)$ such that $v = v_1 + v_2$. By choosing these v_i as test functions in the previous equation and summing over $i \in \{1,2\}$, we obtain:

$$\int_\Omega \nabla\,(u_1^\star + u_2^\star)(x) \cdot \nabla v(x)\mathrm{d}x = \int_\Omega \nabla \quad (u_1^\star + u_2^\star)(x) \cdot \nabla(v_1 + v_2)(x)\mathrm{d}x$$
$$= \int_\Omega f\,(x)\left(v_1(x) + v_2(x)\right)\mathrm{d}x = \int_\Omega f\,(x)v(x)\mathrm{d}x.$$

By the uniqueness of the solution of the variational formulation (3.1), we can deduce that $u = u_1^\star + u_2^\star$.

By Theorem 3.6, we know that the two iterative sequences constructed are weakly convergent and that the sum of their weak limits is equal to the solution of the variational formulation in the entire domain. But according to this theorem, we cannot directly use these two iterative sequences for numerical approximation because the convergences here are weak convergences. So we need to combine the finite element method for numerical approximation, which is introduced in the next subsection.

3.2 Finite Element Discretization

Now we assume that the domains Ω_1 and Ω_2 are polygonal. Let $\left(\mathcal{T}_h^1\right)_h$ be a regular family of triangular meshes of Ω_1 and $\left(\mathcal{T}_h^2\right)_h$ a regular family of triangular meshes of Ω_2 (see Figure 2 for example).

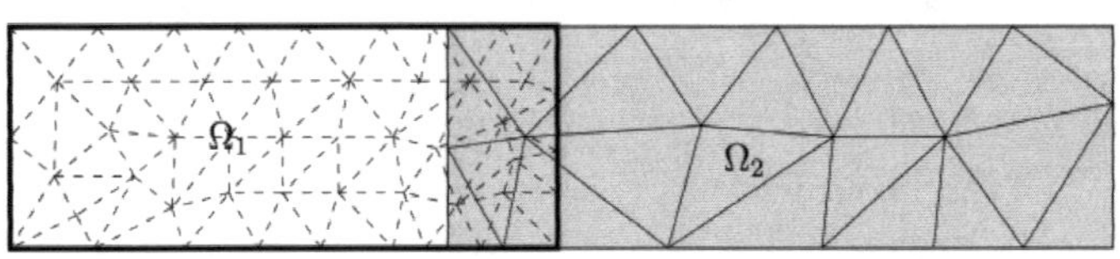

Figure 2. Meshes $\mathcal{T}_h^1$ of Ω_1 and $\mathcal{T}_h^2$ of Ω_2.

We can write the explicit subspaces $V_{i,h}$ of $H_0^1(\Omega_i)(i=\{1,2\})$ constructed from $\mathcal{T}_h^i$ triangulations and Lagrange finite elements of order k as follows:

$$V_{i,h}=\left\{v_{i,h}\in C_0(\overline{\Omega_i}): v_{i,h}\big|_T\in\mathbb{P}_k, \forall T\in\mathcal{T}_h^i \text{ and } v_{i,h}=0 \text{ on } \partial\Omega_i\right\}.$$

For $u_{i,h}^0\in V_{i,h}(i=\{1,2\})$, the discrete version of problem (35) is written as:

$b\left(u_{1(2),h}^{n+1}-u_{1(2),h}^n, v_{1(2),h}\right)+a\left(u_{1(2),h}^{n+1}+u_{2(1),h}^n, v_{1(2),h}\right)=\int_\Omega f(x)v_{1(2),h}(x)\mathrm{d}x, \forall v_{1(2),h}\in V_{1(2),h}(28)$

It can be shown by the same arguments as before that equations (40) define the sequences $\left(u_{1,h}^n\right)_{n\geq 0}$ and $\left(u_{2,h}^n\right)_{n\geq 0}$ in a unique way. Moreover, the entire sequence $\left(u_{1,h}^n, u_{2,h}^n\right)_{n\geq 0}$ converges strongly in $V_{1,h}\times V_{2,h}$ to a limit $\left(u_{1,h}^\star, u_{2,h}^\star\right)$ because weak convergence and strong convergence are equivalent in finite dimensional spaces.

To prove the error estimate result, we introduce the following conclusion:

Proposition 3.7. *Verify the following formula:*

$$\int_\Omega \nabla\left(u_{1,h}^\star-u_1^\star+u_{2,h}^\star-u_2^\star\right)(x)\cdot\nabla\left(v_{1,h}+v_{2,h}\right)(x)\mathrm{d}x=0, \quad \forall v_{1,h}\in V_{1,h}, v_{2,h}\in V_{2,h}.$$

Proof. By passing to the limit in equations (40) in the same way as for the continuous case, we obtain:

$$\int_\Omega \nabla\left(u_{1,h}^\star+u_{2,h}^\star\right)(x)\cdot\nabla v_{1(2),h}(x)\mathrm{d}x = \int_\Omega f(x)v_{1(2),h}(x)\mathrm{d}x, \quad \forall v_{1(2),h}\in V_{1(2),h}.$$

Moreover,we have:

$$\int_\Omega \nabla\left(u_{1,2}^\star+u_{2,1}^\star\right)(x)\cdot\nabla v_{1,2}(x)\mathrm{d}x=\int_\Omega f(x)v_{1,2}(x)\mathrm{d}x, \quad \forall v_{1,2}\in H_0^1(\Omega_{1,2}) \tag{29}$$

Let $v_{1,h}\in V_{1,h}, v_{2,h}\in V_{2,h}$. Now $V_{i,h}\subset H_0^1(\Omega_i)$, so we can choose $v_i=v_{i,h}$ as the test function in (29) which gives:

$$\int_\Omega \nabla\left(u_{1,h}^\star+u_{2,h}^\star\right)(x)\cdot\nabla\left(v_{1,h}+v_{2,h}\right)(x)\mathrm{d}x=\int_\Omega f(x)\left(v_{1,h}+v_{2,h}\right)(x)\mathrm{d}x,$$

and

$$\int_\Omega \nabla\left(u_1^\star+u_2^\star\right)(x)\cdot\nabla\left(v_{1,h}+v_{2,h}\right)(x)\mathrm{d}x=\int_\Omega f(x)\left(v_{1,h}+v_{2,h}\right)(x)\mathrm{d}x.$$

By subtracting these two equations we obtain the desired formula.

So for all $v_{1,h}\in V_{1,h}, v_{2,h}\in V_{2,h}$:

$$\int_\Omega \nabla\left(u_{1,h}^\star+u_{2,h}^\star\right)(x)\cdot\nabla\left(v_{1,h}+v_{2,h}\right)(x)\mathrm{d}x=\int_\Omega \nabla\left(u_1^\star+u_2^\star\right)(x)\cdot\nabla\left(v_{1,h}+v_{2,h}\right)(x)\mathrm{d}x,$$

that is:

$$a\left(u_{1,h}^\star+u_{2,h}^\star, v_{1,h}+v_{2,h}\right)=\tilde{l}\left(v_{1,h}+v_{2,h}\right):=a\left(\underbrace{u_1^\star+u_2^\star}_{\text{known}}, v_{1,h}+v_{2,h}\right).$$

Since the bilinear form a is symmetric, we have:

$$J\left(u_{1,h}^\star+u_{2,h}^\star\right)=\min_{(v_{1,h},v_{2,h})\in V_{1,h}\times V_{2,h}}J\left(v_{1,h}+v_{2,h}\right)$$

which can be written as:

$$\frac{1}{2}\int_{\Omega}\left|\nabla\left(u_{1,h}^{\star}-u_1^{\star}+u_{2,h}^{\star}-u_2^{\star}\right)\right|^2(x)\mathrm{d}x-\frac{1}{2}\int_{\Omega}|\nabla(u_1^{\star}+u_2^{\star})|^2(x)$$
$$=\min_{(v_{1,h},v_{2,h})\in V_{1,h}\times V_{2,h}}\left(\frac{1}{2}\int_{\Omega}\left|\nabla\left(v_{1,h}+v_{2,h}-u_1^{\star}-u_2^{\star}\right)\right|^2(x)\mathrm{d}x-\frac{1}{2}\int_{\Omega}|\nabla(u_1^{\star}+u_2^{\star})|^2(x)\right).$$

Since the term $\frac{1}{2}\int_{\Omega}|\nabla(u_1^{\star}+u_2^{\star})|^2$ is independent of $(v_{1,h},v_{2,h})$, then we obtain:

$$\int_{\Omega}\left|\nabla\left(u_{1,h}^{\star}-u_1^{\star}+u_{2,h}^{\star}-u_2^{\star}\right)\right|^2(x)\mathrm{d}x$$
$$=\min_{(v_{1,h},v_{2,h})\in V_{1,h}\times V_{2,h}}\left(\int_{\Omega}\left|\nabla\left(v_{1,h}+v_{2,h}-u_1^{\star}-u_2^{\star}\right)\right|^2(x)\mathrm{d}x\right).$$

Then by the conclusion of finite element method and proposition 3.7, we can deduce the the error estimate result:

Theorem 3.8. *We assume that the domains Ω_1 and Ω_2 are polygonal. Let $(\mathcal{T}_h^1)_h$ be a regular family of triangular meshes of Ω_1 and $(\mathcal{T}_h^2)_h$ a regular family of triangular meshes of Ω_2. Let $u\in H_0^1(\Omega)$ be the solution of the variational formulation (22), and $u_{i,h}\in V_{i,h}, i=\{1,2\}$ be solutions of (29) by the finite element method $\mathbb{P}_k$. Moreover, we assume that $u_1^{\star}\in H^{k+1}(\Omega_1)$ and $u_2^{\star}\in H^{k+1}(\Omega_1)$. Then we have the following error estimate:*

$$\|u-u_{1,h}^{\star}-u_{2,h}^{\star}\|_{H^1(\Omega)}\le Ch^k\left(\|u_1^{\star}\|_{H^{k+1}(\Omega_1)}+\|u_2^{\star}\|_{H^{k+1}(\Omega_2)}\right), \tag{30}$$

where $h=\max(h_1,h_2)$ and $h_i, i=\{1,2\}$ is the pitch of the $\mathcal{T}_h^i$ mesh.

Proof. For $i=\{1,2\}$, let us consider the interpolation operator $r_h^i\colon H_0^1(\Omega_i)\to V_{i,h}$. For all $k\ge 1$, we have $k+1>N/2=1$. According to Proposition 3.7, we have:

$$\|u_{1,h}^{\star}-u_1^{\star}+u_{2,h}^{\star}-u_2^{\star}\|_{H_0^1(\Omega)}\le\|u_1^{\star}-r_h^1u_1^{\star}+u_2^{\star}-r_h^2u_2^{\star}\|_{H_0^1(\Omega)}.$$

Then, we can obtain:

$$\|u_{1,h}^{\star}-u_1^{\star}+u_{2,h}^{\star}-u_2^{\star}\|_{H_0^1(\Omega)}\le C_1h_1^k\|u_1^{\star}\|_{H^{k+1}(\Omega_1)}+C_2h_2^k\|u_2^{\star}\|_{H^{k+1}(\Omega_2)},$$

where h_i is the pitch of the $\mathcal{T}_h^i$ mesh.
We note that $h=\max(h_1,h_2)$ and use Poincaré inequality, we finally obtain:

$$\|u-u_{1,h}^{\star}-u_{2,h}^{\star}\|_{H^1(\Omega)}\le Ch^k\left(\|u_1^{\star}\|_{H^{k+1}(\Omega_1)}+\|u_2^{\star}\|_{H^{k+1}(\Omega_2)}\right).$$

By Theorem 3.8, when we assume the regularity of the two weak limits in Theorem 3.6, the sum of the two limits of the two convergent sequences obtained by the finite element method can approximate the solution of the variational formulation in H^1 norm, and we can also give the corresponding error estimate. This result shows that for Poisson's equation with the homogeneous Dirichlet boundary condition defined on a complex domain, if we can decompose the domain into two overlapping polygonal domain, then we can construct two iterative sequences obtained by the finite element method. Moreover, if we assume some regularity properties, we can give the numerical solution of Poisson's equation with the homogeneous Dirichlet boundary condition. The advantage of this method is that it avoids gridding complex domains directly.

4. References

[1] H. Brezis, Functional Analysis, Sobolev spaces and partial differential equations, Springer, universitext ed., 2010.
[2] L. C. Evans, partial differential equations, vol. 19 of Graduate Studies in Mathematics, American Mathematical Society, Providence, RI, second ed., 2010.
[3] P. G. Ciarlet, The finite element methods for elliptic problems, North-Holland, Amsterdam (1978).

[4] Houzeaux G, Codina R. A Chimera method based on a Dirichlet/Neumann(Robin) coupling for the Navier-Stokes equations. Comp. Meth. Appl. Mech. Eng. 2003; 192(31-32):3343 – 3377.
[5] Steger J, Benek FDJ. A chimera grid scheme. Advances in Grid GEneration 1983; 5:59–69.
[6] P. Lax, Functional Analysis, Wiley, 2002.
[7] Q. Han and F. H. Lin, Elliptic Partial Differential Equations, American Mathematical Society, 1997.
[8] Ph. Ciarlet, The Finite Element Method for Elliptic Problems (2nd ed.), North-Holland, 1979.
[9] E. Bombieri, [1] Variational problems and elliptic equations, in Mathematical Developments Arising from Hilbert Problems (Browder, F., ed.), Proc. Sympos. Pure Math., Vol. 28, Part 2, American Mathematical Society, 1977, pp. 525–536.
[10] Y. Pinchover and Rubinstein, J.Rubinstein, An Introduction to Partial Differential Equations,

Sanskruti Index of some Chemical Trees and Unicyclic Graphs

H Al-Janabi[1,2*] and G Bacsó[3]

1 Department of Algebra, Institute of Mathematics, Budapest University of Technology and Economics, Hungary
2 Faculty of Computer Science and Mathematics, University of Kufa, Iraq
3 Institute for Computer Science and Control Budapest, Hungary
*E-mail: haneenk.aljanabi@uokufa.edu.iq; tud23sci@gmail.com

Abstract. In chemical graph theory, topological indices are used to estimate the bio-activity of chemical compounds. The molecular graph models molecular compounds. A molecular graph represents the structural formula of a chemical compound relative to the graph. Its vertices represent the atoms of the compound, and its edges represent the chemical bonds. In 2017, Hosamani introduced the Sanskruti index S(G) and defined it as $S(G) = \sum_{uv \in E(G)} ((\delta_u \cdot \delta_v)/(\delta_u + \delta_v - 2))^3$, where δ_u is the sum of the degrees of all neighbors of the vertex u in G. The Sanskruti index displays a good connection with an entropy of octane isomers. In this paper, the Sanskruti index S(G) is computed for some chemical trees and unicyclic graphs.

1. Introduction

Let G be a simple connected graph. $|V(G)|$ is the order of G, its number of vertices denoted by n. $|E(G)|$ is the size of a graph G, its number of edges denoted by m. The degree of a vertex v, indicated by $d(v)$. For a vertex u, we define $\delta u = \sum_{v \in N_u} d(v)$, where N_u denote the neighborhood of vertex u, this means $N_u = \{v \in V(G) \setminus uv \in E(G)\}$. A vertex has degree 0 said to be isolated, and a vertex has degree 1 to be a pendant vertex. A tree T_n is a connected graph without any cycle. A tree of order n with exactly two pendant vertices is said to be a path and denoted by P_n. The tree with $n-1$ pendent vertices is a star of order n, denoted by S_n. A simple connected graph is called unicyclic if it has precisely one cycle [1].

Mathematical chemistry is a part of theoretical chemistry for studying and predicting the molecular structure utilizing mathematical methods without necessarily referring to quantum mechanics. Chemical graph theory is a part of mathematical chemistry that uses graph theory in the mathematical modeling of chemical phenomena to develop the chemical sciences [2]. In mathematical chemistry, numbers encoding specific organic molecules' structural characteristics obtained from the corresponding molecular graph are named graph invariants or topological indices [3].

In 1947, topological indices were utilized in biology and chemistry for the first time when chemist Harold Wiener [4] presented the Wiener index to demonstrate correlations between properties of Physico-chemical for organic compounds of molecular graphs. Topological indices are numerical parameters of a graph that are invariant under the graphs' isomorphism [5]. Many topological indices can be introduced for the connected graph by different and various definitions. The connectivity index is one of the widely used to reflect molecular branching, introduced in 1975 by Randic [6], is defined as follows:

$$R(G) = \sum_{uv\in E(G)} \frac{1}{\sqrt{(u)\cdot d(v)}}$$

For the time being, there are several of these indices that have constructed applications in Mathematical chemistry. Most of these indices depend only on the degree of the vertex of the molecular graph. For example, the atom-bond connectivity index (*ABC*), presented by Estrada et al. [2]. It is defined as:

$$ABC(G) = \sum_{uv\in E(G)} \sqrt{\frac{d(u)+d(v)-2}{d(u)\cdot d(v)}}$$

ABC-index gives a good model for the stability of branched and linear alkanes in addition to the stress-energy of cycloalkanes [2, 7]. Furtula et al. [8] introduced the improvement version of the *ABC*-index and named it the augmented Zagreb index (*AZI*):

$$AZI(G) = \sum_{uv\in E(G)} \left(\frac{d(u)\cdot d(v)}{d(u)+d(v)-2}\right)^3$$

The predictive power is much better than the *ABC*-index in studying of formation heat of octane and heptane [8]. Motivated by previous researches on topological indices and their applications, Hosamani [9] defined the Sanskruti index $S(G)$ for the molecular graph G as follows:

$$S(G) = \sum_{uv\in E(G)} \left(\frac{\delta_u\cdot\delta_v}{\delta_u+\delta_v-2}\right)^3 \quad (1)$$

S-index productivity was examined utilizing the isomers dataset, consisting of the following data: melting point, boiling point, molar refraction, heat capacities, acentric factor, octanol-water partition coefficient, total surface area, and entropy. The S-index correlated with all these properties, and the S-index has a good relationship with the octane isomers entropy, see Figure 1. The following structure-property relationship model was developed for the S-index [9].

$$entropy = 1.7857\, S \pm 81.4286I$$

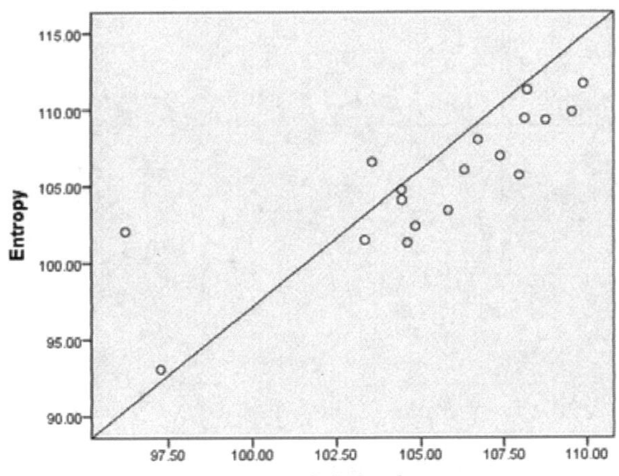

Figure 1. Correlation of S-index with the octane isomers entropy [9].

Sardar et al. computed the Sanskruti index of Polycyclic Aromatic Hydrocarbons [3]. Gao et al. presented explicit formulas for the Sanskruti index of an infinite group of dendrimer nanostars [10]. And, they computed the Sanskruti index for the Circumcoronene series of Benzenoid [11]. In this paper, new theorems are presented with their proofs for the Sanskruti index for many important molecular graphs: caterpillar trees, cycle-caterpillar graphs, starlike trees and sunlike graphs.

2. Results

In this section, important classes of molecular graphs and their Sanskruti index are discussed.

Proposition 1. The Sanskruti index of the star graph S_n is given by:

$$S(S_n) = \frac{(n-1)^7}{(2n-4)^3} \quad Where\ n \geq 3.$$

Proof. In the graph S_n, there are $n-1$ vertices with degree 1 which are $v_1, v_2, \cdots, v_{n-1}$ and one vertex v_n with degree $n-1$. Moreover, every vertex from the vertices $v_1, v_2, \cdots, v_{n-1}$, has one neighbor, which is a vertex $\mathrm{v_n}$, this means $\delta_{v_i} = n-1, 1 \leq i \leq n-1$. And, the vertex v_n has $n-1$ neighbors with degrees 1, this means $\delta_{v_n} = n-1$. Since the graph S_n has $n-1$ edges. So, we apply Equation (1):

$$\begin{aligned} S(S_n) &= \sum_{v_i v_n \in E(S_n)}^{1\leq i\leq n-1} \left(\frac{\delta_{v_i} \cdot \delta_{v_n}}{\delta_{v_i}+\delta_{v_n}-2}\right)^3 \\ &= (n-1)\left(\frac{(n-1)(n-1)}{n-1+n-1-2}\right)^3 \\ &= \frac{(n-1)^7}{(2n-4)^3}. \end{aligned}$$

Proposition 2. The Sanskruti index of the path graph P_n is satisfied with the following consequences: 1. $S(P_3) = 16$; 2. $S(P_4) = 27.39$; 3. $S(P_4) = 43.648$; and 4. $S(P_n) = 18.962n - 64.99$, where $n \geq 6$.

Proof. We can prove the points above (1, 2, and 3) directly using Equation (1). While point (4) can be proven as follows:

In the graph P_n, $n \geq 6$, there are $n-4$ vertices with ($\delta = 4$) which are $v_3, v_4, \cdots, v_{n-3}, v_{n-2}$. And, two vertices ($v_2, v_{n-1}$) with ($\delta = 3$), also two vertices v_1 and v_n with ($\delta = 2$). Moreover, there are $n-1$ edges in P_n which are $v_1v_2, v_2v_3, v_iv_{i+1}, \cdots, v_{n-2}v_{n-1}, v_{n-1}v_n$. So, we can apply Equation (1):

$$\begin{aligned} S(P_n) &= \sum_{v_i v_{i+1} \in E(P_n)}^{1\leq i\leq n-1} \left(\frac{\delta_{v_i} \cdot \delta_{v_{i+1}}}{\delta_{v_i}+\delta_{v_{i+1}}-2}\right)^3 \\ &= \left(\frac{2\cdot 3}{2+3-2}\right)^3 + \left(\frac{3\cdot 4}{3+4-2}\right)^3 + \underbrace{\left(\frac{4\cdot 4}{4+4-2}\right)^3 \cdots \left(\frac{4\cdot 4}{4+4-2}\right)^3}_{(n-5)\ times} + \left(\frac{4\cdot 3}{4+3-2}\right)^3 \\ &+ \left(\frac{3\cdot 2}{3+2-2}\right)^3 = 18.962n - 64.99. \end{aligned}$$

Proposition 3. The Sanskruti index of the cycle graph C_n is given by:

$$S(C_n) = 18.962n, n \geq 3.$$

Proof. In the graph C_n, every vertex v_i, $1 \leq i \leq n$ has two neighbors with degrees 2. So, every vertex v_i has ($\delta_{v_i} = 4$). Moreover, the number of edges is equal to n. So, we can apply Equation (1):

$$\begin{aligned} S(C_n)\& &= \sum_{v_i v_{i+1} \in E(C_n)}^{1\leq i\leq n} \left(\frac{\delta_{v_i} \cdot \delta_{v_{i+1}}}{\delta_{v_i}+\delta_{v_{i+1}}-2}\right)^3, \ where, v_{n+1} = v_1 \\ &= n\left(\frac{4.4}{4+4-2}\right)^3 = 18.962. \end{aligned}$$

2.1. Caterpillar Trees and Cycle-caterpillars

Let G be a labelled graph on n vertices, and let $p_1, p_2, \cdots, p_n$ be non-negative integers. The thorn graph $G(p_1, p_2, \cdots, p_n)$ of the graph G is obtained from G by attaching p_i pendant vertices to the i-th vertex of G, $i = 1, 2, \cdots, n$. The concept of thorn graphs was introduced by Gutman [12]. So, we can define the caterpillar trees are thorn graphs whose parent graph is a path P_n and denoted by $P_n^* = P_n\,(p_1, p_2, \cdots, p_n)$, see Figure 2. In other words, a caterpillar tree is a tree in which all the vertices are within distance 1 of the main path. If we delete all pendant vertices of a caterpillar tree, we reach a path [13].

Remark 4. It is easy to compute the Sanskruti index of caterpillar tree P_n^*, $1 \le n \le 4$, directly. So, the Sanskruti index of P_n^*, $n \ge 5$ is computed in the theorem below.

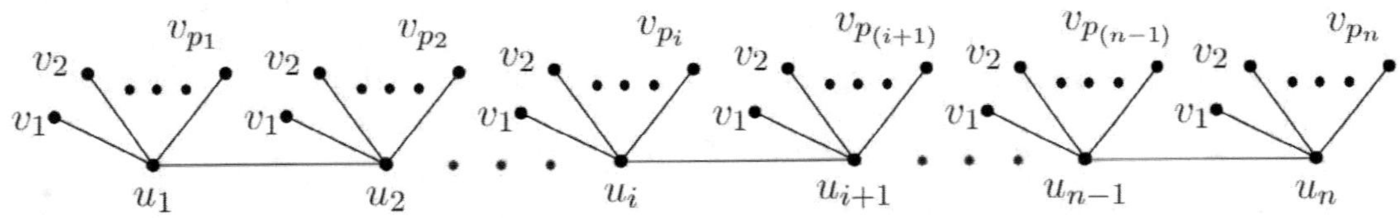

Figure 2. The caterpillar tree $T_n(p_1, p_2, \cdots, p_i, p_{i+1}, \cdots, p_{n-1}, p_n)$ |

Theorem 5. Let $P_n^* = P_n(p_1, p_2, \cdots, p_n)$ be the caterpillar tree, where $p_1, p_2, \cdots, p_n$ are non-negative integers. The Sanskruti index of P_n^* is given by:

$$S(P_n^*) = \sum_{k=1}^{10} A_k,\ n \ge 5$$

Proof. Let $u_1, u_2, \cdots, u_n$ be the vertices of P_n, and let $v_1, v_2, \cdots, v_{p_i}$ be the pendant vertices to the i-th vertex of P_n, where $1 \le i \le n$. The degree of the vertex u_i, $2 \le i \le n-1$, is equal to $(2 + p_i)$, and the degrees of the vertices u_1 and u_n are equal to $1 + p_1$ and $1 + p_n$, respectively. Moreover, there are two types of edges in the caterpillar tree P_n^*: the first type is the edges between the vertices of P_n, this means every edge $u_i u_{i+1}$ has two ends in P_n. And the second type is the edges that have one end u_i belongs to the vertices of P_n and the other end will be pendent vertex v_j, $1 \le j \le p_i$ (see Figure 2).

$$A_1 = \left(\frac{(2+p_1+p_2)(3+p_1+p_2+p_3)}{3+2p_1+2p_2+p_3}\right)^3,\ A_2 = \left(\frac{(3+p_1+p_2+p_3)(4+p_2+p_3+p_4)}{5+p_1+2p_2+p_3+p_4}\right)^3,$$

$$A_3 = \left(\frac{(4+p_{n-3}+p_{n-2}+p_{n-1})(4+p_{n-2}+p_{n-1}+p_n)}{5+p_{n-3}+2p_{n-2}+2p_{n-1}+p_n}\right)^3,$$

$$A_4 = \left(\frac{(4+p_{n-2}+p_{n-1}+p_n)(2+p_{n-1}+p_n)}{3+p_{n-2}+2p_{n-1}+2p_n}\right)^3,$$

$$A_5 = \sum_{i=3}^{n-2} \left(\frac{(4+p_{i-1}+p_i+p_{i+1})(4+p_i+p_{i+1}+p_{i+2})}{6+p_{i-1}+2p_i+2p_{i+1}+p_{i+2}}\right)^3,$$

$$A_6 = p_1\left(\frac{(2+p_1+p_2)(1+p_1)}{1+2p_1+p_2}\right)^3,\ A_7 = p_2\left(\frac{(3+p_1+p_2+p_3)(2+p_2)}{3+p_1+2p_2+p_3}\right)^3,$$

$$A_8 = p_{n-1}\left(\frac{(3+p_{n-2}+p_{n-1}+p_n)(2+p_{n-1})}{3+p_{n-2}+2p_{n-1}+p_n}\right)^3,\ A_9 = p_n\left(\frac{(2+p_{n-1}+p_n)(1+p_n)}{1+p_{n-1}+2p_n}\right)^3,$$

$$A_{10} = \sum_{i=3}^{n-2} p_i\left(\frac{(4+p_{i-1}+p_i+p_{i+1})(2+p_i)}{4+p_{i-1}+2p_i+p_{i+1}}\right)^3.$$

Proof. Let $u_1, u_2, \cdots, u_n$ be the vertices of P_n, and let $v_1, v_2, \cdots, v_{p_i}$ be the attached pendant vertices to the i-th vertex of P_n, where $1 \le i \le n$. The degree of the vertex u_i, $2 \le i \le n-1$, is equal to $(2 + p_i)$, and the degrees of the vertices u_1 and u_n are equal to $(1 + p_1)$ and $(1 + p_n)$ respectively. Moreover, there are two types of edges in the caterpillar tree P_n^*: the first type is the edges between the vertices of P_n, this means every edge $u_i u_i + 1$ has two ends in P_n. And the second type is the edges that have one end u_i belongs to the vertices of P_n and the other end will be pendent vertex v_j, $1 \le j \le$

p_i (see Figure 2). To compute the Sanskruti index of P_n^*, we must find the sum of the degrees of all neighbors of every vertex in P_n^*, this means finding δ for all vertices. The vertex u_1 has p_1 neighbors with degrees 1 and one neighbor u_2 with a degree $(2 + p_2)$. So, $\delta_{u_1} = 2 + p_1 + p_2$. The vertex u_2 has p_2 neighbors with degrees 1, one neighbor u_1 with a degree $(1 + p_1)$ and one neighbor u_3 with a degree $(2 + p_3)$. So, $\delta_{u_2} = 3 + p_1 + p_2 + p_3$. The vertex u_1, $3 \le i \le n-2$, has p_i neighbors with degrees 1, one neighbor u_{i-1} with degree $(2 + p_{i-1})$ and one neighbor u_{i+1} with degree $(2 + p_{i+1})$. So, $\delta_{u_i} = 4 + p_{i-1} + p_i + p_{i+1}$. The vertex u_{n-1} has p_{n-1} neighbors with degrees 1, one neighbor u_n with degree $(1 + p_n)$ and one neighbor u_{n-2} with degree $(2 + p_{n-2})$. So, $\delta_{u_{n-1}} = 3 + p_{n-2} + p_{n-1} + p_n$. The vertex u_n has p_n neighbors with degrees 1 and one neighbor u_{n-1} with degree $(1 + p_{n-1})$. So, $\delta_{u_n} = 2 + p_{n-1} + p_n$. Moreover, the pendant vertex v_j , $1 \le j \le p_i, 2 \le i \le n-1$ has one neighbor u_i with a degree $(2 + p_i)$. So, $\delta_{v_j} = 2 + p_i$. But, the pendent vertex v_r, $1 \le r \le p_1$, has one neighbor u_1 with a degree $(1 + p_1)$. So, $\delta_{v_r} = 1 + p_1$, and the pendant vertex v_s, $1 \le s \le p_n$, has one neighbor u_n with a degree $(1 + p_n)$. So, $\delta_{v_s} = 1 + p_n$. Now, we can apply Equation 1 to the first type of edges $u_i u_{i+1}$, to obtain the terms A_k, $1 \le k \le 5$, and to the second type of edges $u_i v_j$, $1 \le j \le p_i$, to obtain the term A_k, $6 \le k \le 10$.

Caterpillar trees are used in chemical graph theory to represent the structure of hydrocarbon molecules. For example, for positive integer $p_1 = p_n = 3$ and $p_2 = p_3 = \cdots = p_{n-1} = 2$, the caterpillar tree $P_n(3,2,2,\cdots,2,3)$ is the molecular graph of certain hydrocarbon (**straight-chain alkanes**, a single chain with no branches, have the general chemical formula C_nH_{2n+2}), see Figure 3. It is easy to find the Sanskruti index of the molecular graphs of Ethane $P_2(3,3)$, Propane $P_3(3,2,3)$, and Butane $P_4(3,2,2,3)$ using Equation 1. So, Theorem 5 is used to get the formulae for the Sanskruti index of the straight-chain alkanes $C_nH_{2n+2} = P_n(3,2,2,\cdots,2,3)$, $n \ge 5$, as shown in Corollary 6.

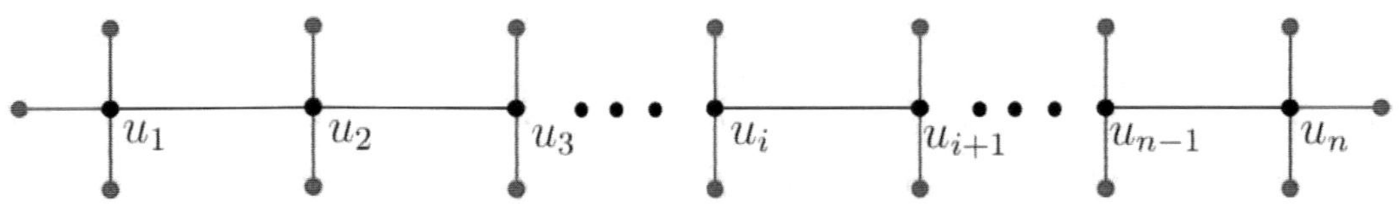

Figure 3. The molecular graphs of hydrocarbon (straight-chain alkanes).

Corollary 6. The Sanskruti index of the straight-chain alkanes $C_nH_{2n+2} = P_n(3,2,2,\cdots 2,3)$, $n \ge 5$, is given by:

$$S(C_nH_{2n+2}) = 199.074\,n - 120.354, n \ge 5$$

Proof. We used Theorem 5 and replaced p_i with $2, 2 \le i \le n-1$, also replaced both p_1 and p_n with 3. □

A unicyclic graph is called **a cycle-caterpillar** if deleting all its pendent (end) vertices will reduce it to a cycle, and denoted by $C_n(p_1, p_2, \cdots, p_n)$. So, the cycle-caterpillars are thorn graphs whose parent graph is a cycle; see Figure 4.

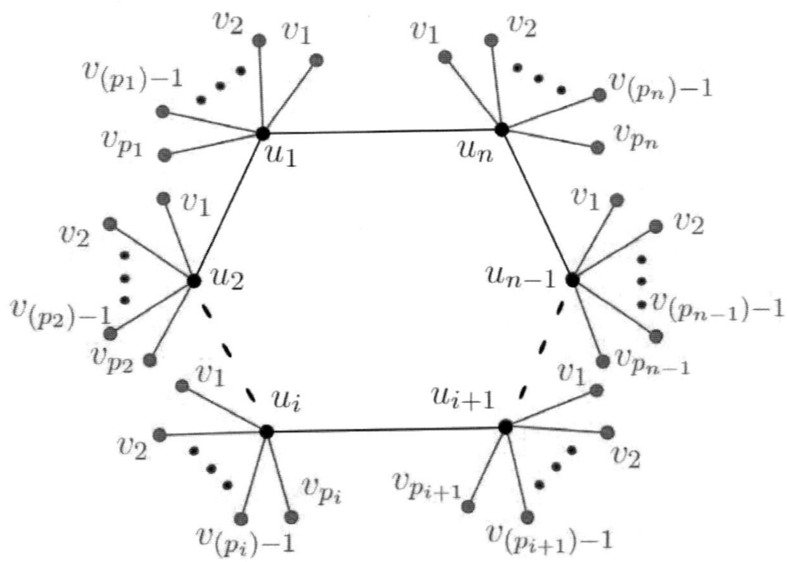

Figure 4. The cycle-caterpillar $C_n(p_1, p_2, \cdots, p_n)$.

Theorem 7. Let $C_n^* = C_n(p_1, p_2, \cdots, p_n)$ be the cycle-caterpillar, and $p_1, p_2, \cdots, p_n$ are non-negative integers. The Sanskruti index of C_n^* is given by:

$$S(C_n^*) = B_1 + B_2$$

Where,

$$B_1 = \sum_{i=1}^{n} \left(\frac{(4+p_{i-1}+p_i+p_{i+1})(4+p_i+p_{i+1}+p_{i+2})}{6+p_{i-1}+2p_i+2p_{i+1}+p_{i+2}}\right)^3, \text{ where } p_{n+1} = p_1$$

$$B_2 = \sum_{i=1}^{n} p_i \left(\frac{(4+p_{i-1}+p_i+p_{i+1})(2+p_i)}{4+p_{i-1}+2p_i+p_{i+1}}\right)^3, \text{ where } p_{n+1} = p_1$$

Proof. Let $u_1, u_2, \cdots, u_n$; un be the vertices of C_n, and let $v_1, v_2, \cdots, v_{p_i}$ be the attached pendant vertices to the i-th vertex of C_n, where $1 \le i \le n$. The degree of the vertex u_i, $1 \le i \le n$, is equal to $(2 + p_i)$. Moreover, there are two types of edges in cycle-caterpillar C_n^*: the first type is the edges between the vertices of C_n, this means every edge $u_i u_{i+1}$ has two ends in C_n. And the second type is the edges that have one end u_i belongs to the vertices of C_n and the other end will be pendent vertex v_j, $1 \le j \le p_i$ (see Figure 4). To compute the Sanskruti index of C_n, we must find the sum of the degrees of all neighbors of every vertex in C_n, this means finding δ for all vertices. The vertex u_i, $1 \le i \le n$, has p_i neighbors with degrees 1, one neighbor u_{i-1} with degree $(2 + p_{i-1})$ and one neighbor u_{i+1} with degree $(2 + p_{i+1})$. So, $\delta_{u_i} = 4 + p_{i-1} + p_i + p_{i+1}$. Moreover, the pendant vertex v_j, $1 \le j \le p_i$, $1 \le i \le n$ has one neighbor u_i with degree $(2 + p_i)$. So, $\delta_{v_j} = 2 + p_i$. Now, we can apply Equation 1 to the first type of edges $u_i u_{i+1}$, to obtain the term B_1, and to the second type of edges $u_i v_j$, to obtain the term B_2.

2.2. *Starlike Trees and Sunlike Graphs*

Let u be an isolated vertex and let $P_1, P_2, \cdots, P_m$ be the paths of orders $n_1, n_2, \cdots, n_m$, where are the positive integers n_i, $1 \le i \le m$. **The starlike tree** $T_m(n_1, n_2, \cdots, n_m)$ is the tree obtained by identifying the end vertex v_i of P_i, with the vertex u for all $i = 1,2, \cdots, m$.

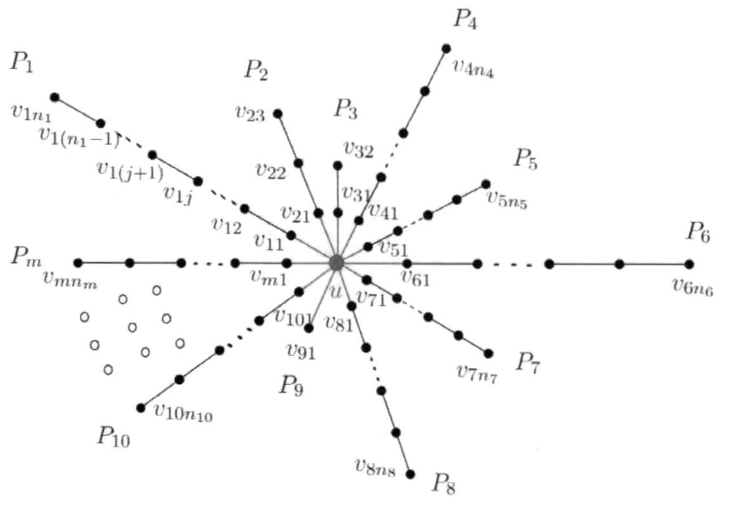

Figure 5. The starlike tree $T_n(n_1, n_2, \cdots, n_m)$.

In other words, a starlike tree is a tree with exactly one vertex u having a degree of at least 3 and has the property that $T_m(n_1, n_2, \cdots, n_m) - u = P_1 \cup P_2 \cup \cdots \cup P_m$, where $|P_i| = n_i$, $n_i \geq 1$, $1 \leq i \leq m$, see Figure 5. The vertex u will be the center of the starlike tree. And v_{ij} means the vertex v at the j-th position of the i path (P_i), $1 \leq i \leq m, 1 \leq j \leq n_i$. Clearly, for $n_i \geq 1, d(v_{in_i}) = 1$. Theorem 8. Let $T_m^* = T_m(n_1, n_2, \cdots, n_m)$ be the starlike tree, and l be the biggest order such that $l \geq n_i, 1 \leq i \leq m$. If $I_1 = \{i: 1 \leq i \leq m, n_i = 1, i.e., |P_i| = 1\}$, $I_2 = \{i: 1 \leq i \leq m, n_i = 2\}, \ldots, I_{l-1} = \{i: 1 \leq i \leq m, n_i = l - 1\}, I_l = \{i: 1 \leq i \leq m, n_i = l\}$, where $|I_1| = t_1, |I_2| = t_2, \cdots, |I_{l-1}| = t_{l-1}, |I_l| = t_l$. Then, the Sanskruti index of T_m^* is given by:

$$S(T_m^*) = \sum_{k=1}^{9} E_k,$$

Where,

$$E_1 = t_1 \left(\frac{(2m-t_1)m}{2m-t_1-2}\right)^3, \; E_2 = t_2 \left(\frac{(2m-t_1)(m+1)}{3m-t_1-1}\right)^3,$$

$$E_3 = (m - t_1 - t_2)\left(\frac{(2m-t_1)(m+2)}{3m-t_1}\right)^3,$$

$$E_4 = 8t_2, E_5 = t_3\left(\frac{3m+6}{m+3}\right)^3, \; E_6 = (m - t_1 - t_2 - t_3)\left(\frac{4m+8}{m+4}\right)^3,$$

$$E_7 = 8(m - t_1 - t_2), \; E_8 = 13.824(m - t_1 - t_2 - t_3),$$

$$E_9 = \sum_{4<r\leq l, s=r-4} 18.962 t_r s$$

Proof. Let the vertex u be the center of T_m^* (it is clear that, T_m^* has m paths). To compute the Sanskruti index of T_m^*, we must find the sum of the degrees of all neighbors of every vertex in P_m^*, this means finding δ for all vertices. Since all neighbors of the center u have degrees 2 except for the vertices v_{i1} that belong. To I_1, $1 \leq i \leq m$ and $|I_1| = t_1$. Then, $\delta_u = m - t_1$. Moreover, the vertex $v_{i1}, i \in I_1$, has only one neighbor, which is the vertex u, this means $\delta_{v_{i1}} = m = \delta_{v_{in_i}}$. Also, the vertex $v_{i1}, i \in I_2$, has two neighbors, which are the vertices u, and v_{in_i}, this means $\delta_{v_{i1}} = m + 1$. Since every vertex v_{ij}, $1 \leq i \leq m, i \in I_3 \cup I_4 \cup A, 2 \leq j \leq n_i - 1$, has degree 2, where $A = \cup_{4<r\leq l} I_r$. Then, the vertex v_{i1} has two neighbors, which are the vertices u and v_{i2}, this means $\delta_{v_{i1}} = m + 2$. Furthermore, the vertex $v_{i2}, i \in I_2$, has only one neighbor, which is the vertex v_{i1}, this means $\delta_{v_{i2}} = 2 = \delta_{v_{in_i}}$. In general, $\delta_{v_{in_i}} = 2$, where v_{in_i} is the end vertex in the path P_i with order $\mathrm{n_i}$. Also, the vertex $v_{i(n_i-1)}, i \in I_3$,

has two neighbors, which are the vertices v_{in_i} with degree 1 and the vertex $v_{i(n_i-2)}$ with degree 2, this means $\delta_{v_{i(n_i-1)}} = 3$. Finally, every vertex v_{ij}, $i \in I_4 \cup A, 2 \le j \le n_i - 2$, has two neighbors, which are the vertices $v_{i(j-1)}$ with a degree 2 and the vertex $v_{i(j+1)}$ with a degree 2, this means $\delta_{v_{ij}} = 4$. Now, we can apply Equation 1 to the first edges uv_{i1}, $1 \le i \le m$ to obtain the terms E_1, E_2 and E_3:

$$\sum_{uv_{i1},\ i\in I_1}^{1\le i\le m} \left(\frac{\delta_u \cdot \delta_{v_{i1}}}{\delta_u + \delta_{v_{i1}} - 2}\right)^3 = E_1,\ \sum_{uv_{i1},\ i\in I_2}^{1\le i\le m} \left(\frac{\delta_u \cdot \delta_{v_{i1}}}{\delta_u + \delta_{v_{i1}} - 2}\right)^3 = E_2,$$

$$\sum_{uv_{i1},\ i\in I_3 \cup I_4 \cup A} \left(\frac{\delta_u \cdot \delta_{v_{i1}}}{\delta_u + \delta_{v_{i1}} - 2}\right)^3 = E_3.$$

And, we apply Equation 1 to the second edges $v_{i1}v_{i2}$, $1 \le i \le m, i \notin I_1$, to obtain the terms E_4, E_5, and E_6. If $i \in I_2$, then the second edges will be $v_{i1}v_{in_i}$, as shown in term E_4 below:

$$\sum_{v_{i1}v_{i2}=v_{i1}v_{in_i},\ i\in I_2}^{1\le i\le m} \left(\frac{\delta_{v_{i1}} \cdot \overbrace{\delta_{v_{i2}}}^{=\delta_{v_{in_i}}}}{\delta_{v_{i1}} + \underbrace{\delta_{v_{i2}}}_{=\delta_{v_{in_i}}} - 2}\right)^3 = E_4,\ \sum_{v_{i1}v_{i2},\ i\in I_3}^{1\le i\le m} \left(\frac{\delta_{v_{i1}} \cdot \delta_{v_{i2}}}{\delta_{v_{i1}} + \delta_{v_{i2}} - 2}\right)^3 = E_5,$$

$$\sum_{v_{i1}v_{i2},\ i\in I_4 \cup A}^{1\le i\le m} \left(\frac{\delta_{v_{i1}} \cdot \delta_{v_{i2}}}{\delta_{v_{i1}} + \delta_{v_{i2}} - 2}\right)^3 = E_6.$$

Also, we apply Equation 1 to the end edges $v_{in_i-1}v_{in_i}$, $1 \le i \le m, i \in I_3 \cup I_4 \cup A$ (if $i \in I_2$, then the end edges will be the second edges $v_{i1}v_{i2}$, as shown in term E_4 above), to obtain the term E_7:

$$\sum_{v_{i(n_i-1)}v_{in_i},\ i\in I_3 \cup I_4 \cup A}^{1\le i\le m} \left(\frac{\delta_{v_{i(n_i-1)}} \cdot \delta_{v_{in_i}}}{\delta_{v_{i(n_i-1)}} + \delta_{v_{in_i}} - 2}\right)^3 = E_7.$$

Furthermore, we apply Equation 1 to the edges $v_{i(n_i-2)}v_{i(n_i-1)}$, $1 \le i \le m, i \in I_4 \cup A$ (if $i \in I_3$, then the edges $v_{i(n_i-2)}v_{i(n_i-1)}$ will be the second edges $v_{i1}v_{i2}$, as shown in term E_5 above), to obtain the term E_8:

$$\sum_{v_{i(n_i-2)}v_{i(n_i-1)},\ i\in I_3 \cup I_4 \cup A}^{1\le i\le m} \left(\frac{\delta_{v_{i(n_i-2)}} \cdot \delta_{v_{i(n_i-1)}}}{\delta_{v_{i(n_i-2)}} + \delta_{v_{i(n_i-1)}} - 2}\right)^3 = E_8.$$

We calculated all edges of the path P_i, $|P_i| = n_i = 4$, as shown in terms E_3, E_6, E_7 and E_8 above. Also, we calculated four edges from every path P_i, $|P_i| = n_i = 5$, which are the first edges, second edges, last edges, and one previous. If $|P_i| = n_i = 5$, then one edge still from P_i. If $|P_i| = n_i = 6$, then two edges still from P_i. Generally, $r - 4$ still from the path P_i, where $|P_i| = n_i = r$. Now, we apply Equation 1 to the edges $v_{ij}v_{(i+1)(j+1)}$, $1 \le i \le m, 2 \le j \le n_i - 3, n_i > 4$, this means $i \in A$, to obtain the term E_9:

$$\sum_{v_{ij}v_{(i+1)(j+1)},\ i\in A}^{1\le i\le m} \left(\frac{\delta_{v_{ij}} \cdot \delta_{v_{(i+1)(j+1)}}}{\delta_{v_{i(n_i-2)}} + \delta_{v_{i(n_i-1)}} - 2}\right)^3 = E_9.$$

Where,

$$\left(\frac{\delta_{v_{ij}} \cdot \delta_{v_{(i+1)(j+1)}}}{\delta_{v_{i(n_i-2)}} + \delta_{v_{i(n_i-1)}} - 2}\right)^3 = 18.962, \forall\ i, j.$$

And, t_r is the number of paths P_i of order $n_i = r(|I_r| = t_r)$, $1 \le i \le m, 4 < r \le l$.

Let C be a cycle on m vertices with the vertex set $V(C) = \{u_1, u_2, \cdots, u_m\}$ and let $P_1, P_2, \cdots, P_m$ be the paths of orders $n_1, n_2, \cdots, n_m$, where n_i is the positive integers, $1 \le i \le m$. The **sunlike graph** $U_m(n_1, n_2, \cdots, n_m)$ is the graph obtained by identifying the end vertex of P_i with the i-th vertex of C for all $i = 1,2, \cdots, m$, see Figure 6. In other words, a sunlike graph has the property that

$U_m(n_1, n_2, \cdots, n_m) - C_m = P_1 \cup P_2 \cup \cdots \cup P_m$, where $|P_i| = n_i, n_i \geq 1, 1 \leq i \leq m$. If the path P_i of order n_i is identifying with the vertex u_i of C_m, then we denote the vertex u_i by $u_{i(n_i)}$. Also, we define $u_i(1)$ is the smallest vertex, this means the identification path P_i has order 1. Moreover, the vertex v_{ij} means the vertex v at the j-th position of the i path (P_i), $1 \leq j \leq n_i$. If $n_i \geq 1, j = n_i$, then $d(v_{in_i}) = 1$.

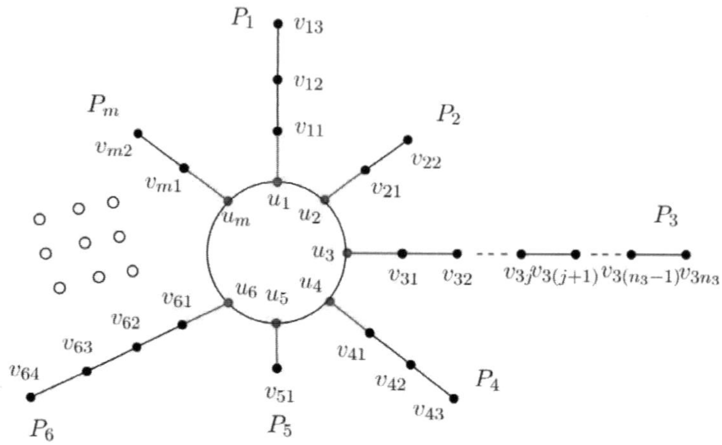

Figure 6. The sunlike graph $C_m(3,2,n_3,3,1,4,\cdots,2)$.

Corollary 9. Let $U_m^* = U_m(n_1, n_2, \cdots, n_m)$ be the sunlike graph, and l be the biggest order such that $l \geq n_i, 1 \leq i \leq m$. If $I_1 = \{i: 1 \leq i \leq m, n_i = 1, i.e., |P_i| = 1\}$, $I_2 = \{i: 1 \leq i \leq m, n_i = 2\}$, ..., $I_{l-1} = \{i: 1 \leq i \leq m, n_i = l-1\}$, $I_l = \{i: 1 \leq i \leq m, n_i = l\}$, where $|I_1| = t_1, |I_2| = t_2, \cdots, |I_{l-1}| = t_{l-1}, |I_l| = t_l$. Then, the Sanskruti index of U_m^* is given by:

$$S(U_m^*) = \sum_{k=1}^{12} H_k$$

Where,

$$H_1 = 18.087t_1,\ H_2 = 17.576t_2,\ \ H_3 = 48.084(m - t_1 - t_2).$$

$$H_4 = 15.625t_3,\ H_5 = 15.625t_3,\ H_6 = 23.323(m - t_1 - t_2 - t_3)$$

$$H_7 = 8(m - t_1 - t_2),\ H_8 = 13.824(m - t_1 - t_2 - t_3),\ H_9 = \sum_{4<r\leq l,\ s=r-4} 4.096t_r s$$

$$H_{10} = 68.083f_1,\ H_{11} = 79.934f_2,\ H_{12} = 95.533(m - f_1 - f_2)$$

While f_1 is the number of edges $u_i u_{i+1}$ that have two smallest vertices $u_i(1)$ and $u_{i+1}(1)$. And, f_2 is the number of edges $u_i u_{i+1}$ that have precisely one smallest vertex $u_i(1)$ or $u_{i+1}(1)$.

Proof. The proof method is the same as the proof method of Theorem 8, except for changing the vertex u in the starlike tree to the number of vertices $u_1, u_2, \cdots, u_m$ in this graph (sunlike), and we calculate the new edges $u_i u_{i+1}$ using Equation 1. Since m was the degree of vertex u in the starlike tree, but the degrees of $u_1, u_2, \cdots, u_m$ are 3. Therefore, the calculation of the first and second edges (mentioned in the proof of Theorem 8) will change depending on the degree of vertex u. So, if $u_i = u_i(1)$ (the identification path P_i has order 1), then $\delta_{u_i} = d(u_{i-1}) + d(u_{i+1}) + d(v_{i1}) = 7$. And, $\delta_{u_i} = 8$ where $u_i = u_i(n_i)$, $n_i \geq 2$, $1 \leq i \leq m$. Moreover, the vertex v_{i1}, $i \in I_1$, has only one neighbor, which is the vertex $u_i(1)$ with degree 3, this means $\delta_{v_{i1}} = 3 = \delta_{v_{in_i}}$. Also, the vertex v_{i1}, $i \in I_2$, has two neighbors, which are the vertices u_i, and v_{in_i} with degree one, this means $\delta_{v_{i1}} = 4$. Since every vertex v_{ij}, $1 \leq i \leq m$, $i \in I_3 \cup I_4 \cup A$, $2 \leq j \leq n_i - 1$, has degree 2, where $A = \bigcup_{4<r\leq l} I_r$. Then, the vertex v_{i1} has two neighbors, which are the vertices u_i and v_{i2}, this means $\delta_{v_{i1}} =$

4. Furthermore, the vertex v_{i2}, $i \in I_2$, has only one neighbor, which is the vertex v_{i1}, this means $\delta_{v_{i2}} = 2 = \delta_{v_{in_i}}$. In general, $\delta_{v_{in_i}} = 2$, where v_{in_i} is the end vertex in the path P_i with order n_i. Also, the vertex $v_{i(n_i-1)}$, $i \in I_3$, has two neighbors, which are the vertices v_{in_i} with degree 1 and the vertex $v_{i(n_i-2)}$ with degree 2, this means $\delta_{v_{i(n_i-1)}} = 3$. Finally, every vertex v_{ij}, $i \in I_4 \cup A$, $2 \leq j \leq n_i - 2$, has two neighbors, which are the vertices $v_{i(j-1)}$ with a degree 2 and the vertex $v_{i(j+1)}$ with a degree 2, this means $\delta_{v_{ij}} = 4$. Now, we can apply Equation 1 to the first edges $u_i v_{i1}$, $1 \leq i \leq m$, to obtain the terms H_1, H_2 and H_3:

$$\sum_{u_i(1)v_{i1},\ i\in I_1}^{1\leq i\leq m} \left(\frac{\delta_{u_i(1)} \cdot \delta_{v_{i1}}}{\delta_{u_i(1)} + \delta_{v_{i1}} - 2}\right)^3 = H_1,\ \sum_{u_i v_{i1},\ i\in I_2}^{1\leq i\leq m} \left(\frac{\delta_{u_i} \cdot \delta_{v_{i1}}}{\delta_{u_i} + \delta_{v_{i1}} - 2}\right)^3 = H_2$$

$$\sum_{u_i v_{i1},\ i\in I_3\cup I_4\cup A}^{1\leq i\leq m} \left(\frac{\delta_{u_i} \cdot \delta_{v_{i1}}}{\delta_{u_i} + \delta_{v_{i1}} - 2}\right)^3 = H_3$$

The method used to obtain $H_4, H_5, \cdots, H_9$ is the same as the method used to obtain $E_4, E_5, \cdots, E_9$ in the proof of Theorem 8. Equation 1 is applied to the second edges $v_{i1}v_{i2}$, $1 \leq i \leq m$, $i \notin I_1$, to obtain the terms H_4 and H_5. Also, it is applied to the end edges $v_{i(n_i-1)}v_{in_i}$, $1 \leq i \leq m$, $i \in I_3 \cup I_4 \cup A$, to obtain the term H_7. Moreover, Equation 1 is applied to the edges $v_{i(n_i-2)}v_{i(n_i-1)}$, $1 \leq i \leq m$, $\in I_4 \cup A$, to obtain the term H_8 . Also, it is applied to the edges $v_{ij}v_{(i+1)(j+1)}$, $1 \leq i \leq m$, $2 \leq j \leq n_i - 3$, $n_i > 4$, this means $i \in A$, to obtain the term H_9 . Finally, we apply Equation 1 to the edges of the cycle C_n which are $u_i u_{i+1}$, $1 \leq i \leq m$, to obtain the terms H_{10}, H_{11} and H_{12}:

$$\sum_{u_i(1)u_{i+1}(1),i\in A}^{1\leq i\leq m} \left(\frac{\delta_{u_i(1)} \cdot \delta_{u_{i+1}(1)}}{\delta_{u_i(1)} + \delta_{u_{i+1}(1)} - 2}\right)^3 = H_{10}$$

$$\sum_{u_i(1)u_{i+1}}^{1\leq i\leq m} \left(\frac{\delta_{u_i(1)} \cdot \delta_{u_{i+1}}}{\delta_{u_i(1)} + \delta_{u_{i+1}} - 2}\right)^3 = \sum_{u_i u_{i+1}(1)}^{1\leq i\leq m} \left(\frac{\delta_{u_i} \cdot \delta_{u_{i+1}(1)}}{\delta_{u_i} + \delta_{u_{i+1}(1)} - 2}\right)^3 = H_{11}$$

$$\sum_{u_i u_{i+1}}^{1\leq i\leq m} \left(\frac{\delta_{u_i} \cdot \delta_{u_{i+1}}}{\delta_{u_i} + \delta_{u_{i+1}} - 2}\right)^3 = H_{12}$$

3. Conclusion

In this paper, we computed the Sanskruti index of some chemical trees: caterpillar trees (especially: the tree $P_n(3, 2, 2, \cdots, 2, 3)$ which is the molecular graph of certain hydrocarbon (straight-chain alkanes)), cycle-caterpillars, starlike trees and sunlike graphs. These results may also hold in other families of molecular graphs. Moreover, there are several research avenues that may naturally extend the results of this paper.

4. References

[1] Li X and Yongtang Shi 2008 A Survey on the Randic Index *(MATCH Communications in Mathematical and in Computer Chemistry* vol 59) pp 127–156

[2] Estrada E Torres L Rodriguez L and Gutman I 1998 An Atom-bond Connectivity Index: Modelling the Enthalpy of Formation of Alkanes *Indian Journal of Chemistry (*vol 37A) issue 10 (India: NISCAIR-CSIR) pp 849–855

[3] Muhammad SS Soheil Z and Mohammad RF 2017 Computing Sanskruti Index of the Polycyclic Aromatic Hydrocarbons *(Geology Ecology and Landscapes* vol 1) issue 1 (London: Taylor & Francis) pp 37–40

[4] Wiener H 1947 Structural Determination of Paraffin Boiling Points *(Journal of the American Chemical Society* vol 69) issue 1 (Washington: ACS Publications) pp 17–20

[5] Ivan G and Oskar EP 1986 *Mathematical Concepts in Organic Chemistry* (Berlin Heidelberg: Springer)

[6] Randic M 1975 Characterization of Molecular Branching *(Journal of the American Chemical Society* vol 97) issue 23 (Washington: ACS Publications) pp 6609–6615

[7] A Ernesto E 2008 Atom-bond Connectivity and the Energetic of Branched Alkanes *(Chemical Physics Letters* vol 463) issues 4–6 (London: Elsevier) pp 422–425
[8] Furtula B Graovac A and Vukičević D 2010 Augmented Zagreb index *(Journal of Mathematical Chemistry* vol 48) (Switzerland: Springer Nature) pp 370–380
[9] Hosamani SM 2017 Computing Sanskruti Index of Certain Nanostructures *(Journal of Applied Mathematics and Computing* vol 54) (Switzerland: Springer Nature) pp 425–433
[10] Gao Y Sardar M Zafar S and Farahani M 2017 Computing Sanskruti Index of Dendrimer Nanostars *(International Journal of Pure and Applied Mathematics* vol 115) issue 2 (Academic Publications Ltd.) pp 399–404
[11] Gao Y Farahani M Sardar M and Zafar S 2017 On the Sanskruti Index of Circumcoronene Series of Benzenoid *(Applied Mathematics* vol 8) issue 4 (Scientific Research) pp 520–524
[12] Ivan G 1998 Distance in Thorny Graph *(Publications de l'Institut Mathématique* vol 63 (77) issue 83 pp 31–36
[13] Mahdieh A and Ali I 2013 Chemical Graphs Constructed from Rooted Product and their Zagreb indices *(MATCH Communications in Mathematical and in Computer Chemistry* vol 70) issue 3 pp 901–919

Acknowledgments
The authors would like to thank the referees for helpful comments and suggestions. First author was supported by the Stipendium Hungaricum PhD at the Budapest University of Technology and Economics.

Journal of Physics: Conference Series **2287** (2022) 012006 doi:10.1088/1742-6596/2287/1/012006

On Finite Subgroups in the General Linear Groups over an Algebraic Number Field

Ying Li

269 South Third Ring Road, Xiushui Road, Shengze Town, Wujiang District, Suzhou City, Jiangsu Province, China
E-mail: liying_louisechina@163.com

Abstract. As is well-known, there are only finitely many isomorphic classes of finite subgroups in a given general linear group over the field of rational numbers. This result can be generalized to any algebraic number field. While the case of field of rational numbers is relatively well-studied, we still do not know much for general algebraic number field cases. In this article, we discuss the finiteness of isomorphic classes of finite subgroups of general linear groups over an algebraic number field. We give a method to calculate a multiplicative bound for the orders of finite subgroups and to classify finite cyclic subgroups.

1. Introduction

For a given positive integer n, it is well-known that, up to isomorphism, there are finitely many finite subgroups of $GL_n(\mathbb{Q})$. It is a result dating back to the era of Minkowski [3]. Decades later, Schur generalized this result by replacing $\mathbb{Q}$ with any algebraic number field [5]. He used the character theory of finite groups to obtain this result.[4]

Many works have been done on the classification of finite subgroups of $GL_n(\mathbb{Q})$. We have even thorough classification results when n is relatively small [1, 2]. However, for other algebraic number fields, we still do not have too many classification results.

The ultimate goal of our project is to determine, up to isomorphism, all finite subgroups of $GL_n(K)$ for a given positive integer n and for an algebraic number field K. In practice, we have obtained the following partial results:

- In Section 2, we give a general method to classify finite cyclic subgroups of $GL_n(K)$.
- In Section 3, we propose another proof of Schur's result [5] when the ring of integers in K is a principal ideal domain, without using the character theory of finite groups.
- In Section 4, we give a general method to calculate a multiplicative bound of finite subgroups of $GL_n(K)$.

The general results proposed in Section 2 and 4 can be applied directly to concrete examples. For example:

1. The order of any finite subgroup of $GL_2(\mathbb{Q}[\sqrt{-1}])$ divides 96, and finite cyclic subgroups of $GL_2(\mathbb{Q}[\sqrt{-1}])$ are precisely C_1, C_2, C_3, C_4, C_6, C_8, C_{12}, up to isomorphism.
2. The order of any finite subgroup of $GL_2(\mathbb{Q}[\sqrt{-2}])$ divides 48, and finite cyclic subgroups of $GL_2(\mathbb{Q}[\sqrt{-2}])$ are precisely C_1, C_2, C_3, C_4, C_6, C_8, up to isomorphism.
3. The order of any finite subgroup of $GL_3(\mathbb{Q}[\sqrt{-1}])$ divides 384, and finite cyclic subgroups of $GL_3(\mathbb{Q}[\sqrt{-1}])$ are precisely C_1, C_2, C_3, C_4, C_6, C_8, C_{12}, up to isomorphism.

The above results are calculated explicitly in Sections 2.3. and 4.3.

The above results can help us classify finite subgroups in $GL_n(K)$ even if we do not know the thorough classification. For example, the symmetric group S_5 is not in $GL_2(\mathbb{Q}[\sqrt{-1}])$, since $|S5|$ contains 5 as a factor whereas we prove that the order of any finite subgroup of $GL_2(\mathbb{Q}[\sqrt{-1}])$ divides 96. For similar reasons, we know that S_5 is not in $GL_2(\mathbb{Q}[\sqrt{-2}])$ or in $GL_3(\mathbb{Q}[\sqrt{-1}])$, either.

2. Finite Cyclic Subgroups of $GL_n(K)$

In this section, we are going to give a method to classify finite cyclic subgroups of $GL_n(K)$. By definition, classifying cyclic subgroups is equivalent to finding all possible orders of elements in $GL_n(K)$. We find that this problem is closely related to the irreducibility of cyclotomic polynomials in different fields. So we start with a brief introduction to cyclotomic polynomials.

2.1. Cyclotomic Polynomials

Let n be a positive integer and $\mathbb{U}_n$ be the set of primitive n-th roots of unity in $\mathbb{C}$. In other words, $\mathbb{U}_n$ consists of the complex numbers of the form $e^{\frac{2\pi ik}{n}}$ with k coprime with n. One may define a polynomial with complex coefficients in the following form:

$$\Phi_n(X) := \prod_{\zeta\in\mathbb{U}n}(X-\zeta).$$

We call Φ_n the n-th cyclotomic polynomial. This polynomial is clearly of degree $\phi(n)$, where $\phi(n)$ denotes the Euler totient function. We list some properties of cyclotomic polynomials that we shall utilize afterwards.

2.1.1. Proposition. Let $\Phi_n(X)$ denote the n-th cyclotomic polynomial.

1. $X^n - 1 = \prod_{d|n}\Phi_d(X)$.
2. For each n, Φn is a polynomial with integer coefficients.
3. Φ_n is irreducible in $\mathbb{Q}[X]$.

Proof: 1. Since $X^n - 1$ and $\prod_{d|n}\Phi_d(X)$ are both monic polynomials, we can prove the equation by showing that they have the same roots. The roots of $X^n - 1$ can be written as $e^{\frac{2\pi ik}{n}}$ where k = 0, 1, · · ·, n - 1. Let d be the greatest common divisor of k and n, thus $\frac{k}{d}$ is coprime with $\frac{n}{d}$, i.e., $e^{\frac{2\pi ik}{n}}$ is a root of $\Phi_{\frac{n}{d}}(X)$. Therefore, $X^n - 1 = \prod_0^{n-1}(X-e^{\frac{2\pi ik}{n}}) = \prod_{d|n}\Phi_d(X)$.

2. First, we are going to prove that Φ_p is a polynomial with integer coefficients when p is a prime number. Since $X^p - 1 = \Phi_1(X)\Phi_p(X) = (X-1)\Phi_p(X)$, we can easily know that $\Phi p(X) = \frac{X^p-1}{X-1} = 1 + X + X^2 + \cdots + X^{p-1}$, which shows that $\Phi_p(X)$ is a polynomial with integer coefficients. Here is a fact we are going to use later: when $f, g, h \in \mathbb{C}[X]$ are monic polynomials and $f = gh$, if $f, g \in \mathbb{Z}[X]$, then h is also a polynomial with integer coefficients. Hence, we assume by induction that for any $m < n$, $\Phi_m(X)$ is in $\mathbb{Z}[X]$. Since $X^n - 1 = \prod_{d|n,d<n}\Phi_d(X)$, $\Phi_n(X)$ is a polynomial with integer coefficients because of the fact mentioned above.

3. This result is well-known and a standard proof can be found in, for example, the Theorem 4.2.6 of [7]

2.1.2. Corollary: Let $\zeta_n \in \mathbb{U}n$ be a primitive n-th root of unity in $\mathbb{C}$. Then, $[\mathbb{Q}[\zeta n] : \mathbb{Q}] = \phi(n)$.
Proof: Since ζ_n is a root of Φ_n, which is irreducible in $\mathbb{Q}[X]$, Φ_n is the minimal polynomial of ζ_n over $\mathbb{Q}$. Then, we can know that $[\mathbb{Q}[\zeta n] : \mathbb{Q}] = \deg \Phi_n = \phi(n)$ by the basic knowledge of the field extension theory.

Property of the order of an element in $GL_n(K)$

In this part, we shall give a property of the order of an element of $GL_n(K)$. To begin with, let ζ_n be a primitive n-th root of unity and $\Phi_n(X)$ be the n-th cyclotomic polynomial. We know that $\Phi_n(X)$ is irreducible in $\mathbb{Q}[X]$. $\Phi_n(X)$ may not be irreducible in K. It is natural to consider the irreducible decomposition of $\Phi_n(X)$ in $K[X]$. In fact, we have the following proposition:

2.1.3. Proposition. Any irreducible factor of $\Phi_n(X)$ in K[X] is of the same degree, namely, $[K[\zeta n] : K]$.

Proof: Let $\Phi_n(X) = \Phi_{n,1}(X)\cdots\Phi_{n,r_n}(X)$ be the irreducible decomposition. We may assume ζ_n^k is a root of $\Phi_{n,1}(X)$, then $\Phi_{n,1}(X)$ is the minimal polynomial of over K. By the basic knowledge of the field extension theory, we can know that $[K[\zeta_n^k] : K] = \deg\Phi_{n,1}$. It is obvious that $K[\zeta_n^k] \subset K[\zeta_n]$. Moreover, since k is coprime with n, there exist $a, b \in \mathbb{Z}$ such that $ak + bn = 1$, and then $\zeta_n = \zeta_n^{ak+bn} = (\zeta_n^k)^a(\zeta_n^n)^b = (\zeta_n^k)^a \in K[\zeta_n^k]$. As a result of $K[\zeta_n^k] \supset K[\zeta_n]$, we can know that $K[\zeta_n^k] = K[\zeta_n]$. Therefore, $\deg\Phi_{n,1}(X) = \cdots = \deg\Phi_{n,r_n}(X) = [K[\zeta_n^k] : K] = [K[\zeta_n] : K]$.

In what follows, we shall denote

$$\phi_K(n) := [K[\zeta_n] : K].$$

Notice that $\phi_{\mathbb{Q}}(n) = \phi(n)$ is nothing but the Euler totient function.

2.1.4. Theorem. Let K be an algebraic number field and n a positive integer. Let $A \in GL_n(K)$ be an element of order d. Then d must be the least common multiple of some integers ℓ, not necessarily distinct, such that the sum of these $\phi_K(n)$ is less than or equal to n.

Proof: Let $\mu_A \in K[X]$ be the minimal polynomial of A with order d. Then $\mu_A \mid (X^d\text{-}1)$ and $\mu_A \nmid (X^d - 1)$ for every $1 \le d' < d$. By the property of cyclotomic polynomials, $X^d - 1 = \prod_{\ell|d}\Phi_\ell(X)$. If we write $\Phi_\ell(X)$ by its irreducible decomposition:

$$\Phi_\ell(X) = \Phi_{\ell,1}(X)\cdots\Phi_{\ell,r_\ell}(X),$$

thus, the irreducible decomposition of $(X^d - 1)$ over $K[X]$ is:

$$X^d - 1 = \prod_{\ell|d}\Phi_{\ell,1}(X)\cdots\Phi_{\ell,r_\ell}(X).$$

Therefore, $\mu_A \mid (X^d - 1)$ implies that μ_A is the product of some irreducible factors of $(X^d - 1)$, which is to say,

$$\mu_A = \prod_{\text{some } \ell|d}\Phi_{\ell,1}(X)\cdots\Phi_{\ell,s_\ell}(X),$$

where $1 \le s_\ell \le r_\ell$. We claim that the least common multiple of these ℓ must be d. In fact, we assume by contradiction that the least common multiple of these integers ℓ is d' where $1 \le d' < d$, and then $\mu_A \mid (X^{d'} - 1)$, which contradicts the fact that $\mu_{A'} \mid (X^{d'} - 1)$ for any $1 \le d' < d$. By the discussion above, since $\deg\Phi_{\ell,i}(X) = \phi_K(\ell)$ for any $1 \le i < s_\ell$, we can know that $\deg\mu_A = \sum S_\ell\,\phi_K(\ell)$ for some $\ell \mid d$ and the least common multiple of these ℓ is d. On the other hand, by the Hamilton-Cayley theorem, $\mu_A \mid \chi_A$, so we can get the inequality that $\deg\mu_A \le deg\,\chi_A = n$, which can give a constraint

$$\sum S_\ell\phi_K(\ell) \le n,$$

for some $\ell \mid d$ and the least common multiple of these ℓ is d.

2.1.5. Corollary. There are only finitely many possible orders of elements in $GL_n(K)$ for a given positive integer n and a given algebraic number field K.

Proof: Let d be the order of an element in $GL_n(K)$. According to Telescopic extension theorem,

$$[K[\zeta_\ell] : \mathbb{Q}] = [K[\zeta_\ell] : \mathbb{Q}[\zeta_\ell]] \cdot [\mathbb{Q}[\zeta_\ell] : \mathbb{Q}] = [K[\zeta_\ell] : K] \cdot [K : \mathbb{Q}],$$

We get this equation:

$$\phi_K(\ell) = [K[\zeta_\ell]:K] = \frac{[K[\zeta_\ell]:\mathbb{Q}[\zeta_\ell]]\cdot\phi(\ell)}{[K:\mathbb{Q}]}.$$

In particular, it implies this equality:

$$\phi_K(\ell) \geq \frac{\phi(\ell)}{[K:\mathbb{Q}]}.$$

Combined with the constraint in the theorem above, this inequality implies that

$$\sum S_\ell \phi_K(\ell) \leq n\,[K:\mathbb{Q}],$$

for some $\ell \mid d$ and the least common multiple of ℓ is d Obviously, there are only finitely many positive integers ℓ whose Euler totient function $\phi(\ell) \leq n[K:\mathbb{Q}]$. Therefore, as the least common multiple of some ℓ satisfying $\phi(\ell) \leq n[K:\mathbb{Q}]$, d has only finitely many possible values.

Some examples

In this section, we are going to classify finite cyclic subgroups for $GL_2(\mathbb{Q}[\sqrt{-1}])$, $GL_3(\mathbb{Q}[\sqrt{-1}])$, $GL_2(\mathbb{Q}[\sqrt{-2}])$ and $GL_3(\mathbb{Q}[\sqrt{-2}])$ by using the method mentioned above to show how the mechanism works. For this purpose, we stretch a famous result from [7]:

2.1.6. Theorem. Let d > 1 be a positive integer and we denote

$$A'_d = \left\{\left(\frac{-1}{p}\right) p : p \text{ is an odd prime number dividing } d\right\},$$

where $\left(\frac{-1}{p}\right)$ is the Legendre symbol. Then, we define

$$A_d = \begin{cases} A'_d \cup \{-1\} & if\ d \equiv 4 \bmod 8; \\ A'_d \cup \{-1, 2\} & if\ d \equiv 0 \bmod 8; \\ A'_d & otherwise. \end{cases}$$

Let m be a square-free integer and ζ_d denote a primitive d-th root of unity. Then $\sqrt{m} \in \mathbb{Q}[\zeta_d]$ if and only if m is a nontrivial product of distinct elements in A_d.

2.1.7. Example. Let $K = \mathbb{Q}[\sqrt{-1}]$ and $n = 2$. Let d be the order of an element in $GL_2(\mathbb{Q}[\sqrt{-1}])$. Considering the equation we have shown above:

$$\phi_K(\ell) = \frac{[K[\zeta_\ell]:\mathbb{Q}[\zeta_\ell]]\cdot\phi(\ell)}{[K:\mathbb{Q}]},$$

since $[K:\mathbb{Q}] = 2$ in this case, we can show by *2.3.1 Theorem* that

$$[K[\zeta_\ell]:\mathbb{Q}[\zeta_\ell]] = \begin{cases} 1 & if\ \sqrt{-1} \in \mathbb{Q}[\zeta_\ell]; \\ 2 & if\ \sqrt{-1} \notin \mathbb{Q}[\zeta_\ell]; \end{cases}$$

which indicates that, in this case,

$$\phi_K(\ell) = \begin{cases} \phi(\ell) & if\ 4 \nmid \ell \\ \frac{1}{2}\phi(\ell) & if 4 \mid \ell \end{cases}$$

Thus, $\phi_K(\ell) \leq n = 2$ indicates that when $4 \nmid \ell, \phi(\ell) \leq 2$, and when $4 \mid \ell, \phi(\ell) \leq 4$.
Then, by calculation, we find that ℓ has seven possible values: 1, 2, 3, 4, 6, 8, 12. Since d is the least common multiple of some ℓ, with the constraint

$$\sum S_\ell \phi_K(\ell) \leq n,$$

The possible values of d are 1, 2, 3, 4, 6, 8, and 12, which means that in $GL_2(\mathbb{Q}[\sqrt{-1}])$, there are seven kinds of finite cyclic subgroups whose orders are possibly 1, 2, 3, 4, 6, 8, or 12. Next, we show that these orders are obtainable.
In fact, we can indeed find possible elements of those different orders to show their existence.

Table 1. Restrict the possible orders of elements

$d = 1$	$\begin{pmatrix} 1 & 0 \\ 0 & 1 \end{pmatrix}$
$d = 2$	$\begin{pmatrix} -1 & 0 \\ 0 & -1 \end{pmatrix}$
$d = 3$	$\begin{pmatrix} 1 & 3 \\ -1 & -2 \end{pmatrix}$
$d = 4$	$\begin{pmatrix} \sqrt{-1} & 0 \\ 0 & \sqrt{-1} \end{pmatrix}$
$d = 6$	$\begin{pmatrix} -1 & -3 \\ 1 & 2 \end{pmatrix}$
$d = 8$	$\begin{pmatrix} 1 & \sqrt{-1}-1 \\ 1 & -1 \end{pmatrix}$
$d = 12$	$\begin{pmatrix} \sqrt{-1} & -3\sqrt{-1} \\ \sqrt{-1} & -2\sqrt{-1} \end{pmatrix}$

As shown in Table 1, using the same method, we can restrict the possible orders of elements of $GL_3(\mathbb{Q}[\sqrt{-1}])$ to be 1, 2, 3, 4, 6, 8, and 12. They are the same with those of $GL_2(\mathbb{Q}[\sqrt{-1}])$. To show the existence, if A is a 2×2 matrix of order d in $GL_2(\mathbb{Q}[\sqrt{-1}])$, then

$$\begin{pmatrix} 1 & 0 \\ 0 & A \end{pmatrix}$$

would be a 3×3 of order in d in $GL_3(\mathbb{Q}[\sqrt{-1}])$

2.1.8. Example. Let $K = \mathbb{Q}[\sqrt{-2}]$ and d be the order of an element in $GL_n(\mathbb{Q}[\sqrt{-2}])$. By 2.3.1 Theorem above, we can get the following equation:

$$[K[\zeta_\ell]: \mathbb{Q}[\zeta_\ell]] = \begin{cases} 1 & if\ \sqrt{-2} \in \mathbb{Q}[\zeta_\ell]; \\ 2 & if\ \sqrt{-2} \notin \mathbb{Q}[\zeta_\ell]; \end{cases}$$

which indicates that, in this case,

$$\phi_K(\ell) = \begin{cases} \phi(\ell) & if\ 8 \nmid \ell \\ \frac{1}{2}\phi(\ell) & if\ 8 \mid \ell \end{cases}$$

As shown in Table 2, we now use the result of the preceding section to classify finite cyclic subgroups of $GL_n(\mathbb{Q}[\sqrt{-2}])$ for n = 2, 3.

When $n = 2$, $\phi_K(\ell) \leq 2$ indicates that when $8 \nmid \ell$, $\phi(\ell) \leq 2$, and when $8 \mid \ell$, $\phi(\ell) \leq 4$. Then, by calculation, we find that ℓ has six possible values: 1, 2, 3, 4, 6, 8. Since d is the least common multiple of some ℓ, with the constraint

$$\sum S_\ell \phi_K(\ell) \leq n,$$

the possible values of d are 1, 2, 3, 4, 6, and 8, which means that in $GL_2(\mathbb{Q}[\sqrt{-2}])$, there are six kinds of finite cyclic subgroups whose orders are possibly 1, 2, 3, 4, 6, or 8. We list possible matrices whose orders are exactly those possible values we find.

2.1.9.

Table 2. The possible values and matrices

$d = 1$	$\begin{pmatrix} 1 & 0 \\ 0 & 1 \end{pmatrix}$
$d = 2$	$\begin{pmatrix} -1 & 0 \\ 0 & -1 \end{pmatrix}$
$d = 3$	$\begin{pmatrix} 1 & 3 \\ -1 & -2 \end{pmatrix}$
$d = 4$	$\begin{pmatrix} 0 & 1 \\ -1 & 0 \end{pmatrix}$
$d = 6$	$\begin{pmatrix} -1 & -3 \\ 1 & 2 \end{pmatrix}$
$d = 8$	$\begin{pmatrix} \sqrt{-2} & 1 \\ 1 & 0 \end{pmatrix}$

When $n = 3$, like the case when $n = 2$, $\phi_K(\ell) \leq 3$ indicates that when $8 \nmid \ell$, $\phi(\ell) \leq 3$, and when $8 \mid \ell$, $\phi(\ell) \leq 6$. Then, by calculation, we find that ℓ has six possible values: 1, 2, 3, 4, 6, 8. Since d is the least common multiple of some ℓ, with the constraint

$$\sum S_\ell \phi_K(\ell) \leq n,$$

the possible values of d are 1, 2, 3, 4, 6, and 8, which means that in $GL_3(\mathbb{Q}[\sqrt{-2}])$, there are six kinds of finite cyclic subgroups whose orders are possibly 1, 2, 3, 4, 6, or 8. If A is a 2×2 matrix of order d in $GL_2(\mathbb{Q}[\sqrt{-2}])$, then

$$\begin{pmatrix} 1 & 0 \\ 0 & A \end{pmatrix}$$

would be a 3×3 matrix of order d in $GL_3(\mathbb{Q}[\sqrt{-2}])$.

3. Finite Subgroups of $GL_n(K)$

In this section, we are going to prove that up to isomorphism, there are only finitely many finite subgroups of $GL_n(K)$ for a given positive integer n and an algebraic number field K, whose ring of integers O_K is a principal ideal domain. First, we will show that any finite subgroup of $GL_n(K)$ is isomorphic to a subgroup of $GL_n(O_K)$. Then after modulo a well-chosen prime element x in O_K, we find any finite subgroup of $GL_n(O_K)$ is isomorphic to a subgroup of $GL_n(O_K/(x))$, while the latter, as a finite group, has only finitely many finite subgroups.

Algebraic number fields whose rings of integers are principal ideal domains are important in practical mathematical uses. First of all, it is easier to calculate with elements instead of with ideals. More importantly, the structure of modules over a principal ideal domain is relatively much clearer than that over arbitrary commutative rings, as can be seen from *Section 3.1*. Therefore, these algebraic number fields are of special interests among all algebraic number fields. Some interesting facts or conjectures about this kind of algebraic number fields are listed below:

1. For an algebraic number field K, O_K is a principal ideal domain if and only if it is a unique factorization domain.
2. If K is an imaginary quadratic number field, i.e., if $K = \mathbb{Q}[\sqrt{-d}]$ with d a square-free positive integer, then O_K is a principal ideal domain if and only if d is one of

   $$1, 2, 3, 7, 11, 19, 43, 67, 163$$

   This interesting fact, discovered and conjectured by Carl Friedrich Gauss in 1801, turned out to be very difficult and was finally proven almost 150 years later.

3. Gauss also conjectured that there are infinitely many real quadratic number fields, i.e., number fields of the form $\mathbb{Q}[\sqrt{d}]$ with d a positive integer, such that their rings of integers are principal ideal domains. This conjecture remains open nowadays.

We are not using these interesting facts in our article. We list them to show the importance of principal ideal domains in the studying of general algebraic number fields.

3.1. Free Modules over a Principal Ideal Domain

A general result on free modules over a principal ideal domain is that submodules of a free module of rank n are free and of rank no greater than n. We will give a proof of this result.

We begin by giving the definition of an A-module. It is a generalization of vector spaces over a field. But the properties of A-modules are more complicated.

3.1.1. Definition. Let A be a commutative ring. A module M over A is a set endowed with two operations:

$$+ : M \times M \to M;$$

$$\cdot : A \times M \to M;$$

satisfying:

$(+ - axioms)$ $(M, +)$ is an abelian group;

$(\cdot - axioms)$

- $\forall \lambda, \mu \in A, x \in M, \lambda(\mu x) = (\lambda\mu)x;$
- $\forall x \in M, 1 \cdot x = x;$
- $\forall \lambda, \mu \in A, x \in M, (\lambda + \mu)x = \lambda x + \mu x;$
- $\forall \lambda \in A, x, y \in M, \lambda(x + y) = \lambda x + \lambda y.$

3.1.2. Definition (Basis). Let M be *an* $A\text{-}module$. $B \subset M$ *is called an* $A\text{-}basis$ *of* M, *if the following conditions hold:*

- $\forall x \in M, \exists x_1, \cdots, x_n \in B, \lambda_1, \cdots, \lambda_n \in A,$ such that $x = \sum_1^n \lambda_i x_i;$
- If $x_1, \cdots, x_n \in B$ are distinct and for $\lambda_1, \cdots, \lambda_n \in A,$

$$\sum_{i=1}^n \lambda_i x_i = 0 \Longrightarrow \lambda_i = 0, \forall i$$

The definition of an A-basis of an A-module is similar to that of a vector space. The difference is that an A-module does not necessarily have a basis. If an A-module M admits a basis, then M is called free. A good example of a free A-module is:

$$A^n = \{(x_1, \cdots, x_n) : x_i \in A\}$$

where $\forall \lambda \in A$, $(x_1, \cdots, x_n) \in A^n, \lambda(x_1, \cdots, x_n) := (\lambda x_1, \cdots, \lambda x_n)$. Apparently, A^n is an $A-module$. There exists a basis of A^n, called the canonical basis of A^n, namely, $B := \{e_1, \cdots, e_n\}$, where e_i is an element in A^n which has 1 in its i-th position and 0 elsewhere. Therefore, we can draw the conclusion that A^n is a free $A\text{-}module$.

3.1.3. Definition. Let M, N be $A\text{-}modules$. $\phi : M \to N$ is called a homomorphism of $A\text{-}modules$ (or A-linear map), if the following conditions hold:

$$\forall_x \in M, \lambda \in A, \phi(\lambda x) = \lambda\phi(x);$$

$$\forall_{x,y} \in M, \varphi(x + y) = \phi(x) + \phi(y).$$

Furthermore, $\phi : M \to N$ is an $A\text{-}isomorphism$, denoted as $M \cong N$, if it is a bijective homomorphism.

Let M be a free $A\text{-}module$ with a basis B. Assume that the $|B| = n$, then clearly $M \cong A^n$ as $A\text{-}modules$.

It is well-known in linear algebra that any basis of a given vector space has the same cardinal. The same result holds for a free $A-module$. In fact, we are going to prove in the following paragraphs that $A^m \cong A^n$ if and only if $m = n$. Given this result, we define the rank of a free A-module as the cardinal of any of its bases.

3.1.4. Lemma 3.1.1. Let $\phi : A^m \to A^n$ be a homomorphism of A-$modules$, and a be an ideal of A. If $x = (x_1,\cdots, x_m)$ is in a^m, then $\phi(x) \in a^n$.
Proof: Let $\phi(x) = (\phi_1(x),\cdots,\phi_n(x))$ and $B := \{e_1,\cdots,e_m\}$ be the canonical basis of A^m. For any j, notice that $\phi_j(x) = \phi_j(\sum x_i e_i) = \sum x_i \phi_j(e_i) \in a$ since $x_i \in a, \forall i$. We conclude that $\phi(x) \in a^n$.

3.1.5. Theorem 3.1.1. Let A be a commutative ring. Then $A^m \cong A^n$ as A-$modules$ if and only if $m = n$.
Proof: If $A^m \cong A^n$, there exists a map $\phi : A^m \to A^n$ which is an isomorphism of $A - modules$. Since A is a commutative ring, let m be a maximal ideal of A. Then, we can define another map:

$$\psi : (A/m)^m \to (A/m)^n$$

as $(\bar{x}_1,\cdots,\bar{x}_m) \mapsto \overline{(\phi_1(x_1,\cdots, x_m)}, \cdots, \overline{\phi_n(x_1,\cdots, x_m))}$. We need to check the following conditions:
Well-definedness: Let $(\bar{x}_1,\cdots,\bar{x}_m) = (\bar{y}_1,\cdots,\bar{y}_m)$, then $x_i - y_i \in m$, $\forall i$. Therefore, by the lemma above, $\phi_j(x_i) - \phi_j(y_i) = \varphi j(x_i - y_i) \in m$, i.e , $\phi_j(x_i) = \phi_j(y_i), \forall i, j$, which means that

$$\overline{(\phi_1(x_1,\cdots, x_m)},\cdots,\overline{\phi_n(x_1,\cdots, x_m))} = \overline{(\phi_1(y_1,\cdots, x_m)},\cdots,\overline{\phi_n(y_1,\cdots, y_m))}.$$

- ψ is an $A/m - linear$ map. It is direct from the fact that ϕ is an $A - linear$ map.
- ψ is injective. In fact, let $\bar{x} = (\bar{x}_1,\cdots,\bar{x}_m) \in \ker \psi$, then $\phi_j(x_1,\cdots, x_m) \in m, \forall j$. If ξ is the inverse map of ϕ, then $x_i = \xi_i(\phi(x)) \in m$ by the above lemma. Therefore, $\bar{x}_i \in m, \forall i, i.e., \bar{x} = 0$ in $(A/m)^m$.
- ψ is surjective. This follows directly from the fact that φ is surjective.

Therefore, we construct an isomorphism of $A/m-$ vector spaces: : $(A/m)^m \cong (A/m)^n$. Since any basis of a vector space has the same number of elements, we conclude that $m = n$.
Then, we can deduce that if M is a free $A -$ module and B_1, B_2 are two $A -$ bases such that $|B_1| < +\infty, |B_2| < +\infty$, then $|B_1| = |B_2|$, which is defined to be the rank of M.

3.1.6. Definition 3.1.4. Let M be an $A - module$, $N \subset M$ is called a submodule if N itself is an A-$module$.
Specially, when $M = A$ is an $A -$ module, the submodule of M is the ideal of A.

3.1.7. Theorem 3.1.2. Let $M \subset A_n$ be a submodule, then M is a free $A - module$ with rank $r \leq n$.
Proof: We argue by induction on n.
- $n = 1$.

– If $M \cong \{0\}$, M is a free $A -$ module with rank $r = 0$ where its basis is empty.
– If $M \ncong \{0\}$, M is an ideal of A. Since A is a principal ideal domain, $M := (m) = mA$ where $m \neq 0$. We can construct a map $\phi: A \to mA$ as $x \to mx$. Apparently, ϕ is a homomorphism. Let $x \in ker\phi$, $x = 0$ since $mx = 0$ and $m \neq 0$. Obviously, ϕ is surjective, then $M \cong A$ as $A -$ modules and the rank of M is 1.
Assume this theorem is true for all $k \leq n - 1$. We shall prove that the theorem holds for n. Let M be a submodule of A^n. Let $N = \{(x, 0,\cdots, 0) : x \in A\} \subset A^n$ and $\Gamma := M \cap N$. By induction assumption, Γ is free of rank ≤ 1. We can construct a map $\psi : M \to A^{n-1}$ as $(x_1,\cdots, x_m) \mapsto (x_2,\cdots, x_n)$. Apparently, ψ is a homomorphism as $A -$ modules. Since $\ker \psi = \{(x_1,\cdots, x_n) \in M : x_2,\cdots, x_n = 0\} = M \cap N$, we can know that $M/M \cap N = M/ \ker \psi \cong$

Journal of Physics: Conference Series **2287** (2022) 012006 doi:10.1088/1742-6596/2287/1/012006

$Im\psi \subset A^{n-1}$. By induction assumption, $M/M \cap N$ is free of rank $\leq n-1$. Then, we need to prove that M is a free module. Let $\{\bar{e}_1,\cdots,\bar{e}_k\}$ be a basis of $M/M \cap N$ and $\{f_1,\cdots,f_p\}$ be a basis of $M \cap N$. May take $e1 \in \bar{e}_1,\cdots,e_k \in \bar{e}_k$. We would like to show that $\{e_1,\cdots,e_k,f_1,\cdots,f_p\}$ forms a basis of M.
$\forall x \in M, x \in \bar{x} \in M/M \cap N, \bar{x} = n_1\bar{e}_1 + \cdots + n_k\bar{e}_k \in M/M \cap N$. Therefore, $x - (n_1e_1 + \cdots + n_ke_k) = 0$, i.e., $x - (n_1e_1 + \cdots + n_ke_k) \in M \cap N$. Since $\{f_1,\cdots,f_p\}$ is a basis of $M \cap N$, there exists $m_1,\cdots,m_p$ such that $x - (n_1e_1 + \cdots + n_ke_k) = (m_1f_1 + \cdots + m_pf_p)$. Hence, $\{e_1,\cdots,e_k,f_1,\cdots,f_p\}$ spans M for $x = n_1e_1 + \cdots + n_ke_k + m_1f_1 + \cdots + m_pf_p$.
While $n_1e_1 + \cdots + n_ke_k + m_1f_1 + \cdots + m_pf_p = 0$, i.e., in $M/M \cap N, n_1e_1 + \cdots + n_ke_k + 0 = 0$, n1 $= \cdots = n_k = 0$ since $\{e_1,\cdots,e_k\}$ is a basis of $M/M \cap N$. Then, because $m_1f_1 + \cdots + m_pf_p = 0$ in $M \cap N$, we can know that $m_1 = \cdots = m_p = 0$.
Therefore, M is a free module of rank less than or equal to n.

From $GL_n(K)$ to $GL_n(O_K)$
Let K be an algebraic number field whose ring of integers O_K is a principal ideal domain. In this situation, we are going to prove that any finite subgroup of $GL_n(K)$ is isomorphic to a subgroup of $GL_n(O_K)$. (Therefore, to classify subgroups of $GL_n(K)$, it suffices to classify subgroups of $GL_n(O_K)$.
Let $G \subset GL_n(K)$ be a finite group. For any $g \in G, x \in K^n$, we have $gx \in K^n$. Regard O_K^n as a subset of K^n. We can define a subset Γ of K^n:

$$\Gamma := \{\textstyle\sum_{i=1}^{m} g_ix_i : g_i \in G, x_i \in O_K^n, m \in \mathbb{N}\}$$

We make the convention of the sum $\sum_{i=1}^{o} = 0$
Apparently, we can know that Γ is an O_K-module. Then, we are going to show that there exists $d \in O_K$ such that $O_K^n \subset \Gamma \subset \frac{1}{d}O_K^n$. Furthermore, we can deduce that Γ is a free O_K-module of rank n. In fact,
$\forall x \in O_K^n$, let $g_i := 1, x_i := x$ for any positive integer i, then $x = g_ix_i \in \Gamma$, which means $O_K^n \subset \Gamma$.
Since G is a finite subgroup of $GL_n(K)$, there are finite elements in G. For O_K is the ring of integers in K, K is the fraction field of O_K. Let d be the least common multiple of the denominators of entries in G whose existence is guaranteed by the fact that O_K is a principal ideal domain and thus a unique factorization domain. $\forall x \in \Gamma$, $x = \sum_1^m g_ix_i$ where $g_i \in G$, $x_i \in O_K^n$ and $m \in \mathbb{N}$. Since the entries of g_i are in $\frac{1}{d}O_K^n$, we conclude that $x \in \frac{1}{d}O_K^n$. That is to say, $\Gamma \subset \frac{1}{d}O_K^n$.
Since Γ is a submodule of $\frac{1}{d}O_K^n$, Γ is a free O_K-module of rank less than or equal to n. Since O_K^n is a submodule of Γ, O_K^n is of rank n less than or equal to the rank of Γ. Then, Γ is a free O_K-module of rank n.
Therefore, there exists an O_K-basis of Γ, namely, $\{e_1,\cdots,e_n\}$. Write every e_i in the coordinate form:

$$e_1 = [a_{11}\ a_{21}\ \cdots\ a_{n1}]$$
$$\vdots$$
$$e_k = [a_{1k}\ a_{2k}\ \cdots\ a_{nk}]$$
$$\vdots$$
$$e_n = [a_{1n}\ a_{2n}\ \cdots\ a_{nn}]$$

where each entry is an element in K. Then, we can define a matrix Q:

$$Q = \begin{bmatrix} a_{11} & a_{12} & \cdots & a_{1n} \\ & \vdots & & \\ a_{k1} & a_{k2} & \cdots & a_{kn} \\ & \vdots & & \\ a_{n1} & a_{n2} & \cdots & a_{nn} \end{bmatrix}$$

Apparently, $\{e_1, \cdots, e_n\}$ is a basis of K^n as $K-$vector space and every column of Q is linearly independent, so we can know that Q is in $GL_n(K)$. Then, we are going to show that for any $g \in G$, there exists $Q^{-1}gQ \in GL_n(\mathcal{O}_K)$ so that we can deduce that G is isomorphic to a subgroup of $GL_n(\mathcal{O}_K)$.

3.1.8. Lemma. For any $g \in G$, $Q^{-1}gQ$ is an invertible map from $\mathcal{O}_K^n$ to $\mathcal{O}_K^n$.
Proof: Let $\{v_1, \cdots, v_n\}$ be the canonical basis of $\mathcal{O}_K^n$ and $\{e_1, \cdots, e_n\}$ be a basis of Γ. We can understand Q as a map from $\mathcal{O}_K^n$ to : $\{v_1, \cdots, v_n\} \mapsto \{e_1, \cdots, e_n\}$.
Surjectivity: $\forall x \in \Gamma$, $\exists s_1, \cdots, s_n \in \mathcal{O}_K$ such that $x = s_1e_1 + \cdots + s_ne_n = s_1Qv_1 + \cdots + s_nQv_n = Q(s_1v_1 + \cdots + s_nv_n)$
Injectivity: Let $x = \lambda_1v_1 + \cdots + \lambda_nv_n$ be an element in $\ker Q$. Then, $Qx = \lambda_1e_1 + \cdots + \lambda_ne_n = 0$. Since $e_1, \cdots, e_n$ is linearly independent, $\lambda_1 = \cdots = \lambda_n = 0$, i.e., $\ker Q = 0$.
Therefore, Q is bijective. Similarly, we can consider the matrix $g \in G$ as a map from Γ to Γ: $\{e_1, \cdots, e_n\} \mapsto \{ge_1, \cdots, ge_n\}$. Obviously, g is bijective since g is an invertible matrix whose inverse is also a map from Γ to Γ. Then, for any $x \in \mathcal{O}_K^n$, $Q^{-1}gQx$ is in $\mathcal{O}_K^n$, which means that $Q^{-1}gQ$ is a map from $\mathcal{O}_K^n$ to $\mathcal{O}_K^n$. Therefore, $Q^{-1}gQ$ is invertible since it is bijective. Then, $Q^{-1}gQ$ is in $GL_n(\mathcal{O}_K)$.
We are ready to prove that G is isomorphic to a subgroup of $GL_n(\mathcal{O}_K)$. Define a map ϕ from G to $GL_n(\mathcal{O}_K)$: $g \mapsto Q^{-1}gQ$. We will prove that φ gives an isomorphism of G to a subgroup of $GL_n(\mathcal{O}_K)$:
Homomorphism: $\phi(I) = Q^{-1}IQ = I$ and $\phi(AB) = Q^{-1}ABQ = Q^{-1}AQQ^{-1}BQ = \phi(A)\phi(B)$.
Injectivity: Let x be in $\ker\phi$. Since $\phi(x) = Q^{-1}xQ = 1$, we can know that $I = QIQ^{-1} = QQ^{-1}xQQ^{-1} = x$, which is to say, $\ker\phi = 1$, i.e., ϕ is injective.
Hence, G is isomorphic to the image of φ as groups. Therefore, in this subsection, we have proven the following theorem:

3.1.9. Theorem. Let n be a positive integer and K be an algebraic number field whose ring of integers $\mathcal{O}_K$ is a principal ideal domain. Then, any finite subgroup of $GL_n(K)$ is isomorphic to a finite subgroup of $GL_n(\mathcal{O}_K)$.

Finite Subgroups of $GL_n(\mathcal{O}_K)$
In this subsection, we are going to prove that, up to isomorphism, there are only finitely many finite subgroups in $GL_n(\mathcal{O}_K)$.
First of all, we define a set of prime numbers:

$$\mathcal{E} = \{p : p \text{ is a prime number and is the order of an element in } GL_n(\mathcal{O}_K)\} \cup \{2\}.$$

This set, as we shall see soon, contains the prime numbers that do not have the properties we need. So we call the prime numbers in this set "*exceptions*".
By *2.2.3 Corollary*, we can see that $\mathcal{E}$ is a finite set. This is an important fact that we will use continuously in what follows.
Now let $p \notin \mathcal{E}$ be a non-exceptional prime number and x be a prime factor of p in $\mathcal{O}_K$. Let us recall the general definition of prime elements: Let A be a commutative ring. $x \in A$ is called a prime element in A, if for any $y \mid x$ in A, $y \sim 1$ or $y \sim x$. Here $a \sim b$ means a is equal to b up to an invertible element. The definition of prime elements is a generalization of prime numbers in integers.

3.1.10. Lemma. $\mathcal{O}_K/(x)$ is a finite field.
Proof: $\mathcal{O}_K/(x)$ is a field since x is a prime element in$\mathcal{O}_K/(x)$. We can define a map ϕ:

$$\phi: \mathbb{Z} \hookrightarrow \mathcal{O}_K \longrightarrow \mathcal{O}_K/(x).$$

Clearly, since $\ker\phi = p\mathbb{Z}$, $\mathbb{Z}/p\mathbb{Z} = \mathbb{Z}/\ker\phi \cong Im\phi \subset \mathcal{O}_K/(x)$. Therefore, $\mathcal{O}_K/(x)$ is an extension of $\mathbb{F}_p$. As a well-known fact, $\mathcal{O}_K$ is a free $\mathbb{Z}$ – module of finite rank $[K : \mathbb{Q}]$ (c.f. [4, p.45]). Then, let $\{e_1, \cdots, e_n\}$ be a $\mathbb{Z}$ – basis of $\mathcal{O}_K$. For any $x \in \mathcal{O}_K$, we can write x as $\sum_1^n \lambda_i e_i$ where $\lambda_i \in Z$. Hence, $\bar{x} = \sum_1^n \lambda_i e_i \in \mathcal{O}_K/(x)$ where $\bar{\lambda_i} \in \mathbb{Z}/\mathbb{Z} \cap (x) = \mathbb{Z}/p\mathbb{Z}$. So, $\mathcal{O}_K/(x)$ can be spanned by $\{\bar{e}_1, \cdots, \bar{e}_n\}$ as the vector space of $\mathbb{Z}/p\mathbb{Z}$. The dimension of $\mathcal{O}_K/(x)$ is no bigger than n over $\mathbb{F}_p$, which implies that $\mathcal{O}_K/(x)$ is a finite field.
Now, we can define a group homomorphism:

$$\tau_x : GL_n(\mathcal{O}_K) \rightarrow GL_n(\mathcal{O}_K/(x))$$

by modulo x for each entry. This homomorphism has the following interesting property:

3.1.11. Proposition. For $g \in ker\,\tau_x$ and a positive integer l such that $p \nmid l$, if $g^l = 1$, then $g = 1$.
Proof: For $g \in \ker\tau_x$, $\tau_x(g) = [1]$, which means that $g = 1 + h$ where x divides all entries of h. It suffices to prove the proposition for l a prime number. In fact, for $l \geq 2$, $l = p_1 \cdots p_r$, and g' is defined to be $g^{p_1 \cdots p_{r-1}}$. Then, $g' \equiv 1^{p_1 \cdots p_{r-1}} = 1 \bmod x$. Since $g'^{p_r} = g^l \equiv 1 \bmod x$, we can know that $g' = 1$ if we assume the proposition for l a prime number is true. By induction on r, $g = 1$.
Now we assume that l is a prime number not equal to p. May assume by contradiction that $h \neq 0$. We can write h as xdh' where $d \in \mathcal{O}_K$ and $h' \in M_n(\mathcal{O}_K)$ such that the greatest common divisor of entries of h' is 1.

$$\begin{aligned} g^l &= (1 + h)^l \\ &= (1 + xdh')^l \\ &= (xdh')^l + \binom{l}{l-1}(xdh')^{l-1} + \cdots + lxdh' + 1. \end{aligned}$$

By $g^l = 1$, we can know that

$$(xdh')^l + \binom{l}{l-1}(xdh')^{l-1} + \cdots + lxdh' = 0.$$

Since $\mathcal{O}_K$ is a domain,

$$(xd)^{l-1}h'^l + \binom{l}{l-1}(xd)^{l-2}h'^{l-1} + \cdots + lh' = 0.$$

Then, from the fact that x divides $(xd)^{l-1}h'^l + \binom{l}{l-1}(xd)^{l-2}h'^{l-1} + \cdots + \binom{l}{2}xdh'^2$ we conclude that $x|lh'$. Therefore, $x|l$ in $\mathcal{O}_K$, since the greatest common divisor of entries of h is 1. Because x divides p, l, p are not coprime in $\mathcal{O}_K$, which implies that l, p are not coprime in $\mathbb{Z}$ according to Bézout's theorem. Therefore, $l = p$, which is contradictory to our assumption of l. Hence, $h = 0$, which is to say, $g = 1$.

3.1.12. Theorem. Let x be a prime factor of non-exceptional prime number p in $\mathcal{O}_K$. For any finite subgroup $G \subset GL_n(\mathcal{O}_K)$,

$$\tau_x : GL_n(\mathcal{O}_K) \rightarrow GL_n(\mathcal{O}_K/(x))$$

gives an isomorphism of G to a subgroup of $GL_n(\mathcal{O}_K/(x))$.
Proof: We define a map $\phi: G \longrightarrow GL_n(\mathcal{O}_K/(x))$ by modulo x for each entry. We are going to show that ϕ gives an isomorphism of G to the image of ϕ. Apparently, ϕ is a group homomorphism. It suffices to show ϕ is injective. For any $g \in \ker\phi$, there exists a positive integer m such that $g^m = 1$. From our choice of p, p does not divide m. Since $\ker\phi = \ker\tau_x \cap G$, by the preceding

proposition, $g = 1$. This proves the injectivity of ϕ. Hence, ϕ gives an isomorphism of G to the image of ϕ.
From this theorem, we conclude that any finite subgroup of $GL_n(O_K)$ is isomorphic to a subgroup of $GL_n(O_K/(x))$, whereas the latter, being a finite group since $O_K/(x)$ is a finite field, has only finitely many subgroups. Taking account of the result of Section 3.2, we have finally proven the following theorem:

3.1.13. Theorem. Let n be a positive integer and K be an algebraic number field whose ring of integers is a principal ideal domain. Then there are finitely many finite subgroups of $GL_n(O_K)$, up to isomorphism.

4. A Multiplicative Bound for the Orders of Finite Subgroups of $GL_n(K)$

The preceding section shows that there are only finitely many finite subgroups of $GL_n(K)$. It is thus interesting to give a multiplicative bound for the orders of finite subgroups of $GL_n(K)$ for a given n and a given algebraic number field K. In this section, we are going to present a general method to calculate a multiplicative bound and apply it to special cases.

4.1. Preliminaries

In this subsection, we will present some tools and theorems that will be used in calculating the multiplicative bound.
Definition 4.1.1. Let A be a finite set of positive integers. A multiplicative bound of A is a positive integer M such that each element of A divides M.
In this case, A is the set of orders of finite subgroups of $GL_n(K)$.

4.1.1. Lemma. Let $\mathbb{F}_q$ be a finite field with q elements and n be a positive integer. Then there are $q^{\frac{n(n-1)}{2}}(q^{n-1}-1)\cdots(q-1)$ elements in the group $GL_n(\mathbb{F}_q)$.
Proof: A matrix is invertible if and only if its rows are linearly independent. For the first row, the only constraint is that it cannot be $(0,\cdots,0)$, so there are q^n-1 choices of the first row. Assume the first k rows are chosen, and to make sure the $(k+1)$-th row is linearly independent of the first k rows, it should not lie in the k-dimensional subspace spanned by the first k rows. So there are q^n-q^k choices of the $(k+1)$-th row. In total, there are $(q^n-1)(q^n-\mathrm{q})\cdots(q^n-q^{n-1}) = q^{\frac{n(n-1)}{2}}(q^n-1)(q^{n-1}-1)\cdots(q-1)$ choices of invertible matrics.

4.1.2. Lemma. Let ℓ be a prime number, then $(\mathbb{Z}/\ell^2\mathbb{Z})^\times$ is a cyclic group.
Proof: Since $|(\mathbb{Z}/\ell^2\mathbb{Z})^\times| = \ell(\ell-1)$, it suffices to find an element of order $\ell(\ell-1)$ in $(\mathbb{Z}/\ell^2\mathbb{Z})^\times$. First, we notice that $(\mathbb{Z}/\ell\mathbb{Z})^\times$ is cyclic since $\mathbb{Z}/\ell\mathbb{Z}$ is a finite field. We can find an integer x' such that the order of x' modulo ℓ is $\ell-1$. The order of x' modulo ℓ^2 is thus $\mathrm{d}(\ell-1)$ in $(\mathbb{Z}/\ell^2\mathbb{Z})^\times$ for some $\mathrm{d} \geq 1$. Then $x := x'^d$ is of order $\ell-1$ modulo ℓ^2. Then, we find that $y = 1+\ell$ is of order ℓ modulo ℓ^2 by noticing the following congruence equation:

$$(1+\ell)^k \equiv 1 + k\ell \bmod \ell^2.$$

Finally, since ℓ and $\ell-1$ are coprime, xy is of order $\ell(\ell-1)$ modulo ℓ^2. In other words, $(\mathbb{Z}/\ell^2\mathbb{Z})^\times$ is a cyclic group.

4.1.3. Theorem (Dirichlet's arithmetic progression) Let m, n be two coprime integers and $m > 0$. Then there are infinitely many prime numbers of the form $mk + n$ for $k \in \mathbb{N}$.
For a proof of this result, we refer the readers to Serre's textbook [6, pp. 61-76].

4.1.4. Theorem (Chinese remainder theorem) Let m, n be coprime integers and a, b be any given integers. Then there exists $x \in \mathbb{Z}$ such that

$$x \equiv a \bmod m, x \equiv b \bmod n$$

Such x is unique modulo mn.

Proof: Chinese remainder theorem is equivalent to saying that the ring homomorphism

$$\Phi: \mathbb{Z}/mn\mathbb{Z} \longrightarrow \mathbb{Z}/m\mathbb{Z} \times \mathbb{Z}/n\mathbb{Z}$$

$$x \mapsto (x \bmod m, x \bmod n)$$

is bijective. Since both sides have mn elements, it suffices to show that Φ is injective. In fact, let $x \in \ker\Phi$, then $m|x$, $n|x$. But m, n are coprime, which implies that $mn|x$, i.e., $x = 0$ in $\mathbb{Z}/mn\mathbb{Z}$.

4.2. General Method to Calculate the Multiplicative Bound

Let $G \subset GL_n(K)$ be a finite group. Let $p > 2$ be a prime number and $x \in O_K$ be a prime factor of p in O_K. Since the characteristic of $O_K/(x)$ is p, $O_K/(x)$ is isomorphic to $\mathbb{F}_{p^r}$ for some $r \geq 1$. To make the notation pithy, we write $\mathbb{F}_{p^r}$ as $\mathbb{F}_q$. Aside from finitely many exceptional prime numbers p, G can be viewed as a subgroup of $GL_n(O_K/(x))$, i.e., $GL_n(\mathbb{F}_q)$. (2.2.2 Corollary and 3.3.3 Theorem)

Let $\ell \geq 2$ be a fixed prime number, $\nu_\ell(|G|)$ is defined to be the power of ℓ of $|G|$. Then by the theorem of Lagrange, $\nu_\ell(|G|)$ is less than or equal to $\nu_\ell(|GL_n(\mathbb{F}_q)|)$. Therefore, finding the minimum of $\nu_\ell(|GL_n(\mathbb{F}_q)|)$ suffices to calculate the maximum of $\nu_\ell(|G|)$. Since

$$|GL_n(\mathbb{F}_q)| = q^{\frac{n(n-1)}{2}}(q^n - 1)(q^n - 1)(q^{n-1} - 1)\cdots(q - 1),$$

By Fermat's little theorem, $p^{\ell-1} - 1 \equiv 0$ by modulo ℓ, so ℓ is certainly a factor of $|GL_n(\mathbb{F}_q)|$ as n becomes large. However, we do not know clearly the power of ℓ and we want it to be as small as possible. Then we consider what the power of ℓ^2 is in $|GL_n(\mathbb{F}_q)|$. As we know from *4.1.2. Lemma* above, $(\mathbb{Z}/\ell^2\mathbb{Z})^\times$ is a cyclic group, which is to say, there exists an $x \in \mathbb{Z}$ such that $Ord_{\ell^2}(p) = \ell(\ell-1)$. Then, we define x in this way and as long as $p \equiv x$ by modulo ℓ^2, p satisfies $Ord_{\ell^2}(p) = \ell(\ell-1)$. According to Dirichlet's arithmetic progression [6], there are infinitely many prime numbers in $\{k\ell^2 + x\}$, while there are only finitely many exceptions which p cannot be chosen from, then we can find a prime number $p = k\ell^2 + x$ such that $Ord_{\ell^2}(p) = \ell(\ell-1)$.

Then, we can find a way to calculate $\nu_\ell(p^k - 1)$ for any $k \leq n$. Moreover, we demand that $\ell > 2$ in this following case and we will consider $\ell = 2$ later. By a direct expansion, we find

$$\nu_\ell(p^k - 1) = \begin{cases} 1 + \nu_\ell(k) & if\ \ell - 1 | k \\ 0 & if\ \ell - 1 \nmid k \end{cases}$$

As a result, we can calculate $\nu_\ell(|\mathrm{GL_n}(\mathbb{F}_q)|)$ by the equation above. Let m be the greatest common divisor of r and $\ell - 1$, then we can know that $\frac{\ell-1}{\mathrm{m}} | \mathrm{k}$ since $\ell - 1 | \mathrm{rk}$. So let $\mathrm{k} = \mathrm{d}\frac{\ell-1}{\mathrm{m}}$.

$$\nu_\ell(|GL_n(\mathbb{F}_{q^r})|) = \Sigma_{k=1}^{n} \nu_\ell(p^{rk} - 1)$$

$$= \Sigma_{d=1}^{\left\lfloor\frac{mn}{\ell-1}\right\rfloor} \nu_\ell(p^{rd\frac{\ell-1}{m}} - 1)$$

$$= \left\lfloor\frac{mn}{\ell-1}\right\rfloor + \Sigma_{d=1}^{\left\lfloor\frac{mn}{\ell-1}\right\rfloor} \nu_\ell(rd\frac{\ell-1}{m})$$

$$= \left\lfloor\frac{\mathrm{mn}}{\ell-1}\right\rfloor + \Sigma_{\mathrm{d}=1}^{\left\lfloor\frac{\mathrm{mn}}{\ell-1}\right\rfloor} \nu_\ell(\mathrm{rd})$$

$$= \left\lfloor\frac{mn}{\ell-1}\right\rfloor(1 + \nu_\ell(r)) + \Sigma_{d=1}^{\left\lfloor\frac{mn}{\ell-1}\right\rfloor} \nu_\ell(d)$$

$$= \left\lfloor \frac{mn}{\ell-1} \right\rfloor (1 + v_\ell(r)) + v_\ell(\left\lfloor \frac{mn}{\ell-1} \right\rfloor !)$$

Now we consider the situation that $\ell = 2$. In this case, $p = 2^e x + 1$ for $e \in \mathbb{Z}$ and x is not an even number. Then, we know that $p^k = (2^e x + 1)^k = 1 + k2^e x + \binom{k}{2}(2^e x)^2 + \cdots + (2^e x)^k$. We classify the possible values of e into two parts:

When $e > 1$, $v_\ell(p^k - 1) = v_\ell(k) + e$. To make e as small as possible, we consider $e = 2$, which means $p \equiv 5$ by modulo 8. Therefore, $\sum_{k=1}^{n} v_2(p^{rk} - 1) = \sum_{k=1}^{n} (2 + v_2(rk)) = n(2 + v_2(r)) + v_2(n!)$.

When $e = 1$, i.e., $p \equiv 3$ by modulo 4, we can know that $v_2(p^k - 1) = v_2(2kx + 2k(k-1)x^2) = v_2(2kx(1 + (k-1)x)) = v_2(k) + 1 + v_2(1 + (k-1)x)$. To avoid possible explosion of $v_2(1 + (k-1)x)$ as k varies, we can take p as small as we can. Take $p = 3$ when 3 is not in the exceptions, then $v_2(3^k - 1) = 1 + 2v_2(k)$. Therefore, we can get the minimum possible value of $GL_n(\mathbb{F}_{p^r})$:

$$\sum_{k=1}^{n} v_2(3^{rk} - 1) = \sum_{k=1}^{n}(1 + 2v_2(r)) = n(1 + 2v_2(r)) + 2v_2(n!)$$

4.3. Some Examples

In this subsection, we will give three examples to show how sharp the multiplicative bound we calculate above is.

4.3.1. Example. Let $K = \mathbb{Q}[\sqrt{-1}]$, then $O_K = \mathbb{Z}[\sqrt{-1}]$. Let p be a prime number in $\mathbb{Z}$. As we know, if p is a prime element in O_K, then $p \equiv 3$ by modulo 4; if p is not a prime element in O_K, then $p \equiv 1$ by modulo 4 or $p = 2$. Since there is $v_\ell(r)$ in the calculation of $|GL_n(\mathbb{F}_q)|$, we want the value of r as small as possible. We hope that $r = 1$, i.e., $\mathbb{Z}[\sqrt{-1}]/(x) \cong \mathbb{F}_p$ where x is a prime element in $\mathbb{Z}[\sqrt{-1}]$ that divides p. When can this happen?

4.3.2. Claim. If x is a prime factor of p where $p \equiv 1 \, mod \, 4$ or $p = 2$ in $\mathbb{Z}[\sqrt{-1}]$, then $\mathbb{Z}[\sqrt{-1}]/(x) \cong \mathbb{F}_p$.

Proof: We are going to show that for any element $y \in \mathbb{Z}[\sqrt{-1}]$, there exists $k \in \mathbb{Z}[\sqrt{-1}]$ such $y = kx + a$ where $a = 0, \cdots, p-1$. Let $e = g + hi$, $x = c + di$ with c coprime with d and $y = M + Ni$. Let $y = M + Ni = ex + s = (g + hi)(c + di) + s = gc - hd + (gd + hc)i + s$ where $s \in \mathbb{Z}$. Since c is coprime with d, we can find suitable g, h such that $gd + hc = N$. However, we do not know whether $s = 0, 1, \cdots, p-1$. Then, we make s equal $a + mp$ where $m \in \mathbb{Z}$ and $a = 0, 1, \cdots, p-1$. Since $p = x\bar{x}$, we know that $M + Ni = (g + hi)(c + di) + a + mx\bar{x} = x(g + hi + m\bar{x}) + a$. Thus, we represent $M + Ni$ as $kx + a$ for some $k \in \mathbb{Z}[\sqrt{-1}]$ and $a \in \{0, \cdots, p-1\}$. Therefore, we prove that $\mathbb{Z}[\sqrt{-1}]/(x)$ has exactly p elements and it is thus isomorphic to $\mathbb{F}_p$.

We now use the result of the preceding section to give multiplicative bounds of finite subgroups of $GL_n(\mathbb{Q}[\sqrt{-1}])$ for $n = 2, 3$. In what follows, let G be a finite subgroup of $GL_n(\mathbb{Q}[\sqrt{-1}])$.

When $n = 2$. We take a prime number $p \equiv 1$ mod 4 which is not an exception. In this case, $r = m = 1$ in the formula of $v_\ell(|GL_n(\mathbb{F}_{p^r})|)$. For $\ell > 3$, we find that $\left\lfloor \frac{mn}{\ell-1} \right\rfloor = 0$, and thus

$$v_\ell(|G|) \leq v_\ell(|GL_n(\mathbb{F}_{p^r})|) = \left\lfloor \frac{mn}{\ell - 1} \right\rfloor (1 + v_\ell(r)) + v_\ell\left(\left\lfloor \frac{mn}{\ell - 1} \right\rfloor !\right) = 0$$

For $\ell = 3$, we find

$$v_\ell(|G|) \leq v_\ell(|GL_n(\mathbb{F}_{p^r})|) = \left\lfloor \frac{mn}{\ell-1} \right\rfloor (1 + v_\ell(r)) + v_\ell\left(\left\lfloor \frac{mn}{\ell-1} \right\rfloor !\right) = 1 \times (1 + 0) + 0 = 1.$$

For $\ell = 2$, by taking a prime number $p \equiv 5$ mod 8 which is not an exception and by utilizing the formula of the preceding part, we find

$$\nu_\ell(|\mathrm{G}|) \leq n\big(2+\ \nu_2(r)\big) + \nu_2(n!) = 2\times 2+1 = 5.$$

Taking account of all the calculations above, we conclude that $|G|$ divides $2^5 \times 3^1 = 96$.
When $n = 3$, we also take a prime number $p \equiv 1 \bmod 4$ which is not an exception. In this case, $r = m = 1$ in the formula of $\nu_\ell\big(|GL_n(\mathbb{F}_{p^r})|\big)$. For $\ell > 3$, we find that $\lfloor\frac{mn}{\ell-1}\rfloor = 0$, and thus

$$\nu_\ell(|\mathrm{G}|) \leq \nu_\ell\big(|GL_n(\mathbb{F}_{p^r})|\big) = \left\lfloor\frac{mn}{\ell-1}\right\rfloor\big(1+\nu_\ell(\mathrm{r})\big) + \nu_\ell\left(\left\lfloor\frac{mn}{\ell-1}\right\rfloor!\right) = 0$$

For $\ell = 3$, we find that

$$\nu_\ell(|\mathrm{G}|) \leq \nu_\ell\big(|GL_n(\mathbb{F}_{p^r})|\big) = \left\lfloor\tfrac{mn}{\ell-1}\right\rfloor\big(1+\nu_\ell(\mathrm{r})\big) + \nu_\ell\left(\left\lfloor\tfrac{mn}{\ell-1}\right\rfloor!\right) = 1\times(1+0)+0 = 1.$$

For $\ell = 2$, y taking a prime number $p \equiv 5 \bmod 8$ which is not an exception and by utilizing the formula of the preceding part, we find

$$\nu_\ell(|\mathrm{G}|) \leq n\big(2+\ \nu_2(r)\big) + \nu_2(n!) = 3\times 2+1 = 7.$$

Taking account of all the calculations above, we conclude that $|G|$ divides $2^7 \times 3^1 = 384$.

4.3.3. Example. $K = \mathbb{Q}[\sqrt{-2}]$, then $O_K = \mathbb{Z}[\sqrt{-2}]$. Let p be a prime number in $\mathbb{Z}$. As we know, if p is not a prime element in $\mathbb{Z}[\sqrt{-2}]$, i.e., p can be written as $a^2 + 2b^2$, then $p \equiv 1, 3$ by modulo 8.
We now use the result of the preceding section to give multiplicative bounds of finite subgroups of $GL_n(\mathbb{Q}[\sqrt{-2}])$ for $n = 2$. Let G be a finite subgroup of $GL_n(\mathbb{Q}[\sqrt{-1}])$. Let $x \in \mathbb{Z}$ such that $Ord_{\ell^2}x = \ell\,(\ell - 1)$. We take a prime number $p \equiv 1, 3$ by modulo 8 and $p \equiv x$ by modulo ℓ^2 which is not an exception. In this case, $r = m = 1$ in the formula of $\nu_\ell\big(|GL_n(\mathbb{F}_{p^r})|\big)$. For $\ell > 3$, we find that $\lfloor\frac{mn}{\ell-1}\rfloor = 0$, and thus

$$\nu_\ell(|\mathrm{G}|) \leq \nu_\ell\big(|GL_n(\mathbb{F}_{p^r})|\big) = \left\lfloor\frac{mn}{\ell-1}\right\rfloor\big(1+\nu_\ell(\mathrm{r})\big) + \nu_\ell\left(\left\lfloor\frac{mn}{\ell-1}\right\rfloor!\right) = 0$$

For $\ell = 3$, we find

$$\nu_\ell(|\mathrm{G}|) \leq \nu_\ell\big(|GL_n(\mathbb{F}_{p^r})|\big) = \left\lfloor\tfrac{mn}{\ell-1}\right\rfloor\big(1+\nu_\ell(\mathrm{r})\big) + \nu_\ell\left(\left\lfloor\tfrac{mn}{\ell-1}\right\rfloor!\right) = 1\times(1+0)+0 = 1.$$

For $\ell = 2$, since $p \equiv 1, 3$ by modulo 8, we take $p \equiv 3$ by modulo 4 in this case. To avoid possible explosion of $\nu_2(1 + (k - 1)x)$ as k varies, we can take p as small as we can. Since p is not among the exceptions which are orders of elements in $GL_2(\mathbb{Q}[\sqrt{-2}])$, while the elements in $GL_2(\mathbb{Q}[\sqrt{-2}])$ has only six possible orders: $1, 2, 3, 4, 6, 8$, we can take $p = 11$ in this case.
By the formula above, we find that $\nu_2(11^k - 1) = \nu_2(k) + 1 + \nu_2(5k - 4)$. Hence,

$$\nu_\ell(|\mathrm{G}|) \leq \nu_\ell(|GL_n(\mathbb{F}_{11})|) = \sum_{k=1}^{n}\big(1+\nu_\ell(k)+\ \nu_\ell(5k-4)\big) = n + \nu_2(n!) + \sum_{k=1}^{n}\nu_2(5k-4) = 4.$$

Taking account of all the calculations above, we conclude that $|G|$ divides $2^4 \times 3^1 = 48$.

5. Conclusion

For a given positive integer n, it is well-known that, up to isomorphism, there are finitely many finite subgroups of $\mathrm{GL_n}(\mathbb{Q})$. This result is generalized to any algebraic number filed by Schur. In this article, by a series of isomorphisms, we propose another proof of Schur's result when the ring of integers in K is a principal ideal domain, without using the character theory of finite groups. Then, we give a general method to calculate an upper bound for the order of finite subgroups of GLn(K), for a given n and a given algebraic number field K. We also give examples of $\mathrm{GL_2}(\mathbb{Q}[\sqrt{-1}])$, $\mathrm{GL_3}(\mathbb{Q}[\sqrt{-1}])$ and $\mathrm{GL_2}(\mathbb{Q}[\sqrt{-2}])$ to show how sharp the upper bound is.

Journal of Physics: Conference Series **2287** (2022) 012006 doi:10.1088/1742-6596/2287/1/012006

Moreover, we give a general method to classify finite cyclic subgroups of $GLn(K)$, where K is an algebraic number field, by considering the irreducible decomposition of n-th cyclotomic polynomial $\Phi_n(X)$ in $K[X]$ where K is an algebraic number field. To specify, we classify finite cyclic subgroups for $GL_2(\mathbb{Q}[\sqrt{-1}])$, $GL_3(\mathbb{Q}[\sqrt{-1}])$, $GL_2(\mathbb{Q}[\sqrt{-2}])$ and $GL_3(\mathbb{Q}[\sqrt{-2}])$ as examples.

6. References

[1] P. M. Gudivok, A. A. Kiriyuk, V. P. Rudko, and A. I. Tsitkin, Finite Subgroups of the Group GL(n, Z), Translated from Kibernetika, No. 6, pp. 71-82, November-December, 1982. Original article submitted June 30, 1982.

[2] G. Mackiw, Finite Groups of 2×2 Integere Matrices, Mathmatics Magazines, Vol. 69, No. 05, Dec. 1996.

[3] H. Minkowski, Zur Theorie der positiven quadratischen Formen, J.Crelle 101 (1887), 196–202.

[4] J¨urgen Neukirch, Algebraic Number Theory, 1999, Springer-Verlag Berlin Heidelberg.

[5] I. Schur, Uber eine Klasse von endlichen Gruppen linearer Substitutionen, Sitz. Preuss. Akad. Wiss. Berlin (1905), 77–91.

[6] J.P. Serre, A Course in Arithmetic, 1996, Springer.

[7] S. H. Weinstraub, Galois Theory, 2009, Spring-Verlag New York.

Application of WKB Method in Approximating Wave Functions in Square Potential Wells

Haohui Guo

Shanghai High School International Division, China
E-mail: haohuiguo2003@outlook.com

Abstract. This paper introduces a widely used approximation method, the Wentzel-Kramers-Brillouin (WKB) method, for solving the Schrodinger equation, sets up a theoretical framework to derive the general solutions of the method, and carries out the application of the approximation method in specific cases to verify the validity and practicality of the method. Throughout the study, numerical methods, such as the Simpsons method, and programming tools, such as Python, are used to process sophisticated calculations. The final approximated solutions for the Schrodinger equation are able to demonstrate the state of a particle in an isolated quantum system. All the presented studies are based upon the knowledge and skillsets of a high school student.

1. Introduction

It has been almost a century since the Schrodinger equation was formulated to describe the wave function – which demonstrates the probability of finding a particle – of a quantum mechanical system. Solving the Schrodinger equation is one of the essential problems in quantum mechanics since non-linear second order ordinary differential equation (ODE) has, in general, no analytic solutions. Therefore, the solutions are limited to a list of very few known potential functions, such as inverse power-law potentials, infinite and finite square wells, and parabolic potentials, and due to this fact, all physical systems are modelled with potentials from this list [1]. To overcome this limitation, many techniques have been developed to derive approximate solutions of the Schrodinger equation, and the Wentzel-Kramers-Brillouin (WKB) method is one of the most commonly used practices [2], [3], [4]. It is a power tool and it has many applications, for example, solving waves in an inhomogeneous plasma as shown in the textbook *Plasma Waves* (D.G. Swanson, Academic, San Diego, CA 1989). Instead of starting with a simplified potential and adding small turns, which leads to perturbation theory, the WKB approximation makes an assumption of a slowly varying potential.

Exact solutions for defined potentials are still an ultimate goal in quantum physics and related fields of science. Many researches have been conducted on the justification of the validity of the WKB method, but many of them focus on more complex potential functions. For beginners in the field of quantum mechanics, an examination of the WKB method in simpler potential functions, with numerical codes that allow them to reproduce results, could help them to get a better grip of the essence of the WKB method. Hence, the purpose of this paper is to verify the validity of the WKB method in unsophisticated potential functions numerically. Specifically, we derive the solutions of Schrodinger equations with three well known potentials where their quantized energy levels and wave functions are determined using the WKB method. Three potential wells are studied in this paper: infinite square well, finite symmetric square well, and finite asymmetric square well. Graphical results are demonstrated and analyzed to verify the validity of the WKB method. All the presented studies are based upon the knowledge and skill sets of a high school student.

We will show that some wave functions obtained via the WKB method are discontinuous at the turning points of the potential. These results are not unexpected due to the fact that our potentials have sudden changes at the turning points, so they cannot be approximated to slowly varying potentials at the turning points. After a little survey over standard textbooks, we found that there is no elaboration on the discontinuity of the waves due to the breakdown of the validity of the WKB method. In this case, we have reasons to assume that the restrictions in applying the standard WKB method are not very well known to students. Therefore, our work presented here is valuable to students to gain the knowledge that in the cases where the potential of interest cannot be approximated to a slowly varying potential, the WKB method would not be an optimal approach.

2. Theoretical framework

The WKB Method is an approximation of the solutions for linear differential equations, including the Schrodinger Equation. Its amplitude is taken to vary slowly in respect to the de Broglie wavelength, thus giving the approximation a semiclassical nature [5].

2.1. General solutions

The general solutions of a wave function using the WKB approximation are separated into two cases: $E > V(x)$ and $E < V(x)$.

2.1.1. $E > V(x)$. The region where $E > V(x)$ is called the classically allowed region. As a result, the wave function is oscillatory as expressed below:

$$\psi(x) = \frac{C_+}{\sqrt{p(x)}} \exp\left[\frac{i}{\hbar}\int^x p(x')dx'\right] + \frac{C_-}{\sqrt{p(x)}} \exp\left[-\frac{i}{\hbar}\int^x p(x')dx'\right] \tag{1}$$

where $p(x)$ is the momentum of a particle moving in a potential well given as $p(x) = \left[2m\left(E - V(x)\right)\right]^{\frac{1}{2}}$, and C_+ and C_- are two normalization constants.

2.1.2. $E < V(x)$. In the case where $E < V(x)$, or the classically forbidden region, the momentum becomes an imaginary value, and the general solution of the wave function is demonstrated as follows:

$$\psi(x) = \frac{C'_+}{\sqrt{|p(x)|}} \exp\left[\frac{1}{\hbar}\int^x |p(x')|dx'\right] + \frac{C'_-}{\sqrt{p(x)}} \exp\left[-\frac{1}{\hbar}\int_x |p(x')|dx'\right] \tag{2}$$

where the exponents become real, thus turning the wave function into an exponentially decaying function.

2.2. Non-Rigid Walls Potential Well

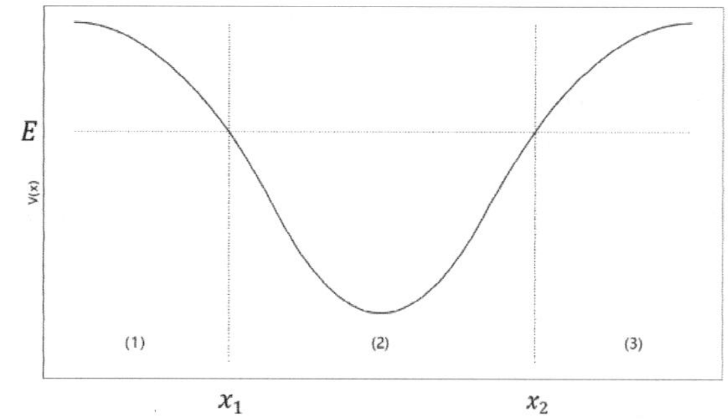

Figure 1. Potential Well with No Rigid Walls

In Figure 1, the potential well is divided into three regions by two turning points x_1 and x_2, where regions 1 and 3 are classified as classically forbidden regions while region 2 is classified as the classically allowed region. According to (1) and (2), the wave function for the potential well can be written as follows:

$$\psi_1(x) = \frac{C_1}{\sqrt{|p(x)|}} \exp\left[-\frac{1}{\hbar}\int_x^{x_1} |p(x')| dx'\right], \; x < x_1 \tag{3}$$

$$\psi_2(x) = \frac{C'_2}{\sqrt{p(x)}} \exp\left[\frac{i}{\hbar}\int_x \; p(x')dx'\right] + \frac{C''_2}{\sqrt{p(x)}} \exp\left[-\frac{i}{\hbar}\int_x \; p(x')dx'\right], \; x_1 < x < x_2 \tag{4}$$

$$\psi_3(x) = \frac{C_3}{\sqrt{|p(x)|}} \exp\left[-\frac{1}{\hbar}\int_{x_2}^{x} |p(x')| dx'\right], \; x > x_2 \tag{5}$$

2.2.1. Connection of ψ_3 and ψ_2. The wave function in region 2 can be further written as

$$\psi_2(x) = \frac{C_2}{\sqrt{p(x)}} \sin\left(\frac{1}{\hbar}\int_x^{x_2} p(x')dx' + \alpha\right), \; x_1 < x < x_2 \tag{6}$$

where α is an unknown phase factor. However, when x approaches to the turning point x_2, the momentum becomes extremely small that the WKB approximation of the wave function fails to hold. As a result, the wave function near the turning point should be solved exactly using the Schrodinger equation. When the distance between x and x_2 is close enough, the potential within this region can be represented by a linear function:

$$V(x) = E + (x - x_2)V' \tag{7}$$

where $V(x_2)$ is substituted by E, and V', the slope of the line at $x = x_2$, is given as $\left.\frac{dV(x)}{dx}\right|_{x=x_2}$. The Schrodinger equation thereby becomes

$$\frac{d^2\psi(x)}{dx^2} - \frac{2mV'}{\hbar^2}(x - x_2)\psi(x) = 0 \tag{8}$$

By applying the change of variable, the equation can be transformed from $y = \left(\frac{2mV'}{\hbar^2}\right)^{\frac{1}{3}}(x - x_2)$ into $\frac{d^2\psi(y)}{dy^2} - y\psi(y) = 0$, which is an airy function. By solving the airy function, we derive the wave function asymptotically [6]

$$\psi(x) = \begin{cases} \frac{A}{\sqrt{p(x)}} \sin\left(\frac{1}{\hbar}\int_x^{x_2} p(x')dx' + \frac{\pi}{4}\right), x \ll x_2, \\ \frac{A}{2\sqrt{|p(x)|}} \exp\left[-\frac{1}{\hbar}\int_{x_2}^{x} |p(x')| dx'\right], x \gg x_2, \end{cases} \tag{9}$$

and the connection formulae that connect the wave function at either side of x_2

$$A = 2C_3, A = C_2, \alpha = \frac{\pi}{4}.$$

Consequently, as we obtain the phase factor α, (6) becomes

$$\psi_2(x) = \frac{C_2}{\sqrt{p(x)}} \sin\left(\frac{1}{\hbar}\int_x^{x_2} p(x')dx' + \frac{\pi}{4}\right), \; x_1 < x < x_2. \tag{10}$$

2.2.2. Connection of ψ_1 and ψ_2. Starting from the turning point x_1, (6) can be written as

$$\psi_2(x) = \frac{D}{\sqrt{p(x)}} \sin\left(\frac{1}{\hbar}\int_{x_1}^{x} p(x')dx' + \beta\right), \; x_1 < x < x_2. \tag{11}$$

Because the approximation fails near the turning point, the wave function in the vicinity of the turning point can be derived from an exact solution using the Schrodinger equation. After processing, similar to the process in 2.2.1, the wave function is expressed as follows:

$$\psi(x) = \begin{cases} \frac{E}{2\sqrt{|p(x)|}} \exp\left[-\frac{1}{\hbar}\int_x^{x_1} |p(x')|dx'\right], x \ll x_1, \\ \frac{E}{\sqrt{p(x)}} \sin\left(\frac{1}{\hbar}\int_x^{x_2} p(x')dx' + \frac{\pi}{4}\right), x \gg x_1. \end{cases} \quad (12)$$

The other group of connection formulae is:

$$E = 2C_1, E = D, \beta = \frac{\pi}{4}.$$

Therefore, after plugging in value of the phase factor, $\psi_2(x)$ becomes

$$\psi_2(x) = \frac{D}{\sqrt{p(x)}} \sin\left(\frac{1}{\hbar}\int_{x_1}^{x} p(x')dx' + \frac{\pi}{4}\right), \ x_1 < x < x_2. \quad (13)$$

2.2.3. Quantization of bound state energy levels. Since (10) is equivalent to (13), we come to an equation:

$$\psi_2(x) = \frac{C_2}{\sqrt{p(x)}} \sin\left(\frac{1}{\hbar}\int_x^{x_2} p(x')dx' + \frac{\pi}{4}\right) = \frac{D}{\sqrt{p(x)}} \sin\left(\frac{1}{\hbar}\int_{x_1}^{x} p(x')dx' + \frac{\pi}{4}\right). \quad (14)$$

The solution must adhere to the following relations:

$$\left(\frac{1}{\hbar}\int_{x_1}^{x} p(x')dx' + \frac{\pi}{4}\right) + \left(\frac{1}{\hbar}\int_x^{x_2} p(x')dx' + \frac{\pi}{4}\right) = (n+1)\pi \quad (15)$$

$$D = (-1)^n C_2 \quad (16)$$

After simplifying (15), the quantized energy levels of the bound states for potential wells with no rigid walls can be calculated using the equation below:

$$\int_{x_1}^{x_2} p(x)dx = \left(n + \frac{1}{2}\right)\pi\hbar, \ n = 0,1,2,3,\cdots. \quad (17)$$

To sum up, the WKB approximated wave function for a particle travelling in a potential well with no rigid walls is divided into three sections:

$$\psi(x) = \begin{cases} \psi_1(x) = \frac{(-1)^n A}{2\sqrt{|p(x)|}} \exp\left[-\frac{1}{\hbar}\int_x^{x_1} |p(x')|dx'\right], \ x < x_1 \\ \psi_2(x) = \frac{A}{\sqrt{p(x)}} \sin\left(\frac{1}{\hbar}\int_x^{x_2} p(x')dx' + \frac{\pi}{4}\right), \ x_1 < x < x_2 \\ \psi_3(x) = \frac{A}{2\sqrt{|p(x)|}} \exp\left[-\frac{1}{\hbar}\int_{x_2}^{x} |p(x')|dx'\right], \ x > x_2 \end{cases} \quad (18)$$

2.3. One Rigid Wall Potential Well

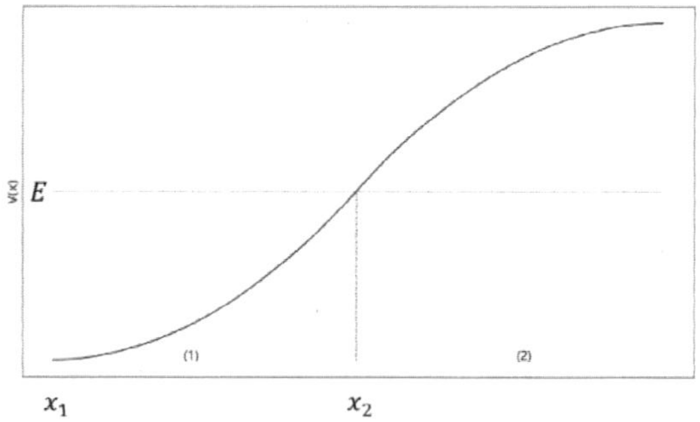

Figure 2. Potential well with one rigid wall

In Figure 2, a particle is travelling in a potential well and an infinite potential wall exists at $\boldsymbol{x = x_1}$. The potential well is divided into classically allowed region 1 $\boldsymbol{x \in (x_1, x_2)}$ and classically forbidden region 2 $\boldsymbol{x \in (x_2, \infty)}$. Similar to the derivation of the wave function in 2.2, the wave function within region 1 exists as an oscillatory form, whereas that of region 2 is presented as an exponentially decaying function:

$$\psi_1(x) = \frac{B}{\sqrt{p(x)}}\sin\left(\frac{1}{\hbar}\int_x^{x_2} p(x')dx' + \frac{\pi}{4}\right) = \frac{A}{\sqrt{p(x)}}\sin\left(\frac{1}{\hbar}\int_{x_1}^{x} p(x')dx'\right),\ x_1 < x < x_2 \quad (19)$$

$$\psi_2(x) = \frac{A}{2\sqrt{|p(x)|}}\exp\left[-\frac{1}{\hbar}\int_{x_2}^{x} |p(x')|dx'\right],\ x > x_2 \quad (20)$$

where the second expression for $\boldsymbol{\psi_1(x)}$ does not have a phase factor because the wave function approaches to zero as it reaches the infinite potential wall at $\boldsymbol{x = x_1}$.
The following equation that can be applied to calculate the bound state energy levels in this scenario can be derived from (19), as demonstrated in 2.2.3:

$$\int_{x_1}^{x_2} p(x)dx = \left(n + \frac{3}{4}\right)\pi\hbar,\ n = 0,1,2,3,\cdots. \quad (21)$$

2.4. Two rigid walls potential well

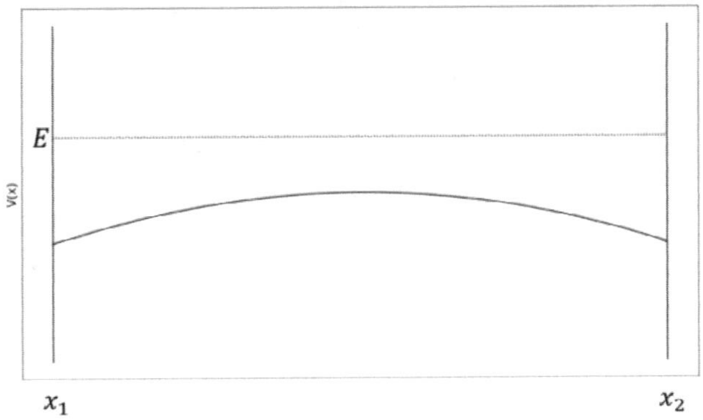

Figure 3. Potential well with two rigid walls

As shown in Figure 3, in a potential well restricted by two rigid walls at $x = x_1$ and $x = x_2$ respectively, the wave function for the particle is expressed as follows:

$$\psi(x) = \frac{A}{\sqrt{p(x)}}\sin\left(\frac{1}{\hbar}\int_{x_1}^{x} p(x')dx'\right) = \frac{B}{\sqrt{p(x)}}\sin\left(\frac{1}{\hbar}\int_x^{x_2} p(x')dx'\right) \quad (22)$$

where neither of the expressions has a phase factor because the wave function has to vanish at both $x = x_1$ and $x = x_2$.
Similar to the quantization of energy levels in previous two types of potential wells, the quantization of bound state energy levels for potential wells that have two rigid walls can be processed similarly, producing the equation

$$\int_{x_1}^{x_2} p(x)dx = (n + 1)\pi\hbar,\ n = 0,1,2,3,\cdots. \quad (23)$$

3. WKB approximation of wave functions for square potential wells

In this section, the application of the WKB method in solving Schrodinger equations with different square potential wells will be discussed. Specifically, to demonstrate that the WKB method is an effective and straightforward approximation, we will show that the solution of the wave function obtained using the WKB method matches exact the solution obtained by solving the Schrodinger equation with an infinite potential well. In finite symmetric and asymmetric potential wells with one rigid wall where calculations for exact solutions of wave functions often require great efforts, we will show that the implementation of the WKB method in these cases will significantly relaxes the effort required to attain the results. The WKB method presented in this section is realized by Python.

3.1. Variables

As shown in table 1, we assign the following values in Python to demonstrate the application of the WKB method in the aforementioned cases:

Table 1. Variables

Notation	Description
$\boldsymbol{x_0}$	Location of the infinite potential wall
$\boldsymbol{x_1}$	Location of the first potential wall
$\boldsymbol{x_2}$	Location of the second potential wall
$\boldsymbol{A}$	Normalization constant
$\boldsymbol{E_n}$	Energy at energy level *n*

3.2. Infinite potential well

Table 2 demonstrates the input values for the initial variables:

Table 2. Assigned values of variables

Variable	Value
n	0,1,3,5
m	$9.11 \times 10^{-34}\ [kg]$
x_1	0
x_2	2
$V(x_1)$	∞
$V(x_2)$	∞

3.2.1. Exact solution by Schrodinger Equation. For a particle of mass m in an infinite potential well, the potential function is expressed as

$$V(x) = \begin{cases} 0, & x_1 < x < x_2 \\ \infty, & x \le x_1 \text{ or } x \ge x_2 \end{cases} . \tag{24}$$

and is visualized as following:

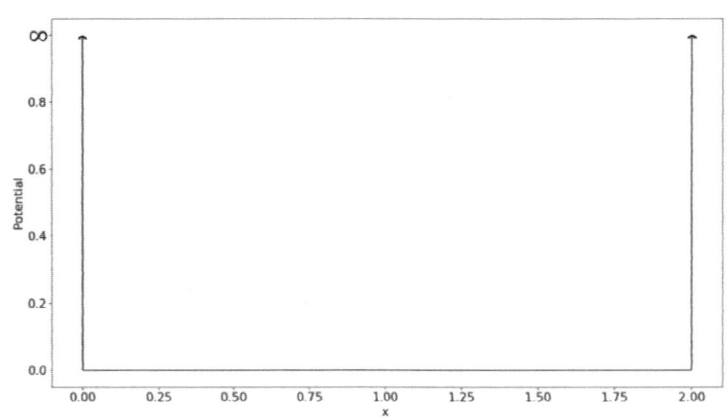

Figure 4. Infinite potential well

In Figure 4, because the potential within the potential well is zero, the time independent Schrodinger equation becomes:

$$\frac{d^2\psi(x)}{dx^2} + k^2\psi(x) = 0, \tag{25}$$

where k is defined as:

$$k = \left[\frac{2mE_n}{\hbar^2}\right]^{\frac{1}{2}}, \; n \in Z^+. \tag{26}$$

The general solution for (25) is

$$\psi(x) = Asin(kx) + Bcos(kx), \; x_1 < x < x_2. \tag{27}$$

Because it is impossible to find a particle outside of an infinite potential well, the wave function vanishes at both walls:

$$\psi(x_1) = \psi(0) = 0 = B. \tag{28}$$

As a result, (27) becomes

$$\psi(x) = Asin(kx), \; x_1 < x < x_2. \tag{29}$$

By substituting the value of B and inserting the value of x into (29), the solution becomes

$$\psi(x_2) = \psi(2) = 0 = Asin(2k). \tag{30}$$

After solving (29), we obtain the expression for k:

$$k = \frac{(n+1)\pi}{2} n \in Z. \tag{31}$$

Substituting (31) into (29) leads to

$$\psi(x) = Asin\left(\frac{(n+1)\pi x}{2}\right), \; n \in Z, \; x_1 < x < x_2, \tag{32}$$

where A is the amplitude of the wave function that serves to normalize the function. In order to calculate the value of A, we need to apply the fact that the probability of finding the particle within the region $x_1 < x < x_2$ must be 100%. To express it mathematically,

$$\int_{x_1}^{x_2} |\psi(x)|^2 dx = 1, \tag{33}$$

where $|\psi(x)|^2$ represents the probability density of the particle.

The value of A can be obtained by substituting (32) to (33):

$$A = \sqrt{\frac{2}{x_2 - x_1}} = \sqrt{\frac{2}{2-0}} = 1. \tag{34}$$

Finally, the wave function of a particle travelling in an infinite potential well becomes

$$\psi(x) = \begin{cases} sin\left(\frac{(n+1)\pi x}{2}\right), x_1 < x < x_2 \\ 0, \; x \leq x_1 \; or \; x \geq x_2 \end{cases}, \; n \in Z. \tag{35}$$

3.2.2. Approximate solution by WKB method.
In an effort to approximate the solution for the wave function, we have to calculate E_n and substitute it into $p(x)$. Subsequently, by normalizing the function (the probability of finding the particle within the confined region is 100%), we can derive the normalization constant A and thereby obtain the final approximate solution of the wave function.
First, since the infinite potential well is a type of a well with two rigid walls, we use (23) to solve for E_n. To express the left side of the equation in terms of E_n, $p(x)$ is equally divided into 10,000 segments. By applying Simpsons method in Python, we can numerically simplify the integral to represent it in terms of E_n. Finally, we can solve for E_n based on the equation and substitute it into $p(x)$.

As demonstrated in the first expression of $\psi(x)$ in (22), we integrate $p(x)$ from x_1. To solve for A, we normalize the wave function as shown in (33). Substituting the wave function into the equation yields

$$\int_{x_1}^{x_2} \frac{A^2}{p(x)} \sin^2\left(\frac{1}{\hbar}\int_{x_1}^{x} p(x')dx'\right) = 1. \tag{36}$$

By dividing $|\psi(x)|^2$ into 10,000 segments and applying Simpsons method in Python, we can solve for A. The following is an example of the final wave function at ground state ($n = 0$):

$$\psi(x) = \frac{1.287\cdot 10^{-17}}{\sqrt[4]{2.744\cdot 10^{-68}}} \sin(1.571x). \tag{37}$$

After simplifying (37), we find that the coefficient is very close to 1, and the angular frequency of the sine wave approximately equals $\frac{\pi}{2}$. As a result, (37) produces approximately the same result as that of the exact solution in (35) when assigning $n = 0$.

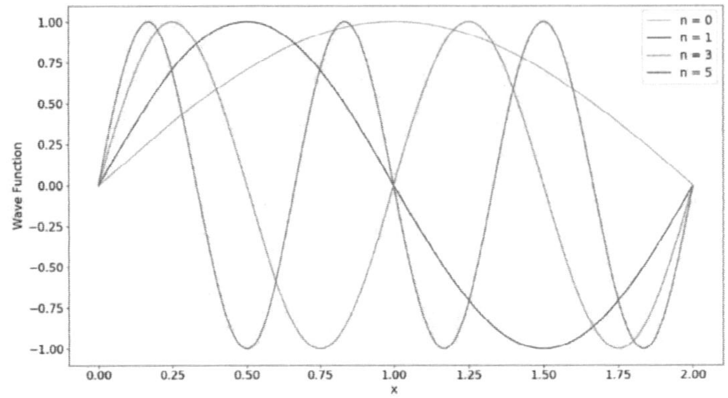

Figure 5. Exact solution of wave function

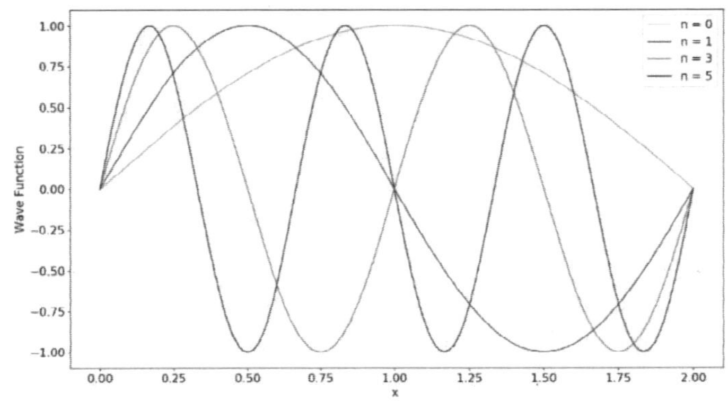

Figure 6. Approximate solution of wave function

As presented in Figure 5 and 6, each containing wave functions for energy levels 0,1,3, and 5, the approximate solutions of the wave functions are virtually the same as the exact solutions, which implies that the WKB method is an effective approximation approach as it produces results that are very close to the exact ones.

3.3. Finite symmetric potential well

Table 3 demonstrates the inputs for the initial variables:

Table 3. Assigned values of variables

Variable	Value
n	0,1,3,5
m	10^{-67} $[kg]$
x_1	-2
x_2	4
$V(x_1)$	2
$V(x_2)$	2

Figure 7 depicts the finite potential well based on the input values:

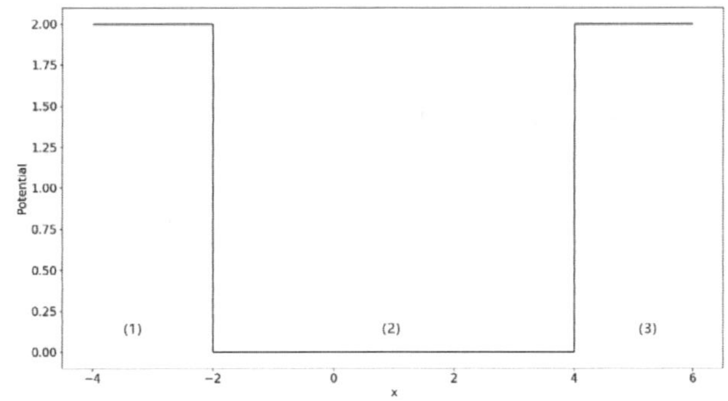

Figure 7. Finite symmetric potential well

The overall calculation process for the wave function in this potential well is similar to that deliberated in section 3.2.2. However, to calculate E_n, we use equation (17) since there are no rigid walls. Moreover, the wave function is divided into three portions: $x < x_1$, $x_1 < x < x_2$, and $x > x_2$, represented by region 1, 2, and 3 respectively. The wave function for each region is indicated in equation (18).

In a finite potential well where bound states exist, $E_n < V$. As a result, $\psi(x) \to 0$ when $x \to \pm\infty$, which becomes normalizable. To normalize the piecewise wave function, its complex square is integrated from -50 to 50. Because when the particle traverses into region 1 or 3, the wave function goes asymptotically infinitesimal, so we believe -50 to 50 is a reasonable and insurable limit. After dividing the region within $-50 < x < 50$ into 1,000 segments and applying Simpsons method to solve the integral (See (33) for equation) numerically, we can derive the value of A.

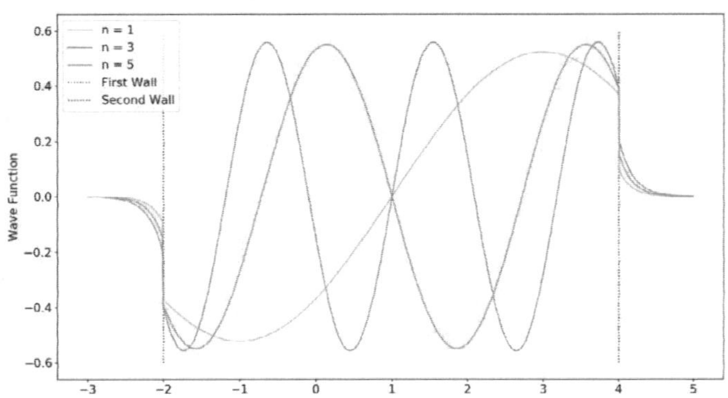

Figure 8. Approximate wave functions at energy levels 1,3,5

As shown in Figure 8, $\psi(x)$ appears as an oscillating wave within region 2, i.e., in the classically allowed region, as $x \to \pm\infty$, there exists finite possibilities of finding a particle outside of the potential well, although the wave function decays and quickly approaches 0 as expected.
Also, notice that at the turning points between the classically allowed and forbidden regions (i.e., $x = -2, x = 4$), the wave function becomes discontinuous. This is because at turning points, the WKB solutions break down as momentum $p(x)$ becomes 0. In order to establish the connection between the WKB solutions in the allowed region as well as in the forbidden region on either side of the breaking down point (turning point), i.e., $p(x) = 0$, an accurate solution is needed, where the potential can be approximated as a linear potential with the slope equals the tangent of the breaking down point (See (7)), hence linking WKB solutions at a turning point [4]. However, such linear approximation only holds valid for a reasonably smooth potential. In a finite symmetric square potential well, the potential changes sharply at turning points, so linear approximation is no longer valid, and thus the discontinuity inevitably persists.

3.4. Finite asymmetric potential well with one rigid wall

In this type of potential well, we will discuss two cases: 1) energy is smaller than both walls and 2) energy is greater than one wall but smaller than the other.

3.4.1. $\boldsymbol{E < V_1}$ and $\boldsymbol{E < V_2}$

Table 4 provides the inputs for the initial variables:

Table 4. Assigned values of variables

Variable	Value
n	1,2,3
m	10^{-68} $[kg]$
x_0	0
x_1	2
x_2	8
V_1	3 [J]
V_2	5 [J]

Figure 9 is the potential well based on the inputs shown in table 4:

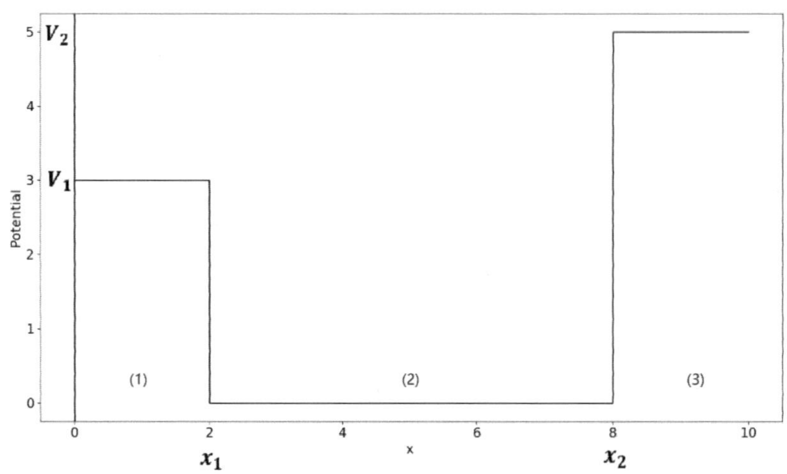

Figure 9. Case 1: Asymmetric potential well with one rigid wall

Since E_n is smaller than both walls, the calculation is the same as that deduced in section 3.3. However, it is noteworthy that the potential walls have different values, and due to the infinite potential wall at $x = 0$, we use 0 rather than -50 as the lower limit of the integral to calculate the normalization constant A in equation (33). The derived wave functions partitioned by regions 1, 2, and 3 are depicted in Figure 10.

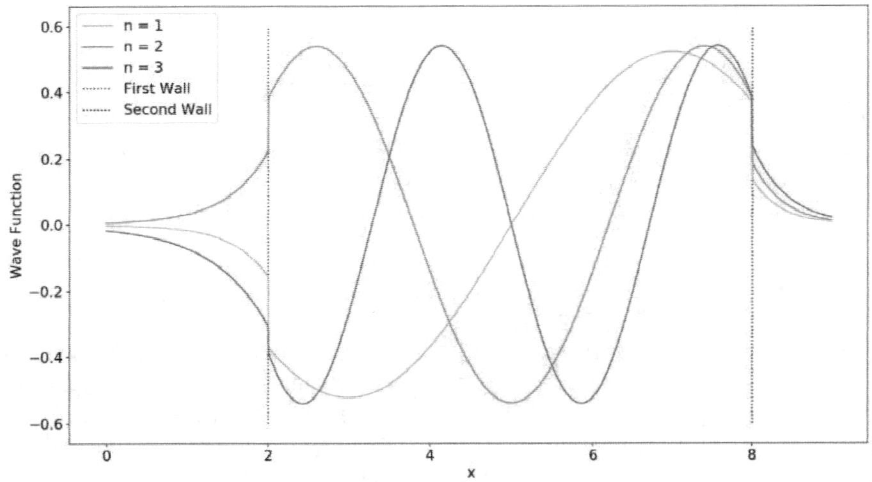

Figure 10. Approximate wave functions at energy levels 1,2,3

In region 2 of Figure 10, we observe that the wave function behaves as an oscillating wave. However, in region 1 and 3, the wave function is decaying tat different rates due to distinct values of potential walls.

3.4.2. $\boldsymbol{V_1 < E < V_2}$

Table 5 is the input for initial variables:

Table 5. Assigned values of variables

Variable	Value
n	4,5,6
m	10^{-68} $[kg]$
x_0	0
x_1	2
x_2	8
V_1	4 [J]
V_2	12 [J]

The potential well is exhibited in Figure 11:

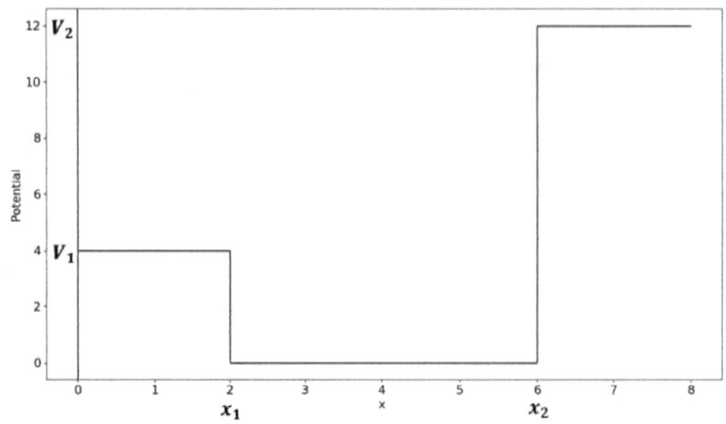

Figure 11. Case 2: Asymmetric Potential Well with One Rigid Wall

Since E_n is greater than V_1, it meets the walls at $x = x_0$ and $x = x_2$, where the potential at $x = x_0$ is infinite. Thus, this potential well can be characterized as that discussed in section 2.3. The wave function is split into two regions: the classically allowed region at $x_0 < x < x_2$ and the classically forbidden region at $x > x_2$, which are represented by equations (20) and (21) respectively. To solve for E_n, we use equation (21), where the lower and upper limits of the integral are x_0 and x_2. To calculate the normalization constant A, we apply equation (33) as expressed below:

$$\int_{x_0}^{50} |\psi(x)|^2 dx = 1. \tag{38}$$

With both E_n and A obtained, we can derive the wave function as depicted in Figure 12:

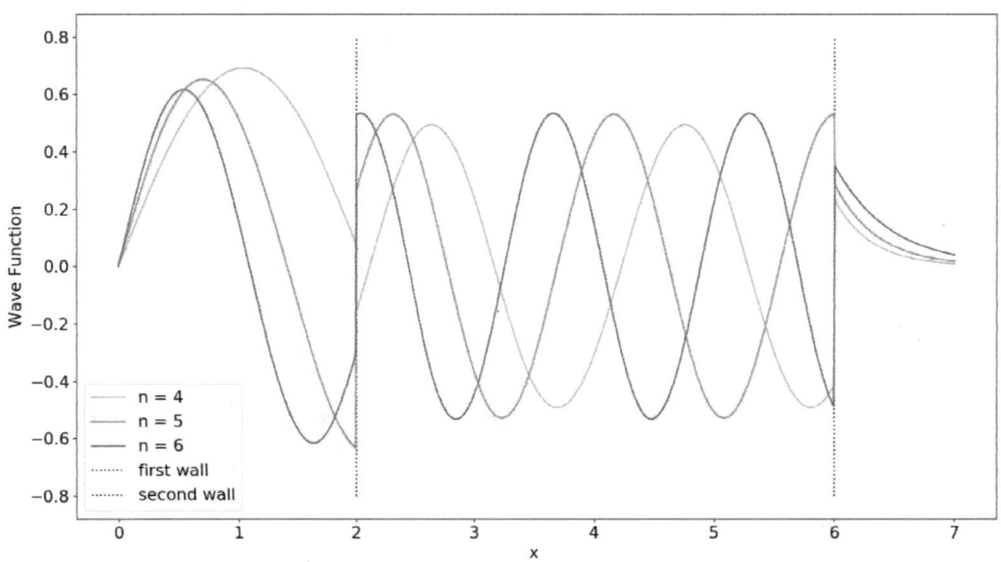

Figure 12. Approximate Wave Functions at energy levels $n = 4,5,6$

As shown in Figure 12, the wave function oscillates at $x_0 < x < x_2$ and decays at $x > x_2$. However, we find that the amplitudes and periods of the wave function differ in regions $x_0 < x < x_1$ and $x_1 < x < x_2$ due to the variation of $p(x)$ with different potentials in these two regions.

We would like to emphasize that there is again a discontinuity showing up in these wave functions from the classically allowed region to the classically forbidden region due to the reason elucidated in the last paragraph of Section 3.3. It is worth pointing out that, the WKB method is a powerful tool dealing with smoothly varying potentials; however, for artificial potentials, such as the symmetric and asymmetric finite potential wells studied here, wave functions derived from the WKB method cannot faithfully establish a particle's quantum state especially in the forbidden region due to the sudden change in the potential.

4. Conclusion

We present detailed derivations for solving Schrodinger equations with three well known potentials using the WKB method. Although the WKB method is not a universal approach for an arbitrary potential, for smoothly varying potentials without sudden change, it does provide a clear route for good approximations. Our results are straightforward and self-explanatory. The necessary coding technique involved in this work is integration with the Simpsons method. Our work indicates that with a proper and simple training, even a high school student is capable of taking on one of the challenging problems in modern physics – solving Schrodinger equations. This would significantly enhance high school students' interest in quantum physics and nurture many well-motivated future physicists.

5. References

[1] Nicolai, Hermann. "Supersymmetry and spin systems." *Journal of Physics A: Mathematical and General* 9.9 (1976): 1497.
[2] Benbourenane, J., and H. Eleuch. "Exactly solvable new classes of potentials with finite discrete energies." *Results in Physics* 17 (2020): 103034.
[3] Dutt, Ranabir, Avinash Khare, and Uday P. Sukhatme. "Exactness of supersymmetric WKB spectra for shape-invariant potentials." *Physics Letters B* 181.3-4 (1986): 295-298.
[4] Lu, Fenghua, et al. "WKB approximation for a deformed Schrodinger-like equation and its applications to quasinormal modes of black holes and quantum cosmology." *Nuclear Physics B* 937 (2018): 502-532.
[5] Tosiek, Jaromir, Ruben Cordero, and Francisco J. Turrubiates. "The Wentzel–Kramers–Brillouin approximation method applied to the Wigner function." *Journal of Mathematical Physics* 57.6 (2016): 062103.
[6] N. Zettili, Quantum Mechanics – Concepts and Applications (Wiley, 2009)

Acknowledgments
The first author thanks Dr. Tian Li for his guidance throughout the completion of the paper.

A Direct Proof of the Prime Number Theorem using Riemann's Prime-counting Function

Zihao Liu

International Department, The Affiliated High School of SCNU, China
E-mail: travor_lzh@163.com

Abstract. In this paper, we develop a novel analytic method to prove the prime number theorem in de la Vallée Poussin's form: $\pi(x) = \mathrm{li}(x) + \mathcal{O}(xe^{-c\sqrt{\log x}})$. Instead of performing asymptotic expansion on Chebyshev functions as in conventional analytic methods, this new approach uses contour-integration method to analyze Riemann's prime counting function $J(x)$, which only differs from $\pi(x)$ by $\mathcal{O}(\sqrt{x}\log\log x/\log x)$.

1. Introduction

The prime number theorem [1][2] has been a popular topic in analytic number theory since the 19th century. Its first proofs were independently given by Hadamard and de la Vallée Poussin in 1896 [3] using analytic methods. Since then, mathematicians such as Apostol [1], Levinson [4], Newman [5][6], Selberg [7], Stein [8], and Wright [9] have explored different approaches (e.g., elementary methods, Tauberian theorems, and contour integrations) to prove this theorem. Instead of analyzing $\pi(x)$ directly, present proofs, regardless elementary or analytic, first attacked either of the Chebyshev functions

$$\vartheta(x) = \sum_{p\le x} \log p \tag{1}$$

$$\psi(x) = \sum_{n\le x} \Lambda(n) = \sum_{k\le \log_2 x} \vartheta(x^{1/k}) \tag{2}$$

to derive asymptotic formulae, and estimates for $\pi(x)$ were later obtained using elementary methods, such as partial summation, to $\vartheta(x)$. Although Landau [1][10] derived the prime number theorem by proving

$$\lim_{x\to\infty} \frac{1}{x}\sum_{n\le x} \mu(n) = 0 \tag{3}$$

$$\lim_{x\to\infty} \sum_{n\le x} \frac{\mu(n)}{n} = 0 \tag{4}$$

he still justified the equivalence of (3) and (4) to the prime number theorem with Chebyshev functions. Hence, it is an interesting task to investigate a method that avoids the use of (1) and (2).

In this paper, we propose a proof of the prime number theorem that avoids the use of Chebyshev functions. In particular, we first introduce Riemann's prime counting function $J(x)$ to our problem and relate it to $\pi(x)$. Then, by inverse Mellin transform, an integral representation is obtained for $\pi(x)$. In the subsequent part of this paper, we evaluate the contour integral and obtain the prime number theorem in the version of de la Vallée Poussin. That is, there exists a positive constant c such that

$$\pi(x) = \mathrm{li}(x) + \mathcal{O}(xe^{-c\sqrt{\log x}}) \tag{5}$$

where $\mathrm{li}(x)$ is the logarithmic integral:

$$\mathrm{li}(x) = \lim_{\varepsilon\to 0^+}\left[\int_0^{1-\varepsilon} + \int_{1+\varepsilon}^{x} \frac{\mathrm{d}t}{\log t}\right] \tag{6}$$

In essence, the main contribution of this paper is a new proof of the prime number theorem. The significance of this proof is that it is an analytic method that avoids the use of Chebyshev functions.

2. Riemann's prime counting function $J(x)$

Let a_n be defined by

$$a_n = \begin{cases} 1/k & n = p^k, p \text{ prime} \\ 0 & \text{otherwise} \end{cases} \tag{7}$$

In his paper on $\pi(x)$, Riemann [11] defined $J(x)$ as the summatory function for a_n:

$$J(x) = \sum_{n\le x} a_n = \sum_{1<p^k\le x} \frac{1}{k} \tag{8}$$

This function has a remarkable property that allows us to connect it to the standard prime counting function $\pi(x)$:

Theorem 1. *For $x \ge 2$, we have*

$$J(x) = \pi(x) + O(\sqrt{x}\log\log x) \tag{9}$$

Proof. It follows from (8) that

$$J(x) - \pi(x) = \sum_{2\le k\le \log_2 x} \frac{\pi(x^{1/k})}{k} \le \sum_{2\le k\le \log_2 x} \frac{\pi(\sqrt{x})}{k}$$

$$\le \sum_{2\le k\le \log_2 x} \frac{\sqrt{x}}{k} \ll \sqrt{x}\log\log x$$

By (9), asymptotic formula for $J(x)$ can be ported to $\pi(x)$ with ease, so we can study $\pi(x)$ simultaneously when we analyze $J(x)$.

3. Integral representation for $J(x)$ and $\pi(x)$

Riemann [11] had shown that

$$J(x) = \frac{1}{2\pi i}\int_{k-i\infty}^{k+i\infty} \frac{x^s}{s}\log\zeta(s)\mathrm{d}s \tag{10}$$

for noninteger $x > 2$ and $k > 1$ using Fourier analysis. However, because (10) involves infinite integrals, it would be difficult to analyze its asymptotic behavior directly, so we wish to convert (10) into a quantitative form. In other words, we prove a special case of Perron's formula [3][12].

Lemma 1. *For $\Re(s) > 1$, we have*

$$\sum_{n=1}^{\infty} \frac{a_n}{n^s} = \log\zeta(s) \tag{11}$$

Proof. Since $\Re(s) > 1$, the left hand side converges absolutely, we can change the order of summation safely:

$$\sum_{n=1}^{\infty} \frac{a_n}{n^s} = \sum_p \sum_{k=1}^{\infty} \frac{1}{kp^{ks}} = \sum_p \log\frac{1}{1-p^{-s}}$$

$$= \log\prod_p \frac{1}{1-p^{-s}} = \log\zeta(s)$$

Lemma 2. *For $k > 1$ and noninteger $x > 2$, if we set*

$$J(x) = \frac{1}{2\pi i}\int_{k-iT}^{k+iT} \frac{x^s}{s}\log\zeta(s)\mathrm{d}s + R_1(x,k,T) \tag{12}$$

Then

$$R_1(x,k,T) \ll \frac{x^k}{T}\sum_{n=1}^{\infty} \frac{1}{n^k|\log x/n|} \tag{13}$$

Proof. If we define

$$h(y) = \begin{cases} 0 & 0 < y < 1 \\ 1 & y > 1 \end{cases} \tag{14}$$

then $J(x)$ can be rewritten into

$$J(x) = \sum_{n=1}^{\infty} a_n h\left(\frac{x}{n}\right) \tag{15}$$

The right hand side converges certainly because the summand vanishes once $n > x$. Due to Titchmarsh [3], for $k > 1$ we have

$$h(y) = \frac{1}{2\pi i}\int_{k-iT}^{k+iT} \frac{y^s}{s} \mathrm{d}s + O\left(\frac{y^k}{T|\log y|}\right) \tag{16}$$

Since $\log\zeta(s)$ converges absolutely whenever $\Re(s) = k > 1$, we plug (16) and (11) into (15). Finally, $|a_n| \le 1$ allows us to deduce (13).

Lemma 3. *Using the same notation, we have*

$$R_1(x,k,T) \ll \frac{x^k}{T(k-1)} + \frac{x\log x}{T} + \frac{x}{T|x-N|} \tag{17}$$

where N denotes the nearest *integer to x.*

Proof. To handle the logarithms in the denominator in (13), we partition the sum into

$$\sum_n = \sum_{n<x/2,n>2x} + \sum_{x/2\le n\le N-1} + \sum_{N+1\le n\le 2x} + \frac{x^k}{Tn^k|\log x/N|}$$

In the first sum, it is evident that $|\log x/n| > \log 2$, so we have

$$\frac{x^k}{T}\sum_{n<x/2,n>2x} \frac{1}{n^k|\log x/n|} \ll \frac{x^k}{T}\sum_{n<x/2,n>2x}\frac{1}{n^k} \le \frac{x^k}{T}\sum_{n=1}^{\infty}\frac{1}{n^k}$$
$$= \frac{x^k}{T}\zeta(k) \ll \frac{x^k}{T(k-1)}$$

in which the last inequality follows from the fact that $\zeta(s) \sim 1/(s-1)$ when $s \to 1$ [13].
When $x/2 \le n \le 2x$, the logarithm function satisfies that

$$\left|\log\frac{x}{n}\right| \gg \frac{|x-n|}{x} \tag{18}$$

so we have

$$\frac{x^k}{T}\sum_{x/2\le n\le N-1}\frac{1}{n^k|\log x/n|} \ll \frac{1}{T}\sum_{x/2\le n\le N-1}\frac{1}{|\log x/n|} \ll \frac{x}{T}\sum_{x/2\le n\le N-1}\frac{1}{x-n}$$
$$\ll \frac{x}{T}\int_{x/2}^{N-1}\frac{\mathrm{d}t}{x-t} = \frac{x}{T}\log\frac{x/2}{x-N+1} \ll \frac{x\log x}{T}$$

in which the last part follows from the fact that $N - 1/2 \le x \le N + 1/2$. Due to symmetry, (18) can be applied to the sum over $N + 1 \le n \le 2x$ to obtain the same estimate. For the last part, using (18), we have

$$\frac{x^k}{Tn^k|\log x/N|} \ll \frac{1}{T|\log x/N|} \ll \frac{x}{T|x-N|} \tag{19}$$

Combining all these estimates gives (17).

Theorem 2. *For half-odd integer* $x > 2$ *and* $k = 1 + 1/\log x$, *we have*

$$\pi(x) = \frac{1}{2\pi i}\int_{k-iT}^{k+iT}\frac{x^s}{s}\log\zeta(s)\mathrm{d}s + R_2(x,T) \tag{20}$$

with

$$R_2(x,T) \ll \frac{x\log x}{T} + \sqrt{x}\log\log x \tag{21}$$

Proof. Plugging $k = 1 + 1/\log x$ into (17) gives

$$R_1(x, 1 + 1/\log x, T) \ll \frac{x\log x}{T} + \frac{x}{T|x-N|} \tag{22}$$

when x is half-odd, we have $|x - N| = 1/2$, so the second term of (22) can be removed. Plugging (22) into (12), we get

$$J(x) = \frac{1}{2\pi i}\int_{k-iT}^{k+iT} \frac{x^s}{s}\log\zeta(s)\mathrm{d}s + O\left(\frac{x\log x}{T}\right) \tag{23}$$

Finally, we replace $J(x)$ with (23) in (9) to obtain the result.

4. Evaluation of the integral

To handle the integral, we consider a rectangular region R with vertices $1 - \delta \pm iT$ and $k \pm iT$. This implies

$$\frac{1}{2\pi i}\int_{k-iT}^{k+iT} \frac{x^s}{s}\log\zeta(s)\mathrm{d}s = \frac{1}{2\pi i}\oint_{\partial R} \frac{x^s}{s}\log\zeta(s)\mathrm{d}s \tag{24}$$

$$+\frac{1}{2\pi i}\left[\int_{k-iT}^{1-\delta-iT} + \int_{1-\delta-iT}^{1-\delta+iT} + \int_{1-\delta+iT}^{k+iT}\right] \tag{25}$$

where ∂R denotes the boundary of R in counterclockwise direction and δ is chosen such that $\zeta(s) \neq 0$ in R and on ∂R. This means that the only singularity of the integrand in R is at $s = 1$. However, because $s = 1$ is a logarithmic singularity, circumvention is needed to evaluate it.

Lemma 4. *If we define*

$$f(r) = \frac{1}{2\pi i}\oint_{(r+)} \frac{x^s}{s}\log\frac{1}{s/r-1}\mathrm{d}s \tag{26}$$

where $(r+)$ denotes any counterclockwise path that encloses only the singularity at $s = r$, then

$$f'(r) = \frac{x^r}{r} \tag{27}$$

for $r \neq 0$.

Proof. Since

$$\frac{\partial}{\partial r}\log\frac{1}{s/r-1} = \frac{\partial}{\partial r}[\log r - \log(s-r)] = \frac{1}{r} + \frac{1}{s-r}$$

we have

$$f'(r) = \frac{1}{2\pi i}\oint_{(r+)} \frac{x^s}{s}\left[\frac{1}{r} + \frac{1}{s-r}\right]\mathrm{d}s$$

$$= \lim_{s\to r}\frac{x^s}{s}\left[\frac{s-r}{r} + 1\right] = \frac{x^r}{r}$$

Lemma 5. *Using the same notation, for all $\eta > 0$ we have*

$$f(1 + i\eta) = \mathrm{li}(x^{1+i\eta}) - i\pi \tag{28}$$

$$f(1 - i\eta) = \mathrm{li}(x^{1-i\eta}) + i\pi \tag{29}$$

Proof. When $\eta > 0$, because (26) implies $\lim_{a\to\infty} f(a) = 0$, we have

$$f(1 + i\eta) = f(1 + i\eta) - f(-\infty + i\eta) = \int_{-\infty+i\eta}^{1+i\eta} f'(r)\mathrm{d}r$$

$$= \underbrace{\int_{-\infty+i\eta}^{1+i\eta} \frac{x^r}{r}\mathrm{d}r}_{u=r\log x} = \int_{(-\infty+i\eta)\log x}^{(1+i\eta)\log x} \frac{e^u}{u}\mathrm{d}u$$

By Cauchy's integral theorem [13], we can deform the path of integration of the last integral into a line segment connecting $-\infty$ and $-\varepsilon < 0$, a clockwise circular arc connecting $-\varepsilon$ and ε, and finally a line segment connecting ε and $(1 + \eta i)\log x$. That is, we have

$$f(1+i\eta) = \int_{-\infty}^{-\varepsilon} + \int_{\varepsilon}^{(1+\delta i)\log x} + \int_{-\varepsilon}^{\varepsilon} \tag{30}$$

for all $\varepsilon > 0$. As $\varepsilon \to 0^+$, the latter integral becomes

$$\lim_{\varepsilon\to 0^+} \int_{-\varepsilon}^{\varepsilon} \frac{e^u}{u} du = \lim_{\varepsilon\to 0^+} \int_{\pi}^{0} \frac{e^{\varepsilon e^{i\theta}}}{\varepsilon e^{i\theta}} (i\varepsilon e^{i\theta}) d\theta = i \lim_{\varepsilon\to 0^+} \int_{\pi}^{0} e^{\varepsilon e^{i\theta}} d\theta$$
$$= -i \int_{\pi}^{0} d\theta = -i\pi$$

in which the interchanging of the limit operation follows from the fact that $e^{\varepsilon e^{i\theta}}$ converges to 1 uniformly:

$$|e^{\varepsilon e^{i\theta}} - 1| \le \sum_{n=1}^{\infty} \frac{\varepsilon^n}{n!} = e^{\varepsilon} - 1 \tag{31}$$

As $\varepsilon \to 0^+$, the first two integrals become

$$\lim_{\varepsilon\to 0^+} \underbrace{\left[\int_{-\infty}^{-\varepsilon} + \int_{\varepsilon}^{(1+\eta i)\log x} \frac{e^u}{u} du\right]}_{t=e^u} = \lim_{\varepsilon\to 0^+} \left[\int_{-\infty}^{e^{-\varepsilon}} + \int_{e^{\varepsilon}}^{x^{1+i\eta}} \frac{dt}{\log t}\right] \tag{32}$$

By (6), it is obvious that the right hand side evaluates to $\mathrm{li}(x^{1+i\eta})$. By plugging everything into (30), we see that (28) is true. A symmetric argument can be applied to prove (29).

Theorem 3. *Using the same notation, we have*

$$\frac{1}{2\pi i} \oint_{\partial R} \frac{x^s}{s} \log\zeta(s) ds = \mathrm{li}(x) \tag{33}$$

Proof. Since $\zeta(s) \sim 1/(s-1)$ when s is near one [13], we can transform the integrand:

$$\frac{1}{2\pi i} \oint_{\partial R} \frac{x^s}{s} \log\zeta(s) ds = \frac{1}{2\pi i} \oint_{(1+)} \frac{x^s}{s} \log \frac{1}{s-1} ds \tag{34}$$

By (26), we see that the right hand side is exactly $f(1)$, which can be calculated by evaluating its Cauchy principal value from (28) and (29):

$$f(1) = \lim_{\eta\to 0^+} \frac{f(1+i\eta)+f(1-i\eta)}{2} = \mathrm{li}(x) \tag{35}$$

Since Theorem 3 allows us to evaluate (24), we now move our focus to (25). To begin with, we extract some properties of $\zeta(s)$ necessary for the proof.

Lemma 6 (de la Vallée Poussin). *Let $s = \sigma + it$. Then there exists a fixed constant $A > 0$ such that* whenever $1 - A/log|t| \le \sigma \le 2$ *and* $|t| \ge 2$ *we have* $\zeta(s) \ne 0$ *and*

$$\frac{\zeta\prime}{\zeta}(s) \ll \log|t| \tag{36}$$

Moreover, in the same region we have

$$|\log\zeta(s)| \ll \log|t| \tag{37}$$

Corollary 1. *When $s \in \partial R$, we have*

$$|\log\zeta(s)| \ll \log T \tag{38}$$

Proof. Proof for (36) is already covered in the third chapter of [3], so we move onto proving (37). Since $|\log\zeta(2+it)| \le \log\zeta(2)$, we can rewrite $\log\zeta(s)$ into

$$\log\zeta(s) = \int_2^{\sigma} \frac{\zeta\prime}{\zeta}(u+it) du + O(1)$$
$$\ll \int_2^{\sigma} \log|t||du| = |\sigma - 2|\log|t|$$

It is evident that σ is bounded, so (37) directly follows.

Theorem 4. *If we choose $\delta = A/log\ T$ and write*

$$\frac{1}{2\pi i}\int_{k-iT}^{k+iT}\frac{x^s}{s}\log\zeta(s)ds = \mathrm{li}(x) + R_3(x,T) \tag{39}$$

then

$$R_3(x,T) \ll x\log^2 T\exp\left(-\frac{A\log x}{\log T}\right) \tag{40}$$

Proof. By Corollary 1, it is evident that the integral over the horizontal segments of ∂R satsifies

$$\left|\int_{k\pm iT}^{1-\delta\pm iT}\frac{x^s}{s}\log\zeta(s)ds\right| \ll \frac{\log T}{T}\int_k^{1-\delta}x^u du \ll \frac{x\log T}{T\log x} \tag{41}$$

Using Corollary 1, we can also establish an upper bound for the left vertical segment of ∂R:

$$\left|\int_{1-\delta-iT}^{1-\delta+iT}\frac{x^s}{s}\log\zeta(s)ds\right| \ll x^{1-\delta}\log T\int_{-T}^{T}\frac{du}{|1-\delta+it|} \tag{42}$$

$$\ll x^{1-\delta}\log^2 T = x\log^2 T\exp\left(-\frac{A\log x}{\log T}\right) \tag{43}$$

wherein the last inequality follows from the definition of δ. Plugging (41), (43) into (25) and applying Theorem 3 to (24) give us the desired result.
With everything prepared, we move our attention back to the prime counting function $\pi(x)$.

5. Proof of the prime number theorem

Applying Theorem 4 to (20), we get

$$\pi(x) = \mathrm{li}(x) + R_2(x,T) + R_3(x,T) \tag{44}$$

If we set $\log T = \sqrt{\log x}$, then we have

$$R_2\left(x, e^{\sqrt{\log x}}\right) \ll x\log x e^{-\sqrt{\log x}} \tag{45}$$

$$R_3\left(x, e^{\sqrt{\log x}}\right) \ll x\log x e^{-A\sqrt{\log x}} \tag{46}$$

If we choose a proper $0 < c < \min(1, A)$, then the logarithms in (45) and (46) can be absorbed into $e^{-c\sqrt{\log x}}$. Plugging them into (44) gives us the final result:
Theorem 5 (de la Vallée Poussin). *There exists a constant* $c > 0$ *such that*

$$\pi(x) = \mathrm{li}(x) + O(xe^{-c\sqrt{\log x}}) \tag{47}$$

Using the fact that $\mathrm{li}(x) \sim x/\log x$, we can also sharpen the remainder of (9):
Corollary 2. *Under Theorem 5, we have for* $x \geq 2$ *that*

$$\pi(x) = J(x) + O\left(\frac{\sqrt{x}\log\log x}{\log x}\right) \tag{48}$$

Proof. Similar to how Theorem 1 is proved, we just need to give a better upper bound for $J(x) - \pi(x)$. That is,

$$J(x) - \pi(x) = \sum_{2\leq k\leq\log_2 x}\frac{\pi(x^{1/k})}{k} \leq \pi(\sqrt{x})\sum_{2\leq k\leq\log_2 x}\frac{1}{k}$$

$$\ll \frac{\sqrt{x}}{\log x}\sum_{2\leq k\leq\log_2 x}\frac{1}{k} \ll \frac{\sqrt{x}\log\log x}{\log x}$$

where the second last $\ll$ follows from the fact that Theorem 5 implies $\pi(x) \ll x/\log x$.

6. Conclusion

In this paper, we present a new proof of the prime number theorem that avoids the use of Chebyshev functions. First, we prove (20) to transform the arithmetic problem into an analytic problem. To evaluate the integral on the right hand side of (20), we set up a counter clockwise rectangular path so that this task is divided in to (24) and (25). Subsequently, differentiation under integral is applied to evaluate (24), and classical zero-free region and upper bound for $\zeta(s)$ are used to give estimates for (25). Finally, choosing a proper T, we deduce again the prime number theorem in the version of de la Vallée Poussin. The most significant contribution of the proof is the evaluation of (24), which gives the main term of Theorem 5. Traditionally, number theorists study the prime number theorem and its generalizations (i.e. prime number theorem for arithmetic progressions [12], prime number theorem for automorphic L-functions [14], or sums of complex numbers over primes [2]) first using the weighted sum

$$\sum_{n\leq x} c_n\Lambda(n) = \sum_{p^k\leq x} c_n \log p \tag{49}$$

and then convert it into sum over primes

$$\sum_{p\leq x} c_p \tag{50}$$

using partial summation. However, introducing the parameter r suggests a possibility to directly evaluate sum over prime powers

$$\sum_{n\leq x} a_n c_n = \sum_{p^k\leq x} \frac{c_{p^k}}{k} = \sum_{p\leq x} c_p + R \tag{51}$$

To study (50) from (49), partial summation is needed for conversion, but studying (50) from (51) only requires us to provide upper estimates for

$$R = \sum_{2\leq k\leq \log_2 x} \sum_{p\leq x^{1/k}} \frac{c_{p^k}}{k} \tag{52}$$

Consequently, the proof technique presented in this paper may inspire new proofs of generalized prime number theorems [14][2] as they are also studied using Perron's formula [3].

7. References

[1] Tom M. Apostol. *Introduction to analytic number theory*. 1976.

[2] Henryk Iwaniec and Emmanuel Kowalski. *Analytic number theory*. Number v. 53 in Colloquium publications. American Mathematical Society, Providence, R.I, 2004.

[3] E. C. Titchmarsh. *The theory of functions*. Oxford science publications. Oxford Univ. Press, Oxford, 2. ed., reprinted edition, 2002. OCLC: 249703508.

[4] Norman Levinson. On the elementary proof of the prime number theorem. *Proceedings of the Edinburgh Mathematical Society*, 15(2):141–146, 1966.

[5] D. J. Newman. Simple Analytic Proof of the Prime Number Theorem. *The American Mathematical Monthly*, 87(9):693–696, 1980.

[6] Zhentao Ma. A Survey on Newmann's proof of the prime number theorem. *Journal of Physics: Conference Series*, 1903(1):012012, April 2021.

[7] Atle Selberg. An Elementary Proof of the Prime-Number Theorem. *The Annals of Mathematics*, 50(2):305, April 1949.

[8] Elias M. Stein and Rami Shakarchi. *Complex analysis*. Number 2 in Princeton lectures in analysis. Princeton University Press, Princeton, N.J, 2003. OCLC: ocm51738532.

[9] E. M. Wright. XVIII.—The Elementary Proof of the Prime Number Theorem. *Proceedings of the Royal Society of Edinburgh. Section A. Mathematical and Physical Sciences*, 63(3):257–267, 1952.

[10] E. Landau. Uber einige neuere grenzwertsatze. *Rendiconti del Circolo Matematico di Palermo*, 34(1):121– 131, December 1912.

[11] H. M. Edwards. *Riemann's Zeta Function*. Academic Press Inc., New York, 1974.

[12] Hugh L. Montgomery and Robert C. Vaughan. *Multiplicative Number Theory I: Classical Theory*. 2006.

[13] E. C. Titchmarsh and D. R. Heath-Brown. *The theory of the Riemann zeta-function*. Oxford science publications. Clarendon Press ; Oxford University Press, Oxford [Oxfordshire] : New York, 2nd ed edition, 1986.

[14] Jianya Liu and Yangbo Ye. Perron's Formula and the Prime Number Theorem for Automorphic L-Functions. *Pure and Applied Mathematics Quarterly*, 3(2):481–497, 2007.

Gradient-induced Model-free Variable Selection Based on Composite Quantile Regression in Reproducing Kernel Hilbert Space

Xin Tan[1], Yan Kong[1] and Mei Yan[2]

1 University of Electronic Science of Technology of China, Chengdu, China
2 Yunnan Normal University, Yunnan, China
E-mail: tanxin_happy@qq.com

Abstract. Variable selection plays an important role to identify truly informative variables in high-dimensional data analysis. In his paper, we propose a variable selection method with composite quantile regression in reproducing kernel Hilbert space (RKHS), which has two main advantages. The first is that our method requires no special model structure assumption and no independence of error term. It is suitable for general non-parametric models and even heteroscedastic models. The second is that the calculation is simple and fast. So, it can also work in high-dimensional situations. Finally, the numerical experiments and real data analysis demonstrate its superior performance in variable selection.

1. Introduction

When analyzing high-dimensional data, we often believed that only a few variables really provide information, which is called the sparsity of data. Therefore, identifying the truly informative variables is regarded as one of the main goals of high-dimensional data analysis and many practical applications (such as health research). In literature, a wide spectrum of variable selection methods had been proposed. For example, the least absolute shrinkage and selection operator (LASSO, [1]), the smoothly clipped absolute deviation (SCAD, [2]), the adaptive LASSO ([3]), the group LASSO ([4]) for the linear model assumption; [5] , [6] , the component selection and smoothing operator (COSSO, [7]), [8] for the additive structure assumption. The above methods all have requirements for the model structure.
In recent years, a type of model-free variable selection method has been proposed, which do not require any special structure of the model. [9] introduced a new variable selection method based on the measurement error model, which can be applied to non-parametric kernel regression. The model-free variable selection method is based on the idea that a variable is truly informative with respect to the regression function if the gradient of the regression function is away from zero. Thus, this type of variable selection method focuses on learning the gradient function. In this paper, we focus on estimating the gradient by directly estimating the quantile function in a reproducing kernel Hilbert space (RKHS, [10]). Based on this idea, [11] and [12] both proposed a model-free variable selection method which identified the unimportant variables by whether the gradient of the regression function is zero. However, their methods are not applicable to the case that the error term is independent of variables X, which can be shown later. To overcome this problem, [13] utilized whether the gradient of the composite quantile function is zero to distinguish important variables. However, their calculations are complex and need iteratively optimize $O(p)$ vectors, which is not suitable for high- dimensional setting. Assuming that the conditional quantiles belong to some smoothing RKHS, we propose an efficient sparse learning procedure, where we first estimate the conditional quantile functions at multiple quantile levels, then

integrate their gradients and finally select informative variables via hard thresholding. We just need to solve m (shown later and independent of p) optimization problems separately, and thus it is computationally efficient for high-dimensional data.
The rest of this article is organized as follows. Section 2 introduces our method and three-step procedure. In section 3, we provide a method to choose parameter λ and v_n. The simulation studies and real applications are contained in Section 5 and 6. Finally, a brief discussion is given to conclude this work.

2. Method

In this paper, we consider the non-parametric model

$$Y = f(\mathrm{X}) + \varepsilon,$$

where $\mathrm{X} \in \mathbb{R}^{\mathrm{p}}$ and ε may not independent of X. Note that X^l is regarded as uninformative if and only if

$$g_\tau^l(\mathrm{x}) \triangleq \partial_l \mathrm{Q}_\tau(\mathrm{X}) = \frac{\partial \mathrm{Q}_\tau(X)}{\partial \mathrm{X}^{\mathrm{l}}} \equiv 0, \text{ for any x and } \tau \in (0,1),$$

where $\mathrm{Q}_\tau(\mathrm{X})$ is the τ-th conditional quantile function of Y given X. Thus, based on this property, we get the following procedure to select informative variables:

Procedure of our method
Step 1: Obtain an estimation $\hat{Q}_{\tau_k}(x)$ in RKHS for any given $\tau_k \in (0,1)$, $k = 1,2,\cdots,m$.
Step 2: Compute $\hat{g}_{\tau_k}^l = \frac{\partial_l \hat{Q}_{\tau_k}(x)}{\partial x^l}$ for $l = 1,2,\cdots,p$ and $k = 1,2,\cdots,m$.
Step 3: Identify the informative variables by checking the norm of $\hat{\boldsymbol{g}}^l \triangleq (\hat{g}_{\tau_1}^l,\cdots,\hat{g}_{\tau_m}^l)$ and the norm is defined as $\parallel \hat{\boldsymbol{g}}^l \parallel_m^2 = \frac{1}{m}\sum_{k=1}^m \parallel \hat{g}_{\tau_k}^l \parallel_n^2$.

For the step 1, we employ the KQR (kernel quantile regression, [14]) method,

$$\widehat{\mathrm{Q}}_{\tau_{\mathrm{k}}} = \mathrm{argmin}_{Q_{\tau_k} \in \mathcal{H}} \frac{1}{\mathrm{n}} \sum_{\mathrm{i}=1}^{\mathrm{n}} \rho_{\tau_{\mathrm{k}}}\left(\mathrm{y}_{\mathrm{i}} - \mathrm{Q}_{\tau_{\mathrm{k}}}(\mathrm{X}_{\mathrm{i}})\right) + \lambda \parallel \mathrm{Q}_{\tau_{\mathrm{k}}} \parallel_{\mathcal{H}}^2, \ \mathrm{k} = 1,2,\cdots,\mathrm{m}, \tag{1}$$

where $\rho_\tau(\mathrm{u}) = \left(\tau - \mathrm{I}_{\{\mathrm{u}<0\}}\right)\mathrm{u}$ and $\mathcal{H}$ is a RKHS associated with kernel $\mathcal{K}$ and $\{(\mathrm{X_i}, \mathrm{y_i})\}_{\mathrm{i}=1}^{\mathrm{n}}$ is an identity independent sample of (X, Y). According to the represent theorem (see details in [10]), the minimizer of equation (1) must have the form

$$\hat{Q}_{\tau_k}(\cdot) = \sum_{i=1}^{n} \alpha_i^{(k)} \mathcal{K}(X_i,\cdot) = K_n(\cdot)^T \boldsymbol{\alpha}^{(k)}, \tag{2}$$

where $\boldsymbol{\alpha}^{(\mathrm{k})} = \left(\alpha_1^{(\mathrm{k})}, \alpha_2^{(\mathrm{k})}, \cdots, \alpha_{\mathrm{n}}^{(\mathrm{k})}\right)^{\mathrm{T}}$ and $\mathrm{K_n}(\cdot) = \left(\mathcal{K}(\mathrm{X}_1,\cdot), \mathcal{K}(\mathrm{X}_2,\cdot), \cdots, \mathcal{K}(\mathrm{X_n},\cdot)\right)^{\mathrm{T}}$.
Next, for the step 2, it follows from Lemma 1 of [13] . This lemma implies that if we want to estimate the gradient of Q_τ within the smooth RKHS, it suffices to estimate Q_τ without loss of information. Thus, if $\widehat{\mathrm{Q}}_{\tau_{\mathrm{k}}}$ is obtained in step 1, $\mathrm{g}_{\tau_{\mathrm{k}}}^l(\mathrm{x})$ can be estimated as $\hat{g}_{\tau_{\mathrm{k}}}^l(\mathrm{x}) = \left(\partial_l \mathrm{K_n}(\mathrm{x})\right)^{\mathrm{T}} \boldsymbol{\alpha}^{(k)}$ where $\partial_l \mathrm{K_n}(\mathrm{x}) = \left(\partial_l \mathcal{K}(\mathrm{X}_1, \mathrm{x}), \partial_l \mathcal{K}(\mathrm{X}_2, \mathrm{x}), \cdots, \partial_l \mathcal{K}(\mathrm{X_n}, \mathrm{x})\right)$.
For the step 3, note that $\mathcal{A}^* = \{\mathrm{l} : \int_0^1 \mathrm{E}[\mathrm{g}_\tau^{\mathrm{l}}(X)]^2 \, \mathrm{d}\tau > 0\}$. Since we don't know the distribution of X, in practice, we apply the empirical norm, i.e., $\parallel \hat{\mathrm{g}}_{\tau_{\mathrm{k}}}^l \parallel_{\mathrm{n}}^2 = \frac{1}{\mathrm{n}}\sum_{\mathrm{i}=1}^{\mathrm{n}} \left(\hat{\mathrm{g}}_{\tau_{\mathrm{k}}}^l(\mathrm{X_i})\right)^2$. Thus, we can estimate the active set as $\hat{\mathcal{A}} = \{l \mid \parallel \hat{\boldsymbol{g}}^l \parallel_m^2 > v_n\}$ where v_n is a parameter.

3. Choosing parameter λ and v_n

Based on some other literatures, for example, [13] and [15] , we set $\lambda = 0.001$ in our experiments. Then we mainly focus on the choice of v_n. Note that the v_n control the variable selection performance, thus we employ the Bayes information criterion (BIC, [16] and [16]) to choose it. Firstly, for a given v_n, we can get $\hat{\mathcal{A}}$ by our proposed method. Secondly, we can get the quantile estimation by step 1 based on sample $\{(X_{\hat{\mathcal{A}},i}, y_i)\}_{i=1}^n$. Thus, we can get the value of BIC

$$BIC(v_n) = \frac{1}{nm}\sum_{k=1}^{m}\sum_{i=1}^{n}\rho_{\tau_k}\left(y_i - \hat{Q}_{\tau_k}(X_{\hat{\mathcal{A}},i})\right) + \frac{\ln n}{2n}|\hat{\mathcal{A}}|, \tag{3}$$

where $|\hat{\mathcal{A}}|$ is the number of elements in $\hat{\mathcal{A}}$. Finally, we take the minimum point of $BIC(v_n)$ as the thresholding parameter $\hat{v}_n$.

4. Numerical experiments

In this section, we study the empirical performance of our proposed method (denoted by GM_cqr), comparing with Median (the same procedure with GM_cqr with $\tau = 0.5$), RF (Random Forest,[18]), COSSO of [7] , GM of [12] and MF of [13] . For COSSO and RF, R packages "cosso" and "randomForest" are used in the calculation. We use average model size (Size, the number of selected informative variables) and the times of correct-fitting, over-fitting and under-fitting (C, O and U) to evaluate the variable selection performance. The following two examples are considered, which were also used in [12] .

Example 4.1. Let $X = (X^{(1)}, X^{(2)}, \cdots, X^{(p)})$ with $X^{(k)} = \left(X_1^{(k)}, X_2^{(k)}, \cdots, X_n^{(k)}\right)^T$ and $X_i^{(k)} = (W_i^{(k)} + rU_i)/(1+r)$ for $r \in [0,1]$, $k = 1,2,\cdots,p$, $i = 1,2,\cdots,n$ where $W_i^{(k)}$ and U_i are independently from $U(-0.5,0.5)$. When $r = 0$, all variables are independent. The true regression function is $g^*(X_i) = 6g_1\left(X_i^{(1)}\right) + 4g_2\left(X_i^{(2)}\right)g_3\left(X_i^{(3)}\right) + 6g_4\left(X_i^{(4)}\right) + 5g_5\left(X_i^{(5)}\right)$ with $g_1(u) = u$, $g_2(u) = 2u - 1$, $g_4(u) = 2u + 1$, $g_4(u) = 0.1\sin(\pi u) + 0.2\cos(\pi u) + 0.3\sin^2(\pi u) + 0.4\cos^3(\pi u) + 0.5\sin^3(\pi u)$ and $g_5(u) = \sin(\pi u)/(2 - \sin(\pi u))$. Generate the response variable by $Y_i = g^*(X_i) + \varepsilon_i$ where ε_i is distributed as $N(0,1)$. Obviously, the first five variables are the active variables in this example.

Example 4.2. Let $X_i^{(k)}$, $i = 1,2,\cdots,n$, $k = 1,2,\cdots,p$ be generate similarly to Example 4.1 except that $W_i^{(k)}$ and U_i are independently drawn from $U(0,1)$. The model is $Y_i = 4X_i^{(1)}X_i^{(2)} + 3\left|X_i^{(3)}\right|\varepsilon_i$ with $\varepsilon_i \sim N(0,1)$. Clearly, the first three variables are the informative variables in this example.

For each example, we consider 12 settings with $r = 0,\ 0.1$, $n = 200,\ 400$ and $p = 20,\ 50,\ 100$. For each setting, we replicate the simulation 50 times. In this paper, we focus on the times of correct fitting to compare the variable selection performance and use bold-font to highlight the methods that has the best performance.

As a summary of the simulation results, our proposed method has much outperformance over the other competitors in most scenarios. The results of examples are given in table 1 and table 2. Note that example 4.1 is an additive model and the error term is independent of X, thus all six methods are applicable and thus are expected to do well. However, our proposed method still has superior performance yet RF and COSSO get worse when the dimension becomes higher.

Table 1. The averaged performance measures of variable selection in Example 4.1.

			$r = 0$				$r = 0.1$			
p	n	Method	Size	C	O	U	Size	C	O	U
20	200	GM_cqr	5.00	**50**	0	0	5.00	**50**	0	0
		MF	5.12	40	3	7	5.06	43	2	5
		RF	5.42	31	19	0	5.74	26	24	0
		Median	5.00	**50**	0	0	5.00	**50**	0	0

		GM	5.00	**50**	0	0	5.00	**50**	0	0
		COSSO	5.42	38	12	0	5.18	45	5	0
	400	GM_cqr	5.00	**50**	0	0	5.00	**50**	0	0
		MF	4.90	45	0	5	5.12	45	1	4
		RF	5.00	**50**	0	0	5.00	**50**	0	0
		Median	5.00	**50**	0	0	5.00	**50**	0	0
		GM	5.00	**50**	0	0	5.00	**50**	0	0
		COSSO	5.18	45	5	0	5.08	47	3	0
50	200	GM_cqr	5.00	**50**	0	0	5.00	**50**	0	0
		MF	5.56	30	19	1	5.42	38	10	2
		RF	7.12	6	44	0	7.44	6	44	0
		Median	5.00	**50**	0	0	5.00	**50**	0	0
		GM	5.00	**50**	0	0	5.00	**50**	0	0
		COSSO	4.92	37	7	6	5.24	36	10	4
	400	GM_cqr	5.00	**50**	0	0	5.00	**50**	0	0
		MF	5.34	40	7	3	5.18	44	5	1
		RF	5.06	47	3	0	5.02	49	1	0
		Median	5.00	**50**	0	0	5.00	**50**	0	0
		GM	5.00	**50**	0	0	5.00	**50**	0	0
		COSSO	5.08	45	4	1	5.40	34	15	1
100	200	GM_cqr	5.00	**50**	0	0	5.00	**50**	0	0
		MF	5.38	29	10	11	5.68	34	13	3
		RF	8.82	4	46	0	9.32	3	47	0
		Median	5.00	**50**	0	0	5.00	**50**	0	0
		GM	5.00	**50**	0	0	5.00	**50**	0	0
		COSSO	5.48	29	19	2	5.76	30	19	1
	400	GM_cqr	5.00	**50**	0	0	5.00	**50**	0	0
		MF	5.16	44	0	6	5.14	44	0	6
		RF	5.22	41	9	0	5.06	47	3	0
		Median	5.00	**50**	0	0	5.00	**50**	0	0
		GM	5.00	**50**	0	0	5.00	**50**	0	0
		COSSO	5.34	40	7	3	5.98	34	16	0

Example 4.2 is a heteroscedasticity model. As shown in table 2, our method has much outperformance over other competitors. Our proposed method can find all the informative variables in the regression function. Yet COSSO may just find the informative variables in the regression function. Besides, the MF can also identify the third informative variable. However, its performance is not as good as GM_cqr and its calculation is complex and runs very slow, especially when p becomes big.

Table 2. The averaged performance measures of variable selection in Example 4.1.

			$r = 0$				$r = 0.1$			
p	n	Method	Size	C	O	U	Size	C	O	U
20	200	GM_cqr	2.82	**43**	0	7	2.70	**37**	0	13
		MF	3.24	24	9	17	3.22	20	8	22
		RF	9.08	0	27	23	8.80	0	29	21
		Median	2.42	5	16	29	2.88	1	10	39
		GM	2.12	0	2	48	2.16	0	2	48
		COSSO	2.48	0	2	48	2.50	2	1	47

	400	GM_cqr	3.00	**50**	0	0	3.00	**50**	0	0
		MF	3.12	28	7	15	3.14	27	7	16
		RF	7.66	0	23	27	7.98	1	25	24
		Median	2.24	2	3	45	2.18	2	2	46
		GM	1.96	0	0	50	1.92	0	0	50
		COSSO	2.54	2	1	47	2.86	0	5	45
50	200	GM_cqr	2.60	**31**	0	18	2.52	**27**	2	21
		MF	3.66	19	13	18	3.52	18	12	20
		RF	18.82	0	14	36	18.50	0	13	37
		Median	2.86	0	5	45	2.94	1	4	45
		GM	2.14	0	0	50	2.52	0	1	49
		COSSO	2.84	1	1	48	3.34	0	1	49
	400	GM_cqr	3.00	**50**	0	0	3.00	**50**	0	0
		MF	3.24	23	9	18	3.20	22	8	20
		RF	12.24	0	17	33	14.50	0	12	38
		Median	2.06	1	0	49	2.18	0	1	49
		GM	1.96	0	0	50	1.92	0	0	50
		COSSO	2.16	0	0	50	2.42	0	0	50
100	200	GM_cqr	2.94	**29**	10	11	2.70	**25**	6	19
		MF	3.52	21	10	19	3.44	18	9	23
		RF	31.40	0	12	38	32.12	0	13	37
		Median	2.24	2	3	45	3.02	1	1	48
		GM	5.76	0	1	49	8.16	0	3	47
		COSSO	3.24	1	0	49	4.42	0	1	49
	400	GM_cqr	3.04	**48**	2	0	3.14	**43**	7	0
		MF	3.16	38	2	10	3.32	28	8	14
		RF	21.40	0	18	32	26.56	0	15	35
		Median	2.46	0	4	46	2.60	2	2	46
		GM	19.94	0	14	36	26.62	0	15	35
		COSSO	1.96	0	0	50	2.04	0	0	50

5. Real data analysis

In this section, we apply our proposed method to the residential building data. The residential building data at https://archive.ics.uci.edu/ml/datasets/Residential+Building+Data+Set, studies the construction costs and sale price corresponding to real estate single-family residential apartments in Tehran, Iran. There are 105 covariates with 372 observations and two response variables. We focus on the cost price in this paper. In our analysis, we take log transformation to the response variable. We randomly take 2/3 of the data as training set and the rest as testing set, and then apply the six methods in the simulations. We repeat the simulations 200 times and use their averages to assess the performance of the methods. The results are given in table 3. GM_cqr and Median selected the same less variables and provides smaller prediction error.

Table 3. The number of selected variables and the prediction errors by various selection methods in the residential building dataset.

Method	GM_cqr	MF	Median	GM	COSSO	RF
Number of selected variables	2	9	2	1	105	49
Prediction error (10^{-2})	5.46 (1.41)	17.13 (2.36)	**4.74** **(1.30)**	56.08 (12.86)	6.85 (12.72)	112.92 (7.38)

6. Conclusion

In this paper, we focus on estimating the gradient by estimating the quantile function in RKHS. Our propose method benefits the following two points: (i) Our proposed method is model-free and also applicable to heteroscedastic models which means that there may be informative variables in the error term; (ii) The calculation procedure of our method is provided. And it is also applicable to high-dimensional settings.

7. References

[1] Tibshirani, R 1996 Regression shrinkage and selection via the lasso Journal of the Royal Statistical Society: Series B (Methodological), 58(1):267–88

[2] Fan, J and Li, R 2001 Variable selection via nonconcave penalized likelihood and its oracle properties Journal of the American statistical Association, 96(456):1348–60

[3] Zou, H 2006 The adaptive lasso and its oracle properties Journal of the American statistical association, 101(476):1418–29

[4] Yuan, M and Lin, Y 2006 Model selection and estimation in regression with grouped variables Journal of the Royal Statistical Society: Series B (Statistical Methodology), 68(1):49–67

[5] Shively, T S, Kohn, R, and Wood, S 1999 Variable selection and function estimation in additive nonparametric regression using a data-based prior Journal of the American Statistical Association, 94(447):777–94

[6] Huang, J Z and Yang, L 2004 Identification of non-linear additive autoregressive models Journal of the Royal Statistical Society: Series B (Statistical Methodology), 66(2):463–77

[7] Lin, Y and Zhang, H H 2006 Component selection and smoothing in multivariate nonparametric regression The Annals of Statistics, 34(5):2272–97

[8] Xue, L (2009) Consistent variable selection in additive models Statistica Sinica, 19:1281–96

[9] Stefanski, L, Wu, Y, and White, K 2014 Variable selection in nonparametric classification via measurement error model selection likelihoods Journal of the American Statistical Association, 109(506):574–89

[10] Wahba, G 1999 Support vector machines, reproducing kernel Hilbert spaces, and randomized gacv Advances in kernel methods, 6:69–88

[11] Yang, L, Lv, S, and Wang, J 2016 Model-free variable selection in reproducing kernel Hilbert space The Journal of Machine Learning Research, 17(1):2885–2908

[12] He, X, Wang, J, and Lv, S 2018b Gradient-induced model-free variable selection with composite quantile regression Statistica Sinica, 28(3):1521–38

[13] He, X, Wang, J, and Lv, S 2018a Efficient kernel-based variable selection with sparsistency arXiv preprint arXiv:1802.09246

[14] Huang, J, Horowitz, J L and Wei, F 2010 Variable selection in nonparametric additive models Annals of statistics, 38(4):2282-313

[15] Wang, X and Leng, C 2016 High dimensional ordinary least squares projection for screening variables Journal of The Royal Statistical Society Series B-statistical Methodology, 78(3):589–611

[16] Schwarz, G 1978 Estimating the dimension of a model Annals of statistics, 6(2): 461-64

[17] Watanabe, S 2013 A widely applicable bayesian information criterion Journal of Machine Learning Research, 14(Mar):867–97

[18] Breiman, L 2001 Random forests Machine learning, 45(1):5–32

Acknowledgment

This work was supported by the Doctoral Foundation of Yunnan Normal University (No. 00800205020503127) and the Yunnan Provincial Education Department (No. 2021J0434).

A Stochastic Process for Music: The Example of K-pop Music

S Park[1], I Kim[2,*] and K Ahn[1,3,*]

[1] Department of Industrial Engineering, Yonsei University, Republic of Korea
[2] Department of Orchestral Music, Yonsei University, Republic of Korea
[3] Center for Finance and Technology, Yonsei University, Republic of Korea

[*] E-mail: inhacello@yonsei.ac.kr (I.K.); k.ahn@yonsei.ac.kr (K.A.)

Abstract. This study analyzes music using a stochastic process, particularly, the Vasicek model. This approach interprets note progression in a song as a mean-reverting process, allowing the estimation of three parameters such as the speed of the revision to the mean, long-term level of the mean, and volatility. In addition, the entropy is evaluated for each song to identify the randomness of rise–fall patterns for each music genre. Our results characterize certain types of music and could be used to create new indicators for music classification.

1. Introduction

Mathematics and computation contribute in numerous ways to music research, including music technology, clarification of music theoretical concepts, and creation of educational music tools. Yet, prior studies have concentrated on scale theories, and current challenges are predominantly focused on modeling musical behavior rather than music products [1].

Accordingly, this study aims to uncover the behavior of notes in music by applying the mean-reverting process, which is widely used in mathematics and physics. In particular, the Vasicek model is considered to model the evolution of notes, and entropy is employed to quantify its randomness.

Prior studies have employed numerous approaches to explain music mathematically. Some of these works have used the Gabor analysis, a type of time-frequency analysis, to investigate music signals [2]. Other works have attempted musical segmentation by finding patterns using musical parallelism [3]. However, to the best of the authors' knowledge, there has not been a single study utilizing the Vasicek model and symbolic time-series analysis (STSA) to examine the mean-revision process and time-varying patterns of note progression in music.

2. Theoretical framework

The notes in a song are analyzed through a mean-reverting process, in particular using the Vasicek model [4–8], which is a one-factor model widely adopted to describe the interest rate movements in finance. According to this model, the progression of a note at time $\boldsymbol{t}$ is defined as

$$dN_t = k(\mu - N_t)dt + \sigma dW_t, \tag{1}$$

where $\boldsymbol{k}$, $\boldsymbol{\mu}$, $\boldsymbol{\sigma}$, and $\boldsymbol{W_t}$ are the speed of adjustment, mean reversion level, volatility, and standard Wiener process, respectively.

To calibrate the parameters in equation (1), we rewrite it in the following regression form [9]:

$$y_t = \alpha + \beta x_t + \varepsilon_t, \tag{2}$$

where $\boldsymbol{y_t} = \boldsymbol{dN_t}$, $\boldsymbol{\alpha} = \boldsymbol{k\mu dt}$, $\boldsymbol{\beta} = -\boldsymbol{kdt}$, $\boldsymbol{x_t} = \boldsymbol{N_t}$, and $\boldsymbol{\varepsilon_t} = \boldsymbol{\sigma dW_t}$. We estimate the parameters of equation (2), i.e., $\boldsymbol{\alpha}$, $\boldsymbol{\beta}$, and $\boldsymbol{\sigma}$, by means of the ordinary least-squares method, where volatility is proportional to the standard error of the residuals according to

$$\sigma = \sqrt{\frac{VAR(\varepsilon_t)}{dt}}. \tag{3}$$

3. Method and data

3.1. Entropy

We apply the STSA and detect the dispersion of the probability allocation onto the rise–fall patterns of consecutive note series. The symbolization of consecutive notes is encoded as 1 for the positive change and 0 for the other changes. Subsequently, the size of the rolling window is determined to quantify the subsequence bundles consisting of $\boldsymbol{S}$ binary numbers. Each subsequence bundle is converted from a binary sequence to a new decimal number, $\boldsymbol{X^S}$ [10–15]. Then, the entropy of the random variable $\boldsymbol{X^S}$ is derived as

$$H(X^S) = -\sum_{i=1}^{M-(S-1)} p(x_i^S) \, log_2 \, p(x_i^S), \tag{4}$$

where $\boldsymbol{M}$ is the number of changes across all notes. Finally, the normalized Shannon entropy is given by

$$h(X^S) = \frac{1}{S} H(X^S). \tag{5}$$

Hereafter, "Shannon entropy" or simply "entropy" will be used to refer to a normalized entropy, that is, $\boldsymbol{h(X^S)}$.

3.2. Data

We reference the music chart data from Melon [16], and the top 30 songs from the chart of May 2021 are retrieved to analyze the latest trending K-pop music.[4] The first verses of these songs are converted into numerical data according to the following rules: (i) the sixteenth note length is taken as the unit length 1, and (ii) the C4 note is taken as the base note 0.

Table 1 summarizes the descriptive statistics. The mean note is between A4 and C5. The minimum note is F3, whereas the maximum note is E6. The skewness and excess kurtosis values suggest that the distribution of notes in K-pop music does not deviate significantly from the normal distribution. In particular, R&B/soul music has the highest variability, whereas rock/metal music exhibits the lowest variability.

Table 1. Descriptive statistics.

	Obs.[a]	Mean	Standard deviation	Maximum	Minimum	Skewness	Excess kurtosis
K-pop	350.20	5.83	3.41	16.00	–4.00	0.15	–0.50
Dance	355.75	5.40	3.61	16.00	–1.50	0.55	–0.12
Ballad	381.75	6.52	2.84	12.50	–1.00	–0.37	–0.44
R&B/Soul	466.00	5.70	3.94	14.00	–4.00	0.01	–0.91
Rap/Hip–Hop	349.25	5.20	3.22	14.50	0.00	0.52	–0.92
Rock/Metal	175.33	6.84	2.76	14.00	–4.00	–0.45	0.94

[4] Melon has the largest market share in the music streaming industry of South Korea.

[a] "Obs." indicates the average number of data points in each song.

4. Results and discussion

Figure 1 displays the histograms of the Vasicek parameters $\boldsymbol{k}$, $\boldsymbol{\mu}$, and $\boldsymbol{\sigma}$. Blue bar represents rock/metal music, and yellow represents the others. The dataset consists of 12 dance, 8 ballad, 3 R&B/soul, 4 rap/hip–hop, and 3 rock/metal music songs. In particular, songs belonging to the rock/metal category exhibit consistent values for all three parameters. Rock/metal music is more volatile than other music types and exhibits a higher mean reversion level and a lower speed of adjustment than other genres.

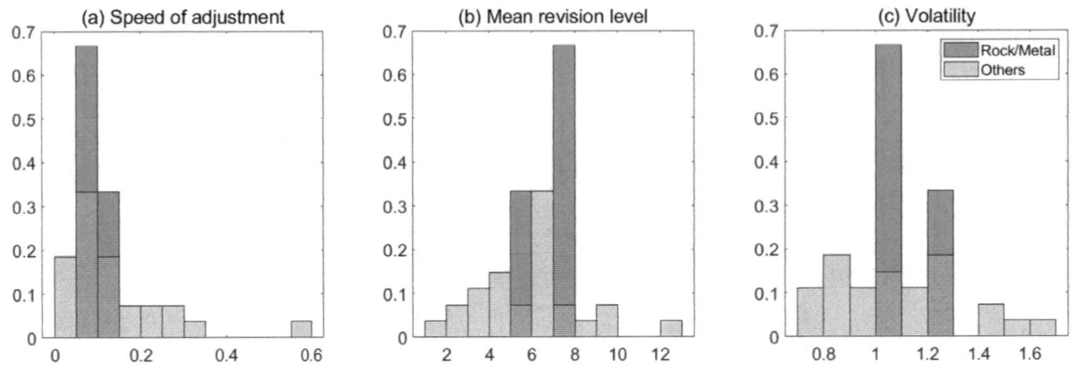

Figure 1. Histogram of the Vasicek parameters for Rock/Metal (see appendix for details).

Table 2 shows the entropy of each genre. R&B/soul and dance music show a higher entropy level than other music types, indicating that their randomness of note progression is higher than that of other genres. In addition, rap/hip–hop and rock/metal music have a lower randomness value, indicating that they have a more consistent rise–fall note pattern. Finally, the results show that the entropy values are robust regardless of the window size.

Table 2. Entropy.

Window size	Dance	Ballad	R&B/Soul	Rap/Hip–Hop	Rock/Metal
3	0.602 (2)	0.595 (3)	0.608 (1)	0.549 (5)	0.574 (4)
4	0.595 (2)	0.590 (3)	0.603 (1)	0.581 (4)	0.567 (5)
5	0.587 (2)	0.584 (3)	0.597 (1)	0.567 (4)	0.559 (5)

5. Conclusion

This study analyzes note progression in music by applying the Vasicek model. Furthermore, the STSA is employed to quantify the unpredictability of the rise–fall patterns in the note progression. Our results characterize certain music genres with three parameters of the Vasicek model and evaluate the randomness of note progression through entropy. Accordingly, our findings suggest the possibility of using a mathematical approach to explain the evolution of music history and propose new indicators for music classification.

Appendix

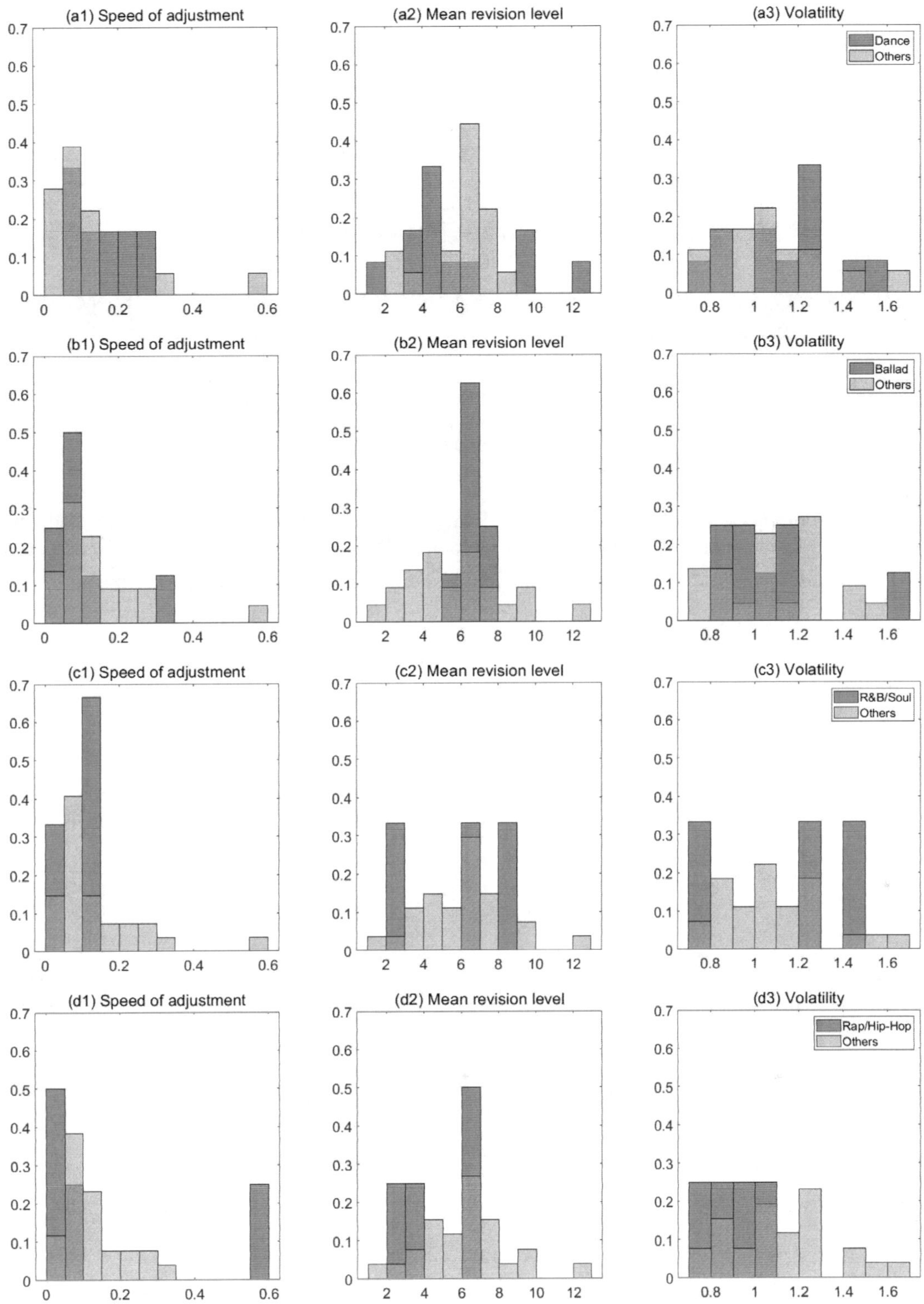

Figure 2. Histogram of the Vasicek parameters for all the details.

References

[1] Volk A and Honingh A 2012 *J. Math. Music.* **6** 73–81
[2] Dörfler M 2001 *J. New Music Res.* **20** 3–12
[3] Cambouropoulos E 2006 *Music Percept.* **23** 249–68
[4] Vasicek O 1977 *J. Financ. Econ.* **5** 177–88
[5] Ryu I, Jang H, Kim D and Ahn K 2021 *Chaos Solit. Fractals* **150** 111070
[6] Ahn K, Forsyth J, Jang H and Kim D 2021 *Fin. Res. Lett.* 102158
[7] Lee G H, Joo K and Ahn K 2020 *J. Phys. Conf. Ser.* **1593** 012037
[8] Ahn K, Choi M Y, Dai B, Sohn S and Yang B 2017 *EPL* **120** 38003
[9] Hull J C and White A D 1996 *J. Deriv.* **3** 26–36
[10] Ahn K, Lee D, Sohn S and Yang B 2019 *Quant. Fin.* **19** 1151–63
[11] Joo K, Jeong M, Seo Y, Suh J H and Ahn K 2021 *Energy Rep.* **7** 4192–204
[12] Yi E, Cho Y, Sohn S and Ahn K 2021 *Chaos Solit. Fractals* **142** 110464
[13] Jang H, Song Y and Ahn K 2020 *Phys. A* **550** 124114
[14] Joo K, Suh J H, Lee D and Ahn K *Energy Strategy Rev.* **30** 100516
[15] Jang S M, Yi E, Kim W C and Ahn K *Entropy* **21** 1116
[16] Melon Chart Finder. Jun 24, 2021 [https://www.melon.com/chart/search/index.htm]

Acknowledgments

This work was supported by (i) the Technology Innovation Program ATC+ (20014125, Development of Intelligent Management Solution for Nuclear Decommissioning Site Characterization) funded by the Ministry of Trade, Industry & Energy (MOTIE, Republic of Korea) and (ii) the Future-leading Research Initiative (Grant Number: 2021-22-0306) funded by Yonsei University.

Drug Level Modelling with Difference and Differential Equations

Shilpa D'Cunha[1] and V R Lakshmi Gorty [2]

1, 2 Basic Sciences and Humanities, Mukesh Patel School of Technology, Management & Engineering, NMIMS University, Mumbai, India
E-mail: shilpa.dcunha@nmims.edu; vr.lakshmigorty@nmims.edu

Abstract. In the current study, drug level in the human body is studied using mathematical modelling, where the difference equations have been used as a means of analyzing the dosage over a prescribed time, reduction in dose half-life, and the total amount of residual drug in the blood-stream for a given time-period. Further, the differential equations approach is investigated for data fitting using Mathematica to extend the application adhering to the continuous nature of the inherent function. The perspective helps to preserve critical resources and render comparative results. In this approach, zero-order, first-order, and fractional order kinetics are explored and compared.

1. Introduction

In the study, a recent real-world problem is selected to observe the given data, comprehend the pattern, and then model the situation. COVID-19 is an infectious disease caused by the most recently discovered coronavirus. The new virus and disease were unknown before the outbreak began in Wuhan, China, in December 2019. COVID-19 is now a pandemic affecting many countries globally. Various drugs are being developed for the COVID-19 pandemic, and mathematical modelling is proving a supportive tool. Bringing some relief for COVID-19 patients, even as the country was reporting its all-time high on each upcoming day of new coronavirus cases, the drugs/vaccines for COVID-19 are now being made available in medical stores/ hospitals. Glenmark Pharmaceuticals on 20 June 2020 launched the antiviral drug Fabiflu, under the brand name Fabiflu, for the treatment of patients with mild to moderate COVID-19. The Mumbai-based drug firm had received manufacturing and marketing approval from the Drugs Controller General of India.

According to the company [1], Fabiflu demonstrated promising clinical evidence, with positive results in mild to moderate COVID-19 cases [2]. Clinical improvement was observed in age groups of 20 to > 90 years. The drug claimed to provide a rapid decrease in viral load within four days, along with quicker symptomatic and radiological improvement. Fabiflu demonstrated up to 88 percent of clinical improvement in patients with mild to moderate COVID-19 [1]. It is a prescription-based medication with recommended dose being 1,800 mg twice daily on day one, followed by 800 mg twice daily up to day 14 [2]. Glenmark declared that mild to moderate COVID-19 patients with co-morbid conditions such as diabetes and heart disease can also use the drug [3]. At the same time, Remdesivir, intravenously administered in hospitals, has been the first treatment to show improvement in COVID-19 trials. It has been approved for emergency use in severely ill patients in the United States, India, and South Korea and received full approval in Japan [1].

Pharma major Cipla Ltd announced the launch of its generic version of antiviral drug Remdesivir for emergency use in the treatment of COVID-19 patients. The new drug was to be sold under the brand name Cipremi. According to a preliminary report from the ACTT-1 (Adaptive COVID-19 Treatment

Trial 1), a randomized clinical trial conducted with Remdesivir in 1063 patients over 60 centers across the US, Europe, and Asia demonstrated a faster time to clinical recovery in hospitalized patients as compared to placebo.

Most of these patients were on oxygen therapy of which some were receiving high flow oxygen or non-invasive ventilation, and some were on a mechanical ventilator. The mortality rates in the study were 7.1 percent in those given Remdesivir and 11.9 percent in those who were given a placebo [2].

In recent years, more costs of drug development and less productivity is a concern for new drug development [4]. To overcome the threat, different approaches like the use of adaptive trial design [5], ample use of biomarkers [6], starting personalized medicines, and the utilization of pharmacokinetics/ pharmacodynamics (PK/PD) modelling and simulation [4] have been adopted. From the early phase of drug development to the last stage, PK/PD modelling is proven to be a good buy [7, 8, 9]. As per the existing literature, the drugs of high half-life and slow-release rate can be better dealt with fractional calculus. The study and analysis of the dosage of the two drugs are investigated and resolved.

2. Modelling of dosage of Fabiflu

Analogous to the study in [10], Table 1 is studied for finding how much amount of Fabiflu should be given to the patients per day, the total amount of absorbed dose of Fabiflu, and such related questions. On Day 1, 1800 mg of dose were given at both times of the day while from the subsequent days 800 mg was administered as in [2]. The pattern of drug absorption is observed, and it is attempted to get a recurrence relation for fitting the data. The recurrence relation is solved for the term identified in it. The relation helped estimate the level of dose in the human body at prescribed day, also the day by which a pre-decided level of the dose is absorbed by the patient.

The 1st dose is represented by a_0, amount of Fabiflu in the body after the first dose is given. a_1 represented the amount of Fabiflu in the body after the second dose. Similarly, a_n represents the amount of Fabiflu in the body after dose $(n+1)$. A steady increase in the amount of dose by 800mg from day 2-14 is observed between the intervals. Pattern development is presented in Table 2. The next value of the dose could thus be predicted having known the previous value. It gives rise to (1). Thus, the difference equation is

$$a_n = a_{n-1} + 800 \tag{1}$$

except for $n = 0,1$ where $a_0 = 1800, a_1 = 3600$.

The total amount of dose absorbed in the body can be calculated just by knowing the number of doses. Solving (1) using the substitution method, we get

$$a_n = 2800 + 800n. \tag{2}$$

Table 1. The total amount of Fabiflu absorbed by the body.

Day	Time	Dose (mg)	Total absorbed (mg)
1	T_1	1800	0
	T_2	1800	1800
2	T_1	800	3600
	T_2	800	4400
3			⋮
⋮			
14	T_1	800	
	T_2	800	

Table 2. Pattern development of Fabiflu dose.

Dose (n)	Total dose given (a_{n-1})	Pattern
1	1800	
2	3600	$a_1 = a_0 + 1800$
3	4400	$a_2 = a_1 + 800$
4	5200	$a_3 = a_2 + 800$
⋮		
n+1	...	$a_n = a_{n-1} + 800$

Table 3. Amount of Fabiflu/Cipremi (mg) in blood-stream over time.

Time (Day 1)	Fabiflu	Cipremi
08:00	1800	200
09:00	1566.99	193.19
10:00	1364.14	186.61
11:00	1187.56	180.25
12:00	1033.83	174.11
13:00	900	168.18
⋮		

Table 4. Reduction in Fabiflu

t	r
1	0.87
2	0.76
3	0.66
4	0.57
5	0.5
⋮	⋮

Thus, after the first two doses, the sequence increased by the constant term which gives rise to a few uncertainties regarding the amount of drug increasing in the patient's body, possible side-effects, and routes of elimination of the drug.
As per [1], Fabiflu (generic name of Fabiflu) is an antiviral drug licensed in Japan in 2014 for the treatment of influenza. The elimination half-life of Fabiflu is estimated to range from 2 to 5.5 hours [1].
Considering the half-life of the drug as 5 hours, the reduction in the amount of Fabiflu in the body over a period of time is analyzed from the data in Table 3.
It is observed that the amount of drug in the blood-stream reduced over a period was not constant. However, the pattern demonstrated a constant ratio in reduction. It is reducing at a rate of 87 percent. Mathematically, it can be demonstrated as in Table 4.

Thus, in general, one can write $r = \left(\frac{1}{2}\right)^{(t/h)}$ where t is the time in hours and h is half-life of the drug.

The pattern seen in the reduction of the amount of drugs in Table 3 is summarized in Table 5. They have represented the initial amount of Fabiflu in the body by a_0, reduction in the amount of Fabiflu after one hour is represented by a_1, and so on. The amount of drug present in the body at the later hour is $r = 0.870550563$ times that of a formal hour and the difference equation for the same can be formulated as:

$$a_n = 0.870550563 * a_{n-1}. \tag{3}$$

Thus, the difference equation formed as:

$$a_n = r * a_{n-1}. \tag{4}$$

To check the relation between the amount of drug reduced at n^{th} hour with n^{th} hour, (4) is solved as

$$a_n = \begin{cases} (r)^n * 1800; & \text{day 1} \\ (r)^n * 800; & \text{day 2 to day 14} \end{cases} \tag{5}$$

Table 5. Day 1 Pattern Development

Hour	Pattern for Fabiflu	Pattern for Cipremi
0		
1	$a_1 = 0.87055*a_0$	$a_1 = 0.96593*a_0$
2	$a_2 = 0.87055*a_1$	$a_2 = 0.96593*a_1$
3	$a_3 = 0.87055*a_2$	$a_3 = 0.96593*a_2$
4	$a_4 = 0.87055*a_3$	$a_4 = 0.96593*a_3$
5	$a_5 = 0.87055*a_4$	$a_5 = 0.96593*a_4$
⋮		
n	$a_n = 0.87055*a_{n-1}$	$a_n = 0.96593*a_{n-1}$

Table 6. Total amount of Fabiflu/Cipremi in blood-stream.

Hour	Fabiflu level	Hour	Cipremi level
0	$a_0 = 1800$	0	$a_0 = 200$
1	$a_1 = r*1800$	1	$a_1 = r*200$
		⋮	
n	$a_n = r^n*1800$	n	$a_n = r^n*200$
⋮			
12	$a_{12} = r^{12}*1800 + 1800$	24	$a_{24} = r^{24}*200 + 100$
n	$a_n = r^n*1800 + r^{n-12}*1800$	n	$a_n = r^n*200 + r^{n-24}*100$
24	$a_{24} = r^{24}*1800 + r^{12}*1800 + 800$	48	$a_{48} = r^{48}*200 + r^{24}*100 + 100$

The graph of the reduced amount of drugs over a period of time is plotted in Figure 1.

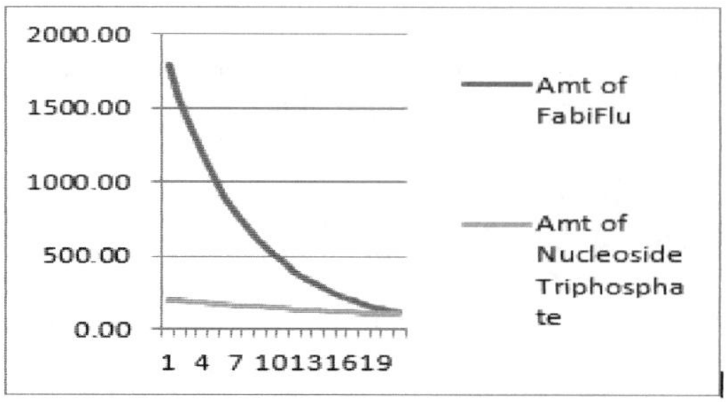

Figure 1. Reduced amount of Fabiflu/Cipremi

The amount of Fabiflu in the body after a certain period won't be the same as (5) as it reveals only the reduced amount of drug. To find the total amount of drugs in the body, Table 6 is tabulated. Continuing in this way, a geometric series is obtained. The series for the sum of first n terms to find a_n indicates the total amount of drug in the body over a certain time n.

$$a_n = \begin{cases} r^n * 1800; & 0 \le n < 12 \\ r^n \left[1000\left(1+\frac{1}{r^{12}}\right) + 800\left(\frac{1-r^{-(n_1+12)}}{1-r^{-12}}\right)\right]; & n_1 \le n < n_1+12, \text{ for } n_1 = 12k \text{ where } 1 \le k < 28 \\ r^n \left[1000\left(1+\frac{1}{r^{12}}\right) + 800\left(\frac{1-r^{-336}}{1-r^{-12}}\right)\right]; & n \ge 336 \end{cases} \quad (6)$$

The total amount of the drug is shown graphically in Figure 2.

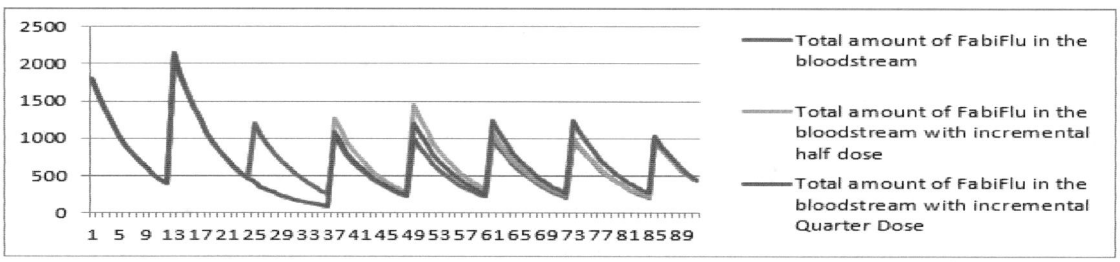

Figure 2. Fabiflu amount for prescribed regular and incremental dosage for missed dosage

On successful completion of the model, the solution of the difference equation is found to be useful in determining the total amount of drug level in the blood-stream for the given time. In the event of a missed dosage an incremental half dose or an incremental quarter dose as shown in Figure 2 is helpful to reinstate the level of the drug in the blood-stream. In the latter case, the drug level is restored gradually with relatively minor upshoot.

2.1 *Differential equation model development:*

An alternative method to the difference equation approach, indicates a possibility of modelling the data through differential calculus using Mathematica to extend the application adhering to the continuous nature of the inherent function. Noyes-Whitney Law [11] for drug release evaluation is given as $\frac{dC}{dt} = k(C_S - C_t)$, where $\frac{dC}{dt}$ dissolution rate, k dissolution rate constant, C_S drug concentration in stagnant layer and C_t drug concentration in the solution bulk at a time as per the Laws in the kinetics of drug release. Fick's law of diffusion [12] diffusion of drug molecule takes place from higher concentration region to lower concentration and diffusion rate is directly proportional to the concentration gradient. Various mathematical models [13] have been used in literature to study drug release profiles, drug diffusion, and drug elimination. In the current study, two such models viz, zero-order model and first-order model are fitted to the reduction (elimination) data of one of the COVID-19 drugs i.e. Fabiflu. The zero-order model defines the process of constant drug levels in the blood throughout the conveyance. In kinetics of zero-order with constant change rate in mass per unit time given as [11]:

$$\frac{dC}{dt} = k. \tag{7}$$

The solution of (7) as a function of time with the initial condition $C(0) = c_0$ is linearly expressed as:

$$C = kt + c_0. \tag{8}$$

The first-order model defines the process where the rate is directly proportional to the drug concentration at any time t [14].
In kinetics of First order, the proportionality of change rate to the existing value (C) is considered [11]:

$$\frac{dC}{dt} = -kC. \tag{9}$$

Classically, the equation of exponential relaxation helps derive the solution considering an initial condition of $C(0) = c_0$:

$$C = c_0 e^{-kt}. \tag{10}$$

As in [15, 16], some diffusional processes do not follow Fick's law. Super-division and sub-division are observed to follow mean-square displacement as non-integer power of time. The fractional calculus approach suits best describe such processes. Classical calculus is a special case of fractional calculus for $\alpha = 1$.
The first-order equation can be converted into fractional terms replacing a derivative of order 1 by a fractional order α and written as [15]:

$$\frac{d^{\alpha}C}{dC^{\alpha}} = -kC. \tag{11}$$

The solution for (11) is given by:

$$C = c_0 E_{\alpha}\left(-kt^{\alpha}\right). \tag{12}$$

For $\alpha = 1$, implies a special case of fractional order resulting in first order. Parameters estimated while applying (7) and (9) to the dataset is depicted in Table 7. R^2, as a measure of goodness of fit, indicates that (9) is better data-fit as shown in Figure 4 than (7) shown in Figure 3.

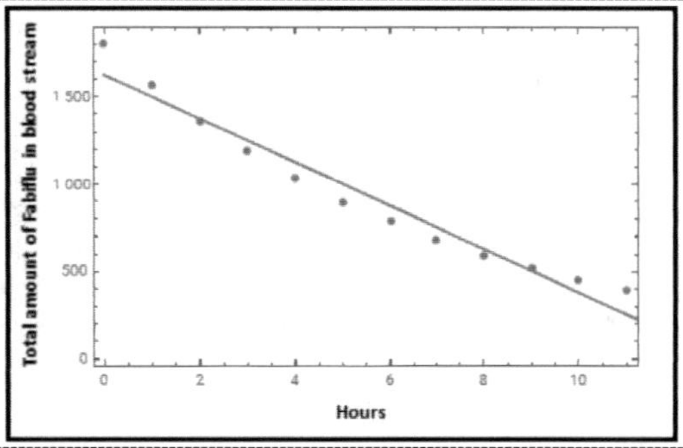

Figure 3. The plot of zero-order kinetic linear equation data fit for Fabiflu level in the blood-stream.

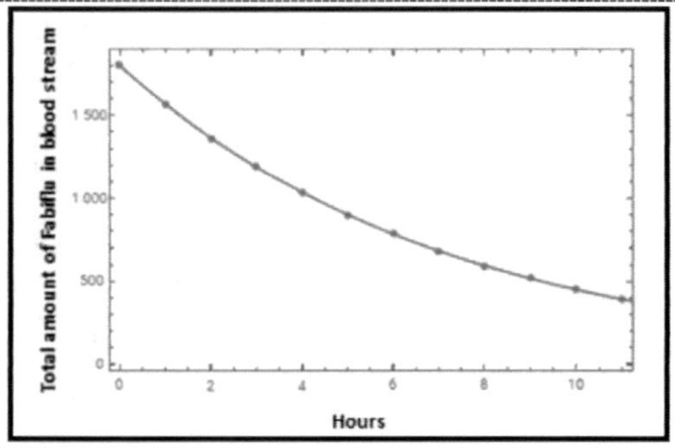

Figure 4. The plot of fractional- order ($\alpha = 1$)/first-order kinetic exponential equation data-fit for Fabiflu level in the blood-stream.

In the difference equation model demonstrated above, a time interval of one hour is considered. However, the level of drug in the blood-stream is a continuous process for which differential equation is advantageous. On application of differential equation model, we can estimate the continuous data function and obtain satisfactory goodness of fit.

Table 7. Parameter estimation (Figure 1).

Parameter	Estimate	Standard-error	R^2
k	-124.53	8.25059	0.95
c_0	1624.11	53.5759	
k	0.13863	$2.6447*10^{-7}$	1
c_0	1800	0.00181664	

Table 8. Total amount of Cipremi absorbed by body.

Day	Dose (mg)	Total absorbed(mg)
1	200	0
2	100	200
.	.	.
10	100	.

3. Modelling for dosage of Cipremi by difference equation

To gain understanding about increasing and decreasing sequence, modelling into difference equation and solving it for n^{th} term, other drugs are taken up for study. For Cipremi, clinical trials used a regimen of 200mg once daily on the first day, followed by 100mg once daily for another 9 days as in [2]. Early data suggested that some patients could benefit from only 5 days of treatment. Table 8 shows the total amount of Cipremi absorbed by the body.
From the Table 8, the difference equation is formed as:

$$a_n = a_{n-1} + 100 \tag{13}$$

for $n = 1, 2, \ldots, 9$ where $a_0 = 200$.
Solving (13) using the substitution method, we get

$$a_n = 200 + 100n. \tag{14}$$

Remdesivir (generic name of Cipremi) is a prodrug that is converted in the body into GS-441524 [17]. It is an antiviral drug developed by Gilead Sciences. It is the main metabolite of the antiviral prodrug remdesivir, which the latter is converted into inside the body. GS-441524 is then phosphorylated to the active nucleotide triphosphate form. A 10mg/kg intravenous dose in non-human primates has a plasma half-life of 0.39 hours. The nucleoside triphosphate metabolite has a half-life of 14 hours in non-human primates. The nucleoside triphosphate metabolite has a half-life of approximately 20 hours in humans [18]. The

calculated ratio is $r = 0.97$. Using the half-life, amount of Cipremi is calculated in Table 3. The pattern can be displayed as in Table 5.
Thus, the difference equation is formed as

$$a_n = 0.965936328 * a_{n-1}. \tag{15}$$

Solving (15) using the substitution method, we get

$$a_n = \begin{cases} (0.965936328)^n * 200; & \text{day1} \\ (0.965936328)^n * 100; & \text{day2 to day10} \end{cases} \tag{16}$$

The graph of the reduced amount of drugs over a period of time is plotted in Figure 1.
To find the total amount of Cipremi in the patient's body Table 6 is prepared.
A geometric series is obtained. The series for the sum of first n terms to find a_n is solved which indicated the total amount of drug in the body over a certain time n.

$$a_n = \begin{cases} r^n * 200; & 0 \le n < 24 \\ r^n \left[100 + 100 \left(\dfrac{1 - r^{-(n_1+24)}}{1 - r^{-24}} \right) \right]; & n_1 \le n < n_1 + 24 \text{ for } n_1 = 24k \text{ where } 1 \le k < 9 \\ r^n \left[100 + 100 \left(\dfrac{1 - r^{-240}}{1 - r^{-24}} \right) \right]; & n \ge 216 \end{cases} \tag{17}$$

Results are plotted graphically in Figure 5.

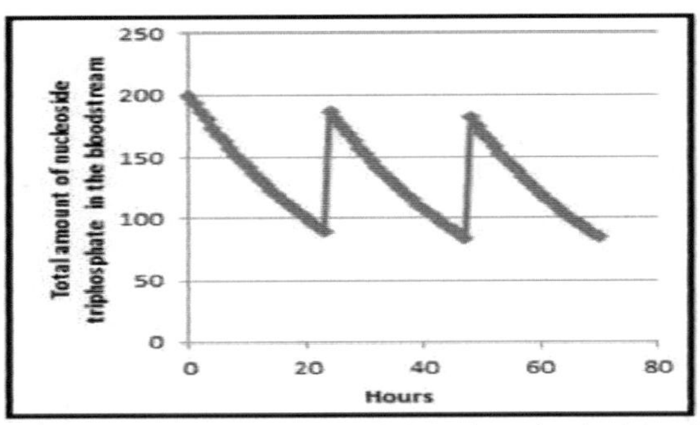

Figure 5. Amount of nucleoside triphosphate in the blood-stream

Thus, the model is developed for the second dosage too. The methodology of difference equations has been used for both medicines. Like Fabiflu, the differential equation approach can be used for fitting data further resulting in analogous inferences.

4. Conclusion

The difference equation to calculate the concentration of drug in the bloodstream was derived in the study by identifying geometric series. It was solved and results were presented graphically to visualize the pattern and estimate the drug level. This model may be useful for recommending remedial actions to restore requisite drug level in cases of possible discrepancies occurring due to missing a dose or overdose. The half-life of the drug was considered as the factor of estimation. The differential equation approach was explored through zero-order and first-order equations to investigate continuous form of the pattern. The data fitting in the study revealed that, the first order kinetics demonstrated better results.

5. References

[1] Joshi S, Parkar J, Ansari A, Vora A, Talwar D, Tiwaskar M and Barkate H 2021 Role of favipiravir in the treatment of COVID-19 *International Journal of Infectious Diseases* **102** 501-508.
[2] Parasher A 2021 COVID-19: Current understanding of its pathophysiology, clinical presentation and treatment *Postgraduate Medical Journal* **97(1147)** 312-320.
[3] Khan S and Al-Balushi K 2021 Combating COVID-19: The role of drug repurposing and medicinal plants *Journal of infection and public health* **14(4)** 495-503.
[4] Rajman I 2008 PK/PD modelling and simulations: utility in drug development *Drug discovery today* **13(7-8)** 341-346.
[5] Chang M, Chow S and Pong A 2006 Adaptive design in clinical research: issues, opportunities, and recommendations *Journal of biopharmaceutical statistics* **16(3)** 299-309.
[6] Biomarkers Definitions Working Group Biomarkers and surrogate endpointsin clinical trials: proposed definitions and conceptual framework *Clin. harmacol.Ther.* **69** 89–95.
[7] Chaikin P, Rhodes G, Bruno R, Rohatagi S and Natarajan C 2000 Pharmacokinetics/pharmacodynamics in drug development: an industrial perspective *The Journal of Clinical Pharmacology* **40(12)** 1428-1438.
[8] Gieschke R and Steimer J 2000 Pharmacometrics: modelling and simulation tools to improve decision making in clinical drug development *European journal of drug metabolism and pharmacokinetics* **25(1)** 49-58.
[9] Colburn W and Lee J 2003 Biomarkers, validation and pharmacokinetic-pharmacodynamic modelling *Clinical pharmacokinetics* **42(12)** 997-1022.
[10] Acker K 2004 Drug Levels and Difference Equations *The Mathematics Teacher* **98(4)** 266-273.
[11] Gouda R, Baishya H and Qing Z 2017 Application of mathematical models in drug release kinetics of carbidopa and levodopa ER Tablets *J. Dev. Drugs* **6(02)** 1-8.
[12] Brahankar D and Jaiswal S 2009 Biopharmaceutics and Pharmacokinetics: A treatise *Vallabh Prakashan.*
[13] Thakkar V, Shah P, Soni T, Parmar M, Gohel M and Gandhi T 2009 Goodness-of-fit model-dependent approach for release kinetics of levofloxacin hemihydrates floating Tablet *Dissolution Technologies* **16(1)** 35-39.
[14] Dash S, Murthy P, Nath L and Chowdhury P 2010 Kinetic modeling on drug release from controlled drug delivery systems *Acta Pol Pharm* **67(3)** 217-223.
[15] Dokoumetzidis A and Macheras P 2009 Fractional kinetics in drug absorption and disposition processes *Journal of pharmacokinetics and pharmacodynamics* **36(2)** 165-178.
[16] Popović J, Atanacković M, Pilipović A, Rapaić M, Pilipović S and Atanacković T 2010 A new approach to the compartmental analysis in pharmacokinetics: fractional time evolution of diclofenac *Journal of pharmacokinetics and pharmacodynamics* **37(2)** 119-134.
[17] Bhagat D, Suryawanshi M and Suryawanshi I 2020 A Systematic Review of Antiviral Drugs for the Treatment of the COVID-19 Pandemic *Manipal Journal of Pharmaceutical Sciences* **6(2)** 109-113.
[18] www.drugbank.ca/drugs

Analysis of the Stability and Hopf Bifurcation of a Three-Dimensional System with Delays

Kutorzi Edwin Yao [1,2] **, Yufeng Shi**[1,2,3]

[1] School of Mathematics, Shandong University, Jinan, 250100, China
[2] Institute for Financial Studies, Shandong University, Jinan, 250100, China
[3] Shandong Big Data Research Association, Jinan, 250100, China

E-mail: ekutorzi@yahoo.com, yfshi@sdu.edu.cn

Abstract. We propose a model of three-dimensional autonomous system with delays. We explore the dynamical behavior of the proposed autonomous system by examining bifurcation diagrams, Lyapunov exponents, equilibrium and stability, and the influence of time delay on Hopf bifurcation. A bifurcation theory is used to analyze and detail the problem. In addition, the explicit algorithm that determines the direction of Hopf bifurcation, along with the stability of bifurcating periodic, has been established. Also, there are specific operating conditions that must be met in order to achieve Hopf bifurcation. In the proposed autonomous system, we analyze the procedures for designing chaotic based systems including parameter selection, discretization of the results, as well as exploring the changing regularity of the bifurcation value. A series of numerical simulations is presented to illustrate the analytical results.

Keywords: Bifurcation, Periodic Solution, Chaotic Attractor, Delay, Numerical simulations

1. Introduction

An autonomous dynamical systems are described by at least three coupled ordinary nonlinear differential equations, that chaos occurs in multiple nonlinear system when numerically computed. As these different values of parameters, variations may occur in the quantitative figure of the solutions for given parameter values. Hopf bifurcation occurs as a entangle conjugate couple of eigenvalues of the linearized system at a certain point becomes purely imaginary. These systems have a unique dynamical appearance, Hopf bifurcation can only occur in systems of dimension two or higher. Lately, researchers started bifurcation analysis and numerical computer simulation for these systems, e.g [1, 2], [3, 4], [5, 6], references therein. In particular, chaos theory has became practicable tools in facilitation secure optical communication devices and secure data encryption.

Over the past one decade, chaos applications have been used in information processing, nonlinear circuits, and etc. Zhang et al studied Hopf bifurcation with chaos entanglement function. The authors reviewed bifurcations of nonlinear dynamical systems in the new chaotic system [7, 8], [9, 10], and [11]. The existence of autonomous systems are extensively encountered in many fields such as engineering and chemistry. Delays are inevitable in autonomous activities; everything takes time, with that we proposed a delayed in Hopf bifurcation in three-dimensional based on chaos entanglement function and the results show there are four requirements that are needed to achieve chaos [12],Kutorzi and Yufeng. The stability problem of the dynamic systems with delays is of more intrigue as part of this study.

In this paper, the priority issue is the stability of dynamic systems with delays. We introduced a time delayed in Hopf bifurcation in three-dimensional based on chaos entanglement function. For autonomous models, chaotic behavior has been developed to displays dynamic behaviors. The systems are considered non-linear because of the multiple feedback between the components of the system. In addition we present a specific range of the operating conditions that are needed to achieve Hopf bifurcation.

The outline of the present paper is organized as follows. In section two and three, we foremost described the characteristics of chaotic three dimensional autonomous system. Significantly, the chaotic dynamics are all established when chaotic dynamics are effectively stabilize, explored in details, and the characteristics are obtained by employing varying subsystems. Furthermore, we study the the algorithm for establishing the direction of hopf bifurcation and as well investigated the stability of bifurcating periodic. In section four, detailed numerical simulations are presented. Finally, we present our findings and conclusion in section five.

2. The Proposed Three-Dimensional Autonomous

Consider 2 linear subsystems: Two-dimensional nonlinear autonomous system and one-dimensional.

$$\begin{aligned} \dot{x} &= ax - by \\ \dot{y} &= cx + ay. \end{aligned} \tag{1}$$

According to [2] where he extend the model to explore the dynamical properties.

$$\dot{z} = dz \tag{2}$$

Where (x, y, z) are steady variable. When $a < 0, c < 0$ and $d < 0$ are both the stable subsystems. Now consider time delay into the system (1) & (2) , the proposed three-order time delayed system could be derived thus and so:

$$\begin{aligned} \dot{x} &= ax - by(t-\tau) \\ \dot{y} &= cx + ay \\ \dot{z} &= dz(t-\tau) \end{aligned} \tag{3}$$

wherein $\tau \in R^+$ is the integral time delay in system (3).

Hopf Bifurcation Analysis,Kutorzi and Shi

2.1. Stability of Fixed Point

According to system (3), equilibria should satisfy

$$\begin{aligned} &ax - by(t-\tau) = 0, \\ &cx + ay = 0, \\ &dz(t-\tau) = 0 \end{aligned} \quad (4)$$

We obtain system (4) has equilibria at the fixed point $\mathbf{E}^\star = (0,0,0)$. Additionally, linearizing system (4) around $\mathbf{E}^\star = (0,0,0)$ expressed as

$$J = \begin{pmatrix} a & -be^{-\lambda\tau} & 0 \\ c & a & 0 \\ 0 & 0 & de^{-\lambda\tau} \end{pmatrix} \quad (5)$$

This leads to the associated characteristic equation

$$D(\lambda,\tau) = \det(\lambda E - C_1 - C_2 e^{-\lambda\tau}) \quad (6)$$

$$\lambda^3 + f_1\lambda^2 + f_2\lambda + f_3 + (f_4\lambda + f_5)e^{-\lambda\tau} = 0 \quad (7)$$

wherein

$$\Big\{ f_1 = a^2 + a + d, f_2 = a^4 + 2a^2 d, f_3 = a^2(1+d^2), f_4 = -bc, f_5 = a^2 bc, \quad (8)$$

The manifold case of (7) with $\tau = 0$, the characteristic equation reduces to

$$\lambda^3 + f_1\lambda^2 + f_2\lambda + f_3 + (f_4\lambda + f_5) = 0 \quad (9)$$

The roots of (9) have negative real parts for every nonnegative integer and the positive equilibrium solution with restriction of time delay is asymptotically stable. With the help of Routh-Hurwitz criterion determinant structure all roots of (9) is stable if and only if the following hypothesis holds:

$$(H1)\Big\{ f_1 > 0, f_2+f_4 > 0, f_3+f_5 > 0, \quad and \quad f_1(f_2+f_4)-(f_3+f_5) > 0 (10)$$

Theorem 1. If (10) is satisfied, then the equilibrium $\mathbf{E}^\star$ of system (3) implies that appropriate parameters are chosen to admits asymptotic stability of the system for $\tau = 0$.

Now we review the effect of the distribution of characteristic roots or delay τ when $\tau > 0$. Hopf bifurcation will occur if at least one eigenvalue has positive real parts, unstable and have negative real parts. Assume that $i\omega(\omega > 0)$ is a root of (7). Then ω must satisfy any nonnegative integer in the following nonlinear equation group

$$-\omega_i^3 - f_1\omega^2 + f_2\omega_i + f_3 + (f_4\omega_i + f_5)(\cos(\omega\tau) - \sin(\omega\tau)_i = 0 \quad (11)$$

thus

$$-\omega_i^3 - f_1\omega^2 + f_2\omega_i + f_3 + f_4\omega\cos(\omega\tau)_i + f_4\omega\sin(\omega\tau)_i + f_5\cos(\omega\tau) - f_5\sin(\omega\tau)_i (12)$$

Hopf Bifurcation Analysis,Kutorzi and Shi

which implies

$$\begin{cases} \omega^3 - f_2\omega = f_4\cos\omega(\omega\tau)_i - f_5\sin(\omega\tau) \\ f_1\omega^2 - f_3 = f_4\omega\sin(\omega\tau) + f_5\cos(\omega\tau) \end{cases} \tag{13}$$

By adding both squares, thus have

$$\omega^6 + (f_1^2 - 2f_2)\omega^4 + (f_2^2 - 2f_1f_3 - f_4^2)\omega^2 + f_3^2 - f_5^2 = 0 \tag{14}$$

For convenience, let $\alpha = \omega^2$, then, (14) can be rewritten as

$$R(\alpha) = \alpha^3 + X\alpha^2 + Y\alpha + Z = 0 \tag{15}$$

where

$$X = f_1^2 - 2f_2, Y = f_2^2 - 2f_1f_3 - f_4^2, Z = f_3^2 - f_5^2 \tag{16}$$

State the following conditions :

$$(H2) \quad X > 0 \quad and \quad Z > 0$$

holds, then it implies that the positive equilibrium of $\mathbf{E}^\star$ is usually asymptotically stable for any $\tau > 0$ and that time delayed system (3) has no periodic solution.

$$(H3) \quad \alpha > 0 \quad R(\alpha) \leq 0, \quad where \quad \alpha = (-X + \sqrt{X_0^2 - 3Y_0})/3$$

According to $(H2)$, (14) has no positive root, denoted any positive (14) by α_0; as well, $\pm\omega_0$

$$\lambda^3 + f_{1,0}\lambda^2 + f_{2,0}\lambda + f_{3,0} + (f_{4,0}\lambda + f_{5,0})e^{-\lambda\tau} = 0 \tag{17}$$

which implies that

$$\sin\omega^*\tau = \frac{(f_{1,0}f_{4,0} - f_{5,0})\omega^{*^3} + (f_{2,0}f_{5,0} - f_{3,0}f_{4,0})\omega^*}{f_{4,0}^2\omega^{*^2} + f_{5,0}^2} \cong f_s \tag{18}$$

$$\cos\omega^*\tau = \frac{f_{4,0}\omega^{*^4} + (f_{1,0}f_{5,0} - f_{2,0}f_{4,0})\omega^{*^2} - f_{3,0}f_{5,0}}{f_{4,0}^2\omega^{*^2} + f_{5,0}^2} \cong f_c \tag{19}$$

Thus, we denote

$$\tau_0^j = \begin{cases} \dfrac{1}{\omega_0}\{2\pi - \arccos f_c + 2j\pi\}, & f_s \geq 0, \\ \dfrac{1}{\omega_0}\{\arccos f_c + 2j\pi\}, & f_c < 0 \end{cases} \tag{20}$$

where $j = 0, 1, 2, 3, ...,$ then $\pm\omega_0$ is a pair of purely imaginary roots of (17) with $\tau_0 = \tau_0^0 = \lim_{j\to+\infty}$. Therefore, we define $\tau_0 = \tau_0^0 = \min\{\tau_0^j\}$.

$$(H4) \quad R'(\alpha_0) \neq 0,$$

thereafter

$$\frac{d(Re\lambda(\tau))}{d\tau}\Big|_{\tau=\tau_0^j,\omega=\omega_0} \neq 0. \tag{21}$$

Hopf Bifurcation Analysis,Kutorzi and Shi

Differentiating (7) with respect to τ , consequently

$$\left(\frac{d\lambda}{d\tau}\right)^{-1} = \frac{(3\lambda^2 + 2f_{1,0}\lambda + f_{2,0})e^{\lambda\tau} + f_{4,0}}{\lambda(f_{4,0}\lambda + f_{5,0})} - \frac{\tau}{\lambda} \tag{22}$$

As a result

$$Re\left(\frac{d\lambda}{d\tau}\right)^{-1}|_{\tau=\tau_0^j,\omega=\omega_0} = \frac{\omega_0^2 R'(\alpha^*)}{(f_{4,0}\omega_0^2)^2 + (f_{5,0}\omega_0)^2} \neq 0. \tag{23}$$

We applied normal form hypothesis and center manifold theorem by Hassard [14]. Accordingly, we have the following result on the analysis of existence of Hopf bifurcation :

Theorem 2. Assume that conditions $(H1), (H2), (H3)$,and $(H4)$ hold. Then, the following is considered.

- The Hopf bifurcation occurs at $\tau = \tau_0$ and the equilibrium solution is unstable for $\tau > \tau_0$
- When $\tau \in [0, \tau_0)$, equilibrium solution $\mathbf{E}^\star$ of the time delayed system (3) are locally asymptotically stable

3. Direction and Stability of the Hopf Bifurcation with Delay

In said section, we investigate explicit algorithm for establishing the direction of hopf bifurcation and the stability of bifurcating periodic solution. Based on the discussion in the last section, we derived some conditions under which the system (3) undergoes a Hopf bifurcation. Without the loss of generality, we considered a pair of purely imaginary eigenvalues of system (3); $A_0 = \{\lambda_1, \lambda_{2,3}\}$ same as the delayed system; $\Lambda_0 = \{\omega_0 i\tau^*, -\omega_0 i\tau^*\}$.

By applying methods from normal form and centre manifold reduction theory [14] ,we to proof more detailed information of Hopf bifurcation. Denote τ_0^j by τ^* and rescaling time $\tau \to t/\tau$.

According to system (7), equilibria should satisfy

$$\begin{aligned} &ax - by = 0, \\ &cx + ay = 0, \\ &dz = 0 \end{aligned} \tag{24}$$

As has already been mentioned in this section, we are able to apply more detailed conditions under which system undergoes Hopf bifurcation at $\mathbf{E}^\star$; that is the four-parameter family of differential equations (1), (2). We will also consider the direction of Hopf bifurcation at the equilibrium and stability of the bifurcating periodic solutions by using the normal form theory and center manifold for functional differential equations.

Normalizing the delay τ by scaling $t \to \tau$ and establish the new parameter $\mu = \tau - \tau^*$, then, the new Hopf bifurcation value is $\mu = 0$. The system (3) is transformed into

$$\dot{U}(t) = L_{\tau^*}(U_t) + F(U_t, \mu) \tag{25}$$

Hopf Bifurcation Analysis,Kutorzi and Shi

Hence $\mu = (x, y, z)^T \in \mathbf{R}^3$, $L_\mu : C \to \mathbf{R}^3$ is the functional differential equation and $f : \mathbf{R} \to \mathbf{R}^3$ is the nonlinear part which are given, respectively, as a result of

$$L_\mu(\varphi) = (\tau^*) \times \left[\begin{pmatrix} a & 0 & 0 \\ c & a & 0 \\ 0 & 0 & 0 \end{pmatrix} \begin{pmatrix} \varphi_1(0) \\ \varphi_2(0) \\ \varphi_3(0) \end{pmatrix} + \begin{pmatrix} 0 & b & 0 \\ 0 & 0 & 0 \\ 0 & 0 & d \end{pmatrix} \begin{pmatrix} \varphi_1(-1) \\ \varphi_2(-1) \\ \varphi_3(-1) \end{pmatrix} \right] \quad (26)$$

$$L_\mu(\varphi) = (\tau^*) \times \begin{pmatrix} a\varphi_1(0) + b\varphi_2(-1) \\ c\varphi_1(0) + a\varphi_2(0) \\ d\varphi_3(-1) \end{pmatrix} + h.o.t. \quad (27)$$

and

$$f(\varphi, \mu) = (\tau^* + \mu) \begin{pmatrix} 0 \\ \varphi_1(0)\varphi_2(0) \\ \varphi_1^2(0) \end{pmatrix} \quad (28)$$

for $\varphi(\varphi_1, \varphi_2, \varphi_3)^T \in C$. Then linearized system (25) at origin is

$$\dot{U}(t) = L_{\tau^*}(U_t) \quad (29)$$

Clearly, from the discussion in section 2, the characteristic equation of (7) has a pair of purely imaginary roots $\Lambda_0 = \{\omega_0 i\tau^*, -\omega_0 i\tau^*\}$, we assume that it has the effect of delay τ; easily cognise that for $\tau = \tau^*$ Let $\Im := C([-1, 0], \mathbf{R}^3)$, following this, we consider the following functional differential equation on $\Im$:

$$\dot{z} = L(\tau^*)(z_t). \quad (30)$$

Using Riesz representation theorem, there exists a 3×3 matrix function $\eta(\theta, \tau)$ of bounded variation for $\theta \in [-1, 0]$. It is obvious $L(\tau^*)$ is a continuous linear function mapping $C([-1, 0], \mathbf{R}^3)$ consequently

$$L(\tau^*)\varphi = \left\{ \int_{-1}^{0} [d\eta(\theta, \tau^*)]\varphi(\theta), \qquad \varphi \in C^1([-1, 0], \mathbf{R}^3) \right. \quad (31)$$

For $\varphi \in C^1([-1, 0], \mathbf{R}^3)$, determine

$$A(\tau)\varphi(\theta) = \begin{cases} \dfrac{d\varphi(\theta)}{d\theta}, & \theta \in [-1, 0)\ , \\ \displaystyle\int_{-1}^{0} d\eta(s, \tau)\varphi(s), & \theta = 0 \end{cases} \quad (32)$$

In fact, we may take

$$L := L_\mu : C([-1, 0], \mathbf{R}^3) \to \mathbf{R}^3 \quad (33)$$

By the Riesz representation theorem,

$$L(\tau^*)(\varphi) = \int_{-1}^{0} d\eta(\theta)\varphi(\theta) \quad (34)$$

where

$$\eta(\theta, \tau^*) = \tau^* \begin{pmatrix} a & 0 & 0 \\ c & a & 0 \\ 0 & 0 & 0 \end{pmatrix} \delta(\theta) - \tau^* \begin{pmatrix} 0 & 0 & 0 \\ 0 & 0 & 0 \\ 0 & 0 & d \end{pmatrix} \delta(\theta + 1), \tag{35}$$

with $\delta(\cdot)$ being the Diarc function, if we know the solution for the $\delta(\cdot)$ function, then we can solve it for every single value. For $\varphi = (\varphi_1, \varphi_2) \in C$, determine

$$A(\mu)\varphi(\theta) = \begin{cases} \dfrac{d\varphi}{d\theta}(\theta), & \theta \in [-1, 0) \ , \\ \displaystyle\int_{-1}^{0} d\eta(\theta, \mu)\varphi(\theta), & \theta = 0 \end{cases} \tag{36}$$

and

$$R(\mu)\varphi = \begin{cases} 0, & \text{for } \theta \in [-1, 0) \ , \\ f(\mu, \varphi) & \text{for } \theta = 0 \end{cases} \tag{37}$$

Thus, the system (3) could be transformed into functional differential equation as follows:

$$\dot{u}_t = A(\mu)u_t + R(\mu)u_t, \tag{38}$$

where $u_t(\theta) = u(t + \theta), \quad \theta \in [-1, 0]$.

Whereas, for $(\Psi_1, \Psi_2) \in C([0, 1], \mathbf{R}^3)$, let us review the operator

$$A^* : C([0, 1], \mathbf{R}^3) \to \mathbf{R}^3, \tag{39}$$

defined by

$$A^*\Psi(s) = \begin{cases} \dfrac{d\Psi}{ds}(s), & \text{for } s \in (0, 1] \\ -\displaystyle\int_{-1}^{0} \Psi(-s)d\eta(s), & \text{for } s = 0, \end{cases} \tag{40}$$

and for $\varphi \in C([-1, 0], \mathbf{R}^3)$ and $\Psi \in C[0, 1], \mathbf{R}^3)$ we depict a bilinear inner product:

$$\begin{aligned} \langle \Psi, \phi \rangle &= (\Psi(0), \phi(0)) - \int_{-1}^{0} \int_{\xi=0}^{\theta} \Psi(\xi - \theta) d\eta(\theta)\phi(\xi) d\xi = (\Psi(0), \phi(0)) \\ &+ \tau^* \int_{-1}^{0} \Psi(\theta + 1) \begin{pmatrix} 0 & 0 & 0 \\ 0 & 0 & 0 \\ 0 & 0 & d \end{pmatrix} \phi(\theta) d\theta, \end{aligned} \tag{41}$$

wherein $\eta(\theta) = \eta(\theta, 0)$, and $A(0)$ and A^* are adjoint operators.

Presuppose that $P(\theta)$ and $P^*(s)$ are generalized eigenvectors of $A(\tau^*)$ and A^* corresponding to $\Lambda_0 = \{i\omega_0\tau^*, -i\omega_0\tau^*\}$, respectively; then, $P^*(s)$ is the adjoint of P and dim P = dim $P^* = 2$. Where $\varphi \in C$ and $\Psi \in C^* = C([0, 1], \mathbf{R}^3)$.By a direct calculation, the lemma follows immediately.

Lemma 4. Let

$$\xi=-\frac{a}{\omega_0 i+b},\zeta=-\frac{c}{\omega_0 i+a},\xi^*=\frac{1}{\omega_0 i+c},\zeta^*=\frac{d}{\omega_0 i+a} \tag{42}$$

then a basis of P with Λ_0 and

$$p_1(\theta)=e^{\omega_0 i\tau^*\theta}(1,\xi,\zeta)^T,p_2(\theta)=\bar{p}_1(\theta),\quad \theta\in[-1,0] \tag{43}$$

is a basis of P^*

$$q_1^*(s)=e^{-\omega_0 i\tau^* s}(1,\xi^*,\zeta^*)^T,q_2(s)=\bar{q}_1(s),\quad s\in[-1,0] \tag{44}$$

Now using condition $\langle q*(s),q(\theta)\rangle=1$, leads to have
Let $\Phi=(\Phi_1,\Phi_2),\Psi^*=(\Psi_1^*,\Psi_2^*)^T$, where

$$\Phi_1(\theta)=\frac{p_1(\theta)+p_2(\theta)}{2},\Phi_2(\theta)=\frac{p_1(\theta)-p_2(\theta)}{2i},\quad \theta\in[-1,0], \tag{45}$$

and

$$\Psi_1^*(s)=\frac{q_1(s)+q_2(s)}{2},\Psi_2^*(s)=\frac{q_1(s)-q_2(s)}{2i},\quad s\in[-1,0], \tag{46}$$

From (41) leads to obtain $(\Psi_1^*,\Phi_2),(\Psi_1,\Phi_2)$, and further notice

$$(q_1,p_1)=(\Psi_1^*,\Phi_1)-(\Psi_2^*,\Phi_2)+[(\Psi_1^*,\Phi_2)-(\Psi_2^*,\Phi_1)]i; \tag{47}$$

that is, leads to

$$(q_1,p_1)=1+\xi\xi^*+\zeta\zeta^*-\zeta\zeta^*\tau^*e^{-\omega_0 i\tau^*}:=D^*. \tag{48}$$

Consequently, we have

$$(\Psi_1^*,\Phi_1)-(\Psi_2^*,\Phi_2)=Re\{D^*\},(\Psi_1^*,\Phi_2)-(\Psi_2^*,\Phi_1)=Im\{D^*\}, \tag{49}$$

Now, we define $(\Psi^*,\Phi)=(\Psi_m^*,\Phi_n)(m,n=1,2)$ and construct a new basis for P^* by $\Psi=(\Psi_1,\Psi_2)^T=(\Psi^*,\Phi)^{-1}\Psi^*$. In addition, we define $f_0=(\xi_0^1,\xi_0^2,\xi_0^3)$, where

$$\xi_0^1=\begin{pmatrix}1\\0\\0\end{pmatrix},\xi_0^2=\begin{pmatrix}0\\1\\0\end{pmatrix},\xi_0^3=\begin{pmatrix}0\\0\\1\end{pmatrix} \tag{50}$$

Let $c\cdot f_0$ be defined by

$$c\cdot f_0=c_1\xi_0^1+c_2\xi_0^2+c_3\xi_0^3 \tag{51}$$

for $c=(c_1,c_2,c_3)^T$, and $c_j\in\mathbf{R}(j=1,2,3)$. Then, the center space of linear equation (29) is determined by $P_{CN}\Im$, where

$$P_{CN}\varphi=\Phi(\Psi,\langle\varphi,f_0\rangle)\cdot f_0,\quad \varphi\epsilon\Im, \tag{52}$$

and $C=P_{CN}\Im\oplus P_s\Im$; $P_s\Im$ we denote the complementary subspace of $P_{CN}\Im$. If

$$A_{\tau^*}\varphi(\theta)=\dot{\varphi}(\theta)+X_0(\theta)[L(\tau^*)(\varphi(\theta)-\dot{\varphi}(0))],\quad \varphi\in B\Im, \tag{53}$$

Hopf Bifurcation Analysis,Kutorzi and Shi

where $X_0 : [-1,0] \to B(X,X)$ and $\begin{cases} 0, & -1 \le \theta \le 0, \\ I, & \theta = 0 \end{cases}$. Then A_{τ^*} is the infinitesimal generator induced by the solutions of (25) and (29) hence could be written as the abstract differential equation operator :

$$\dot{U}_t = A_{\tau^*} U_t + X_0 F(U_t, \mu), \tag{54}$$

Therefore, the solution of (25) could be written in the form of

$$U_t = \Phi \begin{pmatrix} x_1(t) \\ x_2(t) \end{pmatrix} \cdot f_0 + h(x_1, x_2, \mu), \begin{pmatrix} x_1(t) \\ x_2(t) \end{pmatrix} = (\Psi \langle U_t, f_0 \rangle), \tag{55}$$

where $h(x_1, x_2, \mu) \in P_s\Im, h(0,0,0) = Dh(0,0,0) = 0$. Especially plug the solution of (25) on the center manifold is determine by

$$U_t^* = \Phi \begin{pmatrix} x_1(t) \\ x_2(t) \end{pmatrix} \cdot f_0 + h(x_1, x_2, 0). \tag{56}$$

Let $z = x_1 - x_2 i$, by stating that $p_1 = \Phi_1 + \Phi_2 i$, then (56) leads to rewrite as

$$U_t^* = \frac{1}{2}\Phi \begin{pmatrix} z + \bar{z} \\ (z - \bar{z})i \end{pmatrix} . f_0 + W(z, \bar{z}) = \frac{1}{2}(p_1 z + \bar{p}_1 \bar{z}) \cdot f_0 + W(z, \bar{z}), \tag{57}$$

where $W(z,\bar{z}) = h((z+\bar{z})/2, -(z-\bar{z})/2i, 0)$. Furthermore from [17], z satisfies

$$\dot{z} = \omega_0 i \tau^* z + g(z, \bar{z}), \tag{58}$$

where

$$g(z, \bar{z}) = \Psi_1(0) - \Psi_2 i(0)) \langle F(U_t^*, 0), f_0 \rangle. \tag{59}$$

Let

$$W(z, \bar{z}) = W_{20}\frac{z^2}{2} + W_{11} z\bar{z} + W_{02}\frac{\bar{z}^2}{2} + ... \tag{60}$$

and

$$g(z, \bar{z}) = g_{20}\frac{z^2}{2} + g_{11} z\bar{z} + g_{02}\frac{\bar{z}^2}{2} + ... \tag{61}$$

By (57), it can be easily to compute

$$\begin{aligned} \langle F(U_t^*, 0), f_0 \rangle = {} & \frac{\tau^* z^2}{4} \begin{pmatrix} 1 \\ \xi \\ 0 \end{pmatrix} + \frac{\tau^* z\bar{z}}{4} \begin{pmatrix} 2 \\ \xi + \bar{\xi} \\ 0 \end{pmatrix} + \frac{\tau^* \bar{z}^2}{4} \begin{pmatrix} 1 \\ \xi^2 \\ 0 \end{pmatrix} \\ & + \frac{\tau^*}{4} z^2 \bar{z} \begin{pmatrix} \langle 4\omega_{11}^{(1)}(0) + 2\omega_{20}^{(1)}(0), 1 \rangle \\ \langle 2\omega_{11}^{(2)}(0) + \omega_{20}^{(2)}(0) + 2\xi\omega_{11}^{(1)}(0) + \bar{\xi}\omega_{20}^{(0)}(0), 1 \rangle \\ 0 \end{pmatrix} + ..., \end{aligned} \tag{62}$$

Hopf Bifurcation Analysis,Kutorzi and Shi

where $\langle W_{ij}^{(n)}(\theta),1\rangle = (1/\pi)\int_0^{\pi} W_{ij}^{(n)}(\theta)(x)dx, i+j = 2, n = 1,2,3.$ Let $(\Psi_1,\Psi_2,\Psi_3)=\Psi_1(0)-\Psi_2 i(0)$. This leads to the following expressions :

$$\begin{aligned} g_{20} &= \frac{\tau^*}{2}[\Psi_1+\alpha\Psi_2], g_{11} = \frac{\tau^*}{4}[2\Psi_1+(\alpha+\bar{\alpha})\Psi_2],\\ g_{02} &= \frac{\tau^*}{2}[\bar{\Psi}_1+\alpha_2\bar{\Psi}_2], g_{21} = \frac{\tau^*}{2}[\langle 4\omega_{11}^{(1)}(0)+2\omega_{20}^{(1)}(0),1\rangle\Psi_1\\ &+\langle 2\omega_{11}^{(2)}(0)+\omega_{20}^{(2)}(0)+2\alpha\omega_{11}^{(1)}(0)+\bar{\alpha}\omega_{20}^{(1)}(0),1\rangle\Psi_2]. \end{aligned} \tag{63}$$

Since $W_{20}(\theta)$ and $W_{11}(\theta)$ appear in g_{21}, we should compute them. According to (60) we obtain

$$\dot{W}(z,\bar{z}) = W_{20}z\dot{z} + W_{11}(\dot{z}\bar{z}+z\dot{\bar{z}}) + W_{02}\bar{z}\dot{\bar{z}} + ... \tag{64}$$

and

$$A_{\tau^*}W = A_{\tau^*}W_{20}\frac{z^2}{2} + A_{\tau^*}W_{11}z\bar{z} + A_{\tau^*}W_{02}\frac{\bar{z}^2}{z} + ... \tag{65}$$

Along with it, [17] ,$W(z(t),\bar{z}(t))$ also satisfies

$$\dot{W} = A_{\tau^*}W + H(z,\bar{z}), \tag{66}$$

where

$$\begin{aligned} H(z,\bar{z}) &= H_{20}\frac{z^2}{2} + H_{11}z\bar{z} + H_{02}\frac{\bar{z}^2}{2} + ...\\ &= X_0F(U_t^*,0) - \Phi(\Psi,\langle X_0F(U_t^*,0),f_0\rangle)\cdot f_0, \end{aligned} \tag{67}$$

with $H_{ij}\in P_s\Im$ and $i+j=2$. Which lead to (57),(64),(65), and (66), we have

$$\begin{cases} (2\omega_0 i\tau^* - A_{\tau^*})W_{20} = H_{20},\\ -A_{\tau^*}W_{11} = H_{11}. \end{cases} \tag{68}$$

Noticing that A_{τ^*}, comprises two purely imaginary characteristic roots, (66) has the unique solution so as

$$\begin{cases} W_{20} = (2\omega_0 i\tau^* - A_{\tau^*})^{-1}H_{20},\\ W_{11} = -A_{\tau^*}^{-1}H_{11}. \end{cases} \tag{69}$$

For $\theta\in[-1,0]$, it follows from (67)

$$\begin{aligned} H(z,\bar{z}) &= -\Phi(\theta)\Psi(\theta)\langle F(U_t^*,0),f_0\rangle\cdot f_0\\ &= -\left(\frac{p_1(\theta)+p_2(\theta)}{2},\frac{p_1(\theta)-p_2(\theta)}{2i}\right)\cdot(\Psi_1(0),\Psi_2(0))^T\times\langle F(U_t^*,0),f_0\rangle\cdot f_0\\ &= -\frac{1}{2}[p_1(\theta)(\Psi_1(0)-\Psi_2 i(0))+p_2(\theta)(\Psi_1(0)+\Psi_2 i(0))]\times\langle F(U_t^*,0),f_0\rangle\cdot f_0\\ &= -\frac{1}{4}[g_{20}p_1(\theta)+\bar{g}_{02}p_2(\theta)]z^2\cdot f_0 - \frac{1}{2}[g_{11}p_1(\theta)+\bar{g}_{11}p_2(\theta)]z\bar{z}\cdot f_0 + ... \end{aligned} \tag{70}$$

Therefore, $\theta \in [-1, 0]$, leads to

$$H_{20}(\theta) = -\frac{1}{2}[g_{20}p_1(\theta) + \bar{g}_{02}p_2(\theta)].f_0, \tag{71}$$

$$H_{11}(\theta) = \frac{1}{2}[g_{11}p_1(\theta) + \bar{g}_{11}p_2(\theta)].f_0, \tag{72}$$

$$H(z, \bar{z})(0) = F(U_t^*, 0) - \Phi(\Psi, \langle F(U_t^*, 0)\rangle, f_0) \cdot f_0, \tag{73}$$

$$H_{20}(0) = \frac{\tau^*}{2}\begin{pmatrix} 1 \\ \xi \\ 0 \end{pmatrix} - \frac{1}{2}[g_{20}p_1(0) + \bar{g}_{02}p_2(0)] \cdot f_0, \tag{74}$$

and

$$H_{11}(0) = \frac{\tau^*}{4}\begin{pmatrix} 2 \\ \xi + \bar{\xi} \\ 0 \end{pmatrix} - \frac{1}{2}[g_{11}p_1(0) + \bar{g}_{11}p_2(0)] \cdot f_0. \tag{75}$$

With the combination definition of A_{τ^*}, leads to obtain from (69) that

$$\dot{W}_{20} = 2\omega_0 i\tau^* W_{20}(\theta) + \frac{g_{20}p_1(0) + \bar{g}_{02}p_2(0)}{2} \cdot f_0, \quad - \leq \theta < 0. \tag{76}$$

Since $p_1(\theta) = p_1(0)e^{\omega_0 i\tau^*}$, $-1 \leq \theta \leq 0$, leads to

$$W_{20}(\theta) = \frac{i}{2}\left[\frac{g_{20}}{\omega_0\tau^*}p_1(\theta) + \frac{\bar{g}_{02}}{3\omega_0\tau^*}p_2(\theta)\right] \cdot f_0 + Ee^{2\omega_0 i\tau^*\theta}, \tag{77}$$

where

$$E = W_{20}(0) - \frac{i}{2}\left[\frac{g_{20}}{\omega_0\tau^*}p_1(0) + \frac{\bar{g}_{02}}{3\omega_0\tau*}p_2(0)\right] \cdot f_0 \tag{78}$$

From the definition of A_{τ^*} and combine with (69) and (78), leads to

$$\begin{aligned} &2\omega_0 i\tau^*\left[\frac{g_{20}i}{2\omega_0\tau^*}p_1(0) \cdot f_0 + \frac{\bar{g}_{02}i}{6\omega_0\tau^*}p_2(0) \cdot f_0 + E\right] \\ &-L(\tau^*)\left[\frac{g_{20}i}{2\omega_0\tau^*}p_1(\theta).f_0 + \frac{\bar{g}_{02}i}{6\omega_0\tau^*}p_2(0) \cdot f_0 + Ee^{2\omega_0 i\tau^*\theta}\right] \\ &= \frac{\tau*}{2}\begin{pmatrix} 1 \\ \xi \\ 0 \end{pmatrix} - \frac{1}{2}[g_{20}p_1(0) + \bar{g}_{02}p_2(0)] \cdot f_0. \end{aligned} \tag{79}$$

Noticing that

$$\begin{cases} L(\tau^*)[p_1(\theta) \cdot f_0] = \omega_0 i\tau^* p_1(0) \cdot f_0, \\ L(\tau^*)[p_2(\theta) \cdot f_0] = -\omega_0 i\tau^* p_2(0) \cdot f_0. \end{cases} \tag{80}$$

this leads to

$$2\omega_0 i\tau^* E - \tau^* D\Delta E - L(\tau^*)(Ee^{2\omega_0 i\tau^*\theta}) = \frac{\tau^*}{2} \cdot \begin{pmatrix} 1 \\ \xi \\ 0 \end{pmatrix} \tag{81}$$

Hopf Bifurcation Analysis,Kutorzi and Shi

From the previous formulas, this leads to easily obtain

$$E = \frac{1}{2}\begin{pmatrix} 2\omega_0 i + a & e^{-2\omega_0 i\tau^*} & 0 \\ c & 2\omega_0 i + a & 0 \\ 0 & 0 & 2\omega_0 i + de^{-2\omega_0 i\tau^*} \end{pmatrix}^{-1} \times \begin{pmatrix} 1 \\ \xi \\ 0 \end{pmatrix} \tag{82}$$

By the same method, leads to

$$\dot{W}_{11}(\theta) = \frac{1}{2}[g_{11}p_1(\theta) + \bar{g}_{11}p_2(\theta)] \cdot f_0, \quad - \leq \theta \leq 0, \tag{83}$$

and

$$W_{11}(\theta) = \frac{i}{2\omega_0\tau^*}[-g_{11}p_1(\theta) + \bar{g}_{11}p_2(\theta)] \cdot f_0 + F. \tag{84}$$

By the same method, this leads to obtain

$$F = \frac{1}{4}\begin{pmatrix} a & -b & 0 \\ c & c & 0 \\ 0 & 0 & d \end{pmatrix}^{-1} \times \begin{pmatrix} 2 \\ \xi + \bar{\xi} \\ 0 \end{pmatrix} \tag{85}$$

Hence, g_{21} could be represented explicitly.

Each g_{21} is appropriately stated, accordingly, allows us to compute the following results:

$$C_1(0) = \frac{i}{2\omega_0\tau^*}[g_{11}g_{20} - 2|g_{11}|^2 - \frac{|g_{02}|^2}{3}] + \frac{g_{21}}{2}. \tag{86}$$

$$\sigma_2 = \frac{Re(C_1)(0)}{Re(\lambda'(\tau^*)} \tag{87}$$

$$\beta = 2Re(C_1)(0) \tag{88}$$

$$T_2 = \frac{Im(C_1(0)) + \sigma_2 Im(\lambda'(\tau^*)}{\omega_0\tau^*} \tag{89}$$

The general result of Hopf bifurcation theory (see [14]), we recognize that the parameters in (86) demonstrate the properties of Hopf bifurcation which we can describe specifically :- β_2 determines the stability of the bifurcating periodic solutions: the bifurcating periodic solution are stable(unstable) if $\beta_2 < 0(> 0); \sigma_2$ determines the directions of the Hopf bifurcation: if $\sigma > 0(< 0)$, then the direction of the Hopf bifurcation is forward(backward), that is the bifurcating periodic solutions exist when $\tau > \tau_0(< \tau_0)$; and T_2 determines the period of the bifurcation periodic solutions: the period increase(decrease),if $T_2 > 0(T_2 < 0)$.

Theorem 3 System (3) has the following pointcaré normal form :

$$\dot{\omega} = \omega_0 i\tau^*\omega + C_1(0)\omega|\omega|^2 + 0(|\omega|^5), \tag{90}$$

Journal of Physics: Conference Series **2287** (2022) 012012 doi:10.1088/1742-6596/2287/1/012012

Hopf Bifurcation Analysis,Kutorzi and Shi

4. Numerical Results

In the present section, we make some numerical simulations with help of MATLAB to support our analytical results. We choose the following parameter values: $a = -2, b = 6, c = 3$ and $d = -4$. By a direct calculation, we obtain that system (3) is locally asymptotically stable at equilibrium for all $\tau \geq 0$; see Figure 1.

Through discussions in section 3, results in $\tau_0 = 2.005, \tau_1 = 10.984, \tau_2 = 19.935, \cdots$ and $ReC_1^0(0) = -521.13, ReC_1^1(0) = -2853.88, ReC_1^2(0) = -5186.33, \cdots$, which, together with equations (86)-(89), implies $\sigma_2 > 0$ and $\beta_2 < 0$. Therefore, we recognize that the bifurcated periodic solutions are locally asymptotically stable on the center manifold. Hence, equilibrium E_0 is stable when $\tau < \tau_0$. When τ passes through the critical value τ_0, E_0 loses its stability and a Hopf bifurcation occurs. Furthermore, the formulae in section 3, Theorem 1, it follows that system (3) has a stable center manifold in the neighbourhood of equilibrium $\tau_0 = 5.85$. Since $\sigma_2 > 0$ and $\beta_2 > 0$, the direction of the Hopf bifurcation is $\tau > \tau_0$, and these bifurcating periodic solutions from E_0 at τ_0 are stable, as shown in Figure 1. Figure 4 shows their time trajectory planes.

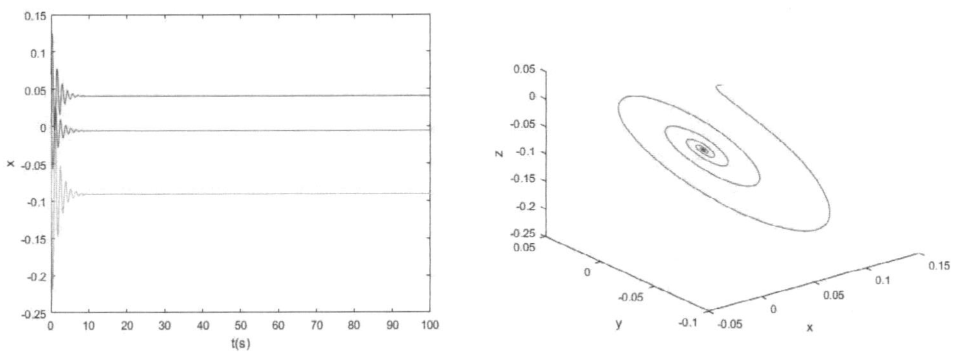

Figure 1. Time history and phase diagram $\tau_0 = 2.005$

5. Conclusion

Our paper proposes a delayed model of three-dimensional autonomous. We determine the direction of hopf bifurcation with the center manifold theorem and the stability of bifurcating periods with respect to the normal form theory. In numerical simulations we report the bifurcating periodic solution is stable or unstable at some bifurcation points, depending on the choice of parameters. The time history and phase diagram of the time-delayed system (3) are shown for a range of parameters. The results could help in appreciating the role of fluctuations and interpreting autonomous systems. Numerical findings are presented to illustrate the analytical results and gives understanding for the direction of Hopf bifurcation of the model. The study will assist in understanding

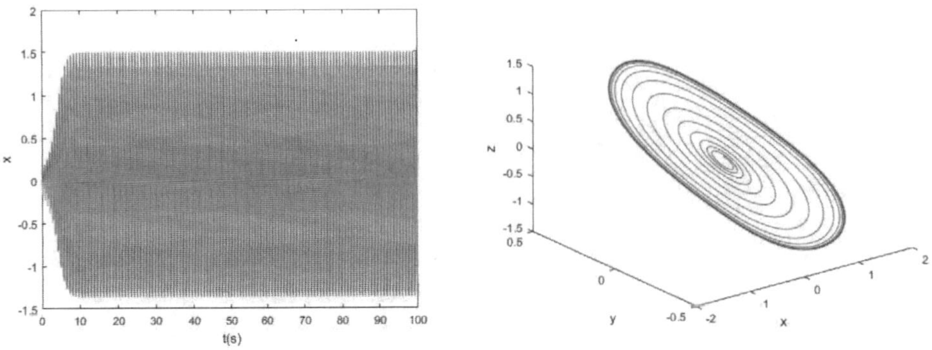

Figure 2. Time history and phase diagram of system

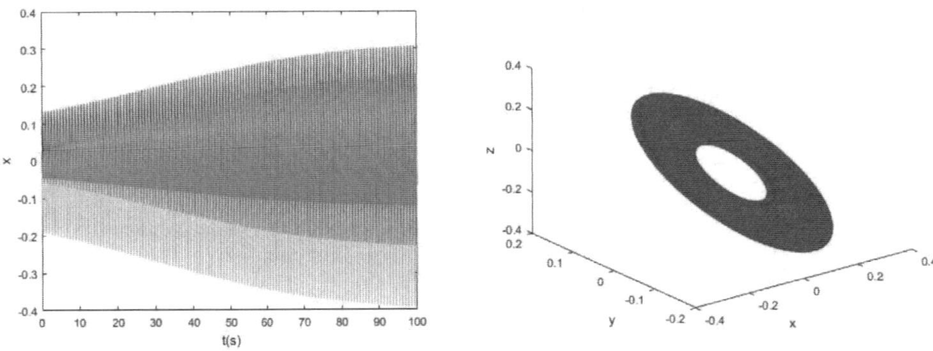

Figure 3. Time history and phase diagram of system

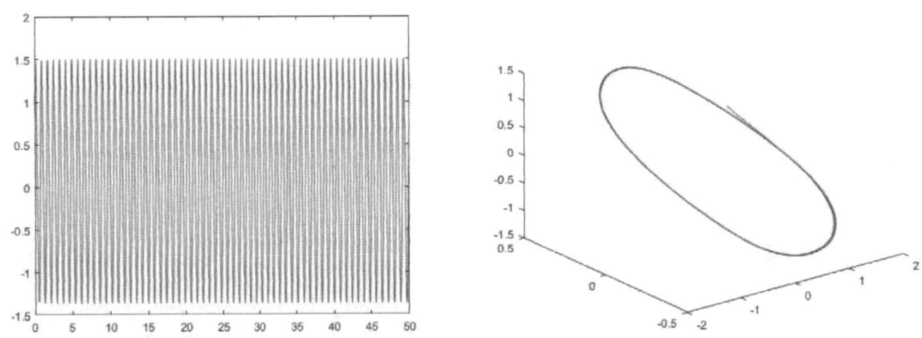

Figure 4. Time trajectories of system and phase diagram

the importance of nonlinear autonomous schemes and explain concepts of fluctuation phenomena.

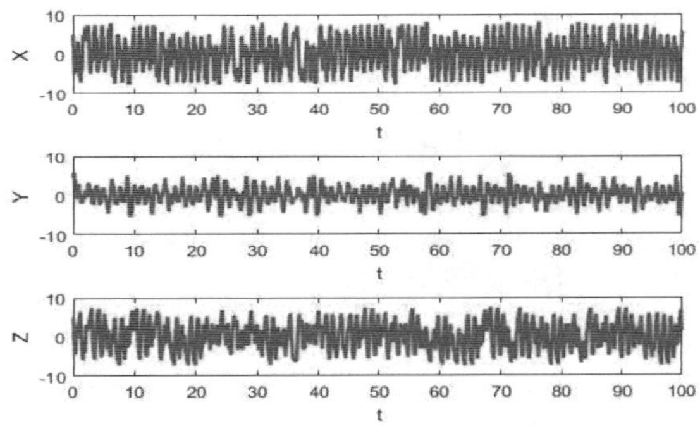

Figure 5. Time series measurement of system (3) $\tau_0 = 1.75$

Acknowledgements

This paper was supported by the National Key R&D Program of China (No. 2018YFA0703900) and the National Natural Science Foundation of China (Grant No. 11871309).

References

[1] F. M. Amaral and L. F. C. Alberto Type-zero saddle-node bifurcations and stability region estimation of nonlinear autonomous dynamical systems, *International Journal of Bifurcation and Chaos* , vol. 22, no. 1, Article ID 1250020, 16 pages, 2012.

[2] J. Hai-Bo, L. Tao, Z. Xiao-Liang, and Z. Li-Ping Bifurcation analysis of the logistic map via two periodic impulsive forces, *Chinese Physics B* , vol. 23, Article ID 010501, 7 pages, 2014.

[3] N. Almazova, G. D. Barmparis, and G. P. Tsironis, Analysis of chaotic dynamical systems with autoencoders *Chaos 31, 103109* (2021); https://doi.org/10.1063/5.0055673

[4] M. M. El-Dessoky, M. T. Yassen, E. Saleh, and E. S. Aly, Existence of heteroclinic and homoclinic orbits in two different chaotic dynamical systems, *Applied Mathematics and Computation*, vol. 218, no. 24, pp. 1185911870, 2012.

[5] X. Chen, W. Huang, V. G. Romanovski, and W. Zhang, Linearizability and local bifurcation of critical periods in a cubic Kolmogorov system, *Journal of Computational and Applied Mathematics*, vol. 245, pp. 8696, 2013.

[6] B. W. Kooi, M. Aguiar, and N. Stollenwerk, Bifurcation analysis of a family of multi-strain epidemiology models, *Journal of Computational and Applied Mathematics* , vol. 252, pp. 148158, 2013.

[7] Zhang Jiangang, Chu Yandong, Du Wenju, Chang Yingxiang, and An Xinlei Hopf Bifurcation Analysis in a New Chaotic System with Chaos Entanglement Function, *Journal of Applied Mathematics*, Volume 2014, Article ID 371509, 13 pages.

[8] Y. Guo, W. Jiang, and B. Niu, Bifurcation analysis in the control of chaos by extended delay feedback, *Journal of the Franklin Institute: Engineering and Applied Mathematic*, vol. 350, no. 1, pp. 155170, 2013.

[9] W. Xu, J. Cao, and M. Xiao, Bifurcation analysis and control in exponential RED algorithm, *Neurocomputing* , vol. 129, pp. 232245, 2014.

[10] J. Paez Chavez, E. Pavlovskaia, and M. Wiercigroch, Bifurcation analysis of a piecewise-linear impact oscillator with drift, *Nonlinear Dynamics*, vol. 77, no. 1-2, pp. 213227, 2014.

[11] S. Li, Q. Wu, and Z. Zhang, Bifurcation and chaos analysis of multistage planetary gear train, *Nonlinear Dynamics* , vol. 75, no. 1-2, pp. 217233, 2014.

[12] Kutorzi Edwin Yao, Yufeng Shi, Hopf bifurcation in three-dimensional based on chaos entanglement function *Chaos, Solitons and Fractals: X*, X 4 (2019) 100027; https://doi.org/10.1016/j.csfx.2020.100027

[13] H. Zhang, X. Liu, X. Shen, and J. Liu, Chaos entanglement: a new approach to generate chaos, *International Journal of Bifurcation and Chaos in Applied Sciences and Engineering* , vol.23, no. 5, Article ID 1330014, 17 pages, 2013.

[14] Hassard, B. D., Kazarinoff, N. D., and Wan, Y. H., Theory and Applications of Hopf Bifurcation, *Cambridge University Press, Cambridge*1st ed.,1981

[15] Y. A. Kuznetsov, Elements of Applied Bifurcation Theory, *Applied Mathematical Sciences, Springer, New York*, NY, USA, 3rd edition , 2004

[16] Z.Wei and Q. Yang, Anti-control of Hopf bifurcation in the new chaotic system with two stable node-foci, *Applied Mathematics and Computation* , vol. 217, no. 1, pp. 422429, 2010.

[17] J.H. Wu, Theory and Applications of Partial Functional-Differential Equations, *Springer, New York,NY,USA,1996.*

Mathematical Model for Combined Radiotherapy and Chemotherapy that Fits with Experimental Data

Hoda Hellal, Hanaa Elabsy and Hesham Elkaranshawy

1 Department of Engineering Mathematics and Physics, Faculty of Engineering, Alexandria University, Egypt
E-mail: hesham_elk@alexu.edu.eg

Abstract. One of the most challenges to effective treatment of cancer is radiation resistance. Published data from a notably experiment performed to examine the ability of a compound therapy to overcome radiation resistance is utilized. The chemotherapeutic drug used was vinblastine. The experiment included control tumors and tumors which were treated with radiation, chemotherapy, or both. In this paper a mathematical model is developed. The model fits with the data reported from the mentioned experiment. Starting with a suggested model for the tumor growth, the model is expanded to include the effect of radiotherapy, chemotherapy, and combined chemotherapy and radiotherapy. By fitting the model with the experimental data, the parameters of the model are obtained in each stage. The final mathematical model acquired can be used with the case with no treatment, with radiotherapy treatment, with chemotherapy treatment, and with combined chemotherapy and radiotherapy treatment. Results capture the behavior of the tumor in all cases in a reasonable way.

1. Introduction

In recent decades, cancer is a major kill source all over the world for humans, as it is a leading cause of death worldwide. For example, nearly 10 million deaths in 2020 [1]. Universally, the cancer burden is expected to rise to approximately 21.4 million cases and 13.5 million deaths by 2030 [2]. Cancer treatments are intermittently diverse and successful, its types are surgery, radiotherapy [3], targeted therapy, hormone therapy [4], viral therapy [5], chemotherapy [6], and immunotherapy [7-9], among others, either separately or in combination of two or more.

Mathematical models have emerged where it is possible to simulate the effects of radiotherapy on tumor growth [3,10,11]. Also, there are many of mathematical models suggested to model the treatment with chemotherapy [6,12,13]. Furthermore, there are a lot of studies that provide mathematical models for treating cancer by radiotherapy and chemotherapy [14-18].

Radiation resistance is a challenge to effective treatment of cancer. A significant experiment was accomplished to test the ability of therapeutic compound to overcome radiation resistance [17,18]. Breast cancer cells were put in a 96-well plate and let grow in a hanging drop for three days. After these three days the resulting spheroid was put in a gravity plate where it stayed for another day. The time was set to zero and the area of the spheroid was measured for the first time. There were control tumors and tumors which were treated with radiation or chemotherapy or both. First the radiation was administered once and after that the chemotherapeutic drug, vinblastine, was administered. The drug stayed in the nutrient solution for six days before it was washed away. The area of the spheroid was measured again on the days 3, 6, 9 and 12. The drug used was vinblastine, a naturally produced from a plant called *Vinca rosea* grows in Madagascar, and it had proved to be effective inhibitors of mammary cancer cell growth. The tumor is assumed to be solid, homogeneous and in an avascular state. In the current research work, it is intended to develop suitable mathematical model that fits with the data reported from the explained experiment. It is aimed to let the model covers reality and considers the influences of treatment. A mathematical model for the growth of the tumor is suggested first and then modified to include the radiotherapy, the chemotherapy, and combined therapy of both. In each step the parameters of the model are obtained by fitting the model with the experimental data. The model can provide an appropriate insight into the interaction between the cancer cells and the therapy. It can help in predicting the foremost effective therapy and strategy and in minimizing the whole drug administered. Hence, it can improve the existing treatment procedures.

2. The Mathematical Model

2.1. *Tumor growth*

2.1.1. Mathematical model. The proposed model for tumor growth is the logistic growth model given by

$$\frac{dT}{dt} = aT(1 - bT) \tag{1}$$

where $T(t)$ represents the number of tumor cells, a is the tumor growth rate, and b is the reciprocal of the tumor carrying capacity. The two parameters a and b are going to be estimated from the experimental data.

2.1.2 Fitting the model to data. In the considered experiment, breast cancer cells were put in a 96-well plate and let grow in a hanging drop for three days, and the resulting spheroid was put in a gravity plate for another day. The time t was set to zero and the area of the spheroid was measured for the first time. For the tumor growth without therapy, the area of the spheroid was measured on days 3, 6, 9, and 12. The Table 1 gives the number of tumor cells over that period [17,18].

Table 1. Number of cells measured over a period of 12 days

Hours	0	72	144	216	288
Number of cells	596.3	2424.5	6986.6	9581.6	10659

To estimate the required parameters, the *Mathematica* commercial program is used to generate a parametric solution for the model in terms of the parameters a and b. This has been performed using the built-in function "$ParametricNDSolveValue$". Using this parametric function, the method of Levenberg-Marquardt is applied using the built-in *Mathematica* function "$FindFit$", to estimate these two parameters. The estimated values for the two parameters are $a = 2.35332 \times 10^{-2}$ h^{-1} and $b = 9.278756 \times 10^{-5}$ cell^{-1}.

2.2. *Radiotherapy*

2.2.1. Mathematical model. To model the effect of radiation on the tumor, the following model is suggested:

$$\frac{dT}{dt} = aT(1 - bT) - \epsilon_r D\, T \tag{2}$$

$$\frac{d\epsilon_r}{dt} = d_r \epsilon_r (1 - \epsilon_r) \tag{3}$$

where D is the amount of radiation administered in Gray, $\epsilon_r(t)$ is the efficacy of the radiotherapy, and d_r is the growth rate of the efficacy. This form for the efficacy allows the variation of the efficacy during the medication period. Parameters a and b have the values estimated in section 2.1.2. The parameter d_r and the initial value for the efficacy $\epsilon_r(0)$ are going to be estimated from the experimental data.

2.2.2. Fitting the model to data. The same procedure for the considered experiment had been performed, but the cells had been treated with radiotherapy this time. The tested cells had been divided into four groups and in the beginning of day zero they were radiated with a dosage of 2, 4, 6 and 8 Gray respectively, where 1 Gray was administered in 2 minutes. The area of the spheroid was measured on days 3, 6, 9, and 12 and Table 2 gives the number of tumor cells over that period [17,18].

Table 2. Number of cells treated with radiotherapy over a period of 12 days

		Number of cells				
Hours		0	72	144	216	288
	2	596.3	2048.9	4636.0	5681.7	7167.2
Radiation 'D'	4	596.3	1801.4	3063.3	3950.6	4194.8
in Gray	6	596.3	2145.8	3130.3	2819.0	4072.0
	8	596.3	1968.3	2319.7	1939.2	2944.5

Mathematica commercial program is used for the curve fitting via the order *FindFit*. Each group gives deferent values for d_r and $\epsilon_r(0)$. The values for d_r are closed to each other, hence, a fixed value for it has been specified, $d_r = 1.8 \times 10^{-3}\ \text{hr}^{-1}$, and the *Mathematica* has been run again to estimate the corresponding value for $\epsilon_r(0)$ for each group. It has been found that the returned values for $\epsilon_r(0)$ can be approximated in the form

$$\epsilon_r(0) = \frac{e_r}{\sqrt{D}} \tag{4}$$

where e_r is a constant which has been estimated to be $e_r = 3.8 \times 10^{-3}\ \text{hr}^{-1}$.

2.3. Chemotherapy

2.3.1. Mathematical model. To model the effect of chemotherapy on the tumor, the following model is suggested:

$$\frac{dT}{dt} = aT(1 - bT) - \epsilon_c C\, T \tag{5}$$

$$\frac{d\epsilon_c}{dt} = d_c \epsilon_c (1 - \epsilon_c) \tag{6}$$

$$\frac{dC}{dt} = -\gamma C \tag{7}$$

where $\epsilon_c(t)$ is the efficacy of the chemotherapy, $C(t)$ is the concentration of the chemotherapy, d_c is the growth rate of the efficacy, and γ is the decay rate of the concentration. Parameters a and b have the values estimated in section 2.1.2. The parameter d_c and the initial value for the efficacy $\epsilon_c(0)$ are going to be estimated from the experimental data. The model given in Eq. 6 for the efficacy allows the variation of it during the medication period. The model of the concentration given in Eq. 7 presents the decay in the concentration of the chemotherapy during the application period. An assumed value of $\gamma = 0.01\ \text{hr}^{-1}$ is given to the decay rate of the concentration. The initial value for the concentration for each group C_0 is specified according to the experimental data, which means that

$$C(0) = C_0 \tag{8}$$

2.3.2. Fitting the model to data. The considered experiment had been progressed with the same steps as in the case explained in section 2.2.2, however, in the time the cells had been treated with the vinblastine chemotherapeutic drug. The tested cells had been divided into five groups and in the beginning of day zero the drug was added with concentrations of 0.3, 1, 3, 10 and 30 μM respectively. Table 3 gives the number of tumor cells over the experimental period [17,18].

Table 3. Number of cells treated with vinblastine over a period of 12 days

		Number of cells				
Hours		0	72	144	216	288
Initial concentration of drug 'C_0' in μM	0.3	596.3	1479.5	2303.3	2134.1	3480.3
	1	596.3	1337.8	2694.8	2730.1	2984.1
	3	596.3	1260.3	2427.1	2538.9	1959.4
	10	596.3	959.1	1639.7	1624.0	1212.8
	30	596.3	924.3	1827.5	1657.2	1405.2

To estimate d_c and $\epsilon_c(0)$, *Mathematica* commercial program is used for the curve fitting via the order *FindFit*. Each group gives different values for d_c and $\epsilon_c(0)$. The values for d_c are closed to each other, hence, a fixed value for it has been specified, $d_c = 1.46 \times 10^{-2}\ \mathrm{hr}^{-1}$, and the *Mathematica* has been run again to estimate the corresponding value for $\epsilon_c(0)$ for each group. It has been found that the returned values for $\epsilon_c(0)$ can be approximated in the form

$$\epsilon_c(0) = \frac{e_c}{{C_0}^{0.95}} \tag{9}$$

where e_c is a constant which has been estimated to be $e_c = 7.6 \times 10^{-3}\ \mathrm{hr}^{-1}$.

2.4. *Combined therapy*

2.4.1. Mathematical model. It is observed that combining radiotherapy with chemotherapy does not simply mean to add the effect of each of them. Hence the combined therapy is not as good as the sum of the single therapies. A kind of efficacy for the combination has to be considered. Hence, the following model is suggested:

$$\frac{dT}{dt} = aT(1 - bT) - \epsilon_1(\epsilon_r D\, T + \epsilon_c C\, T) \tag{10}$$

$$\frac{dC}{dt} = -\gamma C \tag{11}$$

$$\frac{d\epsilon_r}{dt} = d_r\epsilon_r(1 - \epsilon_r) \tag{12}$$

$$\frac{d\epsilon_c}{dt} = d_c\epsilon_c(1 - \epsilon_c) \tag{13}$$

$$\frac{d\epsilon_1}{dt} = d_1\epsilon_1(1 - \epsilon_1) \tag{14}$$

where $\epsilon_1(t)$ is the efficacy of the combined radiotherapy and chemotherapy, and d_1 is the growth rate of the efficacy. Parameters $a, b, d_r, e_r, d_c, \gamma,$ *and* e_c have the values given in sections 2.1.2., 2.2.2., 2.3.1., and 2.3.2. The initial conditions are as follow:

$$T(0) = 596.3,\ C(0) = C_0,\ \epsilon_r(0) = \frac{e_r}{\sqrt{D}},\ \epsilon_c(0) = \frac{e_c}{{C_0}^{0.95}} \tag{15}$$

It can be noticed that these initial conditions are the ones given in the previous sections. The parameter d_1 and the initial value for the efficacy $\epsilon_1(0)$ are going to be estimated from the experimental data. The model given in Eq. 14 for the efficacy allows the variation of it during the medication period.

2.4.2. Fitting the model to data. In this part of the experiment, the cells had been treated with both radiation and vinblastine chemotherapeutic drug. The tested cells had been divided into five groups and in the beginning of day zero the drug was added to each group with concentrations of 0.3, 1, 3, 10, and 30 μM respectively. Each group is divided to four subgroups and each subgroup were radiated with a dosage of 2, 4, 6 and 8 Gray, respectively. The area of the spheroid was measured on days 3, 6, 9, and 12 and Tables 4-7 give the number of tumor cells over that period for each group except the group with 30 μM since the data is missing for that group [17,18].

Table 4. Number of cells treated with radiotherapy and chemotherapy, $C_0 = 0.3\ \mu M$

		Number of cells				
Hours		0	72	144	216	288
Radiation 'D'	2	596.3	1258.1	1695.7	1588.9	1283.0
	4	596.3	1202.1	1992.4	2242.5	2191.6
	6	596.3	1037.9	1842.7	1905.6	1442.5
	8	596.3	1195.5	1606.8	1424.7	1416.8

Table 5. Number of cells treated with radiotherapy and chemotherapy, $C_0 = 1\ \mu M$

		Number of cells				
Hours		0	72	144	216	288
Radiation 'D'	2	596.3	1208.9	1507.9	1436.8	850.9
	4	596.3	1005.3	1663.7	1537.5	1403.4
	6	596.3	1156.8	1669.3	1742.6	1097.7
	8	596.3	967.5	1485.0	1542.5	1136.1

Table 6. Number of cells treated with radiotherapy and chemotherapy, $C_0 = 3\ \mu M$

		Number of cells				
Hours		0	72	144	216	288
Radiation 'D'	2	596.3	1268.2	1708.5	1197.2	892.7
	4	596.3	965,0	1667.3	1492.7	1531.6
	6	596.3	925.6	1594.8	1359.8	764.6
	8	596.3	982.9	1377.4	1371.2	1079.1

Table 7. Number of cells treated with radiotherapy and chemotherapy, $C_0 = 10\ \mu M$

		Number of cells				
Hours		0	72	144	216	288
Radiation 'D'	2	596.3	1134.1	1274.6	1097.1	711.1
	4	596.3	1079.4	1305.0	1383.0	1619.8
	6	596.3	905.6	1566.5	1568.2	1023.8
	8	596.3	1138.6	1443.0	1280.1	905.1

FindFit order for the curve fitting in *Mathematica* commercial program is used to estimate d_1 and $\epsilon_1(0)$. Each subgroup gives deferent values for d_1 and $\epsilon_1(0)$. The values for $\epsilon_1(0)$ are closed to each other, hence, a fixed value for it has been specified, $\epsilon_1(0) = 0.63\ \text{hr}^{-1}$, and the *Mathematica* has been run again to estimate the corresponding value for d_1 for each subgroup. It has been found that the returned values for d_1 can be approximated in the form:

$$d_1 = \frac{e_1}{D^{2.6}\, C^{0.5}} \tag{16}$$

where e_1 is a constant which has been estimated to be $e_1 = 0.1\ \mathrm{hr}^{-1}$. It can be noticed that d_1 is a function of time i.e., it changes with the alteration of time.

3. Results

Figure 1 represents the tumor variation for each subgroup with radiotherapy of 2 Gy, 4 Gy, 6 Gy, and 8 Gy with initial concentration of vinblastine chemotherapeutic drug $C_0 = 3\ \mu M$. The corresponding points of the experimental data in Table 6 are also shown in the figure. Results capture the behavior of the tumor in all cases in a reasonable way. Similar results have been obtained for other values of C_0, however, they are not included because of the space limitation.

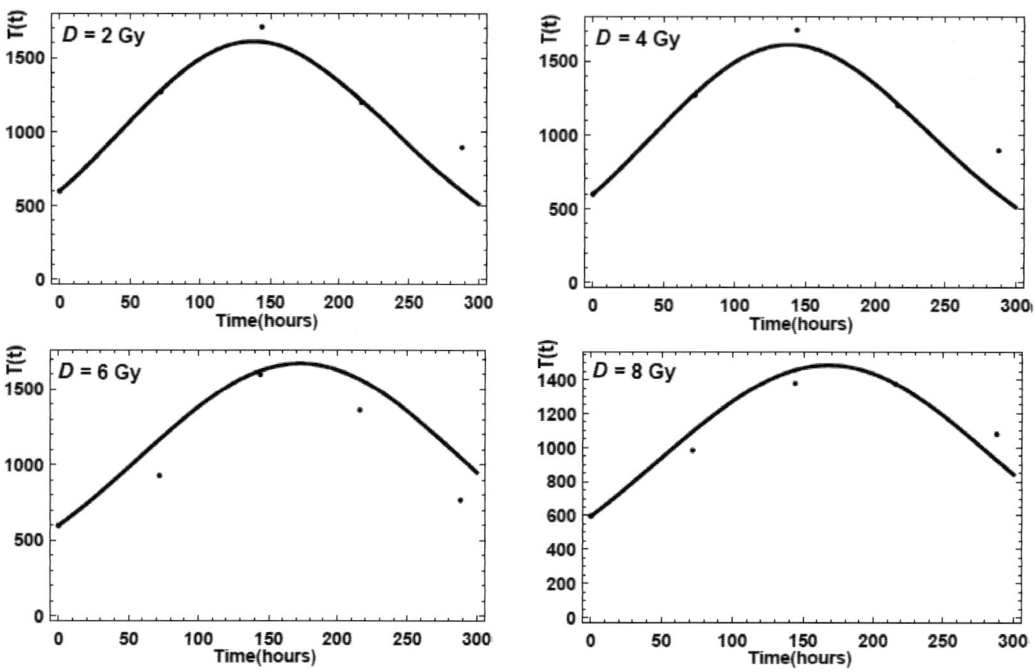

Figure 1. Tumor from mathematical model of combined radiotherapy and chemotherapy (solid lines) assessed with experimental data (big dots), $C_0 = 3, T(0) = 596.3, a = 2.35332 \times 10^{-2}$, $b = 9.278756 \times 10^{-5}$, $d_r = 1.8 \times 10^{-3}, e_r = 3.8 \times 10^{-3}$ $d_c = 1.46 \times 10^{-2}$, $e_c = 7.6 \times 10^{-3}$, and $\gamma = 0.01, \epsilon_c(0) = 0.63, e_1 = 0.1$

4. Discussion and Conclusion

A mathematical model has been constructed to simulate the interaction between the tumor and the therapy. The therapy could be a radiotherapy, a chemotherapy, or a combined chemotherapy and radiotherapy. The model has been designed to simulate an experiment designed to test the ability of therapeutic compound to overcome radiation resistance. The aim has been to cover reality and to consider the influences of treatment.

The construction of the model has started with a simple model for the tumor growth, and the model has been extended to include the effect of radiotherapy, chemotherapy, and combined chemotherapy and radiotherapy. Parameters of the model have been obtained in each stage by fitting the model with the experimental data. Many groups and subgroups had been treated in the cited experiment; hence, some parameters have been assumed as variables which are functions of the radiation dosage, the concentration of the drug, or both. Notably, same expressions for the parameters have been used for all groups. It has been observed that combining radiotherapy and chemotherapy would not result in adding the effect of both, as a result of radiation resistance. Hence, an efficacy for the combination has been included. Results capture the behavior of the tumor during the medication period in all cases in an adequate way.

The mathematical model constructed in this research work has given an insight to the cited experiment. It has highlighted some shadow sites of the experiment. It could be used in predicting the most effective treatment intervention strategies. It can minimize the drug administration, and in general it can enhance the current medication techniques.

References

[1] WHO 2021 https://www.who.int/news-room/fact-sheets/detail/cancer

[2] Mamat M, Subiyanto K A and Kartono A 2013 Mathematical model of cancer treatments using immunotherapy, chemotherapy and biochemotherapy *Appl. Math. Sci* **7** 247–61

[3] Rockne R, Alvord E C, Rockhill J K and Swanson K R 2009 A mathematical model for brain tumor response to radiation therapy *J. Math. Biol.* **58** 561–78

[4] Tao Y, Guo Q and Aihara K 2014 A partial differential equation model and its reduction to an ordinary differential equation model for prostate tumor growth under intermittent hormone therapy *J. Math. Biol.* **69** 817–8

[5] Malinzi J and Soltani M 2019 Mathematical Analysis of a Mathematical Model of Chemovirotherapy: Effect of Drug Infusion Method *Comput. Math. Methods Med.* 7576591

[6] Pinho S, Freedman H and Nani F 2002 Chemotherapy model for the treatment of cancer with metastasis *Math Comput. Model.* **36** 773–803

[7] Elkaranshawy H A, Makhlouf A M and Abouelseoud Y 2020 Using Padé Approximant Method to Solve the Mathematical Model of Tumor-Immune Interactions *Conf. Proc.* ***IEEE*** *Eng. Med. Biol. Soc. (Montreal)* pp 2467-70

[8] Makhlouf A M and Elkaranshawy H A 2021 Sensitivity analysis for a mathematical model of tumor-immune interactions *U.P.B. Sci. Bull.* A **83** 317-26

[9] Makhlouf A M, El-Shennawy L and Elkaranshawy H A 2020 Mathematical modelling for the role of $CD4^{+}$T cells in tumor-immune interactions *Comput. Math. Methods Med.*, Article ID 7187602

[10] Webersinke R 2013 Mathematical Modelling of Tumour Radiotherapy *Technische Universität München (Master thesis)*

[11] Cappuccio A, Herrero M A and Nunez L 2009 Tumour radiotherapy and its mathematical modelling *Contemp. Math.* **402** 77 – 102

[12] Modok S, Hyde P, Mellor H R, Roose T, and Callaghan R 2006 Diffusivity and distribution of vinblastine in three-dimensional tumour tissue: Experimental and mathematical modelling *Eur. J. Cancer* **42** 2404–13

[13] Ledzewicz U, Naghnaeian M and Schättler H 2012 Optimal response to chemotherapy for a mathematical model of tumor-immune dynamics *J. Math. Biol.* **64** 557–77

[14] Ghaffari A, Bahmaie B and Nazari M 2016 A mixed radiotherapy and chemotherapy model for treatment of cancer with metastasis *Math. Methods Appl Sci.* **39** 4603–17

[15] Liuand Z and Yang C 2016 A mathematical model of cancer treatment by radiotherapy followed by chemotherapy Math. and Comput. Simul. **124** 1-15

[16] Barazzuol L, Burnet N G, Jena R, Jones B, Jefferies S J and Kirkby N F 2010 A mathematical model of brain tumour response to radiotherapy and chemotherapy considering radiobiological aspects *J theor. Boil.* **262** 553–65

[17] Bernhardt M 2014 Mathematical Modelling of combined radiation and chemotherapy *Technische Universität München (Master thesis)*

[18] Anastasov N *et al.* 2015 A 3D-microtissue-based phenotypic screening of radiation resistant tumor cells with synchronized chemotherapeutic treatment *BMC Cancer* **15** 466

Accelerated failure time models in analyzing duration of employment

L Basha* and E Gjika

Department of Applied Mathematics, Faculty of Natural Science, University of Tirana, Tirana, Albania
*E-mail: lule.hallaci@fshn.edu.al

Abstract. Parametric accelerated failure time models (AFT), that assess the relationship between the event times and dependent variables constitute an essential class of regression models. In this research we have used the accelerated failure time models: in particular, we have taken into consideration exponential distribution, Weibull distribution, log-logistic model and lognormal distribution, to analyze the duration of employment for employees that have been in their current job or otherwise known as job tenure. A variety of commercial and public companies in Albania were surveyed for this study. Initially the goal of the research is to find the optimal probability distribution to suit the data. Then, the accelerated failure time model is used to assess the impact of the predictors, such as: the employee's age, wages, the employee's age when he has started the job, sex, profession, academic degree, marital status and years of experience prior to this position, in job termination. The log-normal AFT model was the most accurate model for predicting the outcomes of the study, and from this model we can conclude that the employee's age, the employee's age when he has started the job, wages, academic degree and position in the company can affect how long someone stays employed.

1. Introduction

The most common type of model used to predict the duration of socio-economic phenomena is the survival model [1]. There are many researches into the matter of duration of employment and unemployment using survival models [2], [3]. In [4] authors focus on job tenure as a measure of whether a job is a good match for someone after unemployment. The relationship between career flexibility and professional plateau is examined, with focus on how long employees have been employed and their own self-awareness is examined in [5]. Job duration is of uppermost importance to workers, as it can be interpreted as a measure of job-for-life. The median length of time that employees have been with their current employer was 4.2 years in January 2018, according to the Bureau of Labor Statistics. [6] adopt the semi-parametric Cox proportional hazards model, in and [7] determine whether work ability played a decisive role in keeping a job among hospital staff in São Paulo, Brazil, who eventually were laid off within 4 years. By using multiple-record Cox proportional hazards model [8] examine risk variables for health-related job loss over a two-year follow-up period.
In this research, we have surveyed of the feature of 887 employees, in a variety of commercial and public companies, in Albania. In the first stage of the study, we analyze the duration of job tenure and find the parametric probability distribution that best fits the data. Given that the time is a continuous variable we have taken into consideration some continuous distributions such as: lognormal; Weibull; exponential and log-logistic distribution which is mostly used in the literature. Then, variables, which are thought to have a great impact of length of employment are studied. The factors studied are: the employee's age, wages, the employee's age when he has started the job, sex, profession, academic

degree, marital status and years of experience prior to this position. The accelerated failure time (AFT) model, is used to estimate how these factors may effect on job tenure. The procedure of parameter estimation, evaluation and covariate selection was done through R software packages.

2. Materials and methods

When the length of time or survival time, is assumed to follow a known distribution, then we can use a parametric survival model for the data. In the case of parametric models, full maximum likelihood method can be used to estimate parameters and the probability distribution of the variable can be estimated. The probability distributions we have considered are continuous distributions: lognormal; Weibull; exponential and log-logistic distribution. In survival analysis models, explanatory variables are considered to be related with the time it takes for the event to occur. To explore this relationship, researchers proposed the Cox Proportional Hazards model with right censoring [9]. A useful alternative to the Cox PH model is the accelerated failure time (AFT) model [10], [11]. This model supposes that the error term of this linear regression has a specific distribution. This model has the advantage of an intuitive physical interpretation [12] and gives more efficient estimates compare with Cox model [13].

We have denotes with $X_1,.....X_n$ the variable of interest with probability density function f and cumulative distribution function F. The censoring variable is $C_1,.....C_n$ with continuous distribution function G. We suppose that the variable X is independent of C. When taking into account the random right censoring, the variable is not fully observed. We can only notice (T_i,δ_i) where $T_i=\min(X_i,C_i)$ and $\delta_i = I(X_i \le C_i)$, for $i=1,...,n$.

The relationship of logarithm of survival time variable and dependent variables, is linear according to the accelerated failure time model, as

$$\log T = a_0 + \sum_{i=1}^{p} a_i z_i + \sigma\varepsilon \tag{1}$$

where, $z=(z_1,...,z_p)$ are the covariates, $a=(a_1,...,a_p)$ are the coefficients, σ $(\sigma>0)$ is an unknown scale parameter, and ε are random errors that does not depend on z_i.

The AFT model presume that factors interrelate with time directly in the baseline hazard function. The hazard function at time t can be given as:

$$h(t\mid z) = h_0(t\cdot e^{-z^T a})\cdot e^{-z^T a} \tag{2}$$

where $h_0(t)$ is the baseline hazard function. The influence of covariates on hazard in equation (2) depends on time and at the same time is disproportionate. A negative sign on a coefficient for a covariate indicates that it reduces the hazard value, but may also increase the duration of the exposure when the covariate increases [14]. The survivorship, the probability density, the hazard and cumulative hazard function are mathematically equivalent and all the relevant formulas are summarized as below:

$$H(t\mid z) = \int_0^t h(T\mid z)dT = \int_0^t h_0(T\cdot e^{-z^T a})e^{-z^T a}dT = H_0(t\cdot e^{-z^T a}) = -\ln\left[1-F_0(t\cdot e^{-z^T a})\right]$$

$$S(t\mid z) = \exp[-H(t\mid z)] = \exp\left[-H_0(t\cdot e^{-z^T a})\right] = S_0(t\cdot e^{-z^T a}) = 1-F_0(t\cdot e^{-z^T a}) \tag{3}$$

$$f(t\mid z) = h(t\mid z)\cdot S(t\mid z) = h_0(t\cdot e^{-z^T a})\cdot e^{-z^T a}\cdot S_0(t\cdot e^{-z^T a}) = f_0(t\cdot e^{-z^T a})\cdot e^{-z^T a}$$

The general Weibull regression model is $\log T = a_0 + a_1 z_1 + ... + a_p z_p + \sigma X$, where X has standard extreme value distribution: $f_x(x) = \exp(x - e^x)$. The survivor function can be written as:

$$S(t;z_1,...,z_p)=\exp\left\{-\exp[\alpha(\log(t)-a_0-a_1z_1-...-a_pz_p)]\right\} \tag{4}$$

The log-logistic and lognormal regression models are examples of accelerated failure time models as well as the exponential model, which is a special case of Weibull distribution, for $\alpha = 1$. If the event time T has lognormal distribution, we model its dependency from explanatory variables as equation (3), where the variable has standard normal distribution: $X \sim N(0,1)$. The survivor function has the form

$$S(t;z_1,...,z_p)=1-\Phi\left(\frac{1}{\sigma}(\log(t)-a_0-a_1z_1-...-a_pz_p)\right) \tag{5}$$

If the event time T has log-logistic distribution, we model its dependency from explanatory variables as equation (3), where X has standard logistic distribution: $f(x)=\frac{e^{-x}}{(1+e^{-x})^2}$. The survivor function has the form:

$$S(t;z_1,...,z_p)=\frac{1}{1+\exp[\alpha(\log(t)-a_0-a_1z_1-...-a_pz_p)]} \tag{6}$$

3. Study population

In our research, we have looked at how long 887 employees, have been employed in their current job, in a variety of commercial and public companies, in Albania. This study looks at the effect of a given period of time, from January 1991 to December 2018. The time when the employee begins working for the employer, is considered as the start time, while the end time is the time when financial relations are interrupted. At the time the data was received from the company some employees are currently working, while some are no longer employees of the current company. For this purpose, we have established a censoring variable. This variable has a value of 0 for the "still working" variable has a value of 0 if the data was last updated within the past week and a value of 1 otherwise. "still working" variable and a value of 1 otherwise. We have tried to fit a probability model for duration of job tenure. Given that the time is a continuous variable, in our case in months, we have taken into consideration some continuous distributions such as: lognormal; gamma; Weibull; exponential and log-logistic distribution. Some factors that are thought to be important in the time of staying employed with the same employer and the same job position are studied. The accelerated failure time (AFT) model, is used to estimate how these factors may effect on job tenure.

There are in total 887 employees, 534 of them were still in financial relationship with the company, in December 2018, the last month when the data has been update. The other part of the employees has had their employment relationship with the company interrupted. The shortest time that an employee has been in a current job position is one month and the longest time in the same job position is 26 years. In the first stage of this work we have proposed four probability distributions to fit the job tenure time: the lognormal, exponential, Weibull and log-logistic distribution. To estimate and evaluate the fitting we have used R software as a tool. To compare the fitting performance of the probability distributions we have used QQ plot graphical test and the information criteria statistics (AIC and BIC).

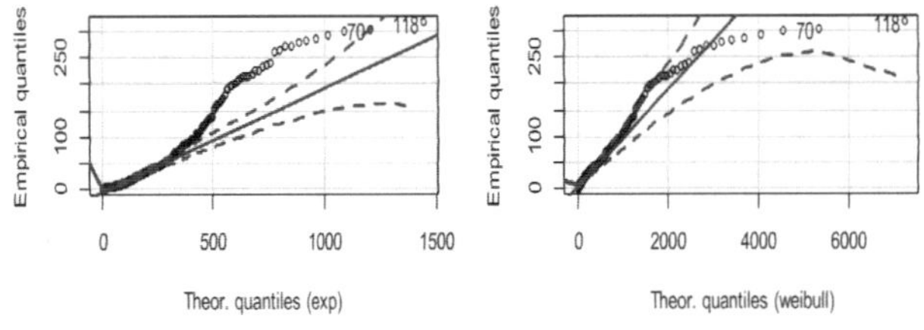

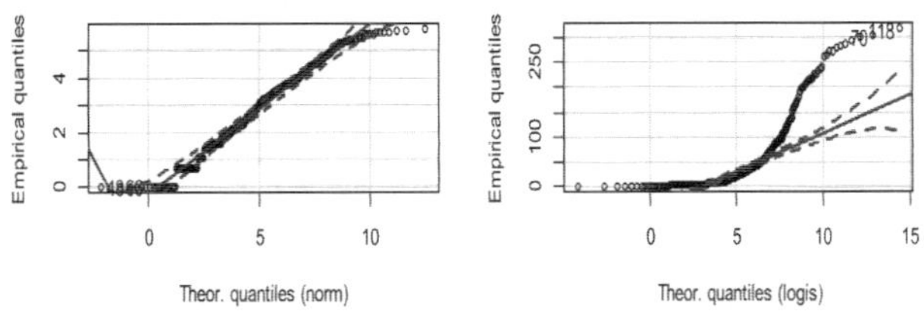

Figure 1. Goodness-of-fit plots for various distributions fitted to job tenure

Plots of empirical quantiles of residuals versus theoretical residuals expected by error distribution (in our case, Weibull, exponential, lognormal and log-logistic distribution) are shown in Figure1. Q-Q plot show a straight line for lognormal distribution, which describe satisfactory the most of the empirical data, and the other proposed distributions such as the exponential, Weibull and log-logistic distribution seem too not describe the real observations.

Table 1. Goodness of fit statistics

	Exponential	Weibull	Lognormal	Log-logistic
AIC	5059	4515	4331	4351
BIC	5064	4521	4336	4358

As it is clearly seen from Table 1, also the goodness-of-fit statistics for the job tenure time are in favor of the lognormal distribution, with parameters: *meanlog: 5.2; sdlog: 2.44.* From a given set of candidate models, we take the one that minimizes the AIC (BIC).
The Kaplan-Meier estimator [15] is a non-parametric maximum likelihood estimator for survival function. This estimator does not make an assumption on the distribution of *T*, but from the curve we can see which parametric distribution is appropriate for the data.

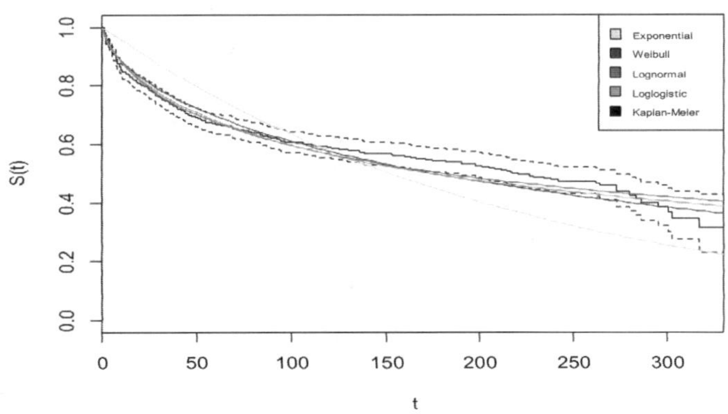

Figure 2. Estimated survival function with Kaplan-Meier method and various distributions fitted to job tenure

Figure 2 shows the estimated survival function using Kaplan-Meier method and the lognormal, exponential, Weibull and log-logistic distributions fit to job tenure, without taking into consideration the covariates. Also, we can see that exponential curve is very different from the Kaplan-Meier curve and again here we can confirm that lognormal distribution "best" fits the data.
In the next step we estimate the relationship between job tenure and all the covariates mention above. A stepwise regression method is conducted, to select the explanatory variables. A forward-selection

rule starts with no explanatory variables and then adds variables, one by one, based on which variable is the most statistically significant, until there are no remaining statistically significant variables. The selection procedure has been done with the principle of minimizing the indicators of goodness-of-fit, in this case we have used the criteria of minimizing Akaike Information Criteria (AIC). This procedure is done using accelerated failure time (AFT), exponential, Weibull, lognormal and log-logistic models. The statistical significance of the explanatory variables was observed at the 5% level after the procedure was completed. The final estimated model yielded the following results, given in Table 2. The first column of the table shows the parameters estimation, with the log-likelihood method. A log likelihood test indicates that the model on the whole is significantly better than one which does not include any covariates. According to the AFT model, factors: *the employee's age, wages, the employee's age when he has started the job, sex, profession, and years of experience prior to this position* in the company are statistically significant with a *p-value* less than 0.05. The *marital status* and *academic degree* are eliminated from the model based on AIC. The parameter estimates from AFT models are interpreted as effects on the time scale, which can either speed up or slow down the survival time. If the acceleration factor is greater than one, then the exposure is beneficial to survival. Male significantly increases the survival time by approximately 30%. One unit change in *the employee's age when he has started the job* at the current position in the company shortens survival time by 16%. The results suggest that the median survival time is decreased by a factor of 2.9 if the job position of an employee is a supervisory, compare with a financier. *Profession, sex* and *the employee's age* extends the duration of financial relationship.

Table 2. Accelerated failure time model

	Value	Std. Error	z	p
(Intercept)	4.45	2.55e-01	17.48	< 2e-16
age	0.138	8.95e-03	15.44	< 2e-16
agebi	-0.165	1.04e-02	-15.78	< 2e-16
ywcp	4.14e-03	5.94e-04	6.97	3.3e-12
salary	-1.16e-05	1.68e-06	-6.87	6.3e-12
genderM	0.265	1.20e-01	2.20	0.027
posfinancier	0.0337	1.53e-01	0.22	0.826
posother	1.03	1.54e-01	6.73	1.7e-11
posspecialist	1.14	1.47e-01	7.80	6.3e-15
possupervisory	1.59e+01	1.67e+03	0.01	0.992
Log(scale)	-0.256	5.57e-02	-4.59	4.5e-06
Factors in Final Model [1] age agebi ywcp salary gender pos				
Deleted	Chi-Sq	d.f.	P	AIC
status	1.16	1	0.2806	-0.84
educ	4.70	2	0.0951	-0.13

When using a model to fit survival data, the survival curves can be adjusted to account for all of the covariates used as predictors. The Kaplan-Meier method is unsuitable for this problem because it is not adjusting for covariates. Figure 3 shows adjusted survival curves for job tenure, where the average value chosen for a covariate being adjusted is used.

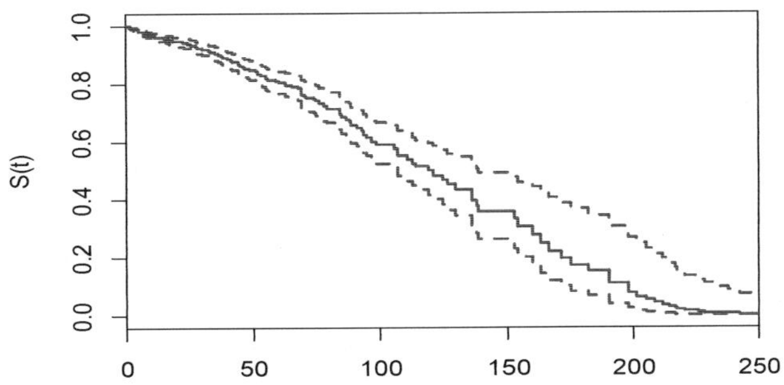

Figure 3. Survival curves adjusted for covariates.

The probability that an employee will continue to work with the current employer or company for at least one year, is 96.2 %. The lower quartile of time is 71 months; this means that, nearly one-quarter of employees leave their company within 71 months from their start date, while half leave within 114 months and 75% of them interrupt their relationship with the company within 160 months from their beginning.

4. Conclusion

The objective of this study was to analyse the length of time the employee has worked in his or her current position or with the current employer, to find the model that best fits the data and to assess the impact of the employee's age, wages, the employee's age when he has started the job, sex, profession, academic degree, marital status and years of experience prior to this position, in job termination. As a result of preliminary analysis on how to model the phenomenon in question, four models were proposed: exponential, Weibull, lognormal and log-logistic. Based on graphical and statistical tests, the lognormal distribution was found to be the best fit for the empirical data, with parameters: *meanlog: 5.2; sdlog: 2.44.*

The results showed that the average work experience of an employee in a company is approximately 187 months, 25% of employees have been in their current position for more than 10 years, and 30% for less than 2 years. According to the AFT analysis, profession, sex and age extends the time of financial relationship, while the employee's age when he has started the job and wages affect positively in the interruption of the relationship, thus in the job tenure. Also, half of the employees leave within 114 months and 75% of them interrupt their relationship with the company within 160 months from their beginning.

5. References

[1] Lawless J L 2003 Statistical *Models and Methods for Lifetime Data* (New Jersey: Wiley Interscience, Hoboken)

[2] LancasterT 1979 Econometric methods for the duration of unemployment. Econometrica vol. **47**, no. 4 pp 939–956.

[3] Landmesser J 2013 The use of methods of analysis of the duration of the labour force survey in Poland,*Warsaw University of Life Sciences*,Warsaw.

[4] Kohara M, Sasaki M and Machikita T 2013 Is longer unemployment rewarded with longer job tenure?Journal *of the Japanese and International Economies*Volume **29** Pages 44-56 https://doi.org/10.1016/j.jjie.2013.06.002

[5] Jiang Z, Hu X and Wang Z 2018 Career adaptability and plateaus: The moderating effects of tenure and job self-efficacy *Journal of Vocational Behavior*Volume **104** Pages 59-71 https://doi.org/10.1016/j.jvb.2017.10.006

[6] Basha L, Gjika E and Puka Ll 2020 Analyzing Job Tenure Factors in Private and Public Companies in Albania by Using Cox Proportional Hazards Model *Advances in Science, Technology and Engineering Systems Journal* vol. **5** no. *2* pp. 254-260 https://dx.doi.org/10.25046/aj050233 DOI: 10.25046/aj050233

[7] Martinez M C and Fischer F M 2019 Work Ability and Job Survival: Four-Year Follow-Up*International journal of environmental research and public health*,**16(17)**, 3143. https://doi.org/10.3390/ijerph16173143

[8] Syddall H E, D'Angelo S, Ntani G, Stevens M, Harris E C, Linaker C H and Walker-Bone K 2020 Work participation and risk factors for health-related job loss among older workers in the Health and Employment after Fifty (HEAF) study: Evidence from a 2-year follow-up period.*PloS one*,**15(9)**, e0239383. https://doi.org/10.1371/journal.pone.0239383

[9] Cox D 1972 Regression models and life – tables (with discussion) *Journal of the Royal Statistical Society* Series B**34** 187–220.

[10] Kalbfleish J D and Prentice R L 2002 *The Statistical Analysis of Failure Time Data 2nd ed.* (New York: Wiley MR1924807)

[11] Yu X and Zhao Y 2018 Jackknife empirical likelihood inference for the accelerated failure time model. *SPRINGER* https://doi.org/10.1007/s11749-018-0601-7

[12] Wei L J 1992 The accelerated failure time model: a useful alternative to the Cox regression model in survival analysis. *Statistics in Medicine*, **11**:1871–1879.

[13] Cox D and Oakes D 1984 *Analysis of Survival Data* (London: Chapman and Hall)

[14] Wang K, Ye X and Jie M 2018 An empirical analysis of post-work grocery shopping activity duration using modified accelerated failure time model to differentiate time-dependent and time-independent covariates. *PLoSONE***13(11):** e0207810. https://doi.org/10.1371/journal.pone.0207810

[15] Kaplan E and Meier P 1958 Nonparametric estimation from incomplete observations. *Journal of the American Statistical Association*, Vol. **53** pp. 457–481

Numerical Simulation Study of the Water Entry Slamming on Oscillating Buoys with Different Shapes

Xuejian Li[1], Xing Luo[2], Ruohong Wang[1] and Yanjun Liu[1,2,3,*]

1 Institute of Marine Science and Technology,Shandong University,Qingdao, China
2 School of Mechanical Engineering,Shandong University,Jinan, China
3 Key Laboratory of High Efficiency and Clean Mechanical Manufacture, Ministry of Education,Shandong University,Jinan, China
E-mail:17865192172@163.com

Abstract. Heaving buoys device is a commonly used form of wave energy utilization. When encountering extreme wind and waves, the float will rise away from the waves, and then fall back to the normal working position will be impacted by water. In this paper, numerical simulation studies are carried out on the slamming characteristics of three types of floats of cone, truncated cone and hemispherical shapes, and the influence of the bottom lift angle of the floating body on the peak acceleration and the peak slamming pressure is analyzed. The CFD method was used to simulate the slamming process of different shapes of floating bodies, and the velocity, pressure and flow field changes during the process of entering the water were obtained, resulting in the main conclusions as follows : The truncated cone-shaped floating body is subject to greater slamming pressure, and the bottom rise angle has little effect on the slamming pressure and speed changes; The slamming pressure on the conical floating body is smaller, and the larger the bottom lift angle, the smaller the slamming pressure; The slamming pressure of the hemispherical floating body is stronger than that of the conical shape, but its speed after entering the water is greater than that of the conical shape, and the kinetic energy loss is minimal.

1. Introduction

When the wave energy generating device encounters extreme wind and waves, the huge wave force causes the movement range of the buoy to exceed the normal working stroke, and it collides with the float and is damaged. In order to prevent damage, a common safety measure is to raise the buoy away from the waves, but after an extreme situation, when the buoy falls back to its normal working position, the bottom of the buoy will be slammed by water.

Slamming is an impact-type hydrodynamic load acting on the hull structure, which specifically involves the coupling process of a gas-liquid-solid three-phase medium. Model tests and numerical methods have gradually become the main means of studying wave slamming. Numerical simulation attempts to obtain accurate numerical solutions through the nonlinear boundary conditions imposed on the free surface in the instantaneous state. Due to problems such as the stability of qualitative algorithms, it is sometimes very difficult to obtain an accurate solution. Numerical simulation methods and theoretical methods are often used to study slamming problems.

Generally speaking, the boundary value problem can be discretely solved by the panel method to obtain the normal derivative of the velocity potential on the surface and the velocity potential on the free surface. Then the velocity component of the upper element of the boundary is calculated, and the position of the free surface and the velocity potential on the free surface can be updated step by step in the time domain. Wu et al. (2004) [1] studied the process of the wedge-shaped body falling into the water freely.They converted the boundary conditions of the free surface to its integral form, and obtained stable and convergent numerical solutions through iterative integration. Since then, the extended coordinate system and auxiliary functions have been widely used in the solution of slamming problems in two-dimensional and three-dimensional situations. For example, Xu, Duan & Wu (2011) [2], Sun & Wu (2013a) [3] and other articles have applied these two methods. Sun & Wu (2013b, 2015) [4, 5] also studied the slamming of non-axisymmetric objects obliquely into the water and the slamming of water surface waves and gravity on the process of wedge-shaped bodies entering the water.

The above-mentioned researches on slamming problems are mostly in the case of wedge-shaped bodies or cones where the wet surface is straight. On the other hand, there have been many research progresses on the slamming of curved objects into water. Scolan & Korobkin (2001) [6] solved the inverse problem of Wagner's problem combined with slice theory to obtain the solution of the slamming of the elliptical paraboloid into water. Tassin et al. (2012) [7] proposed his own algorithm based on Wagner method and boundary element method and used this method to solve the parabolic water entry slamming problem. Sun & Wu (2014) [8] used the potential flow theory combined with the boundary element method to give a similar solution for the expansion of the paraboloid into water. Yves—Marie (2014) [9] demonstrates the feasibility of Wagner's method in solving the problem of parabolic water entry slamming through oblique water entry slamming test of paraboloid in regular waves.

Thanks to the rapid development of computer technology, various numerical simulation methods have been increasingly used in the research of wave slamming, and computational fluid dynamics method is one of them. CFD technology is more and more widely used in various problems of fluid mechanics, and new CFD algorithms are constantly emerging and applied to fluid-solid slamming problems. For example, finite difference method based on CIP technology (Hu & Kashiwagi, 2004) [10], VOF method (Lin, 2007) [11], immersion boundary element method (Wang et al, 2009) [12], SPH method (Liu et al, 2014) [13], Faccio et al. (2015, 2016) [14, 15] created a CFD extension toolkit specially used to accurately calculate the water entry slamming of a wedge.

In this paper, the commercial software Fluent based on the CFD method is used to conduct numerical simulation research on the slamming characteristics of different shapes of floats, and to discuss and analyze the influence of the shape of the floating body, the bottom lift angle and other factors on the peak acceleration and the peak slamming pressure. Analyze the rise of the free liquid surface and the changes of physical quantities in the flow field such as pressure and velocity during the slamming of the floating body, and study the temporal and spatial distribution characteristics of the slamming load. According to the calculation results of the buoy vertical entry into the water, the comparative analysis obtains the influence law of different shape parameters of the floating body during the water entry slamming process. At the same time, the flow field change law is summarized to provide a reference for the design of the buoy.

2. Verification of Numerical Simulation Method for the Water Entry Slamming

In order to prove the feasibility and accuracy of the numerical simulation analysis of float slamming by using Fluent, this paper first selects Zhao's test model and data for comparison and verification. By establishing the same geometric model as Zhao, using Fluent to simulate the water entry slamming, the numerical results are compared with the experimental data.

Model building

Figure 1 is a schematic diagram of Zhao's test model. The wedge-shaped model is 0.5 m long and has a ramp angle of 30°. The weight of the model is 241 kg, and the initial water entry speed is set to 6.15 m/s. There are five pressure sensors distributed on one side of the model to measure the slamming pressure changes during the wedge-shaped body entering the water; accelerometers are used to collect the acceleration changes of the model to calculate the overall slamming load.

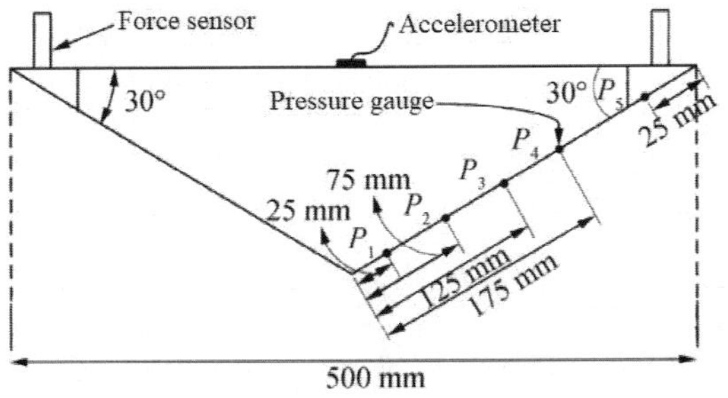

Figure 1. Schematic diagram of Zhao's experimental model.

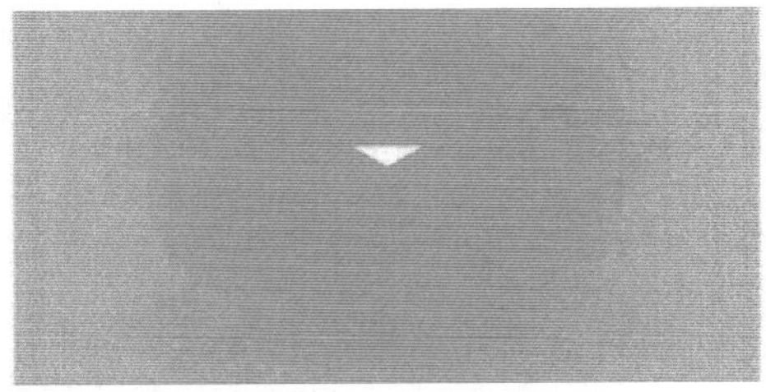

Figure 2. Meshing of the model

Figure 2 shows the meshing of the model. The minimum mesh of 2 mm, the computational domain length 5 m, high 5.2 m, below is water and the water level is 2.7 m. The initial position of the model is 90 mm away from the water surface, and the initial velocity is set to -6.015 m/s, so that the velocity when the model enters the water surface is 6.15 m/s. Choose VOF model for multiphase flow simulation, and use implicit volume force; choose achievable k-ε turbulence model, SIMPLEC pressure algorithm, use six-DOF model to define wedge motion mode, and use elastic smoothing and mesh reconstruction for dynamic mesh method.

2.1. Compare calculation results

2.2.1. Comparison of speed changes during water entry. Figure 3 shows the time history curve of the velocity change of the wedge-shaped body in the process of entering the water. The velocity of the wedge-shaped body at the initial moment of entering the water is 6.15 m/s, then begins to decrease, and the magnitude of the decrease gradually increases. It can be seen that the numerical calculation result is very close to the experimental value, and it is almost completely fitted before 0.015 s, and there is a slight deviation in the later stage. It shows that the numerical method can more accurately simulate the velocity and acceleration changes of the wedge-shaped body in the process of entering the water, and the calculation results have certain accuracy.

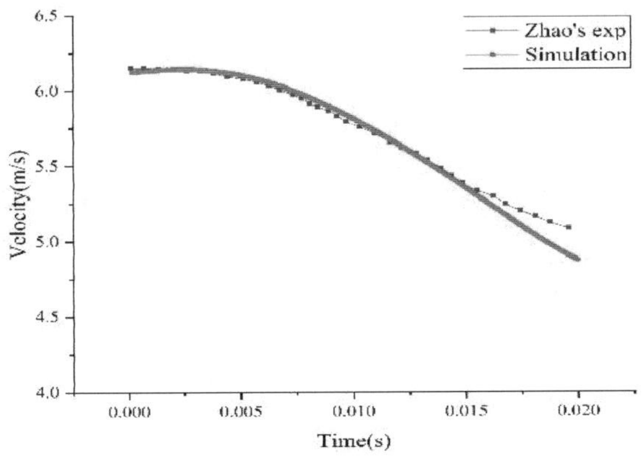

Figure 3. Time history curve of wedge-shaped body entering water velocity.

2.2.2. Slamming pressure comparison at monitoring points. Figure 4, Figure 5, Figure 6, Figure 7 and-Figure 8 show the time history curve of slamming pressure changes at 5 monitoring points during the wedge entering water. The pressures at P1-P5 have peaks in sequence, the peak pressure first increases and then decreases, and the peak pressure at P3 is the largest, about 100,000 Pa. The negative pressure appears before the pressure peak at P4, and the negative pressure at P5 is more obvious.

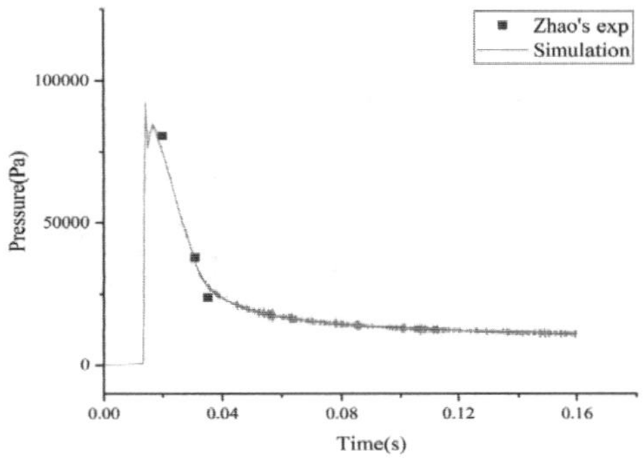

Figure 4. Slamming pressure time history curve at P1

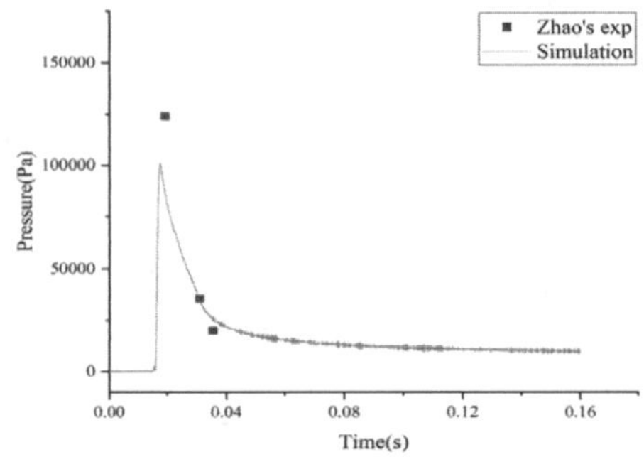

Figure 5. Slamming pressure time history curve at P2.

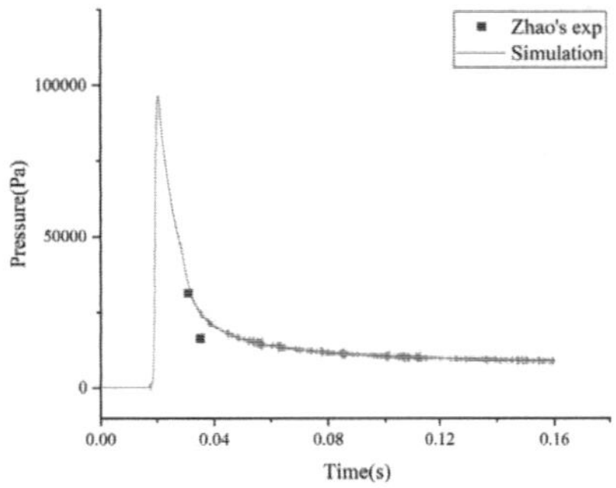

Figure 6. Slamming pressure time history curve at P3.

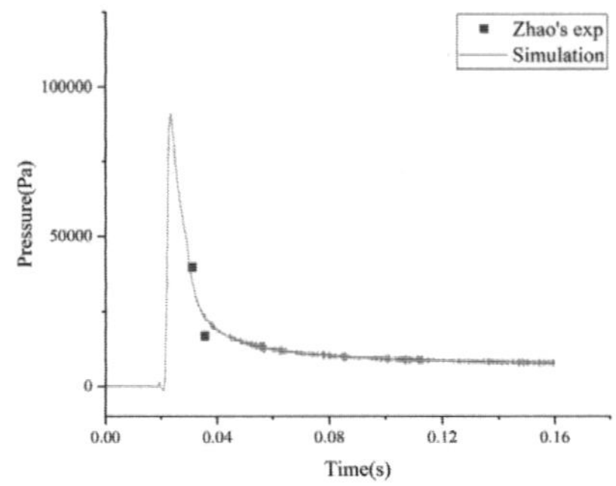

Figure 7. Slamming pressure time history curve at P4.

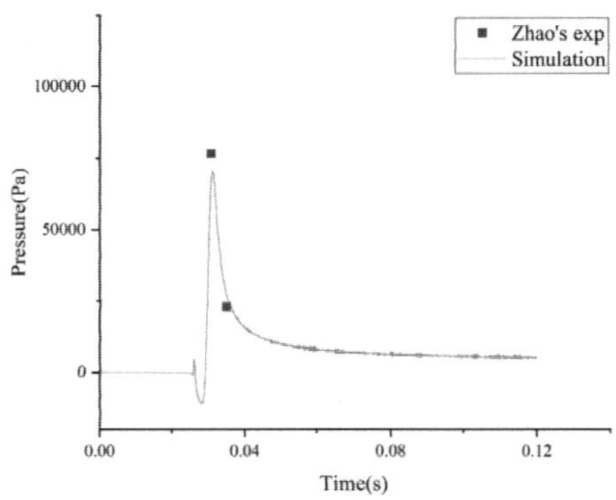

Figure 8. Slamming pressure time history curve at P5.

2.2.3. Slamming pressure comparison at monitoring points. Figure 9 (1)-(8) are the flow field changes at eight different moments in the water entering process. Figure 9 (1) is the initial moment of the wedge-shaped body entering the water. The pressure peak appears and the jet begins to form; In figure 9 (2)-(3), the pressure continues to rise and the jet develops upward; In figure 9 (4)-(8), the wedge-shaped body completely enters the water, the pressure begins to drop, and the jet diverges outward. It can be seen from the figure that the VOF model can clearly show the rise of the free liquid surface and the separation of jets during the wedge-shaped body entering the water.

3. Numerical Simulation Research on Water Entry Slamming Characteristics of Different Shaped Floats

On the basis of verifying the accuracy and effectiveness of the numerical simulation method, numerical simulation of different shapes of wave energy buoys water entry slamming was carried out., The time history curve of the slamming pressure and velocity of the pressure monitoring point during the free entry of the floating body model into the water is calculated, Analyze the influence of the shape of the floating body, the bottom lift angle and other factors on the slamming load.

The geometric modeling and meshing of 30°conical, 45°conical, 30°frusto-conical, 45°frusto-conical and hemispherical floating bodies are carried out. Assuming that the floating body rises to a position of 2.5 m under extreme sea conditions, the falling height is 2.5 m, and the water entry speed is about 7 m/s. The calculation domain is set the same as before, the initial position of the floating body is 90 mm from the water surface, and the initial velocity of the floating body is set as 6.873 m/s.

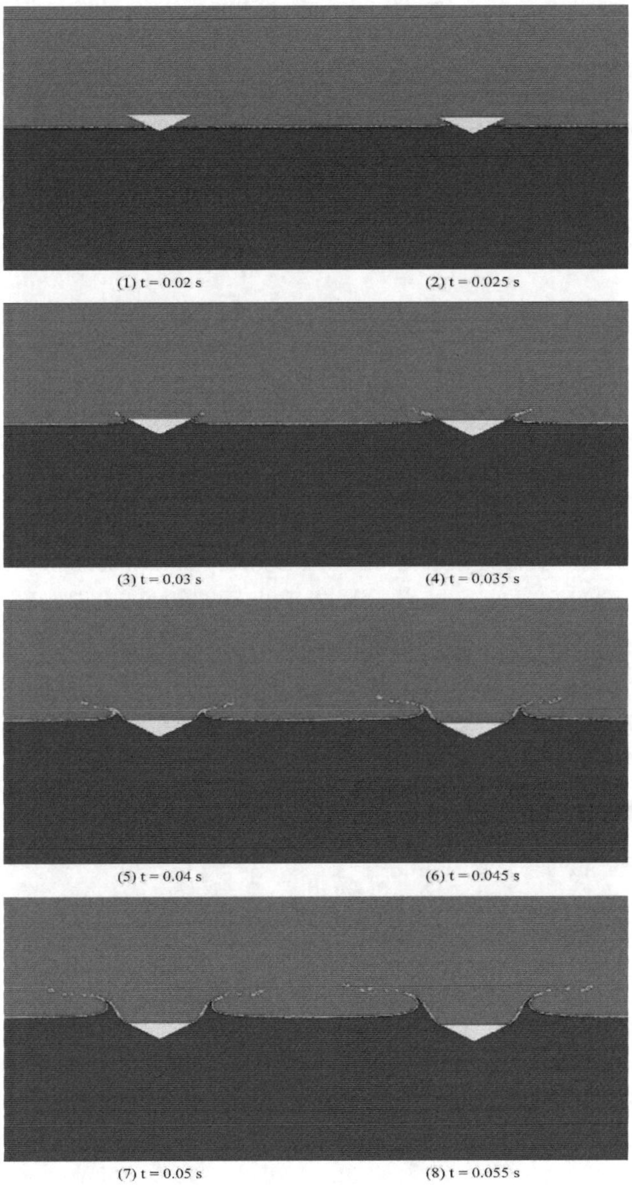

Figure 9. Change of flow field during water entry.

3.1. Water entry slamming of 30° conical floating body

3.1.1. Time history curve of speed change during water entry

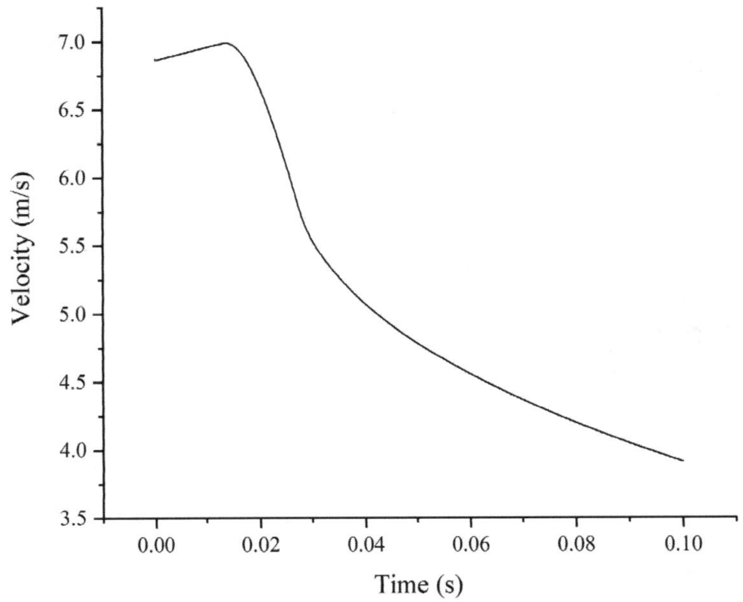

Figure 10. Time history curve of 30°conical floating body entering water speed.

Figure 10 shows the time history curve of the velocity change of a conical floating body with a bottom lift angle of 30° during the slamming process. The velocity of the floating body continues to increase at a constant acceleration during the initial stage of entry into the water, and reaches the maximum value at about 0.018 s, and then rapidly decreases at about After 0.03 s, the decreasing amplitude began to decrease, and the speed change gradually became flat.
Define abbreviations and acronyms the first time they are used in the text, even after they have been defined in the abstract. Abbreviations such as IEEE, SI, MKS, CGS, sc, dc, and rms do not have to be defined. Do not use abbreviations in the title or heads unless they are unavoidable.

3.1.2. Monitoring point slamming pressure time history curve. In order to study the temporal and spatial distribution of slamming load, five pressure monitoring points are set on the surface of the geometric model, and the distribution positions are shown in Figure 11.
Figure 12 shows the time-history curve of the slamming pressure change at the monitoring points P1-P5 during the entry of the floating body into the water. Starting from 0.018 s, the five monitoring points have successively appeared peak pressures, the peak pressure first increases and then decreases, and the peak duration gradually changes Short; P2 has the highest peak pressure, about 144000 Pa; P5 is in the negative pressure zone before the pressure peak.

3.1.3. Changes of jet and flow field shape during slamming. Figure 13 (1)-(8) are the flow field changes at 8 different times during the floating body entering the water, Figure 13 (1) is the initial moment of the floating body entering the water, the pressure peak appears, and the jet begins to form; Figure 13 (2)-(4) The pressure continues to rise and the jet flows upward; the acceleration of the floating body in Figure 13 (5)-(8) decreases, the pressure begins to drop, and the jet diverges outward.

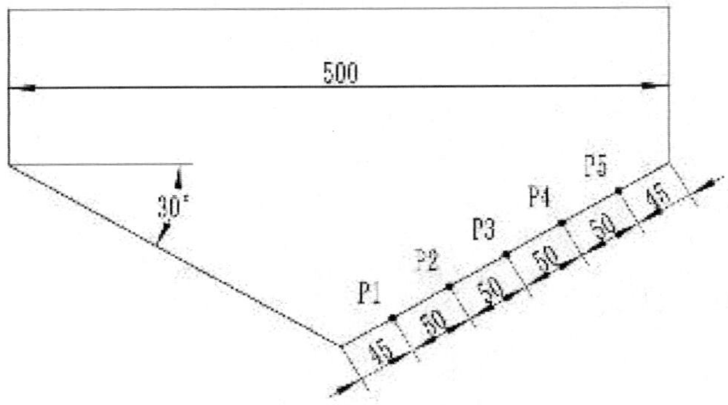

Figure 11. Time Distribution of pressure detection points.

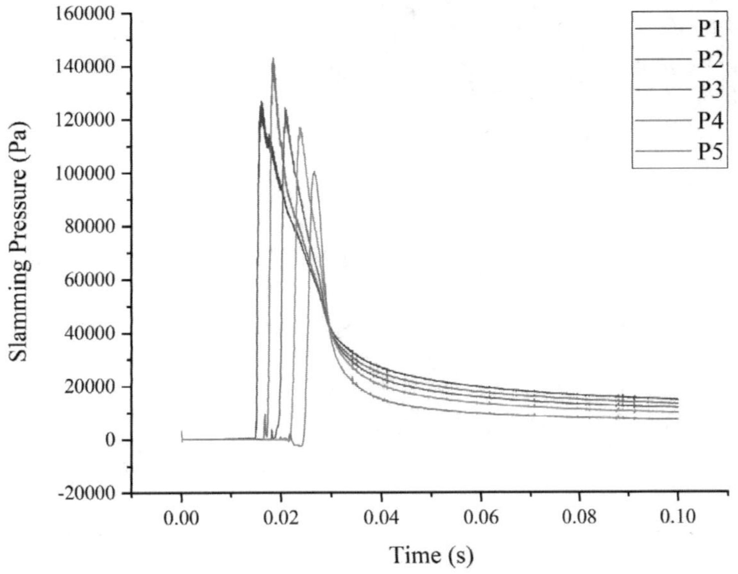

Figure 12. Time history curve of slamming pressure of floating body entering water.

3.2. *Water entry slamming of of 45° conical floating body.*

Time history curve of speed change during water entry. Figure 14 shows the time history curve of the velocity change of a conical floating body with a bottom lift angle of 45° during the slamming process. The velocity of the floating body continues to increase at a constant acceleration during the initial stage of entry into the water, and reaches the maximum value at about 0.018 s, and then rapidly decreases at about after 0.04 s, the decreasing amplitude began to decrease, and the speed change gradually became flat.

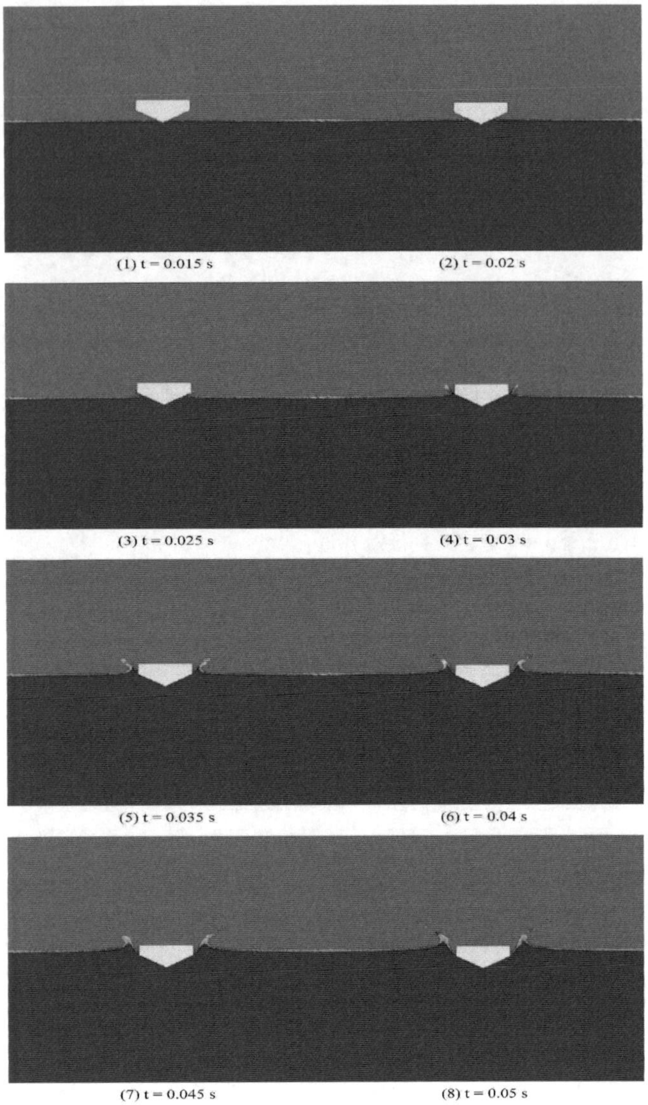

Figure 13. *Change* of flow field during water entry.

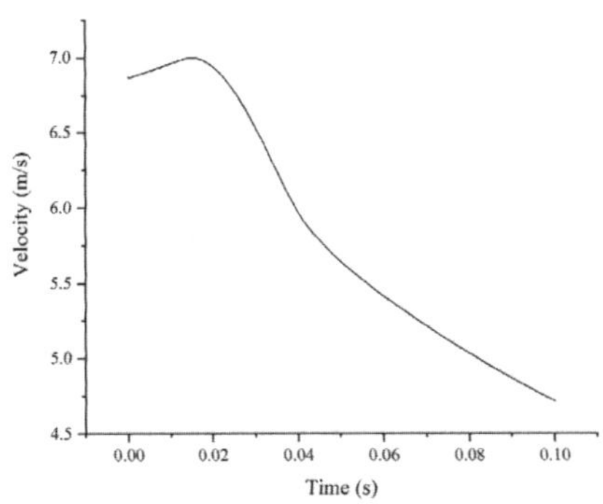

Figure 14. *Time* history *curve of 45°conical floating body entering water speed.*

3.3. *Monitoring point slamming pressure time history curve.* In order to study the temporal and spatial distribution of slamming load, five pressure monitoring points are set on the surface of the geometric model, and the distribution positions are shown in Figure 15.

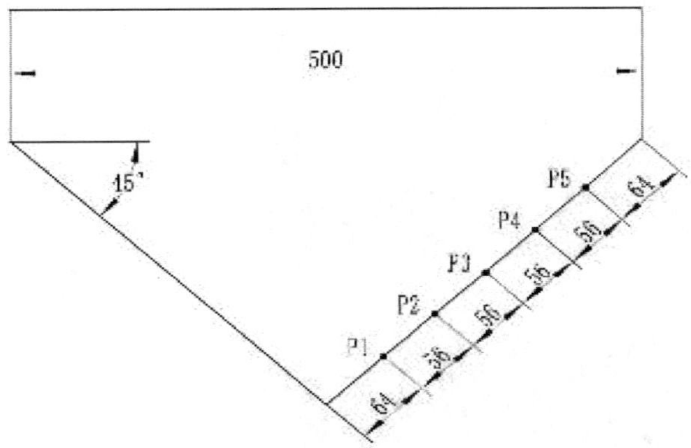

Figure 15. Distribution of pressure detection points.

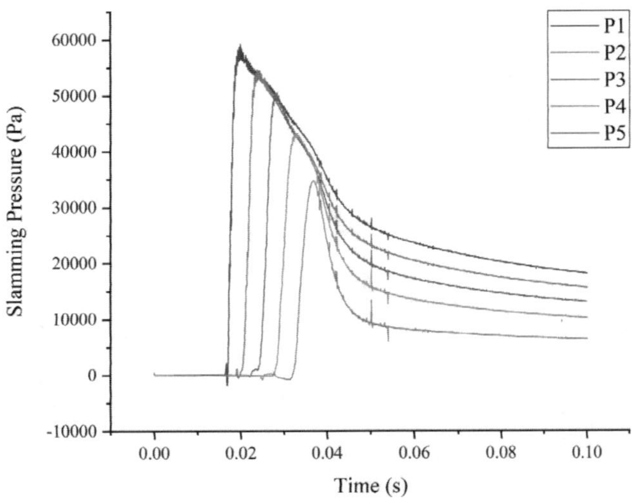

Figure 16. Time history curve of slamming pressure of floating body entering water.

Figure 16 shows the time-history curve of the slamming pressure change at the monitoring points P1-P5 during the entry of the floating body into the water. Starting from 0.018 s, the five monitoring points have successively appeared peak pressures, the peak pressure first increases and then decreases, and the peak duration gradually changes Short; P1 has the highest peak pressure, about 60000 Pa; P5 is in the negative pressure zone before the pressure peak.

3.4. *Changes of jet and flow field shape during slamming.* Figure 17 (1)-(8) are the flow field changes at 8 different times during the floating body entering the water, Figure 17 (1) is the initial moment of the floating body entering the water, the pressure peak appears, and the jet begins to form; Figure 17

(2)-(4) The pressure continues to rise and the jet flows upward; the acceleration of the floating body in Figure 17 (5)-(8) decreases, the pressure begins to drop, and the jet diverges outward.

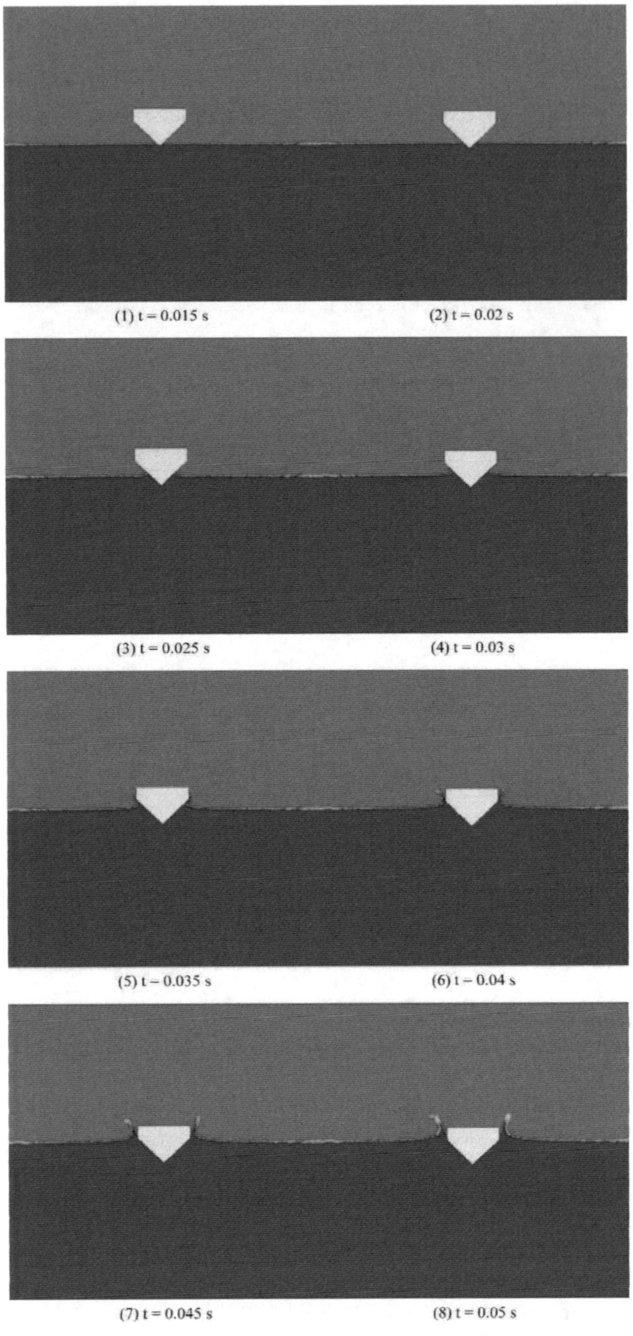

Figure 17. Change of flow field during water entry.

3.5. Water entry slamming of 30° truncated cone-shaped floating body

Time history curve of speed change during water entry. Figure 18 shows the time history curve of the velocity change of a truncated cone-shaped floating body with a bottom lift angle of 30° during the slamming process. The velocity of the floating body continues to increase at a constant acceleration during the initial stage of entry into the water, and reaches the maximum value at about 0.016 s, and then rapidly decreases at about After 0.02s, the decreasing amplitude began to decrease, and the speed change gradually became flat.

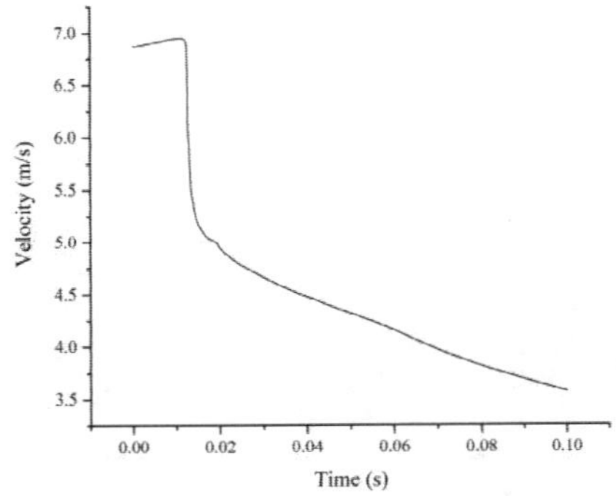

Figure 18. Time history curve of 30° truncated cone-shaped floating body entering water speed.

Monitoring point slamming pressure time history curve. In order to study the temporal and spatial distribution of slamming load, five pressure monitoring points are set on the surface of the geometric model, and the distribution positions are shown in Figure 19.

Figure 20 shows the time-history curve of the slamming pressure change at the monitoring points P1-P5 during the entry of the floating body into the water. P1-P3 monitoring points appeared almost simultaneously at 0.016 s when the slamming pressure peak appeared, the peak pressure gradually decreased but the difference was small, and the peak duration was extremely short; The peak pressure at P1 is the largest, about 3700000 Pa; the slamming pressure peak appears at P4 and P5 later, and the peak pressure is smaller; there is obvious oscillation after the pressure peak at P5.

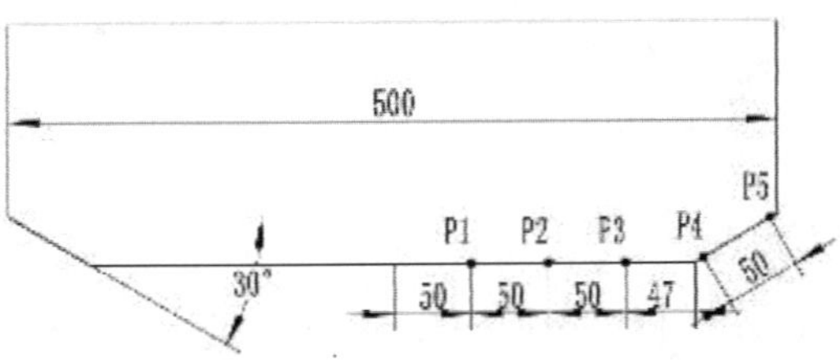

Figure 19. Distribution of pressure detection points.

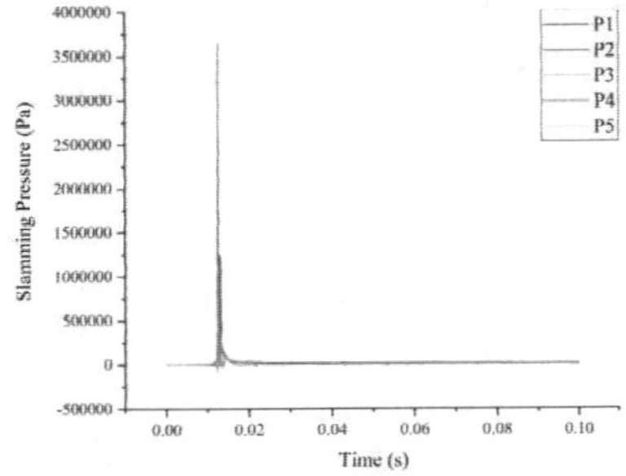

Figure 20. Time history curve of slamming pressure of floating body entering water.

Changes of jet and flow field shape during slamming. Figure 21 (1)-(8) are the flow field changes at 8 different times during the floating body entering the water, Figure 21 (1) is the initial moment of the floating body entering the water, the pressure peak appears, and the jet begins to form; In Figure 21 (2)-(8), the pressure peak has passed, the acceleration of the floating body decreases, and the jet diverges outward.

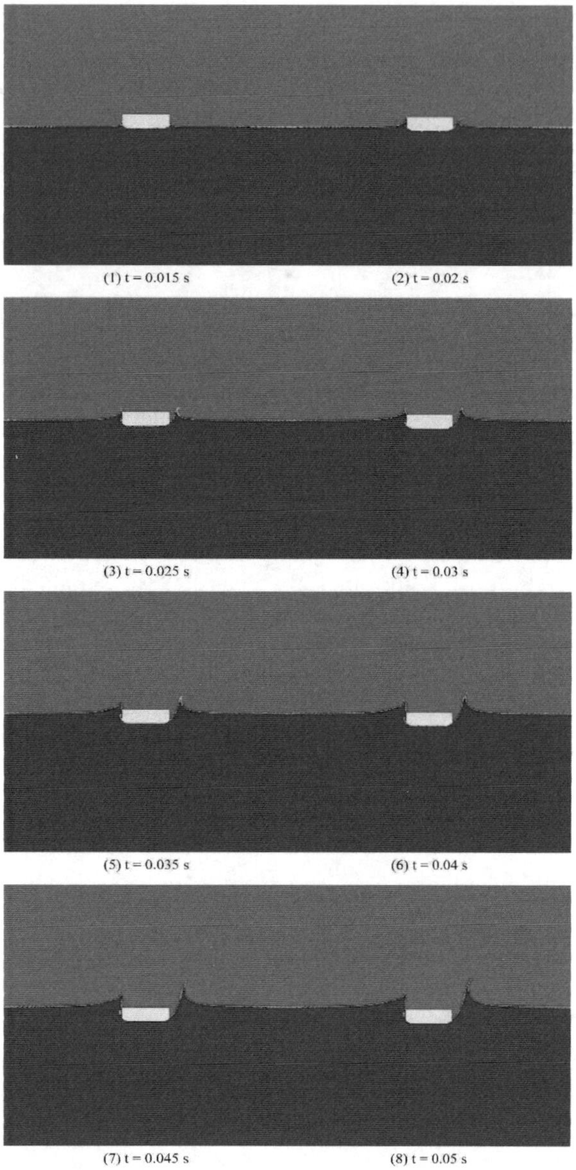

Figure 21. Change of flow field during water entry.

3.6. *Water entry slamming of 45° truncated cone-shaped floating body*

Time history curve of speed change during water entry. Figure 22 shows the time history curve of the velocity change of a truncated cone-shaped floating body with a bottom lift angle of 45° during the slamming process. The velocity of the floating body continues to increase at a constant acceleration during the initial stage of entry into the water, and reaches the maximum value at about 0.016 s, and then rapidly decreases at about after 0.02s, the decreasing amplitude began to decrease, and the speed change gradually became flat.

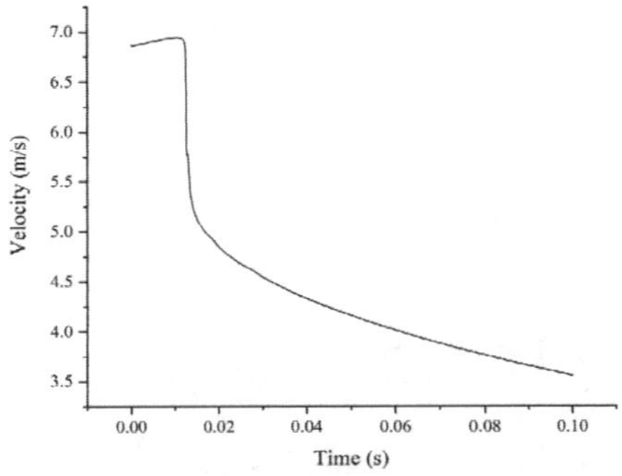

Figure 22. Time history curve of 45°truncated cone-shaped floating body entering water speed.

Monitoring point slamming pressure time history curve. In order to study the temporal and spatial distribution of slamming load, five pressure monitoring points are set on the surface of the geometric model, and the distribution positions are shown in Figure 23.

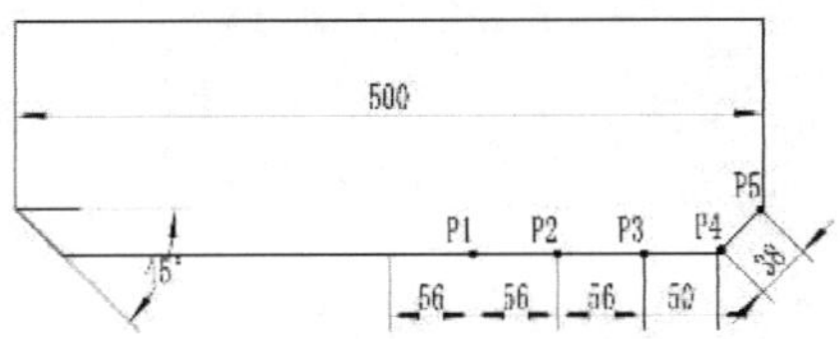

Figure 23. Distribution of pressure detection points.

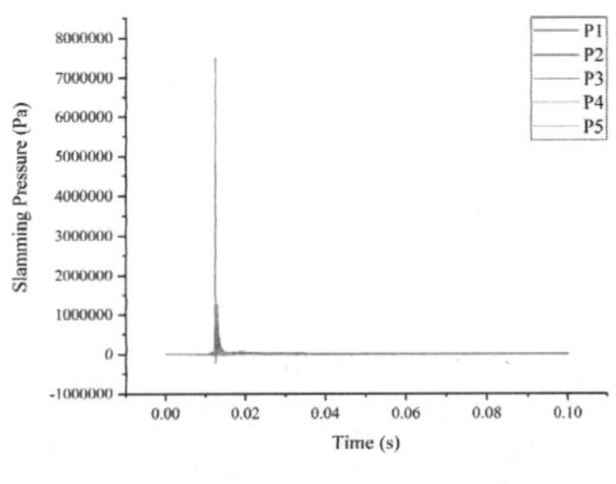

Figure 24. Time history curve of slamming pressure of floating body entering water.

Figure 24 shows the time-history curve of the slamming pressure change at the monitoring points P1-P5 during the entry of the floating body into the water. P1-P3 monitoring points appeared almost simultaneously at 0.016 s when the slamming pressure peak appeared, the peak pressures are almost the same, about 750000 Pa, and the peak duration is extremely short; There is a very short secondary peak after the peak. The secondary peak at P2 is the largest, about 1250000 Pa; the P4 and P5 monitoring points have negative pressure peaks at the same time, but they are relatively small.

Changes of jet and flow field shape during slamming. Figure 25 (1)-(8) are the flow field changes at 8 different times during the floating body entering the water, Figure 25 (1) is the initial moment of the floating body entering the water, the pressure peak appears, and the jet begins to form; In Figure 25 (2)-(8), the pressure peak has passed, the acceleration of the floating body decreases, and the jet diverges outward.

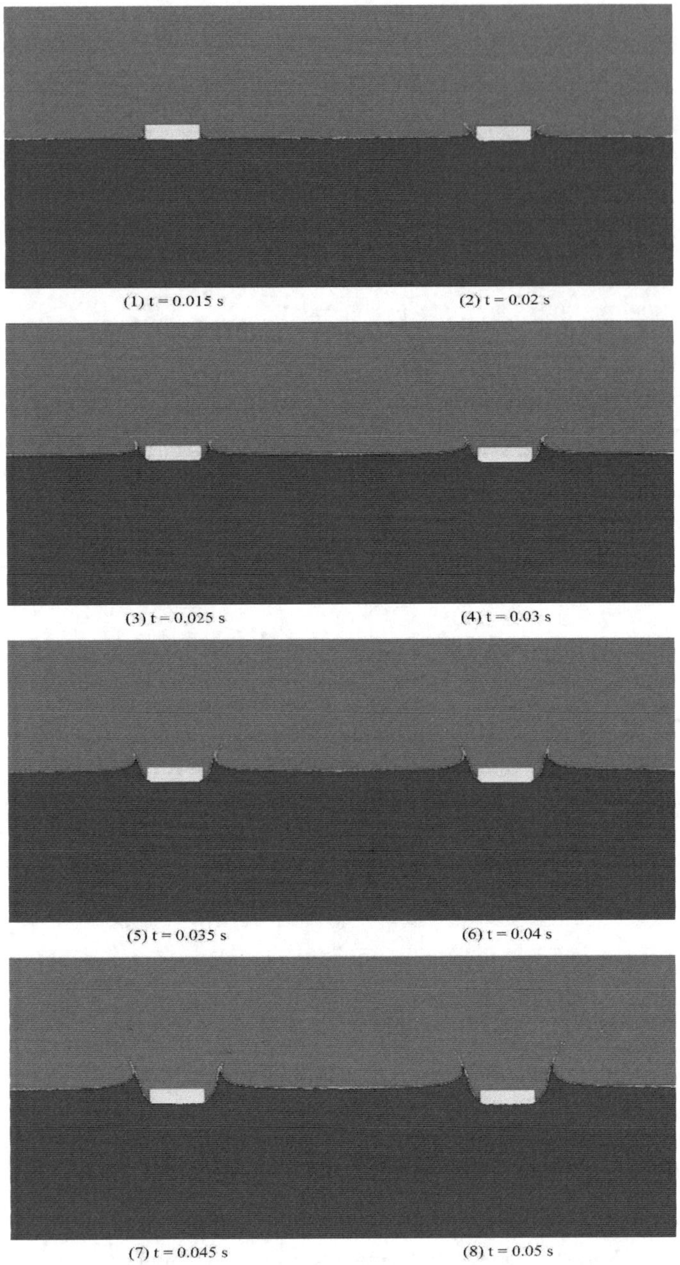

Figure 25. Change of flow field during water entry.

3.7. Water entry slamming of Hemispherical floating body

Time history curve of speed change during water entry. Figure 26 shows the time history curve of the velocity change of a hemispherical floating body during the slamming process. The velocity of the floating body continues to increase at a constant acceleration during the initial stage of entry into the water, and reaches the maximum value at about 0.017 s, and then rapidly decreases at about After 0.05s, the decreasing amplitude began to decrease, and the speed change gradually became flat.

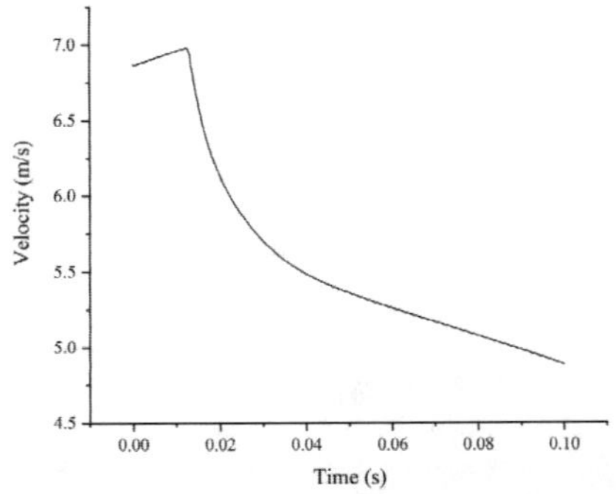

Figure 26. Time history curve of hemispherical floating body entering water speed.

Monitoring point slamming pressure time history curve. In order to study the temporal and spatial distribution of slamming load, five pressure monitoring points are set on the surface of the geometric model, and the distribution positions are shown in Figure 27.

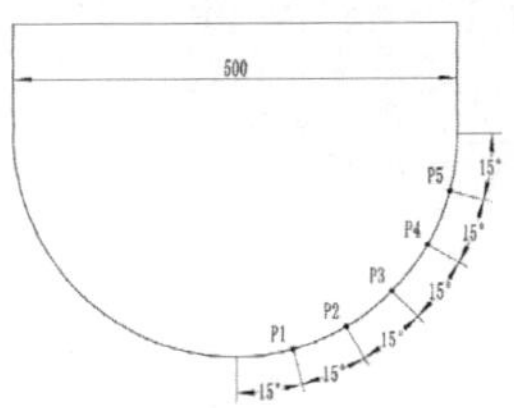

Figure 27. Distribution of pressure detection points.

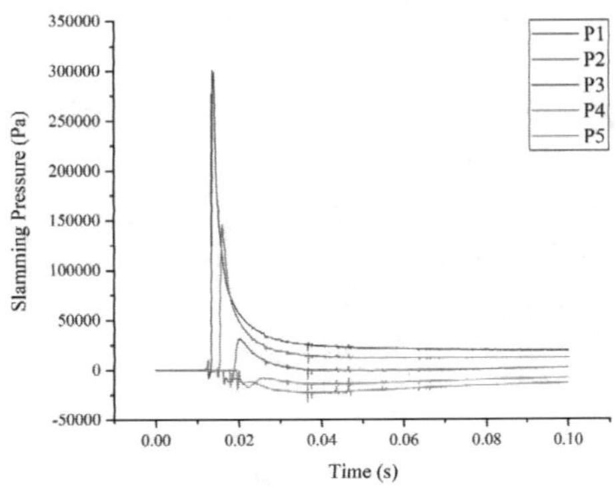

Figure 28. Time history curve of slamming pressure of floating body entering water.

Figure 28 shows the time-history curve of the slamming pressure change at the monitoring points P1-P5 during the entry of the floating body into the water. Starting from 0.017 s, the slamming pressure peaks appear successively at P1-P3, the peak pressure gradually decreases, and the peak duration gradually becomes longer; The peak pressure at P1 is the largest, about 300000 Pa; Negative pressure appears at P3 before the pressure peak; Negative pressure appears at P4 and P5 later, and it stabilizes after a short period of oscillation.

Changes of jet and flow field shape during slamming. Figure 29 (1)-(8) are the flow field changes at 8 different times during the floating body entering the water, Figure 29 (1) is the initial moment of the floating body entering the water, the pressure peak appears, and the jet begins to form; Figure 29 (2)-(3) The pressure continues to rise and the jet flows upward; the acceleration of the floating body in Figure 29 (4)-(8) decreases, the pressure begins to drop, and the jet diverges outward.

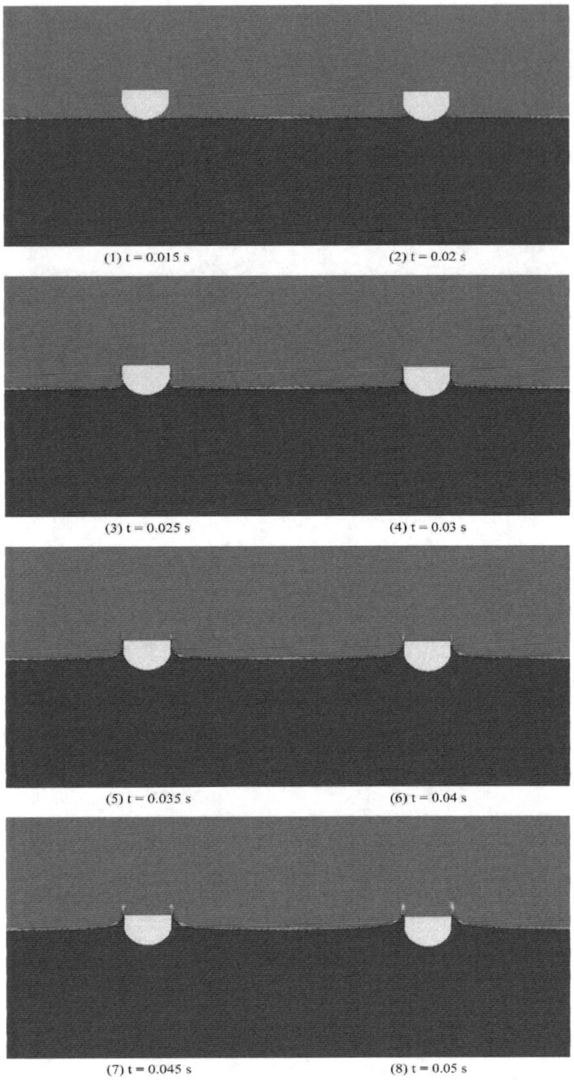

Figure 29. Change of flow field during water entry.

4. Comparison of water slamming characteristics of different shaped floating bodies

Comparison of water entry speed of different shapes of floating bodies. Figure 30 shows the time history curve of the speed change of different shapes of floating bodies entering the water. The speed of the 45° conical floating body decreases more slowly than the 30° conical floating body, that is, the acceleration is smaller, indicating that the slamming load is smaller; There is not much difference in the speed change of the truncated cone-shaped floating body with different bottom elevation angles. Both of them are reduced to about 5 m/s at a great acceleration, and then they become gentle. It shows that when the truncated cone-shaped floating body enters the water, the bottom surface bears the main slamming load, and the smaller conical transition surface cannot reduce the slamming load of the floating body when it enters the water; The slamming load of the hemispherical floating body at the initial stage of entering the water is between the conical shape and the truncated cone shape, and the speed after it stabilizes is the largest, indicating that the kinetic energy loss is the smallest.

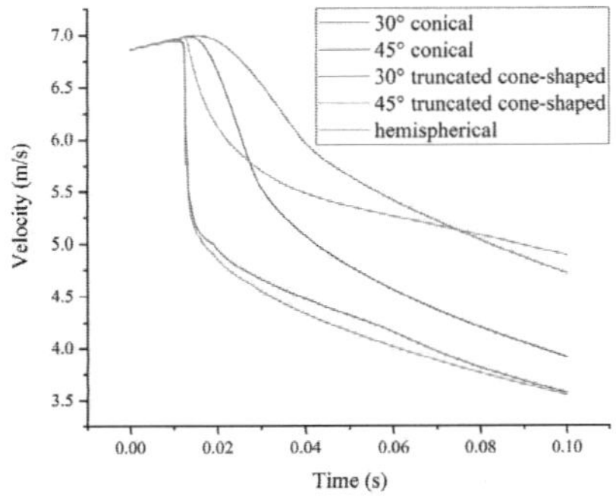

Figure 30. Time history curve of different shapes floating body entering water speed.

Comparison of peak slamming pressure of different shapes of floating bodies. In practical engineering applications, due to the need to coordinate the floating body with the guide column or other parts, the ideal conical or hemispherical floating body is difficult to apply, and the truncated cone-shaped floating body is the most widely used. However, according to Table 1, it can be seen that the peak slamming pressure of the truncated cone-shaped float is much greater than that of the conical and hemispherical shapes, and the bottom lift angle of 30° and the bottom lift angle of 45° are not much different. It shows that when the conical transition surface of the truncated cone-shaped floating body is small, the influence of the bottom rise angle is not large. For the truncated cone-shaped floating body, in order to reduce the bottom slamming pressure, the area of the truncated bottom surface should be reduced as much as possible, and the area of the cone transition surface should be increased. The peak slamming pressure of the 45° conical floating body is less than half of that of the 30° conical floating body, indicating that the bottom rise angle has a greater slamming on the slamming pressure of the conical floating body. The smaller the bottom rise angle, the greater the slamming pressure. The slamming pressure of the hemispherical floating body is stronger than that of the conical shape.

Table 1. Formatting sections, subsections and subsubsections.

Floating body shape	Peak pressure （Pa）
30°conical	143251
45°conical	59141
30°truncated cone-shaped	6348607
45°truncated cone-shaped	7504067
hemispherical	300515

5. Conclusion

The CFD method was used to simulate the slamming process of different shapes of floating bodies, and the velocity, pressure and flow field changes during the process of entering the water were obtained. The truncated cone-shaped floating body is subject to greater slamming pressure, and the bottom rise angle has little effect on the slamming pressure and speed changes; The slamming pressure on the conical floating body is smaller, and the larger the bottom lift angle, the smaller the slamming pressure; The slamming pressure of the hemispherical floating body is stronger than that of the conical shape, but its speed after entering the water is greater than that of the conical shape, and the kinetic energy loss is minimal.

6. References

[1] Wu G X, Sun H, He Y S. Numerical simulation and experimental study of water entry of a wedge in flee fall motion[J]. Journal of Fluids & Structures, 2004, 19(3):277-289.
[2] Xu G D, Duan W Y, Wu G X. Numerical Simulation of Water Entry of A Cone in Free-Fall Motion[J]. Quarterly Journal of Mechanics & Applied Mathematics, 2011, 64(3):265-285(21).
[3] Sun S L, Wu G X. Oblique water entry of a cone by a fully three-dimensional nonlinear method[J]. Journal of Fluids & Structures, 2013a, 42(4):313-332.
[4] Sun S L, Wu G X. Oblique water-entry of non-axisynunetric bodies at varying speed by a fully nonlinear method[J]. Quarterly Journal of Mechanics & Applied Mathematics, 2013b, 66(3):365-393.
[5] Sun, S.Y., Sun, S.L., Wu, G.X. Oblique water entry of a wedge into waves with gravity effect[J]. Fluid Struct. 2015, 52:49-64.
[6] Scolan Y M, Korobkin A A. Three-dimensional theory of water impact, part 1: Inverse Wanger problem[J]. Journal of Mechanics, 2001, 440:293-326.
[7] A. Tassin a,b, N. Jacques a,n, A. E1 Malki Alaoui a, A. Neme a, B. Leble b Hydrodynamic loads during water impact of three-dimensional solids: Modelling and experiments[J]. Journal of Fluids and Structures, 2012, 28:211-23l.
[8] Wu G X, Sun S L. Similarity solution for oblique water entry of all expanding paraboloid[J]. Journal of Fluid Mechanics, 2014, 745:398-408.
[9] Yves-Marie, S. Oblique water entry of a three dimensional body. Int. J. Nav. Archit. Ocean Eng. 2014, 6(4):1197-1208.
[10] Hu C, Kashiwagi M. A CIP-based method for numerical simulations of violent free-surface flows[J]. Journal of Marine Science & Technology, 2004, 9(4):143-157.
[11] Lin P. A fixed-grid model for simulation of a moving body in free surface flows[J]. Computers & Fluids, 2007, 36(3):549-561.
[12] Wang W, Wang Y. An improved free surface capturing method based on Cartesian cut cell mesh for water-entry and -exit problems[J]. Proceedings Mathematical Physical & Engineering Sciences, 2009, 465:1843-1868.
[13] Liu M B, Shao J R, Li H Q. An SPH model for free surface flows with moving rigid objects[J]. Internation Journal for Numercial Methods in Fluid, 2014, 74(9):684-697.
[14] Facci, A. L., Panciroli, R., Ubertim, S., Porfiri, M. Assessment of PIV-based analysis of water entry problems through synthetic numerical datasets. Journal of Fluids and Structures, 2015, 55,:484-500.
[15] Facci, A. L., Porfiri, M., Ubertini, S. Three-dimensional water entry of a solid body: a computational study[J]. Fluid Struct. 2016, 66,:36-53.

Acknowledgments

This project is supported by the 2018 Shandong Major Science and Technology Innovation Project c(No. 2018CXGC0104) "Design and Key Technologies of Supporting Platform for Deep Sea Cage Culture", the 2017 NSFC-Shandong Joint Found Project(U1706230) "Research on Mechanism and Key Technology of Energy Efficiency Enhancement in Floating Hydraulic Wave Power Generation System" and 2017 Marine Renewable Energy Funds Project (No.GHME2017YY01) "Demonstration of Marine Power Supply System for Marine Instruments and Equipment – Wave Energy and Tidal Current Energy Complementary Instruments and Equipment Power Supply System Based on Mooring Platform". At the same time, thanks to the corresponding author Professor Yanjun Liu (Email: lyj111ky@163.com) for his guidance on this paper.

Numerical Simulation Study and Analysis of Water Inrush from Floor Concealed Subsidence Column during Mining

Fu Wei [1,2]

[1] China Coal Technology and Engineering Group Shenyang Research Institute, Fushun 113122, China;
[2] State Key Laboratory of Coal Mine Safety Technology, Fushun 113122, China
E-mail: 2656922560@qq.com

Abstract: In order to clarify the role of mining on the surrounding rock failure and water inrush of the karst collapse column in the working face floor, the concealed karst collapse column in the floor of Geting Coal Mine is taken as the research object, and the FLAC3D numerical simulation software is used to analyze the mining process. The stress field in the vertical direction, the plastic failure zone and the displacement change in the vertical direction are analyzed and studied, and the water inrush process of the hidden floor karst collapse column under different advancing distances is analyzed. The research results show that with the continuous increase of the advancing distance, the plastic failure range of the karst collapse column continues to increase, and the vertical displacement of the floor continues to increase. When the plastic failure zone of the coal floor and the plastic failure zone of the karst collapse column are connected, Water inrush accidents from the karst collapse column of the floor are prone to occur.

1. Introduction

Mine water inrush has become one of the major disasters affecting coal mine safety production. Among them, the water inrush caused by geological structures (faults, collapsed pillars) accounts for 80% [1], among which collapsed pillars are responsible for coal mine water inrush accidents. One of the main geological structures that occurred, due to the existence of the collapse column, not only changed the original coal seam structure, but also reduced the coal recovery rate, and seriously affected the safety production of coal mines. Among the collapsed columns that have been exposed in coal mines in my country, although most of the collapsed columns are not water-conducting or water-free, many mining areas force through the collapsed columns during the actual working face advancement process, and generally, water inrush accidents from the collapsed columns rarely occur.
However, since the 1960s, my country's Fengfeng Jiulong, Luotuoshan coal mines, and Taoyuan coal mines have experienced many karst collapse column water inrush accidents, which have gradually attracted people's attention. For the water inrush problem of the collapse column, domestic and foreign experts Scholars have conducted research from different research perspectives and have achieved certain research results [2-4]. Comprehensive analysis of case data of water inrush accidents from karst collapse column shows that water inrush from karst collapse column has the characteristics of concealment, large amount of water inrush, suddenness, causing serious losses, and expensive treatment. Therefore, water inrush from concealed collapse column on floor Research and analysis of the main control factors are of great significance for preventing coal mine water inrush and ensuring coal mine safety production.

2. Theoretical model of stress and seepage coupling

2.1. Analysis of the influence of stress field on seepage field

Rocks and coal seams are porous media, and their mechanical characteristics are mainly composed of the micro-permeable voids in the rock body and the joints and cracks between the rock bodies. Experimental studies have shown that the change of pore water pressure firstly induces effective stress change, and then it is possible to significantly cause the distribution of fluid pressure in the fracture. The regular formula that the permeability coefficient of the rock mass changes with the stress of the rock mass [5-6]:

$$\Delta K = \frac{\rho g b^3}{12S\mu}\left[1+\Delta\varepsilon\left(\frac{K_n b}{E}+\frac{b}{S}\right)^{-1}\right]^3 \tag{1}$$

It can be seen from the above formula that the stress field affects the permeability coefficient of the rock mass through the volumetric strain in the rock mass, thereby affecting the seepage field in the rock mass.

2.2. Analysis of the influence of seepage field on stress field

In the seepage medium, the water load is expressed by the seepage stress and the seepage volume force. The distribution of the seepage field corresponds to the distribution of the water load one by one. The seepage field changes the stress field distribution by changing the volume strain of the rock mass in the floor. Assuming that the water head distribution in the floor rock body is H(x, y, z), the distribution of seepage pressure P on a certain acting surface is calculated as:

$$P = \gamma(H\text{-}Z) \tag{2}$$

From this, the distribution of the seepage volume force f is calculated as:

$$\begin{Bmatrix} f_x \\ f_y \\ f_z \end{Bmatrix} = -\begin{Bmatrix} \frac{\partial p}{\partial X} \\ \frac{\partial P}{\partial Y} \\ \frac{\partial P}{\partial Z} \end{Bmatrix} = -\begin{Bmatrix} \gamma\frac{\partial H}{\partial X} \\ \gamma\frac{\partial H}{\partial Y} \\ \gamma\left(\frac{\partial H}{\partial Z}-1\right) \end{Bmatrix} \tag{3}$$

2.3. Plastic failure criterion of surrounding rock of coal roof and floor

When the rock mass undergoes shear failure ($f_s < 0$), after the rock mass reaches the yield limit, if there is a relatively large pressure, the rock mass will undergo failure and deformation. The tensile stress yield conditions are:

$$f_s = \sigma_1 - \sigma_3\frac{1+\sin\varphi}{1-\sin\varphi} + 2C\sqrt{\frac{1+\sin\varphi}{1-\sin\varphi}} \tag{4}$$

$$f_t = \sigma_t\text{-}\sigma_s \tag{5}$$

At that time of $f_s < 0$, the coal seam rock mass suffered shear failure; at that time of $f_t > 0$, the coal seam rock mass suffered tensile stress failure.

3. Constructing a numerical model of concealed karst collapse column in the floor

3.1. Mining area engineering hydrogeological conditions

The stratigraphic groups developed in the mining area are: Middle-Lower Ordovician, Benxi Formation, Taiyuan Formation, Lower Permian Shanxi Formation and Quaternary Clay Layers. The mining coal seams are located in the 16th and 17th coal seams of the Taiyuan Formation. The coal seams are located 30~80m above the Ordovician ash aquifer. The 16th and 17th coals are threatened by the Ordovician ash aquifer to varying degrees. The top of the top plate is referred to as mudstone, siltstone, fine sandstone and siltstone, and the bottom plate is referred to as siltstone, mudstone, and limestone.

According to the three-dimensional seismic exploration of Geting Coal Mine, a total of 3 collapsed columns (SX1, SX2, SX3) were found. Except for the larger collapsed column of SX1, the other two are smaller. Located at the edge of the southwest corner of the survey area, it has fallen to the 13th and Ordovician ash, 30m under the 17th coal floor and 60m away from the 16th coal. It is elliptical in the area, with the long axis in the NW direction, with a length of 80m, and the short axis in the NE direction, with a length of 60m and an area of 2436m2. There are altogether 11 fault points, of which 10 are A-level and 1 are B-level. Figure 1 shows the SX1 collapse column.

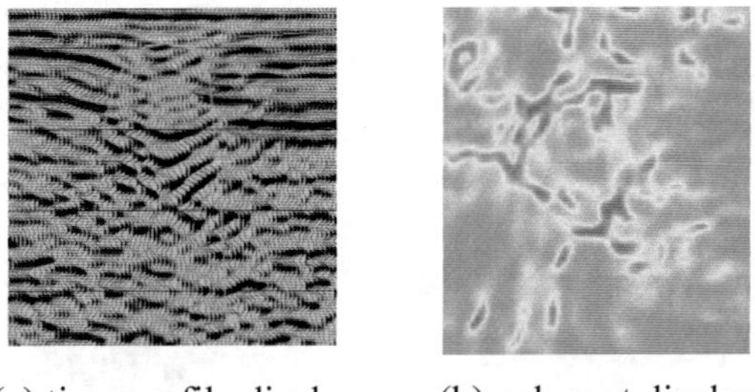

(a) time profile display (b) coherent display

Figure 1. Collapse column of SX1

3.2. Boundary conditions and scheme design of the numerical model

The model is as follows: set the model's front, rear, left, and right to be horizontal constraints, the bottom is a fixed boundary and the vertical orientation is zero; the direction of the working face is along the X axis; the top pressure of the model is replaced by an equivalent load of 10MPa ; Using a fixed boundary water pressure, the ideal water pressure of the aquifer is set to 2.4MPa, and the rock saturation is 1, which changes trapezoidally with the initial water pressure of the karst collapse column. The water pressure of the working face boundary after mining is taken as zero, The size of the model is $300\text{m} \times 150m \times 250m$ that the height of the karst collapse column is 40m and the diameter is 30m. The initial numerical calculation model is shown in Table 1.

Table 1. Mechanical parameters of rock strata.

Parameter / Lithology	Elastic GPa	Poisson's ratio of	Tensile MPa	ρ kg/m-3	porosity η	Permeability coefficient K/10-3ms-1
Fine sandstone	3	0.30	1.6	1600	0.09	0.250
siltstone	32	0.35	3.6	2700	0.17	0.024
marl	15	0.26	2.2	2500	0.16	0.043
Coal seam	2	0.23	1.6	1500	0.24	0.101
mudstone	3	0.30	1.6	1600	0.18	0.028
Fine sandstone	18	0.25	2.1	2400	0.12	0.033
limestone	17	0.34	2.0	2000	0.13	0.050
aquifer	20	0.11	5.1	3000	0.26	0.370
column	2.7	2.7	0.3	1700	0.6	0.440

4. Numerical simulation model results and analysis

4.1. Stress analysis in the vertical direction of surrounding rock

1) From the analysis of Figure 2, it can be seen that the mining on the working face redistributes the stress, forming a stress concentration in front of the working face, and forming a pressure relief zone

and a pressurizing zone on the roof and floor of the collapse column. When the working face advances 20m and the distance from the collapse column is about 60m, the advance distance of the working face has little effect on the stress distribution of the karst collapse column; when the advance distance reaches 80m, the stress area overlaps when it approaches the karst collapse column; At 100m, when pushing directly above the collapse column, a pressure relief zone appears in the collapse column. As a low-stress area, the collapse column is easily affected by mining and is prone to plastic failure. Then the floor collapse column serves as a communication between the working surface and the working surface. The preferred channel of the aquifer, therefore, should be paid attention to in the subsequent coal mining process.

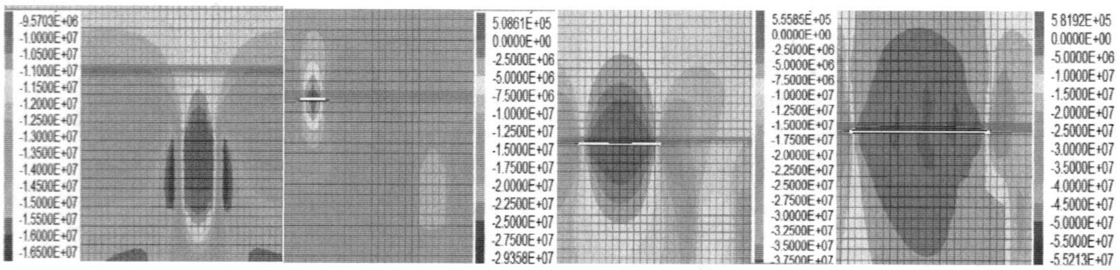

(A) Before mining (b) Advance 40m (c)Advance 80m (d) Advance 100m

Figure 2. Vertical stress diagram of surrounding rock during advancing

2) It can be seen from Figure 3 that the stress distribution of the surrounding rock at the working face before mining is clear. Stress concentration occurs around the collapsed column, and the concentrated stress reaches 16.4MPa, while the internal stress of the collapsed column is relatively small, and the vertical stress of the surrounding rock is only 9.5. MPa. Under the initial stress state of surrounding rock, the vertical stress in the collapsed column is less than the surrounding rock stress in the complete rock formation, indicating that the collapsed column is a natural low stress concentration area. Under the influence of mining, the stress concentration factor continues to increase. When the strength of the rock strata is exceeded, the surrounding rock around the collapsed column of the floor is easily damaged, and the confined water in the aquifer is extremely prone to seepage.

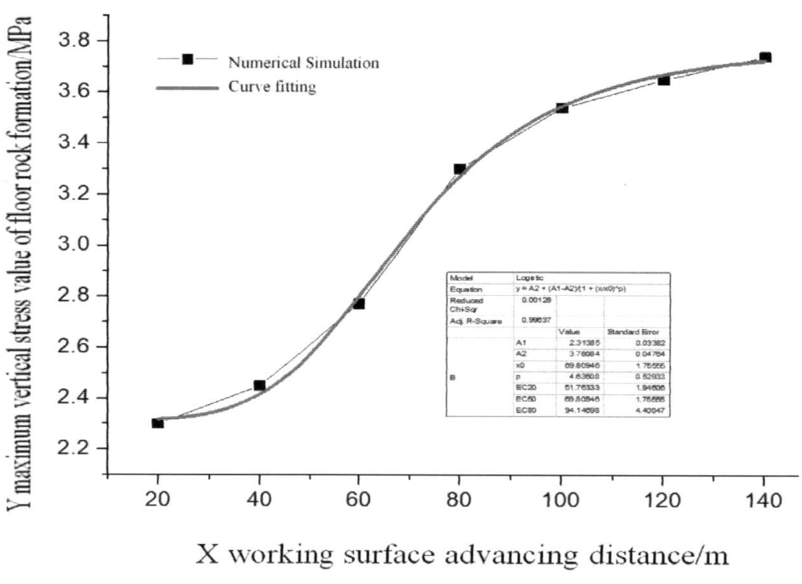

Figure 3. Vertical stress change diagram of surrounding rock during the advancing process

3) With the continuous increase of the advancing distance of the working face, when the mining distance of the working face is getting closer and closer to the collapse column, the stress

concentration factor increases continuously, and the stress concentration range is distributed on the side of the mining face close to the collapse column. When the working face is advanced for 80m, the stress failure zone of the surrounding rock of the working face and the stress zone of the surrounding rock of the collapsed column are superimposed; when the working face is advanced for 100m, the stress of the surrounding rock of the floor and the collapsed column is already Completely coincide, the coal seam floor and the vertical stress failure zone of the surrounding rock of the collapse column coincide, and the possibility of plastic failure of the floor and the collapse column rock formation increases.

4.2. Analysis of plastic failure zone

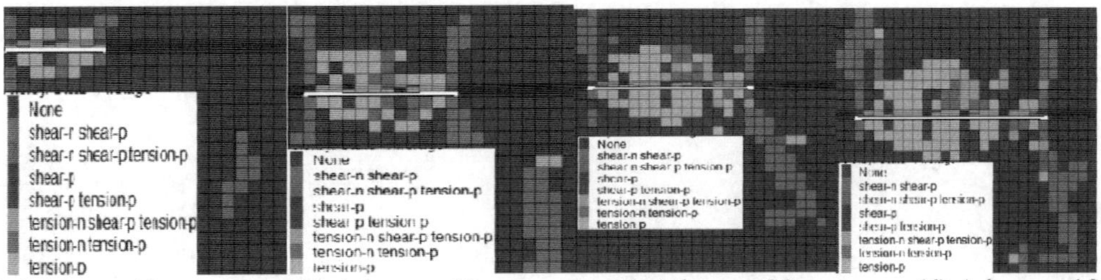

(a) Advance 40m (b) Advance 60m (c) Advance 80m (d) Advance 100m

Figure 4. Diagram of the plastic failure zone during the advancing process

1) When the mining face of the working face does not pass the collapse column, as the advancing distance of the working face continues to increase, the plastic failure depth of the coal seam floor continues to increase, from Figure 4 that the plastic failure depth of the floor reaches the maximum when it reaches the collapse column; when the working face passes the floor After the collapse column, as the working face is far away from the collapse column, the plastic failure depth of the floor gradually decreases. When the advancement distance of the working face reaches 80m, the plastic failure zone of the coal seam floor and the plastic failure zone of the collapse column will butt, and the floor is likely to occur. A water inrush accident occurred in the collapse column.
2) Before mining, a plastic failure zone of about 5m appeared at the junction of the collapse column and the Ordovician aquifer, and with the continuous increase of the distance of the working face, the plastic failure zone of the floor collapse column continued to increase, when the working face was advanced by 60m When left and right, the karst collapse column column communicates with the Ordovician aquifer. And with the continuous increase of the advancing distance of the working face, the water-resistant layer on the upper part of the collapse column is easily damaged under the action of mining and hydraulic fracturing, causing the water in the Ordovician aquifer to gush out.

4.3. Analysis of surrounding rock displacement field change

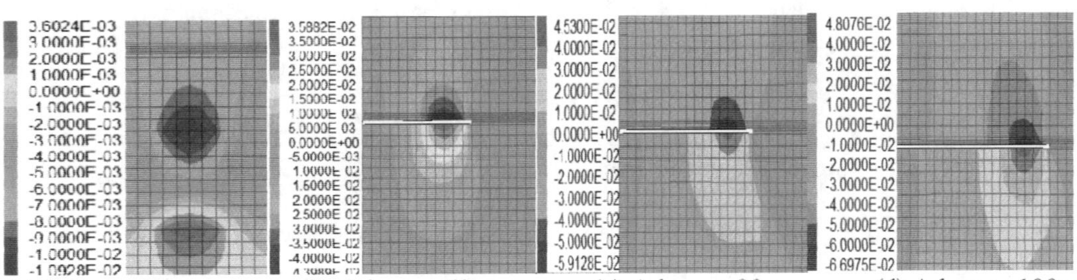

(a)Before mining (b) Advance 60m (c) Advance 80m (d) Advance 100m

Figure 5. Vertical displacement diagram during propulsion

1) From Figure 5 that before the initial excavation , compared with the surrounding rock, the internal displacement of the collapsed column is relatively large, mainly in the vertical direction, and the

vertical displacement of the surrounding rock at the top of the collapsed column is smaller than the bottom of the collapsed column; When the excavation starts, the vertical displacement of the surrounding rock on the top of the working face is greater than the vertical displacement of the surrounding rock on the floor.

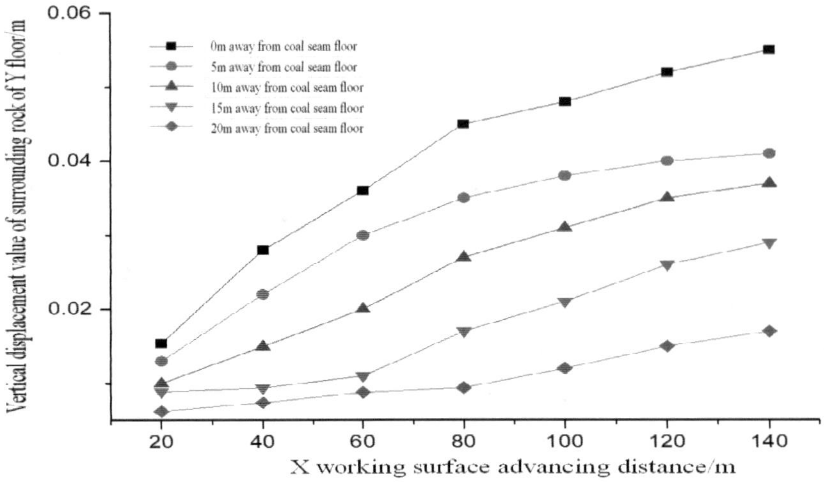

Figure 6. Change of vertical displacement of surrounding rock of floor during advancing process

2) With the continuous increase of the advancing distance of the working face, from Figure 6 that distance to the collapse column is getting closer and closer. When the mining of the working face advances to 80m, the vertical displacement field of the surrounding rock of the floor obviously shifts to the direction of the collapse column. The stress field in the vertical direction of the surrounding rock of the collapsed column is superimposed, and the plastic failure zone is connected.
3) With the continuous increase of the advancing distance of the working face, the vertical displacement of the top and bottom of the working face gradually increases, and the scope of influence continues to expand. Through comparison, it can be seen that the vertical displacement of the roof and floor is mainly concentrated on the newly excavated working face, while the vertical displacement of the surrounding rock of the roof and floor behind the goaf is relatively stable.
4) The vertical displacement value of the surrounding rock located below the goaf floor gradually decreases from top to bottom, and with the continuous increase of the advance distance of the working face, the vertical displacement value of the surrounding rock of the goaf floor gradually decreases. Increase. As for the surrounding rock above the mined-out area, its vertical displacement value gradually decreases from bottom to top .
5) The top rock layer of the goaf is dominated by marl, and its subsidence value gradually decreases from bottom to top, and the bottom rock layer is dominated by fine sandstone and mudstone, and the subsidence value gradually decreases from top to bottom. In addition to its location, the displacement of the rock layer is also related to the lithology and combination of the rock mass. Sandstone with stronger lithology deforms less than marl, and it is also reflected in the front plastic failure zone.

5. Conclusion

By numerically simulating the water inrush from the concealed collapse column of the floor, under the condition that the thickness of the water barrier is 30m, the mining of 17 coal is in danger of water inrush.
1) The surrounding rock on the floor of the mining area and the collapse column near the mining side is prone to shear and tensile failure.
2) Affected by mining, as the advancing distance continues to increase, the stress concentration and range of the surrounding rock gradually increase, the plastic failure range gradually increases, and the vertical displacement range gradually increases. The floor and collapse column should be adjusted before mining. Carry out grouting reinforcement transformation.

6. References

[1] Zhang Tianjun, Zhang Xiufeng, Pang Mingkun, Liu Nan, Zhang Shuo, Gao Hangbiao. Influence of particle loss on pore structure and water inrush behavior of subsided column backfill [J/OL]. *Journal of coal:* 1-10 **[2021-04-08]**. https://doi.org/10.13225/j.cnki.jccs.2020.1110.

[2] ZHANG Shuai. Research on Water Inrush Mechanism and Prevention Technology through Collapse Column in Fully Mechanized Mining Face of Xinzhi Coal Mine [J]. *Coal and Chemical Industry*, **2020, 43(08)**:63-64+70.

[3] Ma Zhigang. Water inrush mechanism and risk evaluation of collapse column area in Shangshe Mine 15501 working face [J]. *Coal Mine Modernization,* **2020(05)**:79-81+84.

[4] Xie Zhigang, Liu Dianqi, Chai Huichan, Ju Qiding, Li Jingying, Rao Jiajian. Prediction of Water Inrush of Hidden Collapse Column in Coal Seam Floor and Evaluation of Grouting Reinforcement before Mining [J]. *Journal of Safety Science and Technology*, **2019,15(05)**:105-110.

[5] Li Guoxin, Liu Nina, Wu Xiaoming, Liu Weiliang, Wang Rumeng, Wang Shaoqiang, Shen Shibao. Application of optimal combination weighting model in water inrush evaluation of floor concealed collapse column [J]. *Safety in Coal Mines*, **2018, 49(11)**:194-198.

[6] Wang Jiawei, Wang Yuhe, Zhao Yazhou, Peng Xiangchao, Wang Mingsen. Simulation and analysis of water inrush from concealed collapse column in working face floor [J]. *Coal Technology*, **2017, 36(11)**:193-195.

Prediction Method of Public Participation in Decision-making Based on Bayes Theorem

Jin Shang[1], Qinyu Feng[2]

1 School of Architecture, University of Sheffield, The University of Sheffield Western Bank, Sheffield, S10 2TN, +44 114 222 2000
2 School of Architecture, University of Sheffield, The University of Sheffield Western Bank, Sheffield, S10 2TN, +44 114 222 2000
Corresponding author's e-mail: JShang5@sheffield.ac.uk

Abstract. Public participation is an activity that citizens participate in the decision-making without government and experts. It could create better informed and more creative decision-making. However, it is difficult to invite all stakeholders to participate in the urban design and planning project. So most of projects could just receive a few decisions from the public. This article would introduce a model based on Bayes theorem. This algorithm model can predict the possibility of other stakeholders' decision-making on a proposal based on the feedback results of a small part of the surveyed participants and big data analysis. This article mainly studies how the algorithm model can be displayed in the actual project through a case of local residents' decision-making and prediction of the building height of the planned area in a planning project.
Key words: Public participation, Decision-making prediction, Bayes theorem

1. Introduction

Public participation is a manifestation of a country's democracy [1]. There are many reasons to organize public participation in the decision-making project. First of all, public participation could create more innovative decisions. Different participants' ideas could open new ways for decision-making. Secondly, public participation could increase the acceptance of public for the decision-making. Because it fully consults the public before making decisions [2]. Thirdly, public participation could increase the openness and integration of government. It could make government policies and decisions more open and transparent [3,4]. Overall, public participation could increase the democracy and wisdom of government decisions.

However, it is not easy for government or urban planners to organize all stakeholders to participate in the decision-making. For example, a community renovation project needs to renovate the public space and infrastructures in this community. There are almost 2000 people in this community, include aging people, young adults and children. They are all stakeholders of this project but it is difficult and impossible to organize 2000 people to participate in every stage of decision-making in this project. So this literature would describe a method to help government and urban planners to predict other stakeholders' decision-making intentions based on Bayes theorem.

2. Related work of Bayes theorem on public participation promotion

The project helps Danish people participate in the supervision and prediction of the water quality of Havelse Water Plant by building a Bayesian network. Bayesian network (Bn) is a decision support system, which is based on probability theory [5–8]. Bayes network is used as a tool for water resources management and forecasting simulation in this case. The main source of the Danish people's domestic water is groundwater. The Havel Water Plant is a large underground water plant in Denmark, with a catchment area covering an area of approximately 26 square kilometers. After local people investigate, most of the water pollution is mainly from pesticides. The goal of the project is to help local residents participate in the control of groundwater quality through the Bayesian network, so that the groundwater is free of pesticides [8].

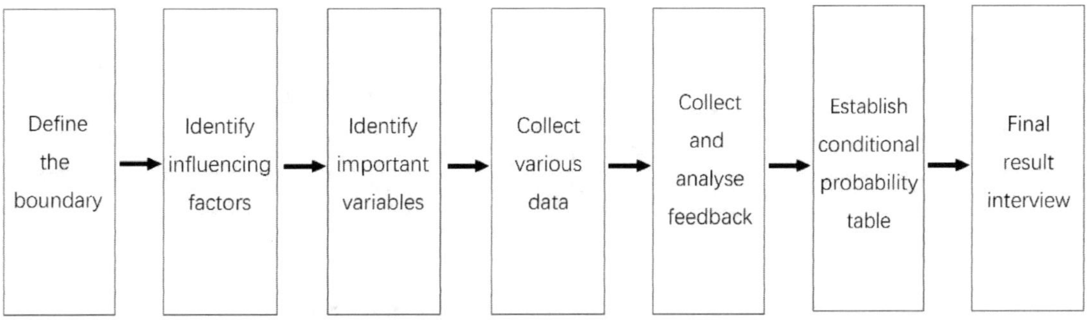

Figure1. 7 steps of building Bayes network

According to figure 1, the project uses 7 steps to build the Bayes network. First of all, the project defines the project background and boundary. The project defines the indicators by defining physical and socio-economic boundaries, areas of interest, alternative actions, etc. Secondly, the project has identified influencing factors, actions to be taken and related indicators. The project also formulated some indicators of public concern, such as groundwater quality. Thirdly, the project identifies important variables and analyzes their possible impact on the results. Fourthly, the project collects various data and conducts data analysis. Fifthly, the project collects and analyzes the suggestions of stakeholders. Sixth, the project establishes a conditional probability table by analyzing and calculating the collected data and determined variables. At the same time, review the list of stakeholders. At last, the project collects feedback from stakeholders on the final results and draws conclusions based on the Bayesian network [8].
So this project builds a Bayesian network and collects and calculates relevant data and public suggestions. In this way, the conditional probability of various events is predicted, so that the public can directly participate in the management and monitoring of water quality. It can be seen that Bayes' law can predict the probability of related events through the control of event variables and data analysis.

3. Methodology

This project would introduce a prediction model that could predict the probability of the public decision-making. The core algorithms of this model is Bayes Theorem. The Bayes' Theorem is a simple mathematical formula, which is mainly for calculating conditional probabilities [9]. It plays an importation role in probabilistic machine learning. According to formula 1, it is the formula of Bayes Theorem. 'P(H)' is the prior probability. It represents the probability data from sample. P(H|D) is the probability of posterior. It represents probability data from other people in the groups of sample. 'P(D|H)' is likelihood. It represents The probability of the event occurring. 'P(D)' is the evidence.

$$P(H|D) = \frac{P(D|H)P(H)}{P(D)} \quad (1)$$

Formula 1 is the formula of Bayes Theorem (Adapted from [10]).
According to figure 2, this system could calculate the probability of all people's choice for some specific urban renovation project based on limited example investigation. At first, system would collect the data for one sample. In this sample, it includes all kinds of stakeholders of the project but the number of examples is limited. After that the system would calculate the prior probability based on big data analysis. And then it would calculate the likelihood based on the sample investigation and get the probability trend at last.

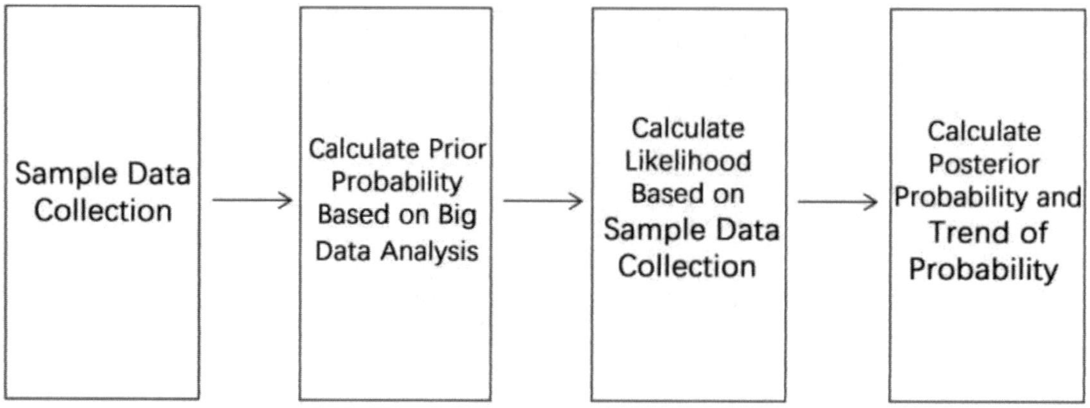

Figure 2. The model of data collection system.

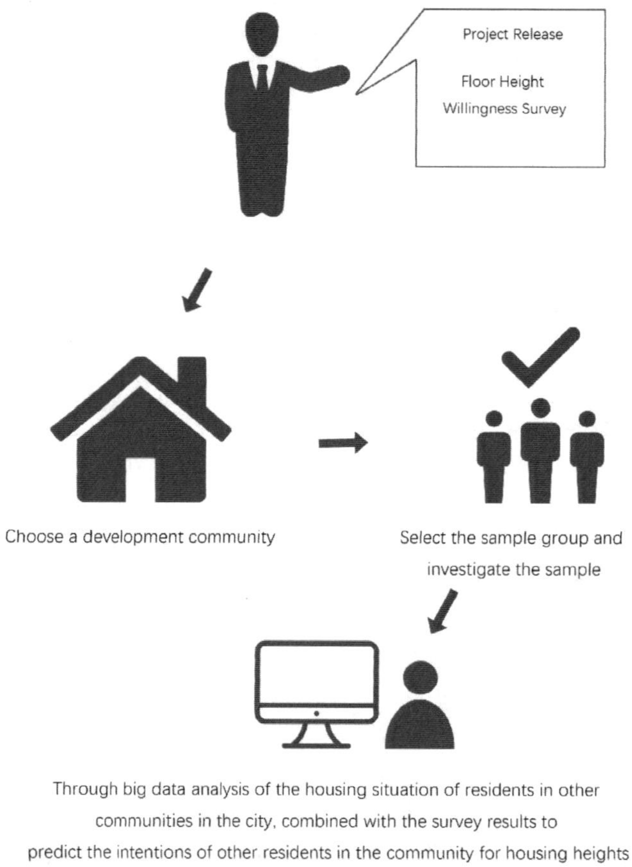

Figure 3. Model of floor height willingness prediction survey

According to formula 2 and figure 3, for the floor height willingness survey, the system collects the stakeholders' data about the desire for community building height renovation, from an example's groups, include young people, children and aging people at first. The number of examples is 'N'. This system collect data mainly from online and offline questionnaire investigation. These questionnaires just have limited choices, like high-rise and mid-rise buildings. There are 'n' residents choosing high-rise building. After that, the system would make a big data analysis for floor choice of residents in this city to analyses the probability (p) of community residents choosing high-rise buildings. And then the system would calculate the likelihood based on the outcomes of the sample investigation based on the second formula. At last, the system would calculate the trend of other community residents choosing high-rise buildings based on the third formula below and provide the results to the urban designers. Overall, this system increases the efficiency of public data investigation and collection in the process of public participation.

01.Algorithm formula 02.Likelihood Calculation 03.Numercial Integration

$$P(p|n,N)=\frac{P(n|p,N)P(p)}{P(n|N)} \qquad P(n|p,N)=\binom{N}{n}p^n(1-p)^{N-n}=\frac{N!}{n!(N-n)!}p^n(1-p)^{N-n} \qquad P(p|n,N)=\frac{P(n|p,N)P(p)}{\sum_p P(n|p,N)P(p)} \tag{2}$$

Formula 2 is the formula of algorithms in system (Adapted from [10]).

4. Conclusion

In summary, this prediction model could help government or urban planners to predict the probability of all stakeholders' decision-making in a limited investigation group. However, it also needs more specific study on the screening of survey groups and determination of the number of groups in specific project. In addition, over time, people's choices may also change. How to predict people's dynamic needs through algorithms still needs further exploration. Therefore, the application of predictive models still needs to consider many complex factors

In the future, the predict method based on Bayes theorem is a useful tool to help people to save time for investigation. It would be widely used in different fields. It needs people to explore more methods to apply it and increase the accuracy of probability prediction.

5. References

[1] Pateman C 2014 Participation and democratic theory *Particip. Democr. Theory* 1–122
[2] Mostert E 2003 The challenge of public participation *Water Policy* **5** 179–97
[3] Langton S 2000 Conditions for meaningful public participation: Precursors of practice *Participatory Processes in Water Management (PPWM)* ed B Barraque pp 179–97
[4] Roberts R 1995 Public involvement: From consultation to participation *Environmental and social impact assessment* ed F Vanclay and D A Bronstein (Chichester: John Wiley & Sons Ltd)
[5] Cowel R G, Dawid A P, Lauritzen S L and Spiegelhalter D J 1999 *Probabilistic networks and expert systems* (New York, NY: Springer)
[6] Korb K B and Nicholson A E 2004 *Bayesian artificial intelligence* (Boca Raton.: Chapmann and Hall)
[7] Pearl J 1988 *Probabilistic reasoning in intelligent systems* (San Francisco, CA: Morgan Kaufmann Publishers)
[8] Henriksen H J, Rasmussen P, Brandt G, von Bülow D and Jensen F V. 2007 Public participation modelling using Bayesian networks in management of groundwater contamination *Environ. Model. Softw.* **22** 1101–13
[9] Joyce J 2003 Bayes' theorem *Stanford Encyclopedia of Philosophy.* https://plato.stanford.edu/entries/bayes-theorem/
[10] Alex R 2021 Bayes theorem and machine learning probability.

Journal of Physics: Conference Series **2287** (2022) 012018 doi:10.1088/1742-6596/2287/1/012018

A Machine Learning Model for Healthcare Stocks Forecasting in the US Stock Market during COVID-19 Period

Prapatchon Jariyapan[1], Jittima Singvejsakul[2] and Chukiat Chaiboonsri[3]

1 King Mongkut's Institute of Technology Ladkrabang, Business School
2 Department of Agricultural Economy and Development, Faculty of Agriculture, Chiang Mai University, Chiang Mai 50200, Thailand
3 Faculty of Economic, Chiang Mai University, Chiang Mai 50200, Thailand
E-mail: Jittimasvsk@gmail.com

Abstract This paper study the nowcasting and forecasting for the healthcare stock price in the united states during the Covid-19 period including the google trend data information. The data is collected in monthly data from 2015 to 2020 which are five interested stock price indexes in the healthcare sector. Empirically, the finding reveals that the Bayesian structural time series analysis can be used to investigate the stock price indexes with the google trend data is becoming useful for the prediction in term of current movement. In term of the machine learning algorithms, the unsupervised learning k-Mean algorithm is employed to cluster the cycle regimes of the stock market which provided three regimes such as Bull market, Sideways and Bear market. There are twenty-nine months stand for bull market, thirty-seven months are predictively provided sideways market and five months are referred as the bear market. Additionally, the supervised learning algorithms by using the Linear Discriminant Analysis (LDA), k-Nearest Neighbors (kNN) and Support vector machine (SVM) are used to investigate the cycle regimes of healthcare stock in next five year. The results indicated that LDA is chosen by the highest coefficient validation which represented the the regimes of stock in the healcare sector of the unites states of America will stay on the sideways periods in the next five years. Thus, the finding in this paper can be the useful information for investor to manage their portfolio especially, in healthcare sector during the Covid-19 period.

1. Introduction

The stock market forecasting is regarded as an important choice for the investors to make their decision. The forecasting in the stock market has been emphasized for many years because it can earn the profit. Interesting, there are many factors that affect the consequence of the return in the stock market including the financial crisis, random walk behavior, trader expectation and the unexpected situation. That the problem from COVID-19 pandemic, happened on 31 December 2019 from the identity of the World Health Organization (WHO) for the first case in China. Many countries around the world face this problem, suffering both the economic and health sector. Thus, the stock index in the many countries has been decreased significantly such as the US stock market fell by 20% as well as the Nikkei (Tokyo stock market) and Shangha (China stock market) has undergo a dropped price in their stock index since mid-March 2020. The composite stock index in US also decline with the concern in the effect of the COVID-19 pandemic on the economy. However, the COVID-19 pandemic is helping the improvement of innovation in biotech toward the biopharma industry and the approach of the healthcare sector reform. The healthcare sector has been the leader return sector during the COVID-19 pandemic, regarding many investors. Interestingly, many people worldwide have experienced with the

COVID-19, the need for vaccine or treatment has become increasing. Then, the increasing effort to handle COVID-19 could also improve the perception of people for the biopharma industry, which in recent years has drawn scrutiny because of rising drug prices. As well as the report from the U.S.A that a more positive view of the healthcare sector before the pandemic began. Therefore, the healthcare stock has outperformed other sector in the stock market. This momentum could continue, especially as the outbreak helps accelerate growth trends in healthcare and, long term, potentially improves sentiment toward the sector. However, the events affect the change of stock price have different characteristics. Unexpected situation that can happen any time, like the COVID-19 pandemic, economic embargoes and the trend of investor expectation may drastically influence the stock price index.
Historically, there several studies are considered the machine learning approach to predict the stock market. The study of A Hybrid Machine Learning System for Stock Market Forecasting which introduced Genetic Algorithm (GA) and Support Vector Machines (SVM) by [1] to predict the stock market price. The accuracy of estimate is found by the different between the actual price and the forecast price. The study found that the hybrid machine learning provided better result than the stand-alone machine learning. [2] study a machine learning model for stock market prediction to maximize investor's gain. The machine learning is proposed to predict the stock market price by using the particle swarm optimization (PSO) and Least square support vector machine learning (LS-SVM). The study showed that the proposed model outperforms in term of the accuracy of the prediction and the PSO has the potential optimization of LS-SVM. In order to study the 21 internal and external shocks of the European market using the Least Absolute Shrinkage and Selection Operator (LASSO) which outperform the traditional regression methods to the problem of limited data and multi-collinearity. The results indicated that the European stock market was affect by gold market before the pandemic announcement by WHO while Germany and France are the most predictors in the stock market after the announcement. [3] study the nowcasting and forecasting for macroeconomic cycle in Thailand by using machine learning algorithms to understand the Thailand economic cycles. The study employed The Bayesian structural time-series (BSTS), k-Mean clustering, Linear Discriminant Analysis (LDA) and k-Nearest Neighbors (kNN) which is the unsupervised and supervised machine learning. The results showed that the finding from these two approaches is different, the supervised machine learning found the expansion period are the predictive sign of Thailand but the unsupervised machine learning or kNN found that the recession period are the signs for Thailand cycles for next three years.
It is undeniable that there is lack of observation in the time series data which is the major concern for analyzing in the economics. The big data is becoming more important which is currently being used in different industries, including the stock market trading. Big data has impacted on the financial sector in term of the changing investment, the huge volume of data is engaged in different way of trading. It helps the trader to make better decision to invest the stocks which are consistent in the process. Consequently, the major stocks in healthcare sector such as ACHC (Acadia Healthcare Company), BIO (Bio-Rad Laboratories), INO (Inovio Pharmaceuticals), QDEL (Quidel Corporation) and SAGE (SAGE Therapeutics) from the Nasdaq stock market in US are employed to statistically explore up-to-date sets of stock market variable with google trend by Nowcasting calculation and forecasting the stock market cyclical movement of monthly data from 2015 to 2020 by using machine learning model. In order to modernize econometric investigation, the machine learning model is used to estimate with the flexibility for handling with some losing information, the huge volumes of data and the decreasing problem of multicollinearity problem. Therefore, the machine learning model including supervised and unsupervised approach are used to investigate the healthcare stocks cyclical stage and validating the data for better forecasting.

2. Data review

For the data review, Table 1 represent the stock price in the healthcare sector from the Nasdaq index to be the observed data for the learning model. The range of data is in the period of 2015 to 2020 which the data were transformed to log return.

Table 1. The log return of stock price in the healthcare sector information

Healthcare stocks	Code	Range	Source
Acadia Healthcare Company	ACHC	2015 – 2020	Nasdaq Index
Bio-Rad Laboratories	BIO	2015 – 2020	Nasdaq Index
Inovio Pharmaceuticals Inc	INO	2015 – 2020	Nasdaq Index
Quidel Corporation	QDEL	2015 – 2020	Nasdaq Index
Sage Therapeutic	SAGE	2015 – 2020	Nasdaq Index

In order to predict the economic cycle in the machine learning model, the behavior factor is needed to estimated. For the technology of databased in the google trends is important one to be used to be the one factor in data prediction which are represent in Table 2.

Table 2. The collected information of behavior factor searching databases from google trend

Behavior factors	Code	Range	Source
Investment	Invest	2015 – 2020	Google Trends
Return on investment	ROI	2015 – 2020	Google Trends
Nasdaq	Nasdaq	2015 – 2020	Google Trends
Health stock	Health stock	2015 – 2020	Google Trends
Stock market and COVID-19	Stock &Covid	2015 – 2020	Google Trends

3. Research methodology

3.1. The Bayesian structural time series (BSTS) analysis

The Bayesian structural time series (BSTS) is proposed [4] for the adjusted non-seasonally in order to predict the united states unemployment benefits with the behavioral indexes from google trends as the effect factor in the BSTS analysis. Furthermore, [5] introduced a Bayesian structural time series Forecasting (BSTSF) model applying the time series data for a sustainable technical analysis in the regression model. The BSTS model can be explain based on the Bayes theorem as follows [6]

$$P(\theta \mid x) = \frac{P(x \mid \theta)P(\theta)}{P(x)}, \tag{1}$$

where x is observed information and θ is the model parameter. $P(x \mid \theta)$ refer the prior and likelihood functions. $P(\theta \mid x)$ is the posterior function. The observed data x given θ is the updated learning for a likelihood. The prior setting is the normal distribution which can be expressed as the Bayesian regression as follows

$$y = \beta_0 + \sum_{p=1}^{P} \beta_i x_i + e \tag{2}$$

where e is the normal distribution with mean = 0 and variance = σ^2. The prior of σ^2 is relied on an inverse-chi square ($Inv-\chi^2$) distribution that can be showed as

$$\sigma^2 \sim Inv-\chi^2(n-p, s^2), \tag{3}$$

n and p are data and parameter sizes, respectively where s^2 is given by

$$s^2 = \frac{1}{n-p}\left(y - X\hat{\beta}\right)^T\left(y - X\hat{\beta}\right) \tag{4}$$

To apply the BSTS model for nowcasting, the Local linear trend without the google trend information is written as

$$\mu_{t+1} = \mu_t + \delta_t + \eta_{\mu,t}, \tag{5}$$

$$\delta_{t+1} = \delta_t + \eta_{\delta,t}. \tag{6}$$

$\eta_{\mu,t}$ and $\eta_{\delta,t}$ are assumed to be normal distributions, $N(0,\sigma^2_{\mu,t})$ and $N(0,\sigma^2_{\delta,t})$, respectively. For the component μ_t is the value of the trend at time t and δ_t is the expected increase in μ between times t and t+1. The model includes the google trend information which is the $\beta^T x_t$ stands for the behavioral factors including in the structural time-series model as represented by following

$$y_t = \mu_t + \beta^T x_t + \varepsilon_t, \tag{7}$$

$$\mu_{t+1} = \mu_t + \delta_t + \eta_{0,t}, \tag{8}$$

$$\delta_{t+1} = \delta_t + \eta_{1,t}. \tag{9}$$

3.2. The k-Means clustering algorithm of the unsupervised machine learning
The k-mean algorithm is specified to be a function of the available data type and the purpose of selective analysis [7]. The Eulidean distance is employed to be the criterion that calculate the hyperspace and two-dimensional space [8]. For the development of k-Means algorithms can be represented as five steps by following

- Step 1: clustering an input data and minimizing variance of the data in order to accept the number of group data and cluster the data into k groups.
- Step 2: taking the first random sampling or the first k instance of k elements by initializing the first k groups.
- Step 3: the arithmetical mean of each cluster formed from the set of data is estimated to be designed as a gradient descent procedure.
- Step 4: the initial cluster is specified class by allocating the k-Means algorithms in each point of the dataset. The nearest cluster assigned each point through a measurement of Euclidean distance.
- Step 5: each point of recorded data is repeatedly assigned by the algorithm

3.3. The supervised machine learning approaches

3.3.1 Linear Discriminant Analysis (LDA). The linear discriminant analysis (LDA) is employed to indicate the linear feature for maximizing the between-class separation of data and minimizing the data within the class, is the basic supervised learning algorithms [9,10]. For the formula of the algorithm can be express by the number of n which is the clusters of the regimes' stock market such as Bull market, Sideways and Bear market. Let μ be the mean for the samples, μ_i and N_i is the mean and number of the observation data in i^{th} class, respectively. Then, Σ_i is the scatter matrix of the observation data, the S_W can be written as follow

$$S_W = \sum_{i=1}^{n} \Sigma_i, \tag{10}$$

where the solution of the between-class scatter matrix, S_B, can be expressed as

$$S_B = \sum_{i=1}^{n} N_i(\mu_i - \mu)(\mu_i - \mu)^T \tag{11}$$

with the condition that $S_W^{-1} S_B$, eigenvalues and eigenvectors are solved.

3.3.2 The k-Nearest Neighbors (kNN) algorithm. The *k*-Nearest Neighbors (kNN) is a non-form artificial approach for both linear regression problem and classification [10]. In order to cluster the data, which was performed as the other feature space, the observation data are allocated. Hence, the mean value of the nearest neighbors in the linear regression is the majority class to group *K* neighbors. The error rate of the misclassified samples from the classification and squared error similar in regression problem can be analyzed by using the *k*-NN algorithm. First, the process of the algorithms is the calculation for the searching distance x_i which is the single sample related to all other points in the feature space. In this stage, the *k* is controlled from the closet neighbors $\{x_j\}_i^k$ as well as the measure distance is generally indicated by the Euclidean distance. Then, the next sage is to assume the result of y_i to be the class membership such as , $y_i \in \{C_1, C_2, ..., C_c\}\ \forall i = 1,2,...,n$, where *c* is a number of class levels from the majority vote of its kNN. The k-NN regression can be expressed a distance-weighted average as following

$$y_i = \frac{1}{k} \sum_{j=1; x_i \in \{x_j\}_i^k}^{k} x_j \tag{12}$$

where y_i, represents the average value of its single nearest neighbor.

3.3.3 The Support vector machine (SVM). For the support vector machine or SVM is the algorithm for applying the regression and classification problem. Generally, in term of the classification the problems are basically modeled as an interval range of data, a binary 0 to 1 which is the logistic regression (Logit model). Hence, the interval of binary 0 to 1 range is assigned to be the position of the hyperplane in the feature space, is the probabilities of class memberships. For the supporting vector is referred to one while the error function is zero, the SVM algorithms are employed to analyze the decision boundary to separate the observation information into two classes by the maximal margin. First, the separate group of the data feature in the space are divided by linearly separation. Then, the point is identified in the input space to support the maximum margin in the vectors. The main properties of the SVM algorithm consists as [11].

$$e = e_{emp} + e_{g'}, \tag{13}$$

where e_{emp} is the training error and $e_{g'}$ is the generalization error.

4. Empirical results

4.1. The Nowcasting prediction by the Bayesian structural time series model

The stock price indexes in the healthcare sector of the united states are represented in Figure 1 in order to explain accurately for the univariate time series. The line prediction shows the movement of the scatter plot which is the predicted rate of the stock market cycles, estimated by its lags, without Google Trend data as the black shade along the line on the left-hand side Figure. However, the estimation of the Nowcasting by BSTS model with google Trend data is presented on the right-hand side of the picture which showed the black shade evidently visualized wider than the prediction without the Google Trend data. According to the power of behavioral factors added into the structural time-series model, the performance of the forecasting is improved in term of the data distribution. Therefore, the efficiency of the forecasting regard to the BSTS model without google trend data is lower than the BSTS estimation with google trend data which is the traditional method. The forecasting by the BSTS model with google trend indexes perform better results than the previous case and become useful for the prediction in term of current movement.

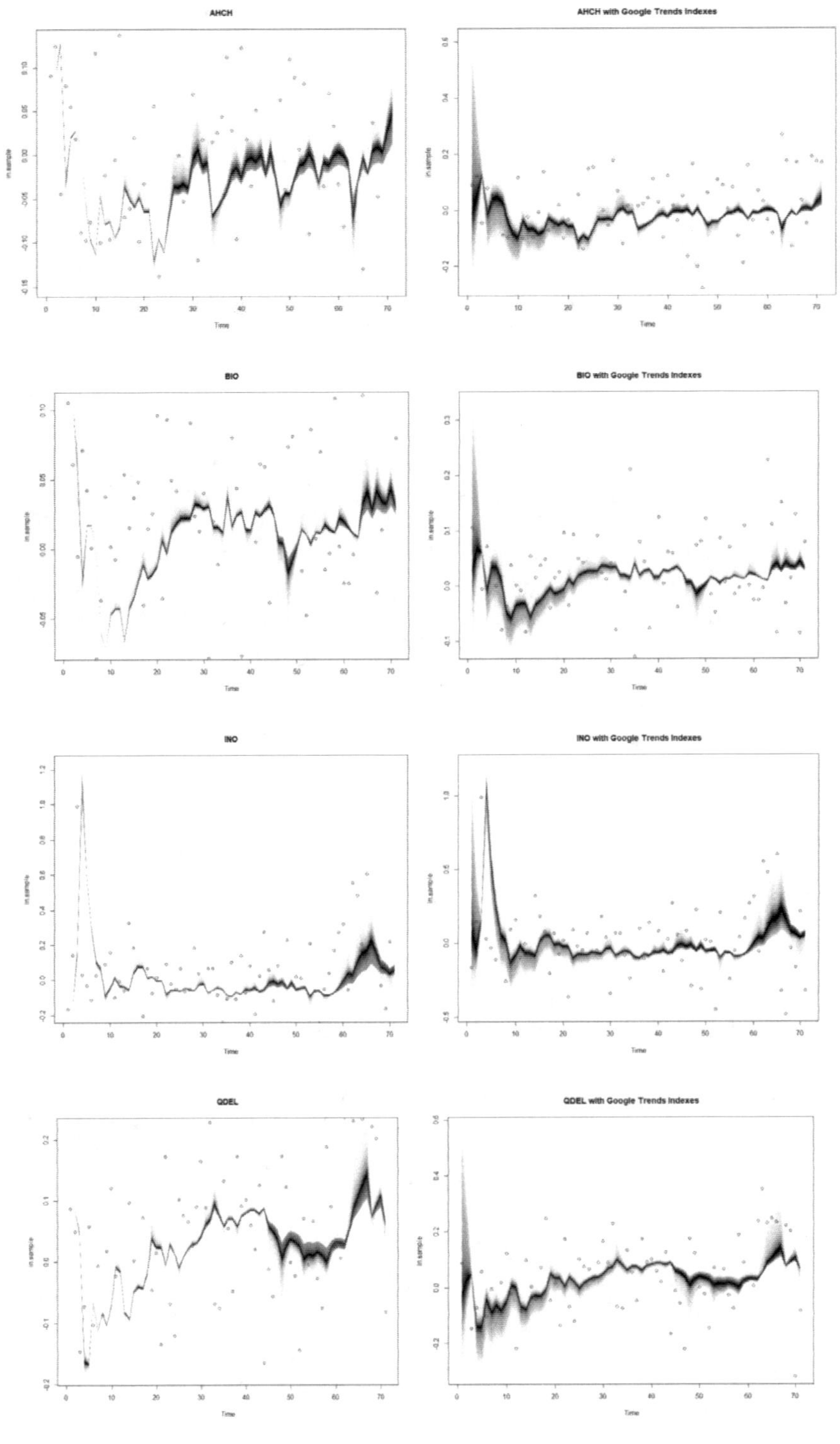
AHCH
AHCH with Google Trends Indexes
BIO
BIO with Google Trends Indexes
INO
INO with Google Trends Indexes
QDEL
QDEL with Google Trends Indexes
Time

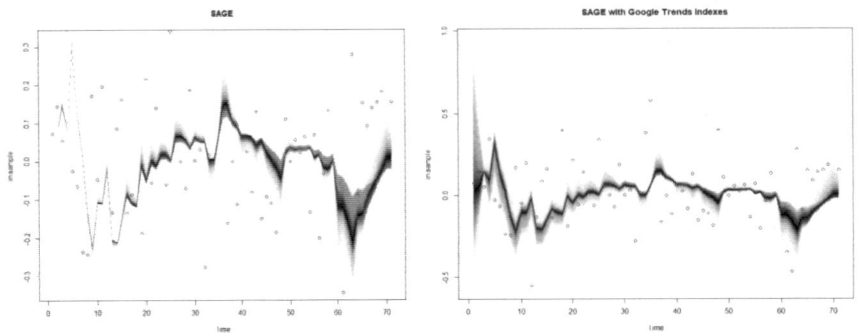

(Source: computing)

Figure 1. The Nowcasting forecasting without google trend and with google trend indexes

4.2. The results of clustering by the k-Means algorithm
In this section the stock market regimes are investigated by the *k*-means algorithm from 2015 to 2020 which cove the Covid-19 situation period. The stock price indexes from the healthcare sector including the google trend data are added to be the important factor that use to find the stock market regimes. As the study of [12] found that the stock market follows three cycle regimes which are the bull market, sideways and bear market by using the optimal trend following to find these sufficient conditions. Consequently, we explore the stock market regimes by clustering into three part which are bull market, sideways and bear market which is showed in Table 3. The results indicate that there are twenty-nine months stand for bull market, thirty-seven months are predictively provided sideways market and five months are referred as the bear market.

Table 3. stock market regimes clustered by the *k*-Means algorithm

Description	Frequencies (Months)
Bull Market	29
Sideways	37
Bear Market	5

Source: Computing

4.3. The stock market regimes forecasting by the machine learning approach
In this part, there are three learning machine algorithms are employed to investigate the stock market regimes in the next five years such as linear discriminant modelling (LDAk-Nearest Neighbors (kNN) and Support Vector Machine (SVM). The Kappa's coefficient is used to validate the machine learning algorithms in order to select the best algorithms for prediction. For the results indicated that LDA that stand for the linear-form learning algorithm is chosen by the highest coefficient validation. Additionally, the accuracies and coefficient are detailed in Table 4.

Table 4. The result of three machine learning approaches by Kappa's coefficients

Algorithm	Cross-validation method	Accuracy	Kappa coefficients
LDA**	Cohen's kappa	0.8138095	0.6425780
k-NN	Cohen's kappa	0.5223810	0.0758920
SVM	Cohen's kappa	0.7847619	0.5868421

Noted: ** indicates the chosen algorithm Source: authors

The data from the stock price indexes and the data set from the google trend data are classifying between the bull market and bear market by the linear learning algorithm. Interestingly, the grouping data is intercept for each other which the algorithm can group the predominant occurences. Consequently, the forecasting result from the LDA found that the stock price indexes in the healcare sector of the unites states of America will stay on the sideways periods in the next five years which are showed in Table 5.

Table 5. The posterior prediction by LDA

Year/Stages	Bull Market	Bear Market	Sideways
2021	0	0	1
2022	0	0	1
2023	0	0	1
2024	0	0	1
2025	0	0	1

Source: Computing

5. Conclusion

The range of time series data observed of stock price indexes from healthcare sector in the united states are covered from the period during 2015 to 2020. That the world economy is currently facing the recession caused by the Covid-19 epidemic occurred in 2020. The machine learning approaches are the tool which can be used for the various type of data for example the big data which is more incentive to be used in the investigation in the real-world problem. Interestingly, the substantiality of the finding in this study reveals the critically predictable trends of the stock market in healthcare sector of the unites states fluctuations through the forecasting, clustering and nowcasting of machine learning approaches. The Bayesian structural time series model is employed to calculate the nowcasting with updating the behavioral information from google trend data which can investigate the misreading up-to-date information which is the insignificant assumption of the traditional econometric methods. For the computational clustering of the stock market cycle regimes, the k-Mean algorithm which is the unsupervised learning can be used to tackle the complexity of data and categorized the cycle regimes of the stock market. Hence, the machines learning approaches can predictively calculate the upcoming regimes situation of stock market in healthcare sector in the united states. For the capability of supervised learning to estimate the enormous amount high dimensional data outperform the traditional method.

The Linear discrimination analysis is the flexibility machine learning model to challenge the parametric estimations. The results reveal that the LDA learning algorithm are evident that the stock market regimes are sideways for the upcoming five years.

To conclude, the machine learning performance can cover the major role of method in econometrics such as a clustering computation, classification, parametric or no parametric predictor. The suggestion provides that the united states' stock market price indexes in healthcare sector do not need the pro-cyclical policies implementation. However, it needs to know the amount stock market in sideways situation which is the useful information for investor to manage their portfolio especially, in healthcare sector during the Covid-19 period.

6. References

[1] Choudhry, R., and Garg, K.: 2008. A Hybrid Machine Learning System for Stock Market Forecasting. *Engineering and Technology*, (39), pp.315-318.

[2] Hegazy, O., Soliman, O.S., and Salam, M.A.: 2013. A Machine Learning Model for Stock Market Prediction. *International Journal of Computer Science and Telecommunications*, (4), pp. 17-23.

[3] Chaiboonsri, C., and Wannapan., S.: 2020. Nowcasting and Forecasting for Thailand's Macroeconomic Cycles Using Machine Learning Algorithms. *International Symposium on Integrated Uncertainty in Knowledge Modelling and Decision Making*, pp 270-282.

[4] Scott, S.L. & Varian, H.R.: 2014. Predicting the present with Bayesian structural time series. *Int. J. Mathematical Modelling and Numerical Optimisation*, Vol. 5, Nos. 1/2, 4-23.

[5] Jun, S.H.: 2019. Bayesian structural time series and regression modeling for sustainable technology management. *Sustainability*, Vol.11, No. 4945; doi:10.3390/su11184945.

[6] Koduvely, H.M. (2015). *Learning Bayesian Models with R*. Packt: Birmingham, UK.
[7] Oyelade, O.J., Oladipupo, O. O., & Obagbuwa, I. C.: 2010. Application of k-Means Clustering algorithm for prediction of students' academic performance. *International Journal of Computer Science and Information Security*, Vol.7, No.1, 292-295.
[8] Li, Y. & Wu, H.: 2012. A clustering method based on K-Means algorithm. *Physics Procedia*, Vol.25, 1104 – 1109.
[9] Bishop, C.: 1995. *Neural networks for pattern recognition*. Oxford University Press.
[10] Ioffe, S.: 2006. *Probabilistic linear discriminant analysis*. A. Leonardis, H. Bischof, and A. Pinz (Eds.). Computer Vision – ECCV 2006. ECCV 2006. Lecture Notes in Computer Science, vol 3954. Springer, Berlin, Heidelberg.
[11] Chakraborty, C. & Joseph, A.: 2017. *Machine learning at central banks*. Staff Working Paper No. 647. Bank of England.
[12] Yu, J., and Zhang, Q.: 2019. Optimal trend-following trading rules under a three-state regime switching model. *American Institute of Mathematical Sciences*, 2(1), pp. 81-100.

Deep Neural Network As a Tool for Appraising Housing Prices: A Case Study of Busan, South Korea

S An[1,2], Y Song[3], H Jang[4], and K Ahn[1,2,*]

[1] Department of Industrial Engineering, Yonsei University, South Korea
[2] Center for Finance and Technology, Yonsei University, South Korea
[3] Department of Geography, Chonnam National University, South Korea
[4] Department of Finance, Accounting and Risk, Glasgow Caledonian University, UK

[*] E-mail: k.ahn@yonsei.ac.kr (K.A.)

Abstract. This study examines whether the number of hidden layers in a deep neural network significantly influences the model accuracy and efficiency for appraising housing prices. We provide empirical evidence that the deep neural network can achieve high accuracy with a small number of hidden layers on our dataset, which contains various hedonic variables. Furthermore, we show that adding layers does not necessarily guarantee the model's accuracy and effectiveness of the computing time.

1. Introduction

Housing market is a crucial part of urban economy, and its price level reflects regional economic factors. At a personal or household level, housing accounts for a large portion of wealth [1]. Therefore, various stakeholders, including governments, investors, and academics, have tried to maintain the stability and effectiveness of the housing market by developing a model for appraising housing prices [2–4].

Various approaches and variables have been introduced to appraise housing prices. Housing prices are determined by various factors, such as housing properties, local amenities, and local demographic characteristics. For instance, transportation accessibility was used as a key variable along with housing and local properties using a hedonic price model (HPM) [5]. Furthermore, recent improvements in the image recognition technology enable us to generate novel variables such as green or sky view index, which can enhance the accuracy and effectiveness of the model [6,7]. Datasets used to estimate housing prices cover various economic, functional, and amenity factors; therefore, they are often nonlinear and complex. However, traditional models such as HPM rely on a simple linear regression model, and whether HPM properly captures the complex nature of variables and their relationships is questionable [8,9].

Considering the complexity and nonlinear property of housing price data, this study develops an appraisal model with a deep neural network (DNN). The DNN has been proved suitable for nonlinear datasets and applied in diverse areas of study, including disease detection and energy consumption prediction [10,11]. Thus, it can be used for appraising housing prices with high accuracy. Moreover, few studies have examined the optimal number of layers, which is important for an accurate and efficient appraisal of housing prices [12,13]. Therefore, this study predicts housing prices using DNN and investigates how many layers are required to capture complex and nonlinear hedonic variables.

2. Method and data

2.1. DNN

DNN comprises an input layer, some hidden layers, and an output layer. We assume the structure of DNN with one to six hidden layers and investigate the optimal number of layers considering the accuracy and efficiency as model performance. Each hidden layer has 64 neurons and the same activation function, with a fully connected ReLU function, is applied, i.e., the most popular activation function in neural networks imposing nonlinearity on the layer structure. We define the ReLU function, which can enhance the model accuracy and prevent gradient from vanishing [14], as

$$\mathrm{ReLU}(x) = \max(0, x),$$

where x stands for an input of the current node. Only if input value x is positive, the activation function conveys value x to nodes in the next layer. Accordingly, ReLU function can significantly reduce the computation cost as it facilitates the calculation of propagating weights by deactivating neurons [15]. Through this feedforward propagation, DNN predicts the target value by aggregating whole weights and inputs of each node and computes the error generated by loss function. Then, DNN starts to update each weight for minimizing the bias through back propagation as follows:

$$w_{n+1} = w_n - \alpha\left(\frac{\partial l}{\partial w_n}\right),$$

where l and α are the loss function and learning rate, respectively. Specifically, we set the learning rate to 0.01 and 'early stopping' to 10% of the maximum number of epochs. For a smooth training procedure, we use the Min–Max scaling instead of the Z-score normalization [16]. The model structure is presented in figure 1. To appraise the model's accuracy, we select mean absolute error (MAE) and R-square, and further estimate the computation time in the model training.

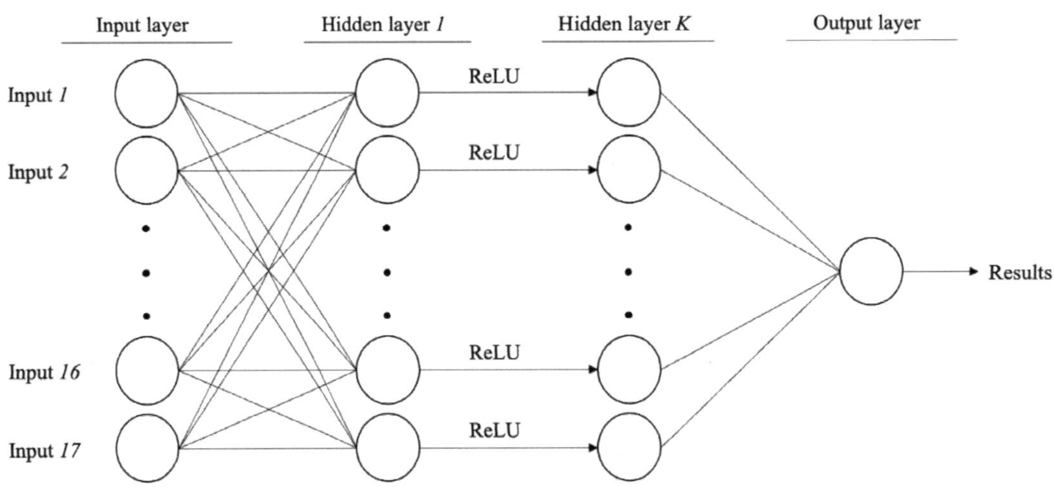

Figure 1. Model structure.

2.2. Data

We utilize the Mendeley data repository (https://data.mendeley.com/datasets/d7grg846wv/3), which was introduced by Song et al. [17], to investigate the impact of extra layers on the DNN model for appraising housing prices. This dataset, which contains 61,152 observations in 2015, comprises four Korean metropolitans: Busan, Daegu, Daejeon, and Gwangju. To explore the impact of a subway network on housing prices in our study, we concentrate on Busan only because it showed the highest modal share of subway, i.e., 17.8%, among the four cities [5]. The prevalent type of housing in Busan is condominium, which is the dominant housing type in South Korea [18]. Herein, 17 variables are used for appraising the housing prices, as shown in table 1.

Table 1. Descriptive statistics.

Variables	Scale	Min.	Max.	Mean	Std.	Skewness	Kurtosis
Higher degree	Ratio	10.16	59.10	35.40	10.77	0.26	−0.89
Year	Date	1962	2015	1999.09	9.23	−0.35	−0.19
Parking	Ratio	0	15	1	0.50	0.94	12.88
Area	Ratio	14.74	250.85	79.27	30.27	1.13	1.93
Households	Ratio	4	5239	871.75	849.79	1.92	5.01
CBD	Ratio	244.28	23900.12	9003.08	4917.75	0.69	−0.03
Floor	Interval	−1	79	10.56	7.90	1.44	3.96
ln_greenspace[a]	Rati	−1.55	8.33	5.21	1.06	−0.36	1.52
ln_waterfront[a]	Ratio	1.75	7.84	6.08	0.92	−0.61	0.10
ln_NET_DIST[a]	Ratio	0.52	9.52	6.99	1.01	−0.99	3.47
SNU	Ratio	0	12	1.17	1.85	3.27	13.49
Dummy_Spring[b]	Nominal	0	1	0.31	0.46		
Dummy_Fall[b]	Nominal	0	1	0.25	0.43		
Dummy_Winter[b]	Nominal	0	1	0.20	0.40		
PopDensity	Ratio	140.41	38720	12978.93	7319.40	0.66	0.15
Heating type[c]	Nominal	0	1	0.91	0.28		
Bus_Count	Ratio	0	46	12.20	6.98	0.80	0.40

[a] This variable is transformed with logarithm.
[b] Control variables of sales period: spring, fall, and winter.
[c] Dummy variable: city gas 0, others 1.

3. Results and discussion

The model accuracy and computation time with additional layers are summarized in table 2. As the number of layers in DNN increases, it becomes more accurate in terms of MAE and its computation requires more time. However, adding more layers does not necessarily guarantee a higher accuracy than others in terms of R-square. For example, the model with four layers (i) achieves the smaller R-square and (ii) reduces the computation time of models with five and six layers by approximately 30%. This suggests that the optimal DNN model can be obtained with few layers when big data are used.

Table 2. Model accuracy and efficiency.

	MAE	R-square	Computation time (s)
Layer 1	0.017	0.836	83.387
Layer 2	0.011	0.895	85.373
Layer 3	0.009	0.915	126.813
Layer 4	0.009	0.913	88.359
Layer 5	0.009	0.918	124.051
Layer 6	0.008	0.924	124.697

DNN optimizes the weight of each layer using various methods such as Adam or RMSProp and minimizes the bias using the stochastic gradient descent algorithm. Consequently, few layers can produce reasonable outcomes in terms of accuracy and efficiency. Most importantly, the performance of DNN is closely related to preventing the overfitting issue. Multiple convolution kernel matrices and training parameters deteriorate the training time and space complexity, resulting in a low speed of network propagation and inaccuracy [19].

Figure 2 shows the average impact of each feature on the model output. We calculate the Shapley Additive exPlanations (SHAP) [20,21] values and compare model predictions with and without each feature. When an individual feature contributes more to appraising housing prices, SHAP value is high. We observe that after removing the subway-related variable, the model's accuracy barely decreases. Moreover, the subway network depreciates the housing prices in contrast to Ahn et al. [5]. Thus, the expected positive effects of transit accessibility do not exist in the Busan metropolitan area.

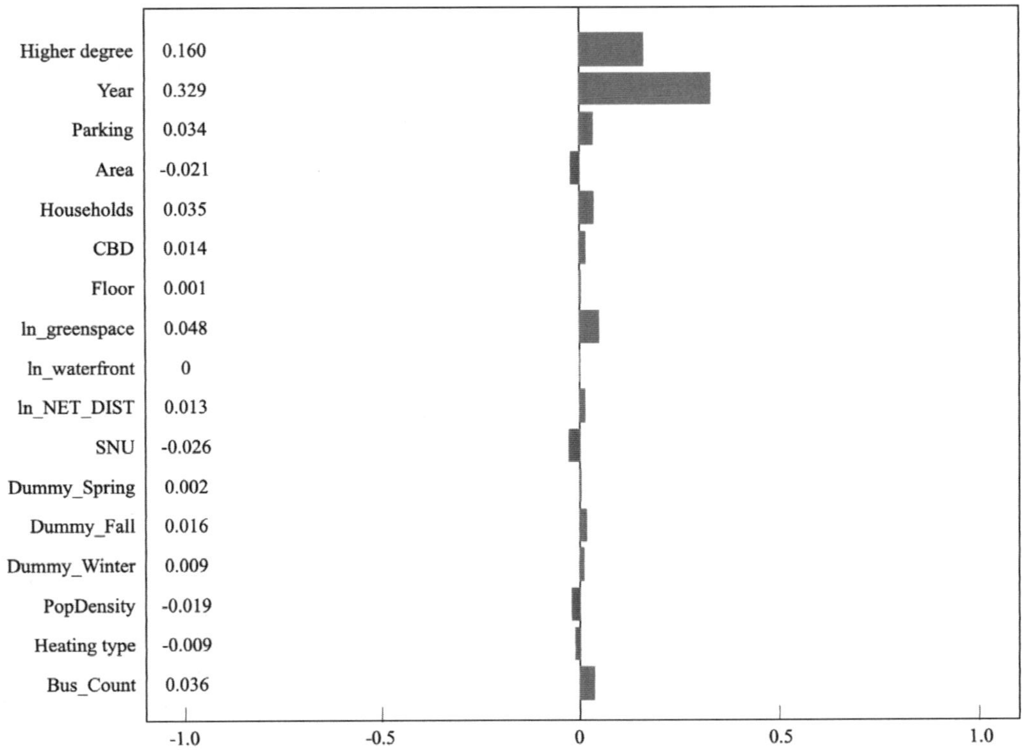

Figure 2. SHAP value plot.

4. Conclusion
This study investigates the number of efficient hidden layers for appraising housing prices in DNN. By comparing the accuracy and computation time, we find that DNN can effectively appraise housing prices with few layers, and an efficient approach involves adding more layers with enough number of neurons.

Acknowledgments
This work was supported by (i) the Technology Innovation Program ATC+ (20014125, Development of Intelligent Management Solution for Nuclear Decommissioning Site Characterization) funded by the Ministry of Trade, Industry & Energy (MOTIE, Republic of Korea) and (ii) the Future-leading Research Initiative (Grant Number: 2021-22-0306) funded by Yonsei University.

References
[1] Glaeser E L, Gyourko J, Morales E and Nathanson C G 2014 Housing dynamics: An urban approach *Journal of Urban Economics* **81** 45–56
[2] Rizi M H 2021 What moves housing markets: A state-space approach of the price–income ratio

International Economics **167** 96–107

[3] Xue C, Ju Y, Li S and Zhou Q 2020 Research on the sustainable development of urban housing price based on transport accessibility: A case study of Xi'an, China *Sustainability* **12** 1497

[4] Pai P P and Wang W C 2020 Using machine learning models and actual transaction data for predicting real estate prices *Applied Sciences* **10** 5832

[5] Ahn K, Jang H and Song Y 2020 Economic impacts of being close to subway networks: A case study of Korean metropolitan areas *Research in Transportation Economics* **83** 100900

[6] Zhang Y and Dong R 2018 Impacts of street-visible greenery on housing prices: Evidence from a hedonic price model and a massive street view image dataset in Beijing *ISPRS International Journal of Geo-Information* **7** 104

[7] Chen L, Yao X, Liu Y, Zhu Y, Chen W, Zhao X and Chi T 2020 Measuring impacts of urban environmental elements on housing prices based on multisource data: A case study of Shanghai, China *ISPRS International Journal of Geo-Information* **9** 106

[8] Hong J, Choi H and Kim W S 2020 A house price valuation based on the random forest approach: The mass appraisal of residential property in South Korea *International Journal of Strategic Property Management* **24** 140–152

[9] Chin T L and Chau K W 2003 A critical review of literature on the hedonic price model *International Journal for Housing Science and its Applications* **27** 145–165

[10] Chien C Y, Hsu S W, Lee T L, Sung P S and Lin C C 2021 Using artificial neural network to discriminate Parkinson's disease from other parkinsonisms by focusing on putamen of dopamine transporter SPECT images *Biomedicines* **9** 12

[11] Kalogirou S A and Bojic M 2020 Artificial neural networks for the prediction of the energy consumption of a passive solar building *Energy* **25** 479–491

[12] Bickel D R 2020 Testing prediction algorithms as null hypotheses: Application to assessing the performance of deep neural networks *Stat* **9** e270

[13] Mayer M, Bourassa S C, Hoesli M and Scognamiglio D 2019 Estimation and updating methods for hedonic valuation *Journal of European Real Estate Research* **12** 134–150

[14] Rahim T, Hassan S A and Shin S Y 2021 A deep convolutional neural network for the detection of polyps in colonoscopy images *Biomedical Signal Processing and Control* **68** 102654

[15] Kulathunga N, Ranasinghe N R, Vrinceanu D, Kinsman Z, Huang L and Wang Y 2021 Effects of nonlinearity and network architecture on the performance of supervised neural networks *Algorithms* **14** 51

[16] Asif M R, Bording T S, Barfod A S, Grombacher D J, Maurya P K, Christiansen A V, Auken E and Larsen J J 2021 Effect of data pre-processing on the performance of neural networks for 1-D transient electromagnetic forward modeling *IEEE Access* **9** 34635–34646

[17] Song Y, Ahn K, An S and Jang H 2021 Hedonic dataset of the metropolitan housing market-Cases in South Korea *Data in Brief* **35** 106877

[18] Jang H, Ahn K, Kim D and Song Y 2018 Detection and prediction of house price bubbles: Evidence from a new city *Lecture Notes in Computer Science* **10862** 782–795

[19] Chen H, Geng L, Zhao H, Zhao C and Liu A 2021 Image recognition algorithm based on artificial intelligence *Neural Computing and Applications* **33** 1–12

[20] Jabeur S B, Mefteh W S and Viviani J L 2021 Forecasting gold price with the XGBoost algorithm and SHAP interaction values *Annals of Operations Research* 1–21

[21] Lundberg S M and Lee S I 2018 A unified approach to interpreting model predictions *International Conferences on Neural Information Processing Systems (CA, USA, December 2017) (Advances in Neural Information Processing Systems* vol 30*)* ed U V Luxburg et al. pp 4766–4775

Information Flow Between Crude Palm Oil and Crude Oil Futures

M Jeong[1,†], S Kim[2,†] and E Yi[3,*]
[1] Department of Industrial Engineering & Center for Finance and Technology, Yonsei University, Seoul 03722, Republic of Korea
[2] Korea Exchange, Seoul 07329, Republic of Korea
[3] Pritzker School of Law, Northwestern University, Chicago, IL 60611, United States
[†] These co-first authors contributed equally to this work.
[*] E-mail: eojinkorea@nlaw.northwestern.edu (E.Y.)

Abstract. This study finds asymmetric information flow from the crude palm oil (CPO) futures to the West Texas Intermediate (WTI) crude oil futures market despite the CPO futures market's low liquidity and small market capitalization. Our finding is robust regardless of the 2019 Coronavirus outbreak and the asymmetric information flow becomes even unilateral considering the exchange rate risk on the Malaysian Ringgit. Finally, we explain the asymmetric information flow from the CPO futures to WTI futures market given that the impact of speculation on market efficiency crowds out that of liquidity.

1. Introduction
Beginning with the Directive on Biofuels for Transport (May 2003), the European Union (EU) attempts to increase biofuel and renewable energy usage for a cleaner environment with the Directive on the promotion of Renewable Energy Sources (June 2009) [1–3]. Since then, the crude palm oil (CPO) market, the leading commodity in edible oil and biofuels, has been growing rapidly. In particular, the economic expansion of China and India (the major CPO importers), that of Indonesia (the largest producer and exporter), and the EU-driven environmental policies have dramatically promoted usage of CPO [4]. However, as CPO production could cause deforestation and contribute to environmental changes, the EU amended its directives and issued new policies, e.g., the Commission Delegated Regulation (EU) 2019/807, to reduce CPO imports from Southeast Asian countries [5,6]. Accordingly, the CPO market keeps getting attention from academics, industrial entities, and regulatory authorities.

The CPO futures market has less liquidity than the West Texas Intermediate (WTI) futures market, the global benchmark of the crude oil market, and it is well documented that markets become more efficient with increased liquidity [7]. From another perspective, after the financialization of the crude oil market in 2000, the pricing mechanism of the WTI futures market was affected by speculation [8]. When the herding behavior, caused by speculative trading activities, becomes widespread, market price is distorted and market efficiency decreases because of information unrelated to fundamentals [9]. Considering that more efficient markets have leadership on information discovery against less efficient markets [10], this study aims to (i) uncover the information flow between the CPO and WTI futures markets and (ii) identify the impact of liquidity and speculative transactions on the information flow particularly with the 2019 Coronavirus (COVID-19) as a natural experimental setting. The COVID-19 outbreak initiated a demand shock lowering the speculative transactions in oil markets [11,12].

Moreover, we investigate the role of the currency risk of Malaysian Ringgit (MYR), the trading currency of CPO futures, due to its impacts on oil prices [13].

2. Methodology and data

2.1. Transfer entropy

Several approaches investigate information flow between different assets: the Granger causality test [14,15], information shares [16], and common factor weights [17]. We used transfer entropy [18], which is free from restrictions on model structure [12,15,19], and calculated the histogram-based transfer entropy defined on equally spaced bins [15], where appropriate bin width h was determined using Freedman–Diaconis' rule [20,21]. Effective transfer entropy was further considered to correct the bias from the finite sample size.

Transfer entropy from system Y to X is defined as follows [18]:

$$TE_{Y\to X} = H\left(X_{t+1}\middle|X_t^{(k)}\right) - H\left(X_{t+1}\middle|X_t^{(k)}, Y_t^{(l)}\right),$$

where $X_t^{(k)} = \{X_t, X_{t-1}, \dots, X_{t-(k-1)}\}$ and $Y_t^{(l)} = \{Y_t, Y_{t-1}, \dots, Y_{t-(l-1)}\}$ are two random processes given by the k and l dimensional delay vectors, respectively. $H(X_{t+1}|\cdot)$ is conditional entropy, implying the degree of uncertainty for predicting X_{t+1} for a given information "$\cdot$". Thus, transfer entropy is the effect of $Y_t^{(l)}$ on predicting X_{t+1}.

Effective transfer entropy is calculated by subtracting the arithmetic mean of the randomized transfer entropy values, $TE_{Y_{(i)}\to X}$ for $i = 1,2,\cdots, M$, from the estimated transfer entropy as follows [22]:

$$ETE_{Y\to X} = TE_{Y\to X}(k,l) - \frac{1}{M}\sum_{i=1}^{M} TE_{Y_{(i)}\to X}(k,l),$$

where $Y_{(i)}$ indicates the randomly shuffled variable Y and M is the number of random shuffling.

2.2. Data

We retrieved the price of CPO futures, particularly for front-month contracts, traded in Bursa Malaysia as a benchmark [23]. For the WTI futures, the price of front-month contracts in the New York Mercantile Exchange was collected. For both commodity futures, daily prices were obtained from Bloomberg, and the dataset covered February 1, 2019, to April 30, 2021. In addition, we separately analyzed the subsample covering from May 2020 to April 2021 (Pandemic period) to examine the impact of low speculative transactions with COVID-19. We excluded the turmoil period (February 2020 to April 2020) in the subsample as the World Health Organization's pandemic declaration caused heterogeneous behavioral bias in both markets [12].

Table 1 shows descriptive statistics for the log-returns of our samples. In the subsample, the skewness of WTI was positive whereas that of CPO was negative, implying that after the pandemic declaration, investors in the CPO futures market are still risk-averse and those in WTI are changed to be risk-loving [24]. The kurtosis of WTI during the whole sample period was extremely high compared to that in the subsample, indicating that many outliers exist, especially in the turmoil period excluded in the subsample. Moreover, after the pandemic, the skewness and kurtosis of CPO became closer to 0 and 3, respectively, similar to those of the Gaussian distribution. This suggests that the CPO futures market became more efficient [25,26] after the pandemic outbreak.[1]

[1] To investigate the market efficiency more accurately, more formal approaches should be considered, such as variance ratio test [27], Hurst exponent [12,28], quantum harmonic oscillator [29–31], etc.

Table 1. Descriptive statistics of log-returns of CPO and WTI futures.

	Min.	Max.	Mean	Std.	Skewness	Kurtosis
Panel A. Overall						
CPO	−0.1025	0.0518	1.200×10^{-3}	0.0183	−0.5135	5.4489
WTI	−0.6017	0.3196	2.422×10^{-4}	0.0488	−3.0554	52.8861
Panel B. Pandemic period						
CPO	−0.0482	0.0509	2.881×10^{-3}	0.0180	−0.1673	3.0381
WTI	−0.0859	0.1861	4.543×10^{-3}	0.0277	0.9168	10.3619

3. Results and discussion

Table 2 summarizes the estimation results of effective transfer entropy between the CPO and WTI futures markets when the CPO futures price is MYR denominated. In Panel A, there exist mutual information flows but net information flow is asymmetric: information flow from CPO to WTI futures dominates that of the other way around. Panel B shows that the cause–effect relationship was not changed and even became stronger after the outbreak of COVID-19. Effective transfer entropy from the CPO to WTI futures is still positive and dominates the other way around regardless of pandemic outbreak. However, the WTI futures market started to provide significantly more information toward the CPO futures market as compared to before due to low speculative trading activities after the COVID-19 crisis.

Table 2. Effective transfer entropy between WTI and CPO futures markets (CPO futures in MYR).

	TE	Std. Err.	*p*-value	ETE
Panel A. Overall (*num. of bins* = 153)				
WTI → CPO	2.3748	0.0525	0.0233	0.0166
CPO → WTI	2.5370	0.0457	0.0000	0.0505
Panel B. Pandemic period (*num. of bins* = 32)				
WTI → CPO	2.1286	0.0786	0.0000	0.0809
CPO → WTI	2.0256	0.0659	0.0000	0.1046

Table 3 shows effective transfer entropy between the CPO and WTI futures markets when the CPO futures price is USD denominated. In all records, the effective transfer entropy in table 3 is lower than that in table 2, and some information flows were not effective anymore with currency risk: evidence for the role of foreign exchange rate in the CPO futures market as a channelling actor for accommodating macroeconomic fluctuations [13,32]. However, transfer entropy from the CPO to WTI futures were effective even with the currency risk regardless of the pandemic outbreak. Put differently, the CPO futures strongly holds its leadership for providing information to the WTI futures regardless of currency channel in the period of high efficiency in the WTI futures market, dominated by low speculative trading activities.

Table 3. Effective transfer entropy between WTI and CPO futures markets (CPO futures in USD).

	TE	Std. Err.	*p*-value	ETE
Panel A. Overall (*num. of bins* = 153)				
WTI → CPO	2.3658	0.0543	0.0167	0.0000
CPO → WTI	2.4998	0.0474	0.0000	0.0193
Panel B. Pandemic period (*num. of bins* = 32)				
WTI → CPO	2.0352	0.0725	0.0000	0.0000
CPO → WTI	1.9504	0.0629	0.0000	0.0113

Prior studies documented that market becomes more efficient with higher liquidity [7] and more efficient markets have leadership on information discovery [10]. However, after the financialization of the crude oil market, the WTI futures market was strongly influenced by speculative trading activities [8]. Speculation causes herding behavior, which results in distorted market prices and lowers market efficiency [9]. Hence, asymmetric information flow from the CPO to WTI futures market can be explained by lower efficiency in the WTI futures market, despite the liquidity of WTI futures being way higher.

4. Conclusion

This study provides evidence that the CPO futures market has leadership on information discovery against the WTI futures market despite having lower liquidity and smaller market capitalization. Moreover, our finding is robust regardless of the COVID-19 outbreak, and asymmetric information flow from the CPO futures to WTI futures market even becomes unilateral considering the currency risk of MYR. Speculation in the WTI futures market causes herding behavior, resulting in price distortion and low market efficiency. Finally, we explain that the impact of speculation on market efficiency crowds out that of liquidity, leading to the asymmetric information flow from the CPO to WTI futures markets.

References

[1] Klessmann C, Lamers P, Ragwitz M and Resch G 2010 Design options for cooperation mechanisms under the new European renewable energy directive *Energy Policy* **38** 4679–4691
[2] Lind A, Rosenberg E, Seljom P, Espegren K, Fidje A and Lindberg K 2013 Analysis of the EU renewable energy directive by a techno-economic optimisation model *Energy Policy* **60** 364–377
[3] Stichnothe H, Schuchardt F and Rahutomo S 2014 European renewable energy directive: Critical analysis of important default values and methods for calculating greenhouse gas (GHG) emissions of palm oil biodiesel *Int. J. Life Cycle Assess.* **19** 1294–1304
[4] Snaith S, Kellard N M and Ahmad N 2018 Open outcry versus electronic trading: Tests of market efficiency on crude palm oil futures *J. Futures Markets* **38** 673–695
[5] Krustiyati A, Janisriwati S, Christine N and Huda M K 2020 Observing European Union rejection of Indonesia's crude palm oil exports from the most favored nation and quantitative restriction principles *J. Adv. Res. L. Econ.* **9** 905–912
[6] Fernández-Tirado F, Parra-López C and Romero-Gámez M 2021 A multi-criteria sustainability assessment for biodiesel alternatives in Spain: Life cycle assessment normalization and weighting *Renew. Energy* **164** 1195–1203
[7] Chordia T, Roll R and Subrahmanyam A 2008 Liquidity and market efficiency *J. Financ. Econ.* **87** 249–268
[8] Joo K, Suh J H, Lee D and Ahn K 2020 Impact of the global financial crisis on the crude oil market *Energy Strategy Rev.* **30** 100516
[9] Froot K A, Scharfstein D S and Stein J C 1992 Herd on the street: Informational inefficiencies in a market with short-term speculation *J. Fin.* **47** 1461–1484
[10] Eom C, Jung W, Choi S, Oh G and Kim S 2008 Effects of time dependency and efficiency on information flow in financial markets *Phys. A* **387** 5219–5224
[11] Norouzi N 2021 Post-COVID-19 and globalization of oil and natural gas trade: Challenges, opportunities, lessons, regulations, and strategies *Int. J. Energy Res.* **45** 14338–14356
[12] Joo K, Jeong M, Seo Y, Suh J H and Ahn K 2021 Shanghai crude oil futures: Flagship or burst? *Energy Rep.* **7** 4197–4204
[13] Zhang Y, Fan Y, Tsai H and Wei Y 2008 Spillover effect of US dollar exchange rate on oil prices *J. Policy Model.* **30** 973–991
[14] Granger C W J 1969 Investigating causal relations by econometric models and cross-spectral methods *Econometrica* **37** 424–438
[15] Jang S M, Yi E, Kim W C and Ahn K 2019 Information flow between Bitcoin and other investment assets *Entropy* **21** 1116

[16] Hasbrouck J 1995 One security, many markets: Determining the contributions to price discovery *J. Fin.* **50** 1175–1199
[17] Gonzalo J and Granger C 1969 Estimation of common long-memory components in cointegrated systems *J. Bus. Econ. Stat.* **13** 27–35
[18] Schreiber T 2000 Measuring information transfer *Phys. Rev. Lett.* **85** 461–464
[19] Yi E, Cho Y, Sohn S and Ahn K 2021 After the Splits: Information Flow between Bitcoin and Bitcoin Family *Chaos, Solitons & Fractals* **142** 110464
[20] Freedman D and Diaconis P 1981 On the histogram as a density estimator: L2 theory *Probab. Theor. Rel.* **57** 453–476
[21] Ahn K, Lee D, Sohn S and Yang B 2019 Stock market uncertainty and economic fundamentals: An entropy-based approach *Quant. Fin.* **19** 1151–1163
[22] Sandoval L 2014 Structure of a global network of financial companies based on transfer entropy *Entropy* **16** 4443–4482
[23] Lean H H and Smyth R 2015 Testing for weak-form efficiency of crude palm oil spot and future markets: New evidence from a GARCH unit root test with multiple structural breaks *Appl. Econ.* **47** 1710–1721
[24] Patton A J 2004 On the out-of-sample importance of skewness and asymmetric dependence for asset allocation *J. Financ. Econ.* **2** 130–168
[25] Fama E F 1965 The behavior of stock-market prices *J. Bus.* **38** 34–105
[26] Fama E F 1970 Efficient capital markets: A review of theory and empirical work *J. Fin.* **25** 383–417
[27] Lo A W and Mackinlay A C 1988 Stock market prices do not follow random walks: Evidence from a simple specification test *Rev. Financ. Stud.* **1** 41–66
[28] Jang H, Song Y and Ahn K 2020 Can government stabilize the housing market? The evidence from South Korea *Phys. A* **550** 124114
[29] Ahn K, Choi M, Dai B, Sohn S and Yang B 2018 Modeling stock return distributions with a quantum harmonic oscillator *Europhys. Lett.* **120** 38003
[30] Lee G H, Joo K and Ahn K 2020 Market efficiency of the crude palm oil: Evidence from quantum harmonic oscillator *J. Phys.: Conf. Ser.* **1593** 012037
[31] Ryu I, Jang H, Kim D and Ahn K 2021 Market efficiency of US REITs: A revisit *Chaos, Solitons & Fractals* **150** 111070
[32] Hadi A R A, Huridi M H, Zaini S M and Zainudin Z 2019 Is Ringgit Really Influenced by Crude Oil Price? Evidence From Commodity and Bank Lending Markets *Contemp. Econ.* **13** 407

Acknowledgments

This work was supported by the NRF grant funded by the Ministry of Science and ICT (Republic of Korea, No. 2021R1A2C1008652) and Technology Innovation Program ATC+ (20014125, Development of Intelligent Management Solution for Nuclear Decommissioning Site Characterization) funded by the Ministry of Trade, Industry & Energy (MOTIE, Republic of Korea).

Equilibrium in Complete Market with Jump-Diffusion Processes

Yunfeng Yang and Zini Wang

School of Science, Xi'an University of Science and Technology, Xi'an 710054, China
E-mail: yangyunfeng_1978@126.com

Abstract. This paper establishes the existence of a solution to the optimization problem. Supposing that risk assets pay continuous dividend regarded as the function of time. It is established that the behaviour model of the stock pricing process is jump-diffusion driven by a count process. We give a characterization of the optimal portfolio by means of the value function and the equivalent martingale measure defined by the utility function. The unique equivalent martingale measure, the unique optimal consumption and portfolio pair and the corresponding wealth process are deduced. We provide a simple characterization of an equilibrium market and discuss existence and uniqueness of equilibrium in the economy.
Keywords: jump-diffusion process; counting process; discount rate; equilibrium; dividend

1. Introduction

The portfolio optimization problem is an important research content in the mathematical finance, which is also important theory and methodology for investors and investment institutions in the asset and risk hedge strategy. In 1976, Merton [1] introduced the variance of the assets returns as the measure of risks. This is the result of a single agent and discontinuous optimization of asset prices. Documents [2-4] studied the extra local equilibrium problem. The equivalent martingale measure technique used in this paper is referred to jump-diffusion model [5-7]. In the continuous time financial market, [8-9] set up a security price equilibrium model which is determined by the law of Supply and demand. Zheng [10] discussed consumption optimization. This paper focuses on the problem of wealth optimization driven by incomplete and discontinuous prices.

We define the wealth optimization model:

$$\sup_{(c_k,\pi_k)\in A_k} E\int_0^T e^{-\int_0^t \beta(u)du} U_k(c_k(t))dt \ ,(k=1,\cdots,m) \tag{1}$$

In this wealth optimization model, $\beta(t)$ is discount rate and process $c_k(t)$ are consumption process. The aggregate endowment can be expressed as process $\varepsilon(t)=\sum_{k=1}^{m}\varepsilon_k(t), 0\le t\le T$,where processes $\{\varepsilon_k(t); 0\le t\le T\},(k=1,\cdots,m)$ are exogenous endowment processes.

2. The financial market

We assume that the price of riskless bond is governed by

$$dB(t) = B(t)[r(t)dt + dA(t)], \tag{2}$$

and the prices of the risky assets satisfy

$$\mathrm{d}S_i(t) = S_i(t-)\big[(\mu_i(t) - \rho_i(t))\mathrm{d}t + \mathrm{d}A(t) + \sigma_i(t)\mathrm{d}W(t) + \varphi_i(t)\mathrm{d}M(t)\big], \quad i=1,2, \tag{3}$$

We suppose risk interest rate, dividend and volatility($r(t)$, $\rho_1(t), \rho_2(t)$, $\sigma_1(t), \sigma_2(t)$) are supposed to be constant.

Process $\{W_t, 0 \le t \le T\}$ is a standard Wiener process, process $M(t) = N(t) - \int_0^t \lambda(s)\mathrm{d}s$ is the compensated martingale and nonexplosive counting process $\{N_t, 0 \le t \le T\}$ with intensity parameter $\lambda(t)$.

We suppose $m_{k0}(t), m_{k1}(t), m_{k2}(t)$ denote the number of shares of riskless bond and two risky assets, the investor k wealth process $X_k^{c_k,\pi_k}(t)$

$$X_k^{c_k,\pi_k}(t) = m_{k0}B(t) + m_{k1}S_1(t) + m_{k2}S_2(t) + \int_0^t \varepsilon_k(u) - c_k(u)du .$$

Let $m_{k1}S_1(t) = \pi_{k1}(t), m_{k2}S_2(t) = \pi_{k2}(t)$, $\pi_k = (\pi_{k1}, \pi_{k2})$. The strategy $(c_k, \pi_k) \in \mathrm{A}_k$ is standard self-financing way.

Lemma 1 The process

$$L(t) = \exp\{-\int_0^t \theta_1(s)\, dW(s) - \frac{1}{2}\int_0^t \theta_1^2(s)ds\} \cdot \exp\{\int_0^t \log\theta_2(s)dN(s) + \int_0^t \lambda(s)(1-\theta_2(s))ds\}$$

is a P martingale, and

$$\theta_1(t) = \frac{(\mu_2(t) - \rho_2(t) - \lambda(t)\varphi_2(t) - r(t))\varphi_1(t) - (\mu_1(t) - \rho_1(t) - \lambda(t)\varphi_1(t) - r(t))\varphi_2(t)}{\sigma_2(t)\varphi_1(t) - \sigma_1(t)\varphi_2(t)},$$

$$\theta_2(t) = \frac{(\mu_2(t) - \rho_2(t) - \lambda(t)\varphi_2(t) - r(t))\sigma_1(t) - (\mu_1(t) - \rho_1(t) - \lambda(t)\varphi_1(t) - r(t))\sigma_2(t)}{\lambda(t)(\sigma_2(t)\varphi_1(t) - \sigma_1(t)\varphi_2(t))}.$$

Lemma 2 Suppose we set up a complete, standard financial market. Consumption process $c_k(t)$ satisfies

$$E^*\int_0^T \frac{c_k(t)}{B(t)}dt = E^*\int_0^T \frac{\varepsilon_k(t)}{B(t)}dt .$$

Then there is a portfolio process $\pi_k(t)$, and process $c_k(t), \pi_k(t)$ satisfy $(c_k, \pi_k) \in \mathrm{A}_k$. In this case, the wealth process can be expressed as

$$X_k^{c_k,\pi_k}(t) = \frac{B(t)}{L(t)} E[\int_t^T \frac{L(u)}{B(u)}(c_k(u) - \varepsilon_k(u))du \,|\, \mathrm{F}_t].$$

Lemma 3 The strategy $(\hat{c}_k, \hat{\pi}_k) \in \mathrm{A}_k$ for optimal problem

$$\sup_{(c_k,\pi_k)\in\mathcal{A}_k} \quad E\int_0^T e^{-\int_0^t \beta(u)\mathrm{d}u}\, U_k(c_k(t))\mathrm{d}t.$$

the wealth process $X_k^{\hat{c}_k,\hat{\pi}_k}(t)$ are satisfy

$$\hat{c}_k(t) = I_k(y_k e^{\int_0^t \beta(u)du}\frac{L(t)}{B(t)}), \qquad X_k^{\hat{c}_k,\hat{\pi}_k}(t) = \frac{B(t)}{L(t)}E[\int_t^T \frac{L(u)}{B(u)}(\hat{c}_k(u) - \varepsilon_k(u))du \,|\, \mathrm{F}_t],$$

$$\pi_{k1}\sigma_1 + \pi_{k2}\sigma_2 = \frac{\theta_1^* + K(t)\theta_1}{L(t)}B(t), \qquad \pi_{k1}\varphi_1 + \pi_{k2}\varphi_2 = \frac{\theta_2^*\theta_2 + K(t)(1-\theta_2)}{L(t)\theta_2}B(t).$$

Lemma 4 If the financial markets are equilibrium, the aggregate endowment

$$\varepsilon(t) = \sum_{k=1}^m I_k(\frac{1}{\delta_k}e^{\int_0^t \beta(u)du}\frac{L(t)}{B(t)}), 0 \le t \le T \quad ,$$

where $\delta_k(t)$ satisfy

$$E\int_0^T \frac{L(t)}{B(t)}[I_k(\frac{1}{\delta_k}e^{\int_0^t \beta(u)du}\frac{L(t)}{B(t)})-\varepsilon_k(t)]dt=0 \ (k=1,\cdots,m)\ .$$

On the contrary, if financial market is satisfying Lemma 3, then this market is an equilibrium market.

3. Main findings

Proposition 1 A standard, complete financial market is an equilibrium market if and only if process $L(t)=\mathrm{e}^{-\int_0^t \beta(u)\mathrm{d}u}B(t)\tau(\varepsilon(t),\vec{\alpha})$, $\vec{\alpha}$ is a solution to the following system equations

$$E\int_0^T \mathrm{e}^{-\int_0^t \beta(u)\mathrm{d}u}\tau(\varepsilon(t),\vec{\alpha})\,[I_k(\tfrac{1}{\delta_k}\tau(\varepsilon(t),\vec{\alpha}))-\varepsilon_k(t)]\mathrm{d}t=0 \qquad (k=1,\cdots,m) \tag{4}$$

Proof

We define a vector $\vec{\alpha}=(\delta_1,\delta_2,\cdots,\delta_m)$. If $\vec{\alpha}$ satisfies either Lemma 4, then $\vec{\alpha}$ cannot be the zero vector. We set

$$f(\vec{\alpha})\triangleq\max_{\{k|\delta_k>0\}}(\delta_k U_k'(\bar{c}_k)). \tag{5}$$

The function

$$I(y,\vec{\alpha})\triangleq\sum_{k=1}^{m}I_k(\frac{y}{\delta_k}),0\le y\le\infty \tag{6}$$

is continuous and strictly decreasing. We define $\tau(\cdot\,\vec{\alpha})$ to be the inverse of $I(\cdot\,\vec{\alpha})$. We can see

$$L(t)=\mathrm{e}^{-\int_0^t \beta(u)\mathrm{d}u}B(t)\tau(\varepsilon(t),\vec{\alpha}). \tag{7}$$

At this point, the optimal consumption process satisfies

$$\hat{c}_k(t)=I_k(\frac{1}{\delta_k}\tau(\varepsilon(t),\vec{\alpha})),0\le t\le T. \tag{8}$$

Proposition 2 Define set $A=\{k|\varepsilon_k(t)=\bar{c}_k,k=1,2,\cdots,m\}$. Suppose $\vec{\alpha}$ satisfies (4), then $k\in A$ if and only if

$$\delta_k\le\frac{\tau(\varepsilon(t),\vec{\alpha})}{U_k'(\bar{c}_k)},0\le t\le T \tag{9}$$

almost surely. Define $\vec{\alpha}^*=(\delta_1^*,\delta_2^*,\cdots,\delta_m^*)$ by

$$\delta_k^*\triangleq\begin{cases}\inf\left(\min\limits_{0\le t\le T}\dfrac{\tau(\varepsilon(t),\vec{\alpha})}{U_k'(\bar{c}_k)}\right), & if\ k\in A,\\ \delta_k, & if\ k\notin A,\end{cases} \tag{10}$$

Then

$$f(\vec{\alpha})=f(\vec{\alpha}^*),\tau(\varepsilon(t),\vec{\alpha})=\tau(\varepsilon(t),\vec{\alpha}^*),$$
$$I_k(\tfrac{1}{\delta_k}\tau(\varepsilon(t),\vec{\alpha}))=I_k(\tfrac{1}{\delta_k^*}\tau(\varepsilon(t),\vec{\alpha}^*)),(k=1,\cdots,m).$$

Proof From (4) and $I_k(y)\ge\bar{c}_k$, we see that parameters $k\in A$ if and only if inequality

$$I_k(\tfrac{1}{\delta_k}\tau(\varepsilon(t),\vec{\alpha}))\ge\bar{c}_k \tag{11}$$

almost surely. (11) is equivalent to (9).

From (10), we can see

$$\delta_k U_k'(\bar{c}_k)\le\delta_k^* U_k'(\bar{c}_k)\le\tau(\varepsilon(t),\vec{\alpha})\le f(\vec{\alpha}),$$

this implies $f(\vec{\alpha})=f(\vec{\alpha}^*)$.

With $\delta_k^* U_k'(\bar{c}_k)\le\tau(\varepsilon(t),\vec{\alpha})$, we can deduce

$$I(\tau(\varepsilon(t),\vec{\alpha}),\vec{\alpha}^*)=\sum_{k\in A}I_k(\frac{\tau(\varepsilon(t),\vec{\alpha})}{\delta_k^*})+\sum_{k\notin A}I_k(\frac{\tau(\varepsilon(t),\vec{\alpha})}{\delta_k})$$
$$=\sum_{k=1}^{m}I_k(\frac{\tau(\varepsilon(t),\vec{\alpha})}{\delta_k})=\varepsilon(t)$$

this implies $\tau(\varepsilon(t),\vec{\alpha})=\tau(\varepsilon(t),\vec{\alpha}^*)$.

If $k\notin A$, $I_k(\frac{1}{\delta_k}\tau(\varepsilon(t),\vec{\alpha}))=I_k(\frac{1}{\delta_k^*}\tau(\varepsilon(t),\vec{\alpha}^*))$ holds with $\tau(\varepsilon(t),\vec{\alpha})=\tau(\varepsilon(t),\vec{\alpha}^*)$ and $\delta_k^*=\delta_k$. If $k\in A$, we obtain

$$\frac{1}{\delta_k}\tau(\varepsilon(t),\vec{\alpha})\geq\frac{1}{\delta_k^*}\tau(\varepsilon(t),\vec{\alpha})\geq U_k'(\overline{c}_k),$$

and

$$I_k(\frac{1}{\delta_k}\tau(\varepsilon(t),\vec{\alpha}))=\overline{c}_k=I_k(\frac{1}{\delta_k^*}\tau(\varepsilon(t),\vec{\alpha}))=I_k(\frac{1}{\delta_k^*}\tau(\varepsilon(t),\vec{\alpha}^*)).$$

That is $\vec{\alpha}^*=(\delta_1^*,\delta_2^*,\cdots,\delta_m^*)$ a solution of the system (4).

Proposition 3 There exists a vector $\vec{\alpha}_0=(\delta_1^0,\delta_2^0,\cdots,\delta_m^0)$ satisfying the system of equations （4）.

Proof We define

$$g_k(\vec{\alpha})\triangleq E\int_0^T \mathrm{e}^{-\int_0^t\beta(u)\mathrm{d}u}\tau(\varepsilon(t),\vec{\alpha})\left[I_k\left(\frac{1}{\delta_k}\tau(\varepsilon(t),\vec{\alpha})\right)-\varepsilon_k(t)\right]\mathrm{d}t,\quad (k=1,\cdots,m). \tag{12}$$

Summing $g_k(\vec{\alpha})$ over k and using (6), we can deduce

$$\sum_{k=1}^{m}g_k(\vec{\alpha})=E\int_0^T \mathrm{e}^{-\int_0^t\beta(u)\mathrm{d}u}\tau(\varepsilon(t),\vec{\alpha})\sum_{k=1}^{m}[I_k(\frac{1}{\delta_k}\tau(\varepsilon(t),\vec{\alpha}))-\varepsilon_k(t)]\mathrm{d}t$$
$$=E\int_0^T \mathrm{e}^{-\int_0^t\beta(u)\mathrm{d}u}\tau(\varepsilon(t),\vec{\alpha})\,[I(\tau(\varepsilon(t),\vec{\alpha}),\vec{\alpha})-\varepsilon(t)]\mathrm{d}t=0. \tag{13}$$

We define set $\mathcal{M}_k\triangleq\left\{\vec{\alpha}=(\delta_1,\delta_2,\cdots,\delta_m)\middle|g_k(\vec{\alpha})\geq 0\right\}$. From the KKM lemma, we have $\bigcap_{k=1}^{m}\mathcal{M}_k\neq\phi$.

Let $\vec{\alpha}_0=(\delta_1^0,\delta_2^0,\cdots,\delta_m^0)\in\bigcap_{k=1}^{m}\mathcal{M}_k$, according to the definition of $\mathcal{M}_k$ and (13), we obtain $g_k(\vec{\alpha}_0)=0$ for arbitrary $k=1,2,\cdots,m$. That is $\vec{\alpha}_0=(\delta_1^0,\delta_2^0,\cdots,\delta_m^0)$ satisfying the system of equations （4）.

Proposition 4 Define function

$$U(c,\vec{\alpha})\triangleq\max_{\substack{c_1\geq\overline{c}_1,\cdots,c_m\geq\overline{c}_m\\ c_1+c_2+\cdots+c_m=c}}\sum_{k=1}^{m}\delta_kU_k(c_k(t)). \tag{14}$$

Then $U'(c,\vec{\alpha})=\tau(c,\vec{\alpha})$ and (8) is a solution to the formula (14).

Proof According to the definition of utility function, we have

$$\sum_{k=1}^{m}\delta_kU_k(c_k)\leq\sum_{k=1}^{m}\delta_k[U_k(\hat{c}_k)+(c_k-\hat{c}_k)U_k'(\hat{c}_k)]$$
$$=\sum_{k=1}^{m}\delta_kU_k(\hat{c}_k)+\sum_{\{k|\hat{c}_k=\overline{c}_k\}}\delta_kU_k'(\hat{c}_k)(c_k-\hat{c}_k)+\tau(c,\vec{\alpha})\sum_{\{k|\hat{c}_k>\overline{c}_k\}}(c_k-\hat{c}_k)$$
$$\leq\sum_{k=1}^{m}\delta_kU_k(\hat{c}_k)+\tau(c,\vec{\alpha})\sum_{k=1}^{m}(c_k-\hat{c}_k)$$

$$=\sum_{k=1}^{m}\delta_k U_k(\hat{c}_k).$$

In other words, (8) is a solution to the formula (14), and we obtain

$$U(c,\vec{\alpha})=\sum_{k=1}^{m}\delta_k U_k(I_k(\frac{1}{\delta_k}\tau(c,\vec{\alpha}))). \tag{15}$$

From (15), we see that

$$\begin{aligned}U'(c,\vec{\alpha})&=\frac{d}{dc}\sum_{k=1}^{m}\delta_k U_k(I_k(\frac{1}{\delta_k}\tau(c,\vec{\alpha})))\\
&=\sum_{k=1}^{m}U_k'(I_k(\frac{1}{\delta_k}\tau(c,\vec{\alpha})))I_k'(\frac{1}{\delta_k}\tau(c,\vec{\alpha}))\tau'(c,\vec{\alpha})\\
&=\sum_{k=1}^{m}\frac{1}{\delta_k}\tau(c,\vec{\alpha})I_k'(\frac{1}{\delta_k}\tau(c,\vec{\alpha}))\tau'(c,\vec{\alpha})\\
&=\sum_{k=1}^{m}\tau(c,\vec{\alpha})\frac{d}{dc}I_k(\frac{1}{\delta_k}\tau(c,\vec{\alpha}))\\
&=\tau(c,\vec{\alpha})\frac{d}{dc}I(\tau(c,\vec{\alpha}),\vec{\alpha})\\
&=\tau(c,\vec{\alpha}).\end{aligned}$$

Proposition 5 A standard, complete market is an equilibrium financial market if and only if each of the following is found

$$r(t)=\beta(t)-\frac{1}{U'(\varepsilon(t),\vec{\alpha})}[U''(\varepsilon(t),\vec{\alpha})(\mu_\varepsilon(t)-\varphi_\varepsilon(t)\lambda(t))+\frac{1}{2}U'''(\varepsilon(t),\vec{\alpha})\sigma_\varepsilon^2(t)$$
$$+(U'(\varepsilon(t-)+\varphi_\varepsilon(t),\vec{\alpha})-U'(\varepsilon(t-),\vec{\alpha}))\lambda(t)], \tag{16}$$

$$\theta_1(t)=-\frac{U''(\varepsilon(t),\vec{\alpha})}{U'(\varepsilon(t),\vec{\alpha})}\sigma_\varepsilon(t), \tag{17}$$

$$\theta_2(t)=\frac{-1}{U'(\varepsilon(t),\vec{\alpha})}(U'(\varepsilon(t-)+\varphi_\varepsilon(t),\vec{\alpha})-U'(\varepsilon(t-),\vec{\alpha}))\lambda(t). \tag{18}$$

Proof We get the following from (17) that $\tau(\varepsilon(t),\vec{\alpha})=e^{\int_0^t\beta(u)du}\frac{L(t)}{B(t)}, 0\le t\le T; \tau(\varepsilon(0),\vec{\alpha})=1$, Ito's lemma provides

$$\tau(\varepsilon(t),\vec{\alpha})=1+\int_0^t\tau(\varepsilon(u),\vec{\alpha})(\beta(u)-r(u))du$$
$$-\int_0^t\tau(\varepsilon(u),\vec{\alpha})\theta_1(u)dW(u)-\int_0^t\tau(\varepsilon(u),\vec{\alpha})\theta_2(u)dM(u). \tag{19}$$

However, since market is an equilibrium market, we can see

$$\tau(\varepsilon(t),\vec{\alpha})=U'(\varepsilon(t),\vec{\alpha}).$$

According to the Ito's lemma and

$$\varepsilon(t)=\varepsilon(0)+\int_0^t\mu_\varepsilon(u)du+\int_0^t\sigma_\varepsilon(u)\mathrm{d}W(u)+\int_0^t\varphi_\varepsilon(u)\mathrm{d}M(u), 0\le t\le T.$$

We see that

$$\tau(\varepsilon(t),\vec{\alpha})=1+\int_0^t U''(\varepsilon(u),\vec{\alpha})(\mu_\varepsilon(u)-\phi_\varepsilon(u)\lambda(u))+\frac{1}{2}U'''(\varepsilon(u),\vec{\alpha})\sigma_\varepsilon^2(u)\mathrm{d}u$$
$$+\int_0^t[U'(\varepsilon(u-)+\varphi_\varepsilon(u),\vec{\alpha})-U'(\varepsilon(u-),\vec{\alpha})]\lambda(u)\mathrm{d}u$$

$$+\int_0^t U''(\varepsilon(u),\vec{\alpha})\sigma_\varepsilon(u)\mathrm{d}W(u)$$

$$+\int_0^t [U'(\varepsilon(u-)+\varphi_\varepsilon(u),\vec{\alpha})-U'(\varepsilon(u-),\vec{\alpha})]\lambda(u)\mathrm{d}M(u)\,. \tag{20}$$

In light of the decompositions (19) and (20), we get (16) (17) and (18).

According to Proposition, choose $\vec{\alpha}_0=(\delta_1^0,\delta_2^0,\cdots,\delta_m^0)$ satisfying the system of equations (4), define $r(t),\theta_1(t),$ and $\theta_2(t)$ by (16) (17) and (18). Then the market is an equilibrium market. The uniqueness of the equilibrium market follows from **Proposition 2.**

4. Concluding remarks

By using the theory of Martingale and point process, we study the optimal equilibrium problem of jump process markets and prove the existence of a solution to the problem. The existence and uniqueness of equilibrium are discussed.

References

[1] Merton, R. C. Option pricing when underlying stock Return are discontinuous. Journal of Economics, Vol.36,no.3, pp. 125-144,1976.
[2] Aase, K. Ruin problems and myopic portfolio optimization in continuous trading. Stochastic Processes and Their Applications, Vol.21,no.2, pp.213-227,1986.
[3] Das, S., & Uppal, R. Systemic risk and international portfolio choice. Journal of Finance, Vol.59,no.6, pp. 2809–2834,2004.
[4] Liu, J., Longstaff, F., & Pan, J. Dynamic asset allocation with event risk. Journal of Finance, Vol.58,no.5, pp. 231–259,2003.
[5] Jeanblanc, M., & Pontier, M. Optimal portfolio for a small investor in a market model with discontinuous prices. Applied Mathematical Optimization, Vol.22,no.3, pp. 287–310,1990.
[6] Xue, X. Martingale representation theory for a class of Levy processes and its applications. Ph.D.Dissertation, 1991.
[7] Bardhan, I., & Chao, X. Martingale analysis for assets with discontinuous returns. Mathematics of Operations Research, Vol.20,no.1, pp. 243–256,1995.
[8] D.Thakoor and M. Bhuruth, Fast quadrature methods for options with discrete dividends, Journal of Computational and Applied Mathematics, vol. 330,Mar. 2018,pp. 1–14.
[9] F.M.Bandi and R.Renòc, Price and volatility co-jumps, Journal of Financial Economics. 119, Jan. 2016,pp. 107–146.
[10] Y.C.Zheng and Y.F. Yang, Wealth optimization models on jump-diffusion model, Journal of Interdisciplinary Mathematics, vol.21, Mar 2018,pp. 201–212.

The Regional Expressions of Secular Stagnation

Zhiye Zhou[1, *], Yating Mao[2], Xinyao Wang[3] and Yuqing Qian[4]

[1]School of Mathematics, University of Liverpool, Liverpool,UK
[2]School of Business, Zhujiang College of South China Agriculture Universit, [3]Guangdong, China
[4]Lee Shau Kee School of Business and Administration, Hong Kong Metropolitan University, Hong Kong, China
[5]St Swithun's School, Winchester, UK
*Corresponding author E-mail: Z.ZHOU75@student.liverpool.ac.uk

Abstract. This The terminology secular stagnation is suggested by Larry Summers in the last decade to demonstrate the globally economic downward trend. Although this standpoint was not paid importance at that time, the unstoppable decrease in world economy of recent years has shown this. Driven by these miscellaneous data, this report aims to search for evidence that can argue the secular stagnation. Thereby, we first analyse and select the objective tabular data from 1970s to 2010s covering all countries to apply the multiple linear regression model. Then the statistical theory is used for both preliminary measurement and subsequent regression diagnostics, also for the estimation method. In data processing, this paper explores the relationship between economic slow-down rate and (the logarithm of) affluence level with the random forest logarithm to find the threshold. Through the data visualization, the economic trend during the corresponding objective period can be observed intuitively. Finally, in the extension of the slow-down rate, we try to excavate the deep interactive terms among all countries. By the Solow growth model, it can be concluded that the gap between developing countries and developed countries is narrower due to the low original capital of poor countries. This phenomenon is also known as the catch-up effect. This report conducts the slow-down rate to verify the secular stagnation and meanwhile provides evidence for the relationship between slow-down and affluence levels.

1. Introduction

1.1. Background

It seems that the growth rate of world economy has already gradually slowed down before the Covid-19 pandemic, which is suggested by Jacobs et al. [1]. It is noteworthy that this phenomenon is more concerning in developed countries, as majority of high-income countries are stuck in a trap that there is an insufficient demand to boost their economies. Moreover, as Younsi and Nafla [2] suppose that the monetary policies of these countries are facing invalid dangers, possibly causing financial instability and adverse effects on economic development. Therefore, people started to take notice of the reliability of a controversial hypothesis called 'secular stagnation'.
In 1938, six years after Great Depression, Alvin Hansen [3], a chair professor at Harvard University, coined a new nation 'secular stagnation', which means that a diminishing growth rate could persist in long run after a great recession due to a lack of technological innovation and a slow population growth. However, his argument was subsequently proved wrong when the World War 2 started, a significant

increase in spending on military defences, and competitions on technology and weapons among countries largely facilitated an economic boom.
His theory was forgotten until 2013, the former Chief Economist of the World Bank Lawrence Summers (2013) re-introduced it after Global Financial Crisis, bringing it back to the public eye again. In the speech given by Summers [4] at International Monetary Fund (IMF), he suggested that the recovery from 2008 Financial crisis was the slowest in history even with such a low interest rates bound. The rise in saving, the slowdown of population and technology constituted an insufficient demand, which then led to secular stagnation.
In responding to this concept, it should be found out that whether the secular stagnation exists or not, and the belief that those richer countries are suffering severer secular stagnation than poorer countries.

1.2. Objective

The main objective of this research about secular stagnation is to find whether there are evidences to support this controversial hypothesis. In this paper, the measurement of affluence level in all countries will be evaluated and the investigation about the relevance between countries' wealth extent and growth slowdown rate will be demonstrated to visualize the secular stagnation. In the first stage, the comparisons between productivity, GDP and GDP per capital give advice on which one is a more suitable parameter to quantify the wealthy degree. In the second stage, a scatterplot is generated to indicate the relationship between the slow down rate and countries' affluence level. By observing the tendency of the diagram, the secular stagnation seems to be substantiated. The terminal stage is bent on excavating deeper into the causes of the distribution of slow-down rate and explaining some internal connections through theories. Furthermore, a reflection about limitations is discussed, following is the conclusion.

2. Literature

This section illustrated other economists' perspectives about secular stagnation, incorporating the critics from Joseph Stiglitz, and the insufficient evidence from Summers' research noted by Roger Farmer. In addition, brief views by Pail Krugman and The Economists are placed as well.
The former Chief Economist of the World Bank Joseph Stiglitz [5] countered the ideology of secular stagnation by noting that the effective fiscal stimulus can fully offset this dire prediction. He argued that the languishment of the economy is not caused by the economy itself, but the noneffective and insufficient fiscal stimulus policy. He gave an example of a large fiscal stimulus in December 2017 and January 2018 and said that the effectiveness of that would have been enlarged a decade earlier with a high unemployment rate. He also asserted that whether the economic growth rate in the future can be as fast as that in the past, it depends on the rate of technology development and proper intervention by government.
According to the debate between Summers and Stiglitz [6], the Professor of Economics at the University of Warwick Roger Farmer [7] gave his own opinion about secular stagnation. He claimed that both Lawrence Summer's and Joseph Stiglitz's theories are expected to be re-examined by new economic model, and he asserted that Summer did not illustrate a fully well-structured equilibrium model that backed his suggestion. However, he agreed with Summer's advocacy that free economy may be failed to adjust back to full employment after a significant shrink of economy. His internally coherent theory explained the reason of the secular stagnation, which incorporated Keynesian economics. The high permanent unemployment is irrelevant with sticky price, but a missing essence of markets.
On top of that, Paul Krugman [8] pointed out that the root of this problem is that whether the policy maker can stimulate people to spend money when they lose the faith of the economy. While Eggers, Ellison and Lee [9] criticized that the theory of secular stagnation ignored the underly improvement of technology and it is always to be revised after recessions, and soon it is to be proved false during a boom in economy.

3. Data and Theory

3.1. Sources

According to Probst [4], the researcher in economic history, the OECD projections foresee the growth in all advanced economies is much slower than that a few months ago. The following Figure 1 display the predications from last May to the latest outlook, with the red arrows indicating the most dramatic downward.

	2017	2018	2019
World	3.7	3.8	3.9
Australia	2.3	2.9	3.0
Canada	3.0	2.1	2.2
Euro area	2.6	2.2	2.1
Germany	2.5	2.1	2.1
France	2.3	1.9	1.9
Italy	1.6	1.4	1.1
Japan	1.7	1.2	1.2
Korea	3.1	3.0	3.0
United Kingdom	1.8	1.4	1.3
United States	2.3	2.9	2.8

	2018	2019	2020
World	3.6	3.3	3.4
Australia	2.9	2.7	2.5
Canada	1.8	1.5	2.0
Euro area	1.8	1.0	1.2
Germany	1.4	0.7	1.1
France	1.5	1.3	1.3
Italy	0.8	-0.2	0.5
Japan	0.7	0.8	0.7
Korea	2.7	2.6	2.6
United Kingdom	1.4	0.8	0.9
United States	2.9	2.6	2.2

Figure 1. Economic changes in the world and some developed countries in recent years [4]

Additionally, economic growth rates of developed countries have been slowed down in recent years. The most representative one is American economy, whose actual economic growth has been below the potential economic growth rate since global financial crisis [10]. Meanwhile, monetary policy becomes an invalid policy, as most interest rates of developed countries are closed to zero, it is no longer an effective method in terms of stimulating demand [11]. The statistics in Figure 2 demonstrate the business cycle of USA, and in Figure 3 the interest rates of developed countries.

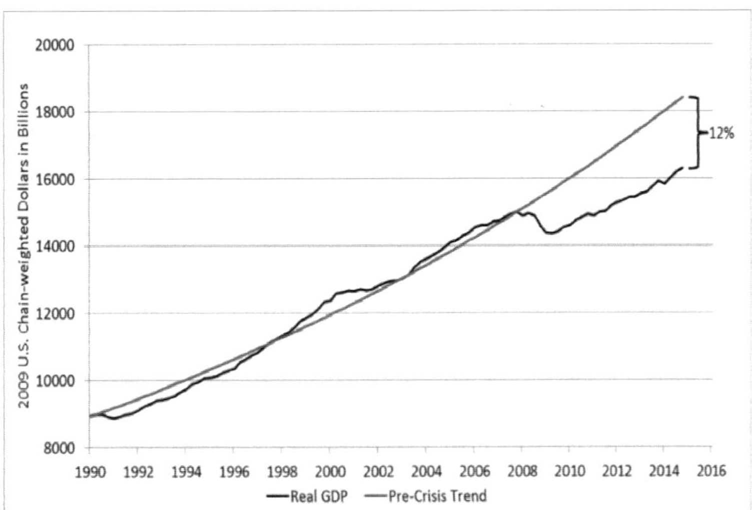

Figure 2. The economic growth rate of the USA [10]

Country	Last	Previous	Reference	Unit
Switzerland	-0.75	-0.75	Mar/21	%
Denmark	-0.5	-0.6	Mar/21	%
Japan	-0.1	-0.1	Apr/21	%
Sweden	0	0	Apr/21	%
Spain	0	0	Apr/21	%
Slovenia	0	0	Apr/21	%
Slovakia	0	0	Apr/21	%
Portugal	0	0	Apr/21	%
Norway	0	0	May/21	%
Netherlands	0	0	Apr/21	%
Malta	0	0	Apr/21	%
Luxembourg	0	0	Apr/21	%
Lithuania	0	0	Apr/21	%
Latvia	0	0	Apr/21	%
Italy	0	0	Apr/21	%
Ireland	0	0	Apr/21	%
Greece	0	0	Apr/21	%
Germany	0	0	Apr/21	%
France	0	0	Apr/21	%
Finland	0	0	Apr/21	%
Euro Area	0	0	Apr/21	%
Estonia	0	0	Apr/21	%
Cyprus	0	0	Apr/21	%
Bulgaria	0	0	May/21	%
Belgium	0	0	Apr/21	%
Austria	0	0	Apr/21	%
United Kingdom	0.1	0.1	May/21	%
Poland	0.1	0.1	May/21	%
Israel	0.1	0.1	Apr/21	%
Australia	0.1	0.1	May/21	%
United States	0.25	0.25	Apr/21	%

Figure 3. Interest rates of developed countries [11]

Lawrence Summer's research made a comparison about the difference between actual growth and economic potential trend within three economies, pointing out that the estimate of GDP potential growth of America in Figure 4 [12] was lowering down after global financial crisis while the disparity between the predict lines and actual growth has been increased. Additionally, the example of Japan and EU economy in Figure 5 and Figure 6 [12] showed that their economy went to a trap as the old monetary policies were unable to stimulate the demand any longer.

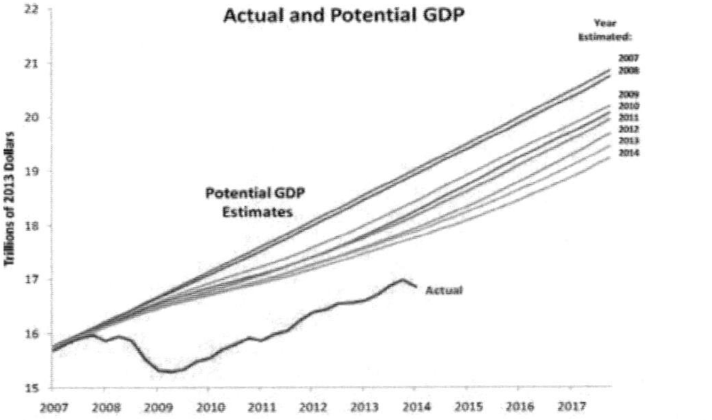

Figure 4. Actual and potential GDP from 2007 to 2014 [12]

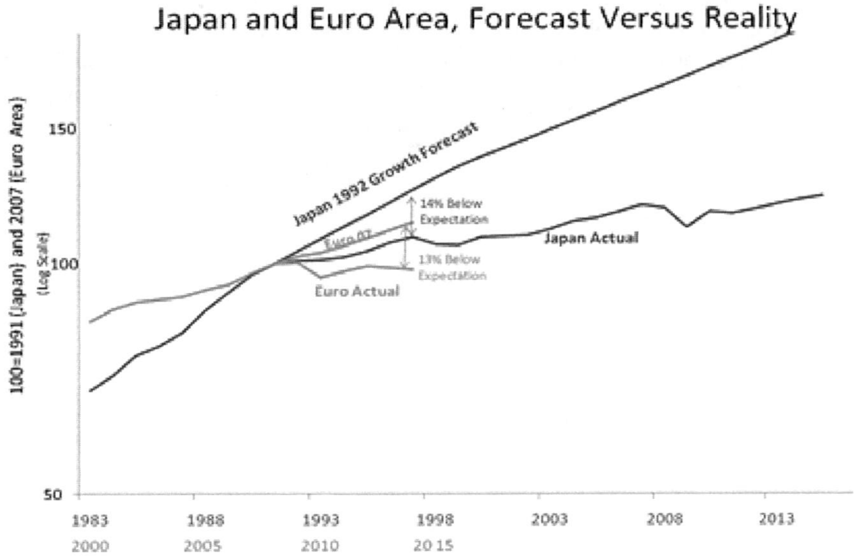

Figure 5. Forecast and reality of Japan and Europe [12]

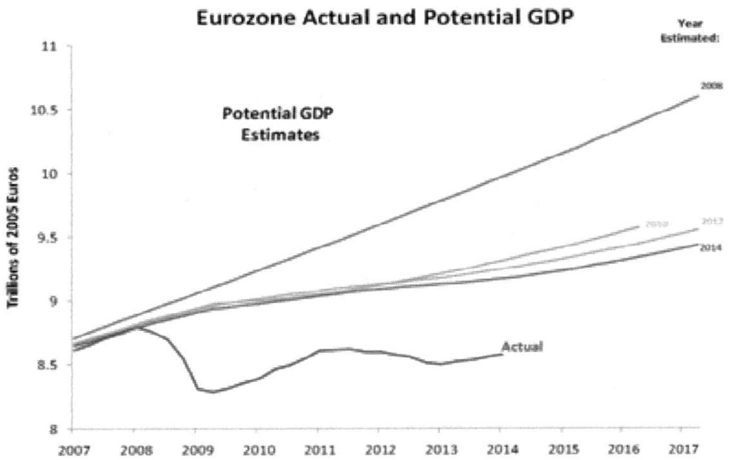

Figure 6. Actual and potential GDP in developed countries [12]

3.2. Theory derivatives

A. The R language theory such as multiple linear regression analysis and diagnostics is used.
B. The statistical theory is used in three sessions, to make comparisons among measurements in data processing; in the form of OLS method in estimation part; in the form of Gauss-Markov theory to judge whether the error is reasonable.
C. The approach of random forest is used for the appearance of slow down rate.
D. The Solow growth model and Summer hypothesis are applied to the extension of research, including catch-up effect and leapfrogging effect.

3.3. Data Processing

3.3.1. GDP per capital and productivity, which is a better measure? Since either GDP per capital or productivity will be used to measure the slow down rate in the final stage, this research begins with finding out the relationship between GDP per capital and time, as well as productivity and time, respectively. Therefore, the multiple lineal regression is applied to these two relationships.
In both experiments, three dummy variables are uniformly constructed, namely D_1, D_2, D_3, which represent for progressive levels of affluence, to divide all countries in the world into three parts. Next, breakdown countries secondarily in the light of the data from World Bank [13]. Additionally, ε_i is used to describe the interactive terms that has impacts on GDP per capital or productivity.
Firstly, for GDP per capital, we have the following regression function:

$$GDP\ per\ capital_i = \alpha + \beta x_i + \gamma_1 D_{1i} + \gamma_2 D_{2i} + \gamma_3 D_{3i} + \varepsilon_i$$

where i is serial numbers of countries, i=1,2,…,182
x is year,
α is the intercept,
β is the slope,
$\gamma_1, \gamma_2, \gamma_3$ are coefficients of dummy variables
It is demonstrated by Dynan and Sheiner [14] that the expression for GDP can be conducted by expenditure-side approach as
GDP = private consumption + gross investment + government investment + government spending + (exports - imports).
According to the formula of GDP calculation, select the following data respectively,

- *Real consumption of households and government* (ccon), which means the personal consumption of households and the consumption of government;
- *Capital stock at Constant 2011 National Prices* (rnna), which means gross growth in investment and government investment, "gross" refers to the fact that GDP measures production and does not consider the various uses of the product. Production can be consumed, invested in fixed assets or inventories, or used to replace depreciated fixed assets [15];
- *Price level of exports* (pl-x), *Price level of imports* (pl-m). Since the price level of the market is affected by the rate of inflation, our exports and imports are measured by the price level.

Then the GDP per capital is

$$\frac{GDP}{population} = \frac{\text{ccon} + \text{rnna} + \text{plx} - \text{plm}}{population}$$

Use R to do the regression analysis, the adjusted-R^2 can be worked out to quantify the relationship between GDP per capital and time as in Figure 7.

```
Residuals:
     Min       1Q   Median       3Q      Max
-0.11081 -0.03361 -0.00621  0.01894  0.92518

Coefficients: (1 not defined because of singularities)
             Estimate Std. Error t value Pr(>|t|)
(Intercept) -2.565e+00  6.690e-02 -38.345   <2e-16 ***
Y            1.309e-03  3.364e-05  38.903   <2e-16 ***
D1                  NA         NA      NA       NA
D2          -1.287e-02  1.448e-03  -8.891   <2e-16 ***
D3           8.893e-02  1.492e-03  59.609   <2e-16 ***
---
Signif. codes:  0 '***' 0.001 '**' 0.01 '*' 0.05 '.' 0.1 ' ' 1

Residual standard error: 0.0608 on 10001 degrees of freedom
  (2371 observations deleted due to missingness)
Multiple R-squared:  0.4003,	Adjusted R-squared:  0.4001
F-statistic:  2225 on 3 and 10001 DF,  p-value: < 2.2e-16
```

Figure 7. R output for the regression of GDP per capital

Besides, the conformity of total GDP is also be studied. However, as the Figure 8 shows below, it seems to be a subordinate choice.

```
Residuals:
     Min       1Q   Median       3Q       Max
-3279992 -1283908  -649195   174002 100776230

Coefficients: (1 not defined because of singularities)
            Estimate Std. Error t value Pr(>|t|)
(Intercept) -76364867    3882430 -19.669  < 2e-16 ***
Y               38723       1957  19.786  < 2e-16 ***
D1                 NA         NA      NA       NA
D2             435018      90648   4.799 1.61e-06 ***
D3            1543908      95205  16.217  < 2e-16 ***
---
Signif. codes:  0 '***' 0.001 '**' 0.01 '*' 0.05 '.' 0.1 ' ' 1

Residual standard error: 4273000 on 12372 degrees of freedom
Multiple R-squared:  0.05106,	Adjusted R-squared:  0.05083
F-statistic: 221.9 on 3 and 12372 DF,  p-value: < 2.2e-16
```

Figure 8. R output for the regression of GDP

Secondly, for productivity, the regression function is:

$$productivity_i = \alpha + \beta x_i + \gamma_1 D_{1i} + \gamma_2 D_{2i} + \gamma_3 D_{3i} + \varepsilon_i$$

where i is serial numbers of countries, i=1,2,…,182

x is year,

α is the intercept,

β is the slope,

$\gamma_1, \gamma_2, \gamma_3$ are coefficients of dummy variables

Scheryer and Pilat [16] pointed out that the productivity can be measured by gross output, that is, the relation is presented as a production function H with gross output Q,

Q = H (A, K, L, M)

where L is labor input,

K is capital input,

M is intermediate input,

A is a parameter of technical change

Here use emp*avh in excel to stand for the expression of productivity. That is,

Productivity= Number of persons engaged (in millions) * Average annual hours worked by persons engaged

By R, the adjusted-R^2 can be observed in Figure 9.

```
Residuals:
   Min      1Q  Median      3Q     Max
 -50697  -22663  -11882    1178 1677292

Coefficients:
              Estimate Std. Error t value Pr(>|t|)
(Intercept) -927020.33   91050.43 -10.181   <2e-16 ***
Y               471.34      45.89  10.270   <2e-16 ***
D1            -7875.86    3195.39  -2.465   0.0137 *
D2            22849.76    2425.74   9.420   <2e-16 ***
D3             4181.60    2518.13   1.661   0.0968 .
---
Signif. codes:  0 '***' 0.001 '**' 0.01 '*' 0.05 '.' 0.1 ' ' 1

Residual standard error: 100200 on 12371 degrees of freedom
Multiple R-squared:  0.02112,   Adjusted R-squared:  0.02081
F-statistic: 66.74 on 4 and 12371 DF,  p-value: < 2.2e-16
```

Figure 9. R output for the regression of productivity

In Figures 7,8,9, R language is used to seek variables linked to the multiple linear regression. For instance, the list of residuals enumerates 5 numbers in the form of boxplot, the quartiles are introduced to describe the median (Q_2), the median of the lowest half (Q_1), and the median of the highest half (Q_3), also with the minimum and the maximum. Likewise, the tabulation of coefficients covers four characteristics of both the intercept and the explanatory variables. 'Estimate' means the estimated value calculated by R, the second column represents for the standard error, 't value' is the test statistic for t-test, the last column is the probability of rejection in hypothesis test. Below are the significance codes, which help to judge what extent of evidence should be rejected. Finally, the multiple R-squared is the initial value and the adjusted R-squared is corrected value. The p-value can also be viewed as the test statistic.

R^2 correlation system test method is used to judge the fitting degree of the regression equation. The value of R^2 is between 0 and 1, and the closer to 1, the better the fitting degree is. We want to assess how useful the model is in explaining the data. This can be assessed by the value of adjusted-R^2.

Let SSR denote the regression sum of squares, which measures how much of the variation in the y values is explained by the regression; SST denote the total sum of squares, which measures the total amount of variation in the y values. The proportion of SST explained by SSR is R^2.

$$SST = \sum_{i=1}^{n} (y_i - \bar{y})^2$$

$$SSR = \sum_{i=1}^{n} (\hat{y}_i - \bar{y})^2$$

$$R^2 = \frac{SSR}{SST}$$

$$0 \le R^2 \le 1$$

By looking at the data result of the model, it can be observed that the p-value is less than 0.01. Therefore, it is known from t-test that the independent variable X is very significant in Y, and it is judged from F-test that the independent variable of the whole model is very significant. Meanwhile, R^2=0.4001 in Figure 7 means that the independent variable and the dependent variable are highly correlated. According to the R output in Figure 8, on the whole, independent variable has a significant correlation with Y, but R^2=0.05083. Therefore, it is judged that the independent variable and the dependent variable have a low correlation and a low goodness of fit.

Hence, in these experiments, it can be obtained that the **GDP per capital** is a better measure as its adjusted-R^2 is closer to 1.

3.3.2. The scatterplot of slow down rate.

Linear regression analysis

Next stage, it is necessary to work out the trend of slow down rate to visualise the secular stagnation by studying the relationship between slow down rate and affluence level. Firstly, a scatterplot should be drawn to explore the distribution of slow down rate in different countries and the relationship between slow down rate and the affluence level. Choose GDP per capital as a measure for these two variables according to the last stage, because it is more appropriate than the productivity.

For the y-axis, that is, the slow down rate, which is defined to be equal to the average growth rate from 1994 to 2017 minus the average growth rate from 1970 to 1993. That is,

Slow down rate = average growth rate from 1994 to 2017 - average growth rate from 1970 to 1993

If the slope is positive, this means that there exists an acceleration in the economic growth, while the negative value means the downward economic growth rate. However, if the GDP per capital is used as x-axis, there will be a strong influential outlier, compelling all scatters clustering to the left. Hence, a transformation is conducted by adding logarithm to the GDP per capital. The influential observation is weakened in this case. Likewise, rank the values of GDP per capital from the smallest to the largest through the time period 1983-1993, where the assumed slow-down has not happened. Therefore, a scatterplot can be drawn as shown in Figure 10 to indicate the distribution of slow-down rate in all countries.

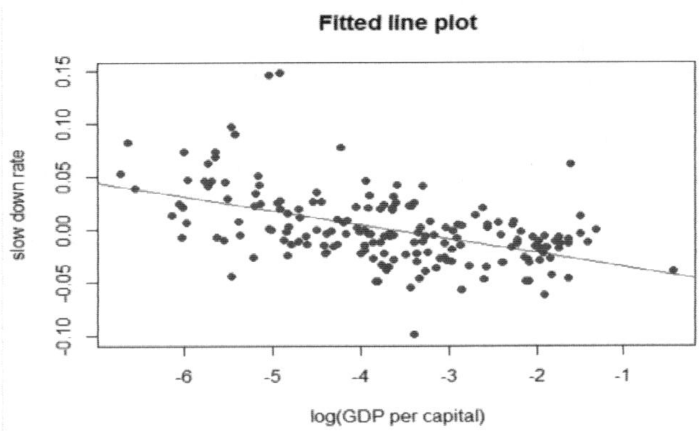

Figure 10. The scatterplot and fitted line of slow-down rate versus the logarithm of GDP per capital

By R, the scatterplot is runed out. It can be obviously observed that there are three outliers, all should be checked. One is on the right-hand-side, this country has the largest GDP but its slow down rate is negative. It can be found from the data table that this country is Qatar, which is an oil produced country [16]. Thus, its export level is much higher than other countries, explaining the high GDP value. Another two outliers are outside the interval [-0.1, 0.1], they are Georgia and Equatorial Guinea, the former [17] accelerated the economic development in the 1990s and the later [18] implemented the economic restructuring in 1987.

From the scatterplot, the tendency can be roughly observed. Next, a fitted linear regression can be worked out, which appears to imply the secular stagnation as its slope is negative. However, the practical meaning of this graph may be illogical because of the constant drop. This tendency derives from the laws of the real world. Therefore, there exist a threshold point that divide the tendency of slow down rate into two parts, with a gradient descent.

Data transformation

To find the threshold, position it relative to the middle of the scatterplot at first according to the distribution of scatters, i.e., within the interval (-5, -4). By the approach of random forest from Prasetiyowati, Maulidevi and Surendro [19], traverse all points in this interval to find a point where the residual sum of squares is the smallest among them. Hence, after calculating all possible points, the country code with the smallest SSR is CHN. Next, take the means of the two segments, one has the logarithm of GDP per capital smaller than CHN and the other has the logarithm of GDP per capital larger than CHN. Then, a transformed fitted line plot with a gradient decent at the threshold can be displayed in Figure 11. The average value of the former segment is 0.03477754 and the average value of the later segment is -0.007733709.

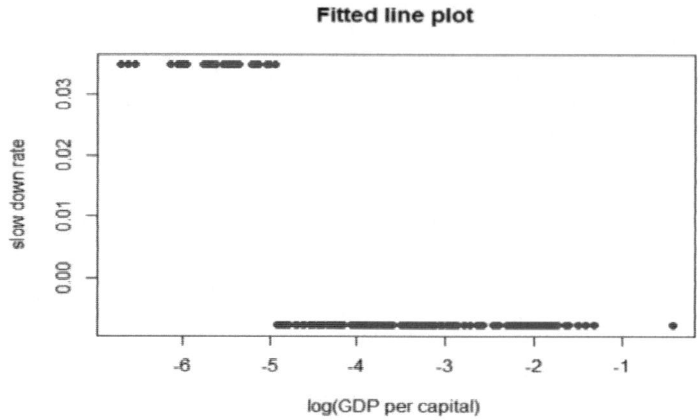

Figure 11. The transformed slow-down rate

It can be summarised from Figure 10 that the world undertakes the economic downward pressure as the global economy grows steadily. Moreover, there's a relationship between slow down rate and affluence level, a richer country suffers from a more severe economic downward. Since the slow down rate is below zero, it can be concluded that the world economy is experiencing **secular stagnation**.

3.3.3. Regression diagnostics

To examine the fitness of the linear regression, run R and we obtain four regression diagnostic diagrams as shown in Figure 12.

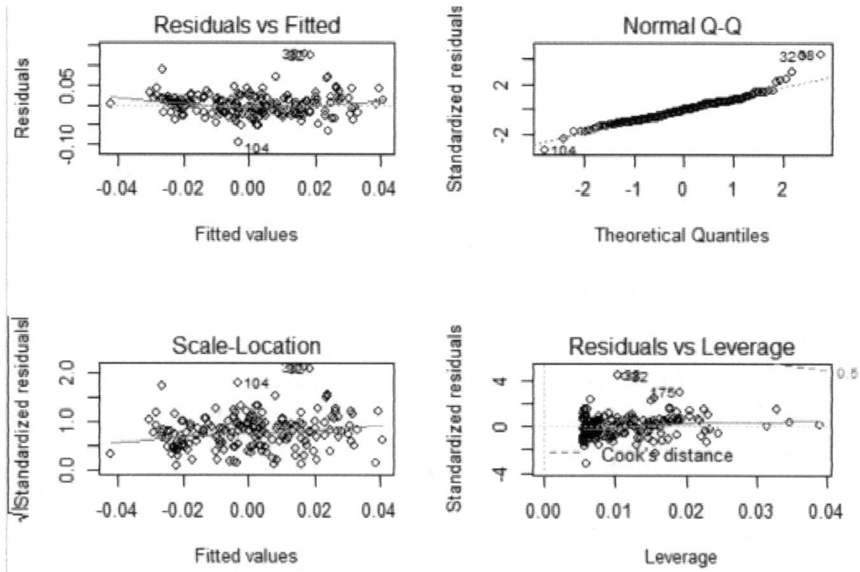

Figure 12. The regression diagnostic diagrams

From Figure 12, the two graphs on the left-hand side are residuals and fitting diagrams, they reveal the relationship between fitted values and true values by residuals and standardized residuals. Both suggest that the points are reasonably close to a straight line, with the scatters evenly distributed on both sides of the horizontal line.

Additionally, the Normal Q-Q figure is used to judge whether the residuals satisfy normal distribution. It shows three outliers and a non-distinct tail that away from the straight line, but most of points are within the interval [-1.96, 1.96]. However, the problem is that the points don't fall on the line at an angle of 45 degrees. This will be explained later.

Finally, the residual versus leverage diagram is displayed. In this Scale-Location graph, several influential observations exist, meaning that they make strong influence on the regression.

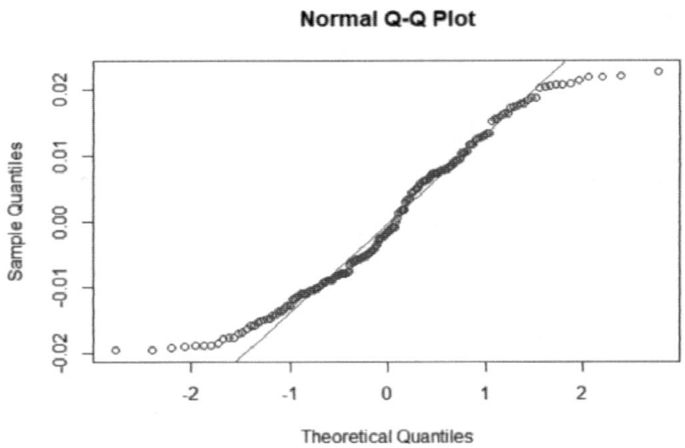

Figure 13. The normal Q-Q plot

Figure 13 is the normal Q-Q plot of the transformed tendency, the degree of the slope of the straight line is about 45^0, which is more appropriate than before despite of some points do not fall on the line. To sum up, the fitted line is reasonable to assumption although some deviations exist.

3.4. Advantages and improvements

3.4.1. Advantages

- The multiple linear regression model is used to conduct regression analysis on GDP per capital and productivity, and the optimal measurement is selected to make the results more reliable.
- Dummy variables are introduced in the experiment to divide countries into three partitions, so that the data can improve the accuracy of the model, also it can reflect the relationship between the affluence level and slow down rate more accurately.
- The least square method is used in curve fitting of slow-down, which can make the parameters fitting result better.

3.4.2. Improvements

The strategies for improvements are based on the potential threatens of the data analysis. In this study, firstly, the limited data in data processing will affect the final results to some extent. One is that the data of GDP in target countries is incomplete in the origination year, which gives rise to an inaccuracy in the process of data analysis. Considering the necessity to analyse the GDP of developed countries, developing countries and poor countries in the same time period to keep conformity, chose from 1983 to 1993, which is before the gradient decent and relatively complete in data. Another limitation is that, when filtrating the data, there exist two countries (CUW and SXM) that only have the data from 2005 to 2017. They are deleted as they break the consistency of the selected data. However, some other countries also have missing data when calculating the GDP, although these are much fewer than that in two countries. For example, data are not available for some years in selected countries in the time period 1983-1993, some countries only have data from 1990 to 1993. Thus, the GDP is ultimately measured by its definition, that is, GDP = Private Consumption + Gross Investment + Government Investment + Government Spending + (Exports - Imports). The indispensable problem is that the time periods from 1983 to 1993 or 1990 to 1993 is too short, so it may be one-sided to judge the opulence degree of a country based on the data within 10 years, let alone the economic changes.

Secondly, the optimality of variables in the regression model is noncommittally remaining a problem. From the perspective of classification, the developed and developing countries can also be divided by other categories, such as exports, technology and people's living standards, etc., leading to enormous different consequences. From the perspective of interactive terms, since the price level of exports and price level of imports are used in the first stage, the price level may be affected by inflation. Adversely, this term is not added to the model as the interactive term is set to be default 0. There is another variable in the excel tabulation needed to be noticed--- the purchasing power parity (PPP). If convert the expression of GDP per capital into the one that with the variable purchasing power parity (PPP), it may be far from the actual result due to the variability of theoretical exchange rate in the calculation. The only thing that ought to be ensured is that each country measures the GDP in the same way with the same monetary unit.

Thirdly, the regression diagnostics exist some defects inevitably. In Residuals vs Fitted diagram, which is a scatterplot of residuals-fitted values (Y-X) for residual analysis. This graph can be used to detect nonlinearity, unequal error variances, and outliers. In the Figure 12, the distribution is basically uniform on both sides of the line, but the scatters seem to be inclined to a central distribution and the line is slightly skew to a horizontal straight line. The error distribution also does not conform to the Gauss-Markov Condition. Similarly, the Y-axis of the scale-location diagram is the extraction of the Y-axis of Residuals vs Fitted diagram and both have the same X-axis. For the scale-location diagram, it shows whether the residuals are uniformly distributed along the range of predictors to check the equivalence of variance hypothesis. Theoretically, there would be a horizontal straight line with evenly distributed data, but in reality, it seems not to be a horizontal line with the far left and right points deviate from it.

The Normal Q-Q plot represents for the normality of the univariate distribution of the data set. If the data is normally distributed, the points will fall on the 45-degree reference line. Otherwise, the points will deviate from the reference line. But in fact, the 45-degree sloped line in the model does not start at the origin and the data does not fall completely in line, with a tail divorced from the precedent trajectory. Therefore, the data does not fit the Y=X line, and the error cannot follow a perfectly normal distribution as well.

For the Residuals vs Leverage diagram, which can check the outliers of the regression model. It can also be used to detect the heteroscedasticity and nonlinearity of data set. The Figure 12 witnesses that some points are too far away from the regression line, which contributes to the imprecision and unpredictability of the model.

Generally, the problems in the regression diagnostics stem from the precedent limited data and the variables that may not be correct. Instead of deleting or keeping the missing values, maybe the preferred method is to do missing value processing by the multiple imputation method, that is, to seek for alternative values. Furthermore, to upgrade the accuracy of the regression model, some algorithm conversion can be used and the variable adjustments can also be applied. Finally, outliers may sometimes affect the result severely, this could be amended by some mathematical transformations.

4. Estimation and results

4.1. Estimation method

First, it should be checked that whether the residuals have a normal distribution, through drawing a histogram of residuals, this consumption can be verified as in Figure 14.

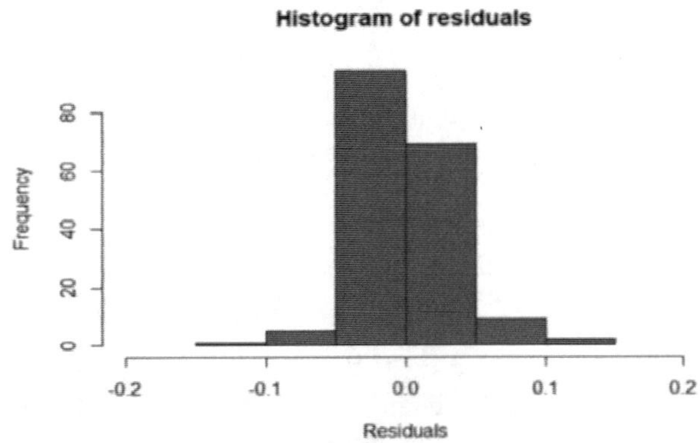

Figure 14. The histogram of residuals

Since the multiple linear regression is used before, naturally the method of Least Square (OLS) is applied to examine how well the model is. Need to minimize the sum of squares of the difference between the true values and the fitted values.

$$\min_{\alpha,\beta} \sum\left(y_i - \alpha - \beta x_i - m(x_i)\right)^2$$

where $m(x_i)$ represents for the dummy variables term and the interactive term.
This gives the estimates $\hat{\alpha}, \hat{\beta}$, where $\hat{\alpha} = \bar{y} - \hat{\beta}\bar{x}$ and $\hat{\beta} = \frac{\Sigma(x_i-\bar{x})(y_i-\bar{y})}{\Sigma(x_i-\bar{x})^2} = \frac{\mathrm{Cov}(x_i,y_i)}{\mathrm{Var}(x_i)}$
As it is tested before, the estimates are as following (Figure 15):

```
Residuals:
     Min       1Q   Median       3Q      Max
-0.11081 -0.03361 -0.00621  0.01894  0.92518

Coefficients: (1 not defined because of singularities)
             Estimate Std. Error t value Pr(>|t|)
(Intercept) -2.565e+00  6.690e-02 -38.345   <2e-16 ***
Y            1.309e-03  3.364e-05  38.903   <2e-16 ***
D1                  NA         NA      NA       NA
D2          -1.287e-02  1.448e-03  -8.891   <2e-16 ***
D3           8.893e-02  1.492e-03  59.609   <2e-16 ***
---
Signif. codes:  0 '***' 0.001 '**' 0.01 '*' 0.05 '.' 0.1 ' ' 1

Residual standard error: 0.0608 on 10001 degrees of freedom
  (2371 observations deleted due to missingness)
Multiple R-squared:  0.4003,    Adjusted R-squared:  0.4001
F-statistic:  2225 on 3 and 10001 DF,  p-value: < 2.2e-16
```

Figure 15. The R output for the regression of GDP per capital

The missing values in D_1 is caused by the fewer number in the richest countries.
The coefficient shows that there is a relation between the year and GDP per capital, also the p-value indicates the strong evidence. The standard error deviation is within a reasonable range. Have R^2= 40.01%, meaning that the regression explains 0.01% of the variation in the data. This is not a low percentage, suggesting a useful regression relationship and the relation is not causal.

4.2. The relationship between slow down rate and affluence levels
After the verification, the fitted line is a piecewise function and there is a continuous linear relationship with a discrete shift between the slow down rate and affluence level. It has an overall decrease in the slow down rate, which implies the affluence level of countries is in direct proportion to the slow down rate. Since the slow down rate of almost all countries is negative without the logarithm, it comes out that the world is confronted with secular stagnation.
However, what the interactive nexuses between poorer countries and richer countries or of slow down at similar rates are needed to be explained, although the hypothesis of the subject has been verified. A

correlative terminology is the Solow growth model, which demonstrates three factors -- technology, capital accumulation and labour force that drive economic growth. It implies that the less capital in poor countries to start with can narrow the gap between rich and poor countries as each additional unit of capital is inclined to yield higher return than in a richer country [20]. This is also the concept of catch-up growth. Hence, the gradient decent can be explained by this model because the difference of slow down rate on both sides of the threshold is stable and will not become wider. Meanwhile, it is suggested by Steinmueller [21] that the development strategy of 'leapfrogging' is supported by information and communication technologies (ICTs), it also narrows the gaps in industrialized and developing countries. The leapfrogging indicates that small and incremental innovations can sustain the leading position of a dominant enterprise [22].

Recently, as the Summers hypothesis says, the productivity improvement for agriculture and manufacturing are close to saturation [23]. Therefore, the underlying rate of technology progress has slowed down. More productivity improvements are needed in service sector for which may require large ICT developments. Since a large amount of technology transfer comes through FDI (foreign direct investment), by searching for the data source traversing all countries from 1970 to 2019, it can be seen in Figure 16 [24] that the countries with high FDI tend to have the similar slow-down rate. But the countries with low FDI do not follow this trait owing to several special countries.

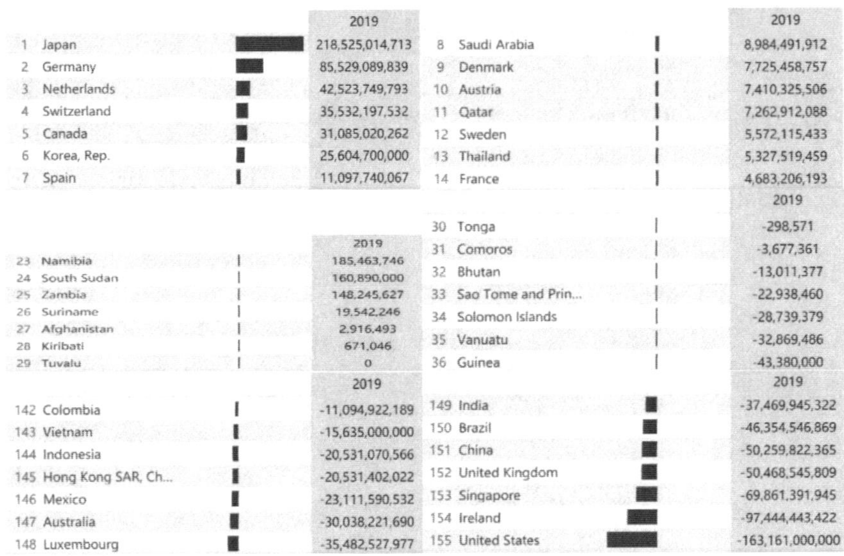

Figure 16. FDI in specific countries [19]

5. Conclusion

This report aims to study the relationship between the slow down rate of economy and affluence level among all countries in the world to explore the forward or backward evidence for secular stagnation. At the beginning, GDP and productivity are compared to choose a better measure for economy, also namely affluence level. After some calculations and comparisons, GDP per capital is more suitable to measure the level of economic development of a country, whether from a theoretical perspective or from specific data analysis. It can be thought of as a "best fit" unit of measure to argue the tendency of economy. Next, the tendency of slow down rate is worked out through seeking the relationship between slow down rate and affluence level. The negative value of slow down rate implies the secular stagnation. Finally, to examine the accuracy of multiple linear regression model and fitted line plot, the regression diagnostics and estimation method are used. The result comes out to be reasonable and appropriate but lack of precision.

However, the problem is that, the absence of data on GDP, the Capital Stock at Constant 2011 National Prices (RNNA), Price Level of Exports (PL-X), and Price Level of Imports (PL-M), which are consistent with the formula of GDP calculation, resulting in an unsatisfactory regression model. Although the model has been analysed and the direction of improvement has been put forward, the problem remains to exist. When using OLS method to test the model, the error of the data was within a reasonable range, so it can be concluded that the world is faced with secular stagnation.

After providing the evidence for secular stagnation of Larry Summers's suggestion, it can be more deeply realized that when the economy reaches to a certain point, the downward tendency may be inevitably always existing. Thus, the root of stagnation needs to be found to solve this global problem more effectively. It is definitely a challenge to find a way of constant growth in economy within a global scope.

6. References

[1] Jan Jacobs, Kazuo Ogawa, Elmer Sterken & Ichiro Tokutsu (2020) 'Public Debt, Economic Growth and the Real Interest Rate: A Panel VAR Approach to EU and OECD Countries', *Applied Economics*, 52(12), pp.1377-1394, [Online]. Available at: https://doi.org/10.1080/00036846.2019.1673301 (Accessed: June 2021)

[2] Moheddine Younsi & Amine Nafla (2019) 'Financial Stability, Monetary Policy, and Economic Growth: Panel Data Evidence from Developed and Developing Countries', *Journal of the Knowledge Economy*, Vol 10, pp.238-260, Springer US [Online]. Available at: https://doi.org/10.1007/s13132-017-0453-5 (Accessed: June 2021)

[3] Alvin H. Hansen (2020) 'Secular stagnation', *TRIMESTRE ECONOMICO*, 87(348), pp.1189-1207, FONDO CULTURA ECONOMICA [Online]. Available at: https://doi.org/10.20430/ete.v87i348.1179 (Accessed: June 2021)

[4] Probst, J. (2019) 'Secular stagnation: it's time to admit that Larry Summers was right about this global economic growth trap', *The Conversation* [Online]. Available at: https://theconversation.com/secular-stagnation-its-time-to-admit-that-larry-summers-was-right-about-this-global-economic-growth-trap-112977 (Accessed: June 2021)

[5] Stiglitz, J. (2018) 'The Myth Of Secular Stagnation', *Social Europe* [Online]. Available at: https://www.socialeurope.eu/the-myth-of-secular-stagnation (Accessed: May 2021)

[6] Aylward, R. et al. (2018) *Debate: Stiglitz vs. Summers on Secular Stagnation - Project Syndicate*. [online] Project Syndicate. Available at: https://www.project-syndicate.org/bigpicture/debate-stiglitz-and-summers-face-off-on-secular-stagnation?barrier=accesspaylog (Accessed: May 2021)

[7] Farmer, R. (2018) 'Secular Stagnation Revisited', *Project Syndicate* [Online]. Available at: https://www.project-syndicate.org/commentary/secular-stagnation-revisited-stiglitz-summers-debate-by-roger-farmer-2018-09?barrier=accesspaylog (Accessed: 11 May 2021)

[8] Paul Krugman (2013) 'Paul Krugman's Triple Doctorate: Economics in the Crisis', *Notas Economicas*, pp.11-17 [Online]. Available at: http://www.uc.pt/feuc/notas-economicas/numeros (Accessed: May 2021)

[9] Eggers, Andrew C., Ellison, Martin & Lee, Sang Seok (2021) 'The economic impact of recession announcements', *Journal of Monetary Economics*, Vol 120, pp.40-52, Elsevier B.V [Online] Available at: https://doi.org/10.1016/j.jmoneco.2021.03.002 (Accessed: June 2021)

[10] Money, Banking and Financial Markets (2015) *A Simple Guide to "Secular Stagnation" — Money, Banking and Financial Markets*. [online] Available at: https://www.moneyandbanking.com/commentary/2015/3/23/a-simple-guide-to-secular-stagnation (Accessed: May 2021)

[11] Tradingeconomics.com (2021) *Interest Rate - Countries - List*. [online] Available at: https://tradingeconomics.com/country-list/interest-rate (Accessed: 14 May 2021)

[12] Teulings, C. & Baldwin, R. (2014) Reflections on the 'New Secular Stagnation Hypothesis'. In: Coen, T. & Richard, B. (Eds.), Secular Stagnation: Facts, Causes and Cures. CEPR Press., London: Centre for Economic Policy Research. pp.27-40 (Accessed: June 2021)

[13] Bank, World. (2021) World Bank Country and Lending Groups – World Bank Data Help Desk. [online] *Datahelpdesk.worldbank.org*. Available at: https://datahelpdesk.worldbank.org/knowledgebase/articles/906519 (Accessed: May 2021)

[14] Karen, D. & Louise, S. (2018) 'GDP as a Measure of Economic Well-being', *Brookings*, page 5, [Online]. Available at: https://www.brookings.edu/research/gdp-as-a-measure-of-economic-well-being (Accessed: May 2021)

[15] Dirk, P. & Paul, S. (2003) 'Measuring productivity', OECD Economic Studies, OECD Publishing, 2001(2), pp.127-170. [online]. Available at: https://doi.org/10.1787/eco_studies-v2001-art13-en (Accessed: May 2021)
[16] Matthew, G. (2013) 'Qatar politics and the challenges of development / Matthew Gray', London: Lynne Rienner Publishers. [Online] Available at: https://ebookcentral.proquest.com/lib/liverpool/detail.action?docID=3329108 (Accessed: May 2021)
[17] Ramaz, P. (2019) 'GEORGIA IN THE EUROPEAN UNION'S EASTERN PARTNERSHIP COUNTRIES: HISTORICAL ASPECTS AND CHALLENGES', *Ecoforum*, 8 (1), Asociatia de Cooperare Cultural-Educationala Suceava [Online]. Available at: https://doaj.org/article/95a754526c8a48e9a642cc99692c7c6f (Accessed: May 2021)
[18] Drew T, C. (2015) 'Long-Term Urban Market Dynamics Reveal Increased Bushmeat Carcass Volume despite Economic Growth and Proactive Environmental Legislation on Bioko Island, Equatorial Guinea', *PLoS ONE*, 10 (7), pp.1-22, [Online]. Available at: https://doi.org/10.1371/journal.pone.0134464 (Accessed: May 2021)
[19] Maria Irmina, P., Nur Ulfa, M., & Kridanto, S. (2021) 'Determining threshold value on information gain feature selection to increase speed and prediction accuracy of random forest', *Journal of Big Data*, 8 (1), pp. 1-22, SpringerOpen [Online]. Available at: https://doi.org/10.1186/s40537-021-00472-4 (Accessed: May 2021)
[20] Md. Saifur, R. & Farihana, S. (2017) 'The nexus between financial integration and real economy: Solow-growth model concept', Research in International Business and Finance, Vol 42, pp. 1244-1253, Elsevier B.V. [Online]. Available at: https://doi.org/10.1016/j.ribaf.2017.07.062 (Accessed: June 2021)
[21] Steinmueller, W.E., (2001). 'ICTs and the possibilities for leapfrogging by developing countries', *International Labour Review*, 140(2), pp.193-194. [Online] Available at: https://doi.org/10.1111/j.1564-913X.2001.tb00220.x| (Accessed: June 2021)
[22] Li, Y. & Cantwell, J. (2021) 'Rapid FDI of emerging-market firms: foreign participation and leapfrogging in the establishment chain', *Transnational Corporations*, 28(1), page 51 [Online], Available at: https://link.gale.com/apps/doc/A662833687/AONE?u=livuni&sid=bookmark-AONE&xid=6f0df375 (Accessed: June 2021)
[23] Wessel, D. (2021) 'Secular stagnation' even truer today, Larry Summers says. [online] Brookings. Available at: https://www.brookings.edu/articles/secular-stagnation-even-truer-today-larry-summers-says/ (Accessed: June 2021)
[24] Knoema. (2021) Search-knoema.com. [online] Available at: https://knoema.com/search?query=FDI&pageIndex=&scope=&term=&correct=&source=Header (Accessed: June 2021)

7. Appendix

```
library(readxl)
pwt92 <- read_excel("R prologue/pwt92.xlsx",
          sheet = "GDP")
View(pwt92)

G <- pwt92$'GDP per capital'
Y <- pwt92$'year'
D1 <- pwt92$'D1(poorest)'
D2 <- pwt92$'D2(middle)'
D3 <- pwt92$'D3(richest)'
states <- as.data.frame(state.x77[c("G", "Y", "D1","D2","D3")])
cor(states)
fit <- lm(G ~ Y + D1 + D2 + D3, data=states)
summary(fit)

library(readxl)
pwt91 <- read_excel("R prologue/pwt91.xlsx",
          sheet = "productivity")
View(pwt91)
```

```
L <- pwt91$'laborinput'
Y <- pwt91$'year'
D1 <- pwt91$'D1(poorest)'
D2 <- pwt91$'D2(middle)'
D3 <- pwt91$'D3(richest)'
states <- as.data.frame(state.x77[c("L", "Y", "D1","D2","D3")])
cor(states)
fit <- lm(L ~ Y + D1 + D2 + D3, data=states)
summary(fit)

library(readxl)
slow_down_rate <- read_excel("R prologue/slow down rate.xlsx")
View(slow_down_rate)

S <- slow_down_rate$'slow down rate'
G <- slow_down_rate$'GDP per capital'
states <- as.data.frame(state.x77[c("S", "G")])
fitted(fit)
residuals(fit)
fit <- lm(S ~ G, data=states)
par(mfrow=c(2,2))
plot(fit)

plot(G,S,Glim=c(2200,500000),Slim=c(-1,1))
plot(G,S, "l", Glim=c(1100,660000),Slim=c(-1,1))
abline(fit,col="red")
res <- residuals(fit)
hist(res, main = paste("Histogram of residuals"), breaks = 8,
   xlab = "Residuals",xlim = range(-0.2,0.2), col="purple")
library(nortest)
ad.test(res)
qqnorm(res, col="blue")
qqline(res, col="red")
plot(G, S, col="blue", pch=19, main = paste("Fitted line plot"),
   xlab = "GDP per capital", ylab="slow down rate")
abline(fit, col="red")
library(randomForest)
randomForest(S~G, data=states,importance=TRUE, ntree=180,na.rm=TRUE)
```

Simulation of Meshing of Herringbone Double Circular-Arc Helical Gears using Multibody Dynamic Analysis Software

C K Lee[1], C L Hsieh[1], J C Ruan[2] and C Y Wang[2]

1 Department of Industrial Engineering and Management, Cheng Shiu University, 840 Chengcing Rd., Kaohsiung, Taiwan (R.O.C.)
2 Department of Mechanical Engineering, Cheng Shiu University, 840 Chengcing Rd., Kaohsiung, Taiwan (R.O.C.)
E-mail: cklee@gcloud.csu.edu.tw

Abstract. This paper aims to complete the simulation of meshing of a pair of herringbone double circular-arc helical gears by using the multibody dynamic analysis software, SolidWorks motion. Three types of edge modifications, including no edge modification, chamfered edge modification, and rounded edge modification, are applied respectively to the pair of gears. The output angular velocities and contact regions corresponding with the three types of edge modifications are analyzed. Moreover, whether the center distance assembly error affects the output angular velocity and contact regions are analyzed. The results obtained by the simulation of meshing show that rounded edge modification is better than chamfered edge modification, and chamfered edge modification is better than no edge modification. The center distance assembly error has no effect on the output angular velocity and has little effect on the contact regions. Therefore, double circular-arc helical gears are not sensitive to center distance assembly errors.

1. Introduction

Compared with involute gears and single circular-arc helical gears, double circular-arc helical gears have much higher contact strength and tooth bending strength, especially the stepped double circular-arc helical gears. The main disadvantage of helical gears is that the axial component force will be generated when the gears mesh. The axial force is caused by the helix angle. The larger the helix angle, the greater the axial force. The application of herringbone gear is the way to eliminate the axial component of force. The herringbone double circular-arc helical gears have the advantages of both double circular-arc helical gears and herringbone gears. Litvin and Lu [1] investigated the generation, geometry, meshing, and contact of double circular-arc helical gears. Litvin and Lu [2] modified the geometry of double circular-arc helical gears for absorption of transmission errors caused by misalignment. Lu et al. [3] investigated the conditions of load share under a load and determined the real contact ratio for aligned and misaligned double circular-arc helical gears. Wang and Chen [4] applied an adjustable bearing to compensate for offset and angular misalignment for double circular-arc helical gears. Litvin et al. [5] proposed a new version of Novikov-Wildhaber helical gears with lower noise and vibration caused by errors of alignment, the possibility of grinding and application of hardened materials, and lower stresses. Yang [6] proposed stepped triple circular-arc gears, of which contact stress was investigated via finite element analysis. Zhang et al. [7] proposed modified double circular-arc helical gears by tooth end relief with helix to reduce transmission errors and stresses. Wang [8] designed a planetary gear reducer with double circular-arc helical gear. Qu et al. [9] investigated the load sharing of double circular-arc helical gears based on finite element analysis. Li et al. [10] proposed a new theoretical approach for the analysis of the impact of the double circular arc gear meshing. Gao et al. [11] proposed a novel point contact gear drive with pinion circular arc teeth and gear parabolic curve teeth, of which the gear stress distribution and the contact deformation were investigated by using finite element analysis. Liang et al. [12] studied the generation and meshing of a double helical gear transmission with curve element constructed tooth pairs. At present, no literature has been found to perform the simulation of meshing and analyze the influence of edge modification on the output angular velocity for a pair of herringbone double circular-arc helical gears by using multibody dynamic software.

2. Mathematical modeling of double circular-arc helical gear

Based on the coordinate transformation theory and the theory of gearing, the mathematical model of double circular-arc helical gear can be created. According to standard GB/T12759-1991, the profile of the basic rack of double circular-arc helical gear is as shown in figure 1. The mathematical modeling of double circular-arc helical gear has three steps.

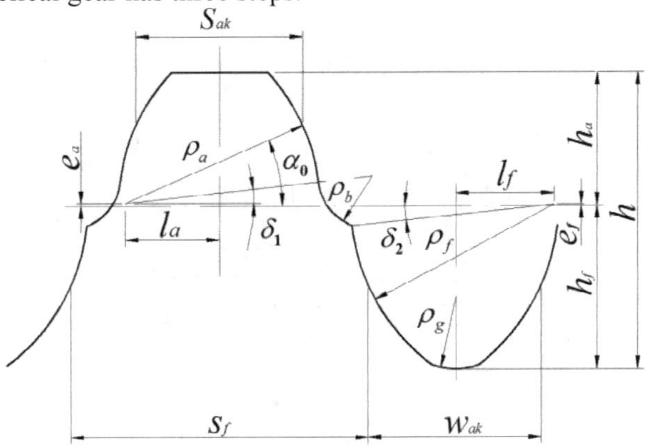

Figure 1. The profile of the basic rack of double circular-arc helical gear in standard GB/T12759-1991.

Step 1: Placing a coordinate system $S_n(x_n, y_n, z_n)$ to the profile and representing the mathematical model of the profile in $S_n(x_n, y_n, z_n)$ by equation (1).

$$\mathbf{r}_n^{(i)}(\theta) = \begin{bmatrix} x_n^{(i)}(\theta) & y_n^{(i)}(\theta) & 0 & 1 \end{bmatrix}^T \tag{1}$$

Step 2: Placing a coordinate system $S_1(x_1, y_1, z_1)$ to the basic rack and representing the mathematical model of the basic rack in $S_1(x_1, y_1, z_1)$ by equation (2).

$$\mathbf{r}_1^{(i)}(\theta, u) = \mathbf{M}_{1n}(u)\mathbf{r}_n^{(i)}(\theta) = \mathbf{Ry}(-\beta)\mathbf{Tz}(u)\mathbf{r}_n^{(i)}(\theta) \tag{2}$$

Step 3: Placing a coordinate system $S_2(x_2, y_2, z_2)$ to the double circular-arc helical gear and representing the mathematical model of the double circular-arc helical gear in $S_2(x_2, y_2, z_2)$ by equation (3). The second equation in equation (3) is obtained by using the theory of gearing.

$$\begin{cases} \mathbf{r}_2^{(i)}(\theta, u, \phi) = \mathbf{M}_{21}(\phi)\mathbf{r}_1^{(i)}(\theta, u) = \mathbf{Rz}(-\phi)\mathbf{Ty}(\rho)\mathbf{Tx}(-\rho\phi)\mathbf{r}_1^{(i)}(\theta, u) \\ u = -\rho\phi\csc\beta + \cot\beta\left[x_n^{(i)}(\theta) + \dfrac{y_n^{(i)}(\theta)y_n'^{(i)}(\theta)}{x_n'^{(i)}(\theta)}\right] \end{cases} \tag{3}$$

3. CAD modeling of herringbone double circular-arc helical gear

To complete the CAD modeling of a pair of herringbone double circular-arc helical gears, three kinds of software are used one by one. Firstly, the mathematical software, Mathematica, is applied to calculate the coordinates of cross-section curves of surfaces of one single tooth groove. Secondly, the CAD system, Pro/Engineer is applied to create the surfaces of one single tooth groove. In Pro/Engineer, the cross-section curves are plotted using the datum curve function. The surfaces of one single tooth groove are created using the boundary blend function. The surfaces are exported to SolidWorks using an IGES format file. Thirdly, the CAD system, SolidWorks, is applied to create the solid model of a pair of herringbone double circular-arc helical gears. In SolidWorks, the excess surfaces are trimmed using the surface trimming function. The discrete surfaces are combined into one single tooth groove surface using the surface knitting function. The whole tooth groove surfaces are created using the surface copying function. The gear blank cylinder is created using the extrusion function. All teeth are formed using the surface cutout function. The solid model of herringbone double circular-arc helical gear is obtained using the solid mirroring function. Figure 2 shows the obtained solid model of a pair of herringbone double circular-arc helical gears. The small gear has 8 teeth. The big gear has 17 teeth. The gear module is 20 mm. The helix angle is 15 degrees. The pressure angle is 24 degrees.

Figure 2. The solid model of a pair of herringbone double circular-arc helical gears.

4. Settings for the simulation of meshing

The simulation of meshing of the pair of herringbone double circular-arc helical gears is completed by the multibody dynamic software, SolidWorks motion, which uses ADAMS as its solver. The settings for the simulation are as follows: In the contact property manager, the contact type is set to solid bodies, the materials are set to steel (greasy), and the contact force is set to impact model. The time function curve of torque applied to the big gear shaft hole is set to STEP5(time,0,0,1,5000) N-mm, indicating that the torque is gradually loaded from 0 to 5000 N-mm in the time range from 0 to 1 second, and then keeps 5000N unchanged. The time function curve of angular velocity applied to the small gear shaft hole is set to STEP5(time,1,0,2,3) deg/s, indicating that the angular velocity is gradually loaded from 0 to 3 deg/s in the time range from 1 to 2 seconds, and then remains 3 deg/s unchanged. The 3D contact resolution is set to using precise contact. The accuracy is set to 0.00001. The integrator type is set to SI2_GSTIFF, which provides better accuracy of velocities and accelerations.

5. Simulation of meshing

5.1. Original design

For the original design, the edges on the end faces are not modified, as shown in Figure 3. The output angular velocity for the design is shown in Figure 4. The angular velocity has a high variation at 11.13

seconds. At this moment, the contact regions on the no edge modification small gear are shown in Figure 5. Edge contact phenomenon can be observed at this moment.

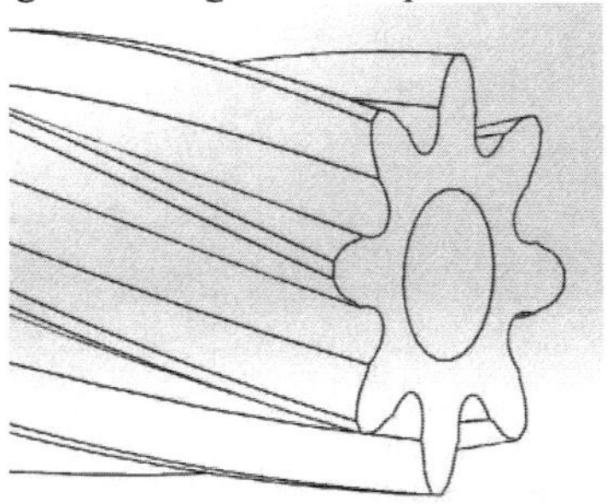

Figure 3. The edges on the end faces are not modified.

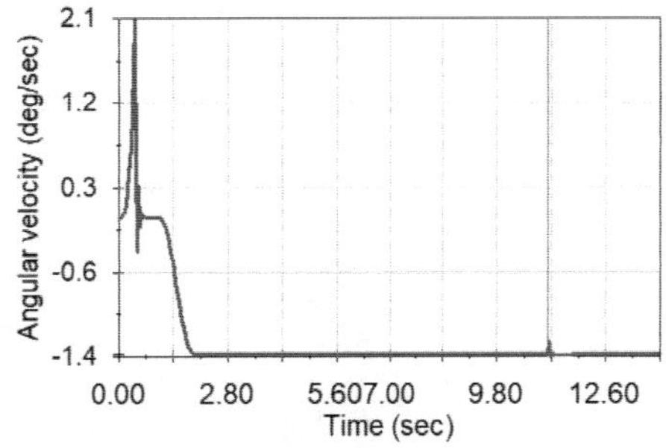

Figure 4. The output angular velocity.

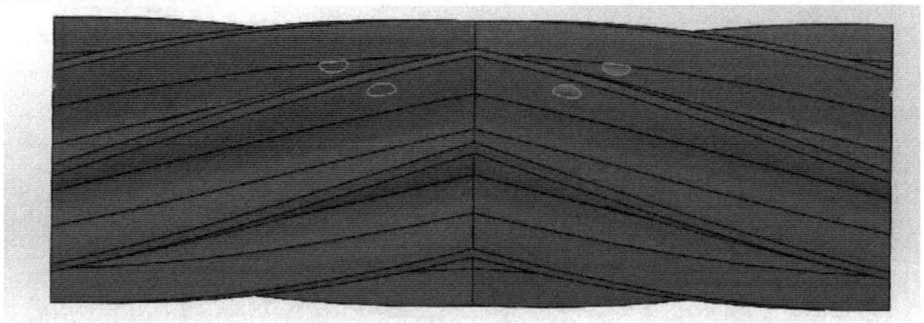

Figure 5. The contact regions on the no edge modification small gear.

5.2. Improved design 1

For the improved design 1, the edges on the end faces are chamfered, as shown in Figure 6. The output angular velocity for the design is shown in Figure 7. The angular velocity has a moderate variation at 11.32 seconds. At this moment, the contact regions on the chamfered edge modification small gear are shown in Figure 8. Edge contact phenomenon can also be observed at this moment.

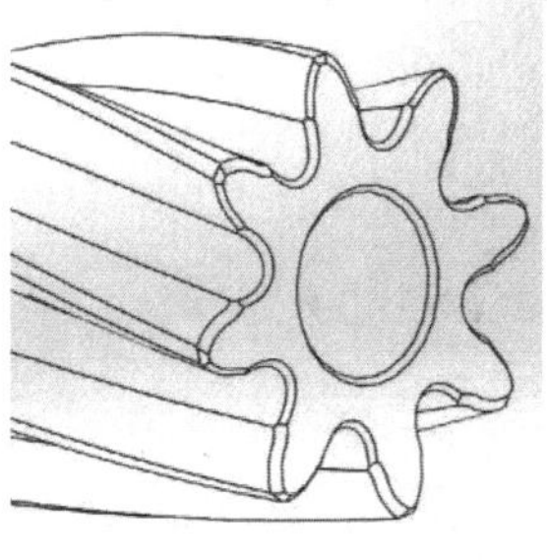

Figure 6. The edges on the end faces are chamfered.

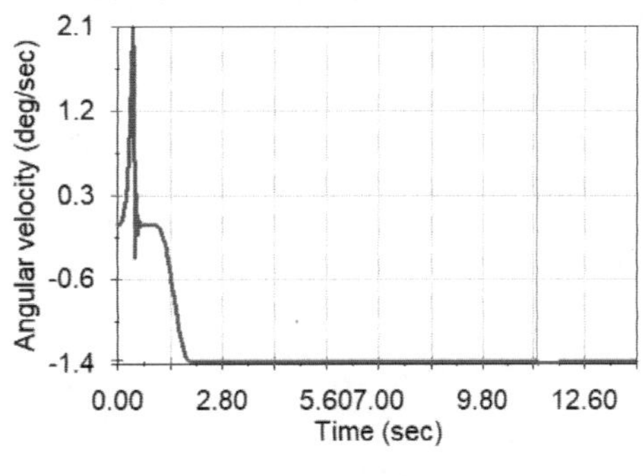

Figure 7. The output angular velocity.

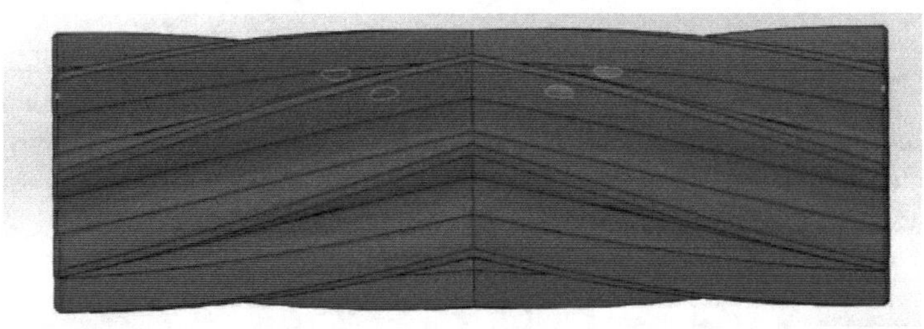

Figure 8. The contact resigns on the chamfered edge modification small gear.

5.3. Improved design 2

For the improved design 2, the edges on the end faces are rounded, as shown in Figure 9. The output angular velocity for the design is shown in Figure 10. The angular velocity has a low variation at 11.48 seconds. At this moment, the contact regions on the rounded edge modification small gear are shown in Figure 11. Edge contact phenomenon can also be observed at this moment.

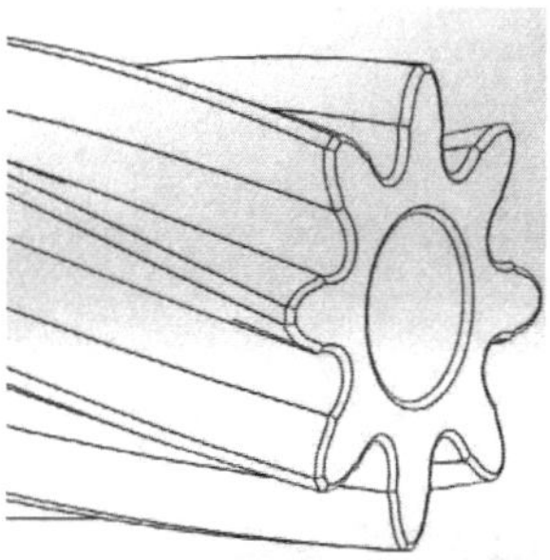

Figure 9. The edges on the end faces are rounded.

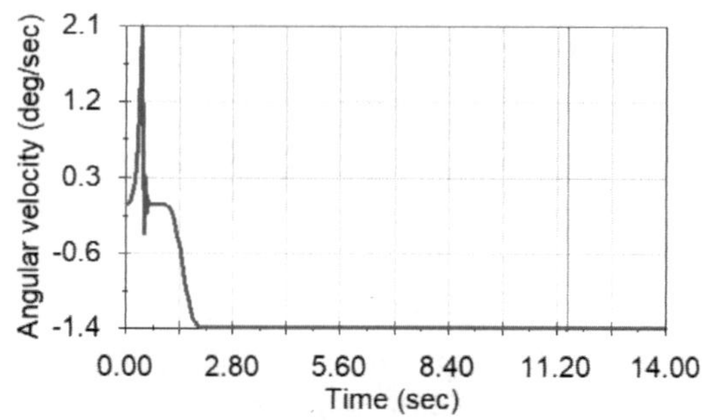

Figure 10. The output angular velocity.

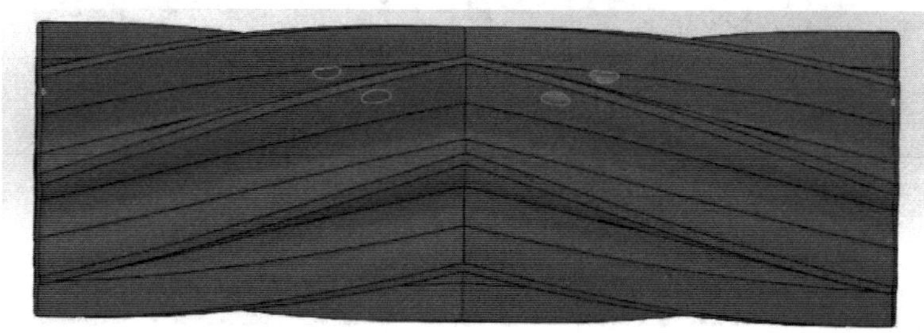

Figure 11. The contact resigns on the rounded edge modification small gear.

5.4. Influence of center distance assembly error

To analyze the influence of center distance assembly error, a center distance assembly error of 0.05 mm is applied intentionally to the gear shafts, as shown in Figure 12. The pair of gears with rounded edge modification is used as the object of analysis. The output angular velocity for the pair of gears is shown in Figure 13. The contact regions on the small gear are shown in Figure 14. It shows that the center distance assembly error has no effect on the output angular velocity and has little effect on the contact regions.

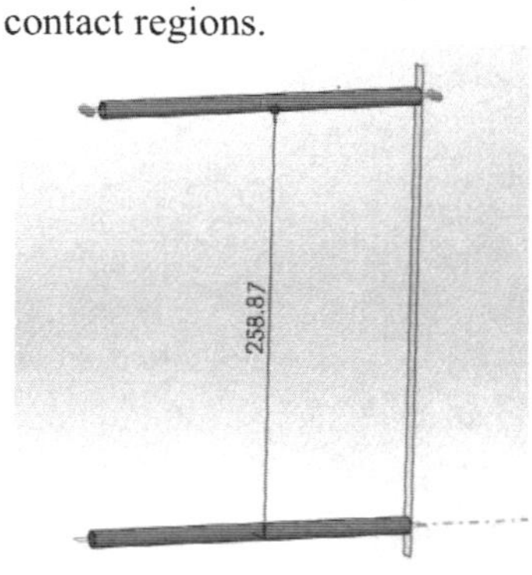

Figure 12. A center distance assembly error of 0.05 mm is applied to the gear shafts intentionally.

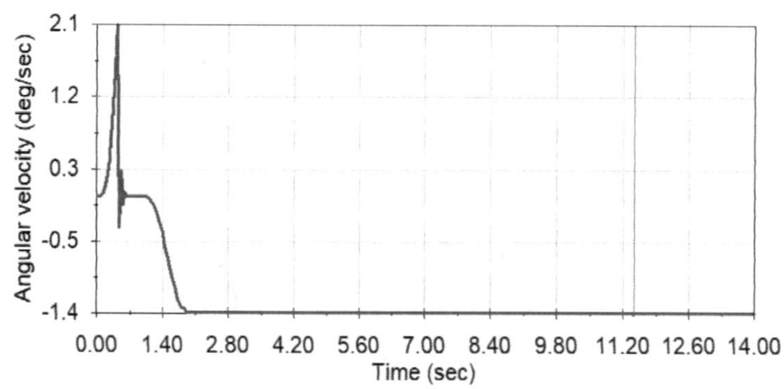

Figure 13. The output angular velocity.

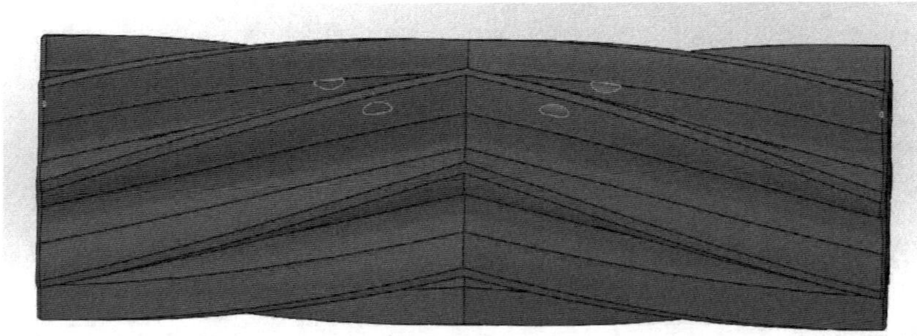

Figure 14. The contact resigns on the small gear having rounded edge modification and center distance assembly error.

6. Conclusion

This paper has completed the simulation of meshing of a pair of herringbone double circular-arc helical gears by using the multibody dynamic analysis software, SolidWorks motion. The gear pair with no edge modification is very sensitive to the edge contact phenomenon. By modifying the sharp edges as chamfered edges, the problem can be improved to a certain extent. By modifying the sharp edges as rounded edges, the problem can be greatly improved. Besides, the center distance assembly error does not influence the output angular velocity of the gear pair and has little influence on the contact regions. Although the edge modification technology can reduce the sensitivity of the gear pair, it needs to spend extra time to process the sharp edges. Therefore, it is suggested to directly apply the longitudinal crowning technology to save the extra edge modification time.

References

[1] Litvin F L and Lu J 1993 *Math. Comput. Model* **18** 31-47
[2] Litvin F L and Lu J 1995 *Comput. Meth. Appl. Mech. Eng.* **127** 57-86
[3] Lu J, Litvin F L and Chen J S 1995 *Math. Comput. Model* **21** 13-30
[4] Wang C Y and Chen C K 2001 *Proc. Inst. Mech. Eng. C.* **215** 759-71
[5] Litvin F L, Fuentes A, Gonzalez-Perez I, Carnevali L and Sep T M 2002 *Comput. Meth. Appl. Mech. Eng.* **191** 5707-40
[6] Yang S C 2009 *Mech. Mach. Theory* **44** 1019-31
[7] Zhang H, Hua L and Han X H 2010 *Mech. Mach. Theory* **45** 46-64
[8] Wang C Y 2012 *AMM* **152-154** 1595-1600
[9] Qu W, Peng X, Zhao N and Guo H 2012 *Structural Engineering and Mechanics* **43** 439-48
[10] Li Y, Wu B and Zhu Lin 2015 *The Open Mech. Eng. J.* **9** 160-7
[11] Gao Y, Chen B, Tan R and Zhang Y 2016 *J. Adv. Mech. Des. Syst. Manuf.* **10** JAMDSM0009
[12] Liang D, Luo T and Tan R 2018 *Math. Probl. Eng.* **2018** 1539849

Simulation of 2D Acoustic Wave Propagation with Absorbing Boundary Condition

Zekun Liu[1] and Tianyou Lu[1]

1 WuHan Sannew School, Wuhan, China
E-mail: 2788660534@qq.com, 377813780@qq.com

Abstract. In this work we design an algorithm to simulate the 2D acoustic wave (P wave) propagation in a medium with different regional velocities. We use the Generalized Conjugate Residual (GCR) algorithm to solve the trapezoidal time integration, where for a given time step of 0.01 seconds, the error is approximately 8% with respect to the reference. We further speed up the simulation by a factor of 900 by implementing a model reduction scheme using Proper Orthogonal Decomposition (POD). Lastly, to suppress the artificial reflections formed when the wave bounces off the fixed boundaries, we created two virtual absorbing layers by tapering the solution of nodes close to the corresponding boundaries after each time iteration. However, we note that the successful implementations of the model reduction and absorbing boundary condition are not compatible with each other and provide explanations for why and how to combine them in future work.

1. Introduction

Earthquakes generate seismic energy waves that propagate the earth's interior and get recorded by geophones when they reach the surface, as shown in Fig. 1. The seismic recordings are then analysed to extract geological information about the Earth. The simulation of seismic wave propagation is widely used in the seismology community for various purposes. Sometimes, the synthetic data generated by simulation is treated as the theoretical replacement of real data to be tested with novel imaging processing techniques. For instance, techniques like the curvelet transform and slant stacking are used on the seismic data to analyse the Earth's mantle [1,2]; however, the real data are usually contaminated with ambient noise and false instrumental response. Thus, synthetic data is first used before any work on the real data. Since different earthquakes generate different seismic data, seismologists wish to generate simulations for different scenarios as quickly and as accurately as possible. To cater to these demands, we focus on speeding up the simulation and mimicking the real-world scenarios by adopting model reduction and absorbing boundary conditions. However, we simplify the Earth's geometry to a rectangular 2D grid due to the limited scope of this work.[3]

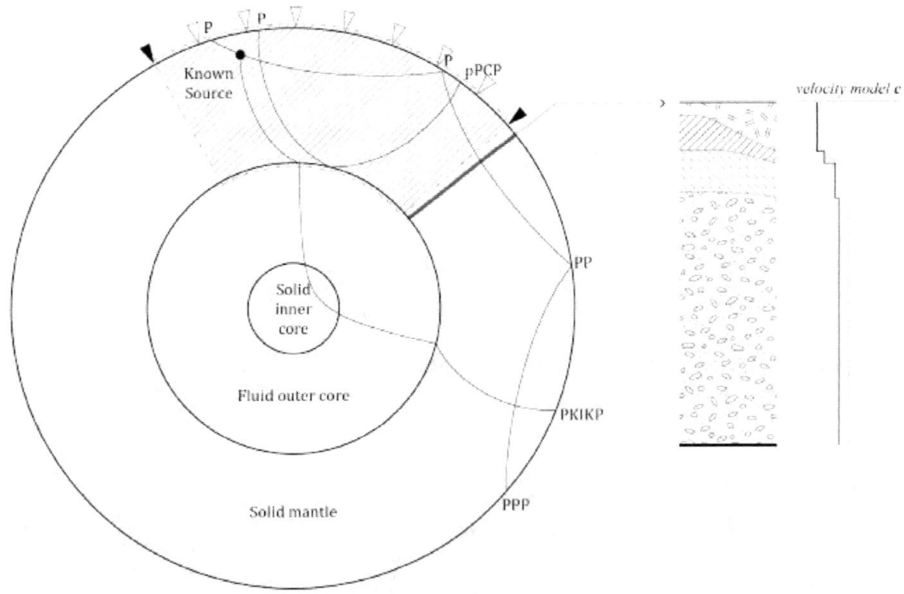

Figure 1. Illustration of P-wave propagation and the velocity model.

2. Problem formulation

Our work starts from the 1D wave equation:

$$\frac{1}{v_p^2}\frac{d^2q}{dt^2} = \frac{\partial^2 q}{\partial r^2} \tag{1}$$

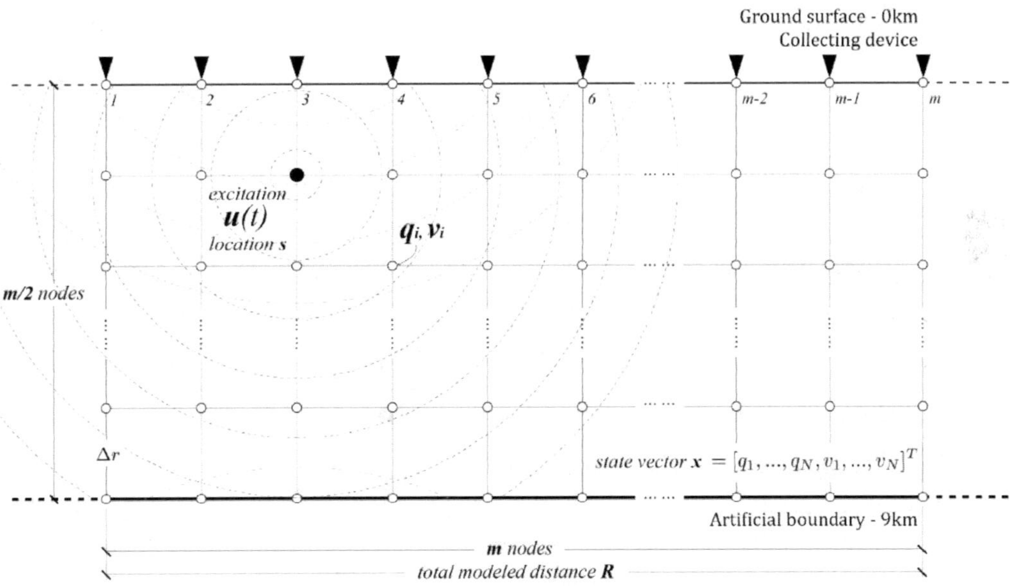

Figure 2. Illustration of model setup.

where, $q \equiv particle\ displacement$, $v_p \equiv P - wave\ velocity$, $r \equiv spatial\ distance$. We take $\frac{d^2q}{dt^2}$ equals $\frac{dv}{dt}$, construct the state vector, and formulate the wave equation as:

$$f(\cdot) = \begin{cases} \frac{dq}{dt} = v \\ \frac{dv}{dt} = v_p^2\left(\frac{\partial^2 q}{\partial r^2} + u\right) = \frac{v_p^2}{\Delta r^2}\left(q_{i-1}^k - 2q_i^k + q_{i+1}^k\right) + \frac{v_p^2}{\Delta r^2} u^k \ (at\ time\ step\ k) \end{cases} \quad (2)$$

where $u \equiv excitation$. The function is synthesized into a linear system, where A is made of the finite difference discretization matrix E and identity matrix I:

$$\frac{dx(t)}{dt} = Ax(t) + bu(t) \quad (3)$$

$$A = \begin{bmatrix} 0 & I \\ \frac{v_p^2}{\Delta r^2} E & 0 \end{bmatrix} \quad (4)$$

We setup the grid as a $m \times \frac{m}{2}$ network, where m is the number of nodes.

- The state of our system is a vector of length m^2 and each node includes displacement q and particle velocity v:

$$x = [q_1, \dots, q_N, v_1, \dots, v_N]^T \quad (5)$$

- The parameters of our components includes:

$$p = \left[v_p\ \Delta r\ R\ s\ \Delta t\right]^T \quad (6)$$

where v_p is the P-wave velocity, Δr is the spatial increment, R is the model distance, s is the location of excitation, and Δt is the time increment.

- The excitation of the system is:

$$u(t) = \cos\left(\frac{2\alpha\pi t}{T}\right), t \in [0, t_0] \quad (7)$$

- Thus, the final set of equations are written explicitly as:

$$f_i(x, p, u) = \frac{dq_i}{dt} = v_i, i \in [1, N] \quad (8)$$

$$f_{N+1}(x, p, u) = \frac{dv_1}{dt} = v_p^2 \frac{-2q_1 + q_2}{\Delta r^2} + v_p^2 \frac{u_1}{\Delta r^2} \quad (9)$$

$$f_{N+i}(x, p, u) = \frac{dv_i}{dt} = v_p^2 \frac{q_{i-1} - 2q_i + q_{i+1}}{\Delta r^2} + v_p^2 \frac{u_i}{\Delta r^2} \quad (10)$$

$$f_{N+N}(x, p, u) = \frac{dv_N}{dt} = v_p^2 \frac{q_{N-1} - 2q_N}{\Delta r^2} + v_p^2 \frac{u_N}{\Delta r^2} \quad (11)$$

- The output y is the displacement of surface nodes as a function of both distance $i \times \Delta r$ and time t, which is known as the synthetic seismic recording at the surface.

$$y(t) = C^T x(t) = [q_1(t), q_2(t), \dots, q_m(t)]^T \quad (12)$$

3. Fundamental numerical methods

3.1. Model parameters

In this study, $\Delta r = 200m$, $\Delta t = 0.01s$. $v_p^{top} = 6600m/s$ for the top half block and $v_p^{bot} = 9900m/s$ for the bottom half block. The total horizontal and vertical distances are 18000m and 9000m respectively, which yield a total number of 4186 simulating points. The acoustic wave velocity v_p is the physical property assigned to each node as constant. The scope of this study is to simulate the first 5 seconds of the wave propagation.

3.2. ODE integration method

We compared the Forward Euler (FE) and Trapezoidal approximation methods, and found that Trapezoidal is preferred. During the FE progression, the system diverges easily because of the accumulated errors, which resemble introducing fake energy into the system. Thus, the largest feasible Δt for FE is 10^{-6}, which is computationally expensive. Trapezoidal on the other hand, increases Δt to 10^{-2}.

The structure of algorithm is shown in Fig. 3. We use GCR iteration inside Newton method to solve for each time interval. Note that, the system is completely linear and does not require 6 Newton method, but it is retained for the future implementation of nonlinear terms, such as energy loss. In our case, GCR takes about 3.3 min to finish, compared to around 23 min for LU decomposition.

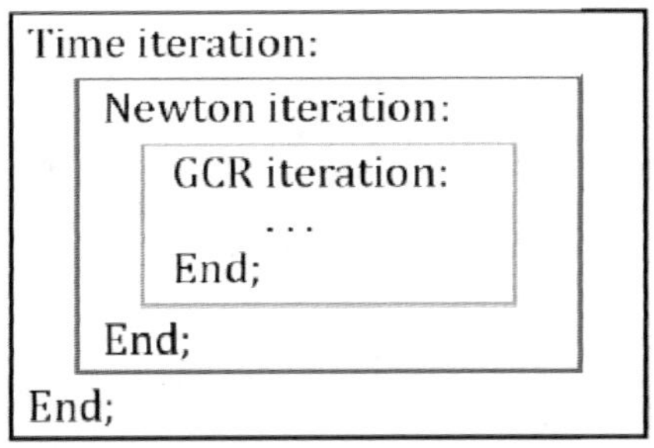

Figure 3. Fundamental ODE structure.

4. The technical challenge

We completed two technical challenges: model order reduction using the Proper Orthogonal Decomposition (POD) method and absorbing boundary conditions.

4.1. POD model order reduction

In seismology, only nodes at the surface matter because that is where geophones are located. Therefore, model reduction is desired to speed up the simulation by preserving only the input-output relationship between the excitation and surface nodes. With POD, singular value decomposition (SVD) is applied to matrix W, constructed with a sufficient number of x's generated by the original model (501 in this case). Projection matrix V_q is constructed by taking the singular vectors corresponding to q largest singular values, in our case $q = 140$. The model can then be reduced with $\hat{A}$, $\hat{b}$ and $\hat{c}$ and can quickly and effectively handle different excitations as shown in Figure 4.

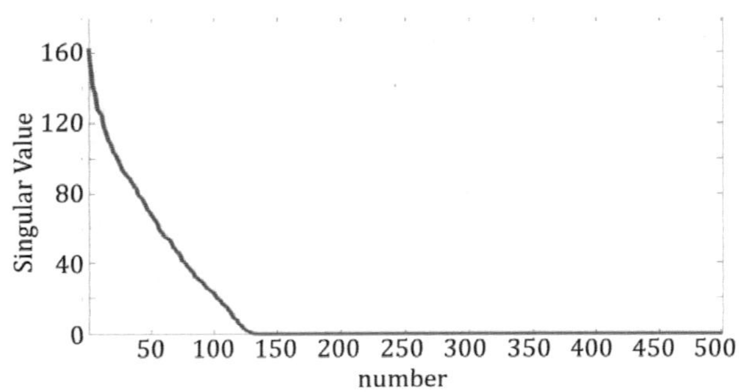

Figure 4. POD: Singular values of SVD of matrix W.

Algorithm 1 Trapezoidal Time Integration Method

Require: initial condition $x(t_s) = x^{init}$

Ensure: x is the solution of $\frac{dx}{dt} = f(x, p, u)$

---

function TRAPEZOIDAL$(f(\cdot), x^{init}, p, u(\cdot), t_s, t_f \Delta t)$
  $x^0 \leftarrow x^{init}$
  $t_l \leftarrow t_s$
  $l \leftarrow 1$
  repeat ▷ Time iteration, index l
    $k \leftarrow 0$
    $\hat{x}^{l,k} \leftarrow \hat{x}^{l-1}$
    $\gamma \leftarrow \hat{x}^{l-1} + \frac{\Delta t}{2} f\left(\hat{x}^{l-1}, u(t_{l-1})\right)$
    repeat ▷ Newton iteration, index k
      $r_0 \leftarrow -[\hat{x}^{l-1} - \frac{\Delta t}{2} f\left(\hat{x}^{l,k}, u(t_l)\right) - \gamma]$
      $j \leftarrow 0$
      repeat ▷ GCR iteration, index j
        $\widetilde{p_j} \leftarrow r_j$
        compute image of residual (r_j): $A\widetilde{p_j} = r_j - \frac{\Delta t}{2\epsilon}\left(f\left(\hat{x}^{l,k} + \epsilon r_j, u(t_l)\right) - f\left(\hat{x}^{l,k}, u(t_l)\right)\right)$
        get Ap_j, p_j by orthonormalizing
        $\alpha_j \leftarrow \langle r_j, Ap_j \rangle$
        $\Delta x_{j+1}^{l,k+1} \leftarrow \Delta x_j^{l,k+1} + \alpha_j p_j$
        $r_{j+1} \leftarrow r_j + \alpha_j A p_j$
        $j \leftarrow j + 1$
      until $\|r_j\|$ small
      $\hat{x}^{l,k+1} \leftarrow \hat{x}^{l,k} + \Delta x^{l,k+1}$
      $k \leftarrow k + 1$
    until $\|\Delta \hat{x}^{l,k}\|$, $\left\|\hat{x}^{l,k} - \frac{\Delta t}{2} f\left(\hat{x}^{l,k}, u(t_l)\right) - \gamma\right\|$ small
    $\hat{x}^l \leftarrow \hat{x}^{l,k}$
    $t_{l+1} \leftarrow t_l + \Delta t$
    $l \leftarrow l + 1$
  until $t_l \geq t_f - t_s$
  return $[x^0\ x^1 \ldots x^l]$

Algorithm 2 POD model order reduction

Require: matrix $W = [\hat{x}^1\ \hat{x}^2 \ldots \hat{x}^m]$ sampled from $x(t_i)$
Ensure: m at least 200-300
  function POD(A, b, c, W, q)
    compute V_L through $SVD(W)$
    $V_q \leftarrow$ column 1 to q of V_L
    $\hat{A} \leftarrow V_q^T A V_q$
    $\hat{b} \leftarrow V_q^T b$
    $\hat{c} \leftarrow V_q^T c$
    return $\hat{A}$, $\hat{b}$, $\hat{c}$

4.2. Absorbing boundary condition

The Earth rarely possesses vertical geological features, so when waves hit the vertical boundaries of our 2D grid, they should pass through instead of being reflected. To address this, we taper the displacement of 10% of nodes on both left and right sides. Those tapered nodes form the absorbing layers. It is implemented by multiplying a tapering matrix T to state $x(t_i)$ at the end of each time iteration, shown in Fig.5.

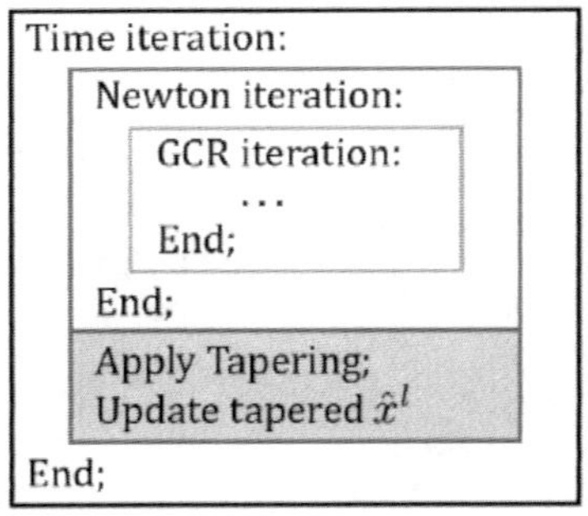

Figure 5. Applying tapering window at each time step.

5. Results

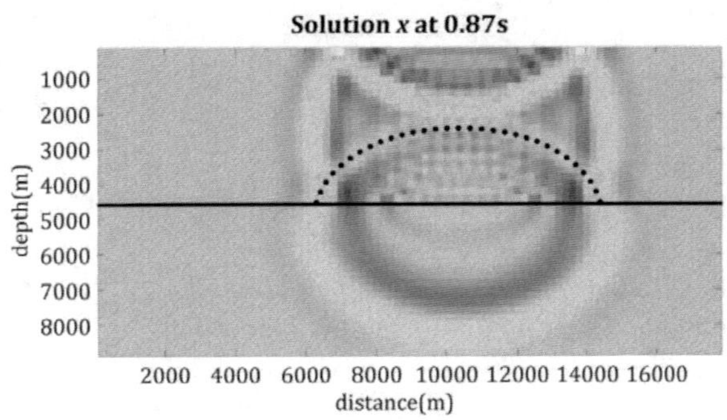

Figure 6. Intermediate reflection at transition interface of v_p

Fig. 6 is a snapshot of results in time, and it shows a clear reflection at the depth of 4500 m, where an intermediate interface is introduced by setting the top and bottom blocks with different p-wave velocities. This is a good sign suggesting that our model is working properly.

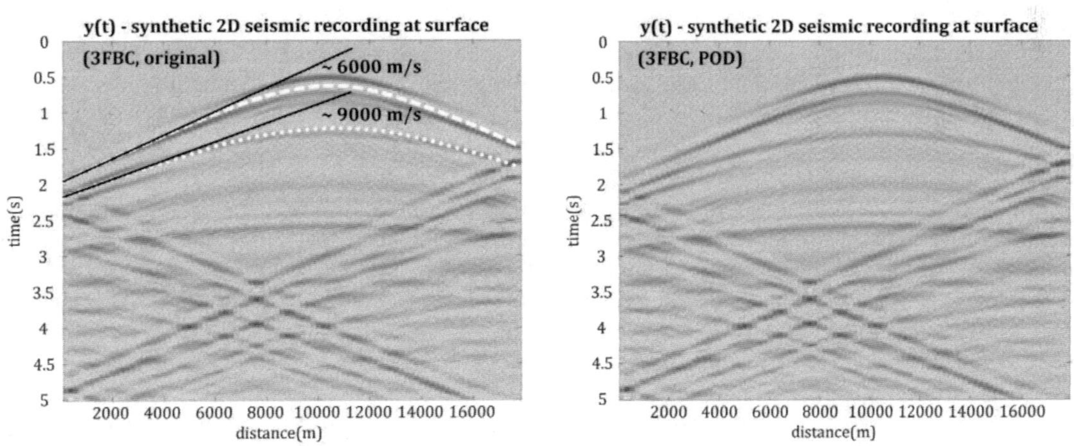

(a) before POD (b) after POD

Figure 7. Results of original (left) & reduced model (right)

Fig. 7 shows a comparison between the original model and the reduced model. Both look very similar, and specifically the L^{-2} error between the two is only 0.0124%. However, the reduced model only takes 0.22s to run, while the original one takes about 25 min, a speed up factor of 900.

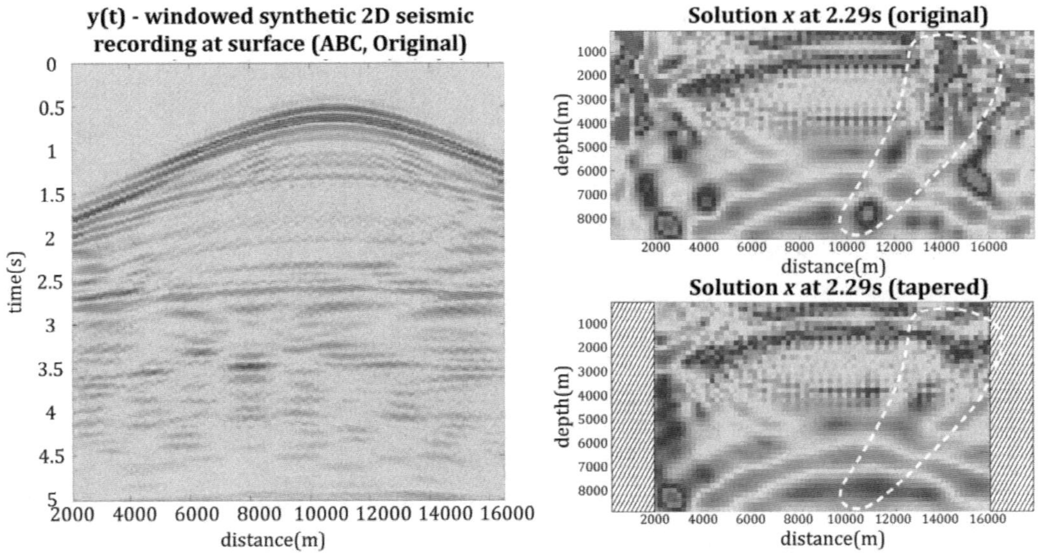

Figure 8. Tapering window effectively suppress reflections

Fig. 8 shows a comparison between the model with and without absorbing boundary condition. It is obvious to see, on the synthetic 2D seismic recording at the surface in the left panel, the straight-line features are being effectively suppressed, which are introduced by the side reflections. Also, in the right panel, the simulation result with absorbing boundary condition (bottom) no longer contains as much side-reflection energy as the one without absorbing boundary condition (top).

6. Technical discussion

In Fig. 8, the reciprocal of the slope where the hyperbola converges is 6000 m/s for the top and 9000 m/s for the bottom, which is close to the true P-wave velocities assigned to the top and bottom blocks ($v_p^{top} = 6600m/s$, $v_p^{bottom} = 9900m/s$). The velocity approximations together with the half-depth reflection suggest that our model is working properly, and the results are valid.

Although our reduced model makes the simulation 900 times faster than the original model, one can easily speed it up more. In this study, we pick 140 largest singular values, and achieve a low error of 0.0124%, which may not be necessary in other applications. Thus, choosing fewer singular values by making q smaller is feasible to further speed up the simulation.

Since we only take 10% of nodes from both sides to be the absorbing layer, the result still experiences faint side reflections, shown in 9. If one increases the size of the layer, the performance of suppression will be better, but it will become more computationally expensive.

Implementing the absorbing boundary condition using tapering window has its limit. The tapering process after each time iteration is manipulating matrix A in GCR, but the model reduction requires A to be unchanged throughout the simulation. Thus, our implementation of absorbing boundary condition is not compatible with the POD model reduction. The address it, more advanced implementation of absorbing boundary condition is needed. For instance, Neumann boundary condition is working directly on matrix A, but it requires one to consider first order spatial derivative, which adds another layer of constrains to the model [4-7].

7. Conclusions

Though this work addresses a physical problem without direct technical concerns, we note several limitations.

- Seismic waves contain energy loss through the conversion from kinetic to thermal energy. We used a constant loss term - 10% reduction along the diagonal of nodal matrix $E-$ to approximate this process. However, in reality, the attenuation should be non-linear terms. Thus, any amplitude analysis should not be performed with this model.
- The current model is tapered manually with sharp suppression in 10% of the nodes at both sides. Reflections still exist from those borders. Our method is not yet perfect given the incompatibility with other techniques. Future work is needed to provide better performance.
- The bottom boundary is an artificial boundary, and the reflection produced is not geologically meaningful. In application, the property of this interface depends on the difference of density across materials from both sides.
- $\Delta t = 0.01s$ for Trapezoidal integration introduces of 8% with respect to the true solution. If more accuracy is needed, one may need to make Δt smaller, but the computation will be more demanding.
- The harmonic solutions of the wave equation are sinusoidal function of both time and space domain. In this work the wave equation is simulated numerically, and we are sampling from the real solution spatially and temporally. Thus, governed by the Nyquist frequency rule, the sampling frequency of both time and space have to be greater than twice the frequency of the excitation u in time and space.

8. Conclusions

In this study, we design a modelling algorithm to quickly and accurately simulate 2D acoustic wave propagation. To solve the time integration of 2D wave equation, we use the Trapezoidal method, which allows us to set a large time step while maintaining a reasonable level of error. We adopt the POD model reduction technique to speed up the simulation by a factor of 900. Also, we implement absorbing boundary conditions to reduce the artificial reflections from vertical boundaries. The simulation results can be used in seismological research to quickly assess the quality of real data. Moreover, they can be treated as constructive a-priori information for inverse problems. It is also worth mentioning that our implementation of absorbing boundary condition is not compatible with POD model reduction. Thus, more work needs to be done in the future to combine the two.

9. References

[1] C. Yu, et al, Mapping Mantle Transition Zone Discontinuities Beneath the Central Pacific With Array Processing of SS Precursors, Journal of Geophysical Research: Solid Earth, 2017.

[2] Z. Zheng, et al, High resolution upper mantle discontinuity images across the Pacific Ocean from SS precursors using local slant stack filters, Geophysical Journal International, 2015.

[3] A. Deuss, et al, The nature of the 660-kilometer discontinuity in Earth's mantle from global seismic observations of PP precursors, Science, 2006.

[4] A. Reynolds, Boundary Conditions for Numerical Solution of Wave Propagation Problems, GEOPHYSICS, 1978.

[5] Luca Daniel, 2.091/6.336/16.910 Introduction to Numerical Simulation MIT, Graduate Course, Fall 2018.

[6] John Dalbey, "How to Write Pseudo-Code", Department of Computer Science and Software Engineering, Cal Poly College of Engineering, 2001: https://users.csc.calpoly.edu/˜jdalbey/SWE/pdl_std.html

[7] John Dalbey, "Examples of Pseudo-Code", Department of Computer Science and Software Engineering, Cal Poly College of Engineering, 2001: https://users.csc.calpoly.edu/˜jdalbey/SWE/pdl_vague.html

Power-based Model for Temperature Prediction in FSW

D Ambrosio[1], V Wagner[1], G Dessein[1], A Tongne[1], M Fazzini[1], C Garnier[1], O Cahuc[2]

[1] Laboratoire Génie de Production, INP/ENI Tarbes, 65016 Tarbes, France
[2] University of Bordeaux, Institute of mechanics and engineering (I2M), CNRS, 33400 Talence, France
Email: dambrosi@enit.fr

Abstract. This paper describes a thermal numerical model accessible to all users for predicting temperature in friction stir welding from the power, material thermal properties, process parameters, tool, and plate dimensions. Starting with the information obtained from the machine, power or torque, the heat flux is modeled as a circular moving source with a diameter equal to that of the shoulder. The model calibrated in a specific setup (CNC machine) successfully predicted without recalibration the weld temperature field in another one (robot). The simple thermal model was applied without recalibration to data available in the literature to test its effectiveness. The results obtained with this model are promising, although more tests are needed to cover all possible varieties of tool geometries and material thickness. If extended over a broader range of configurations (i.e., process parameters and tool-workpiece geometries), it could be a handy tool for all FSW users. The tool may help study the thermal cycles in the heat affected zone that influence final mechanical properties and make it easier to identify optimal parameters if the desired optimal peak temperatures are determined.

1. Introduction

The friction stir welding (FSW) process continues to grow, and more and more industries are becoming interested in it. Thanks to the maturity achieved after more than thirty years of development, today, the process seems much more accessible to all users. However, the difficulties related to the process remain many. Depending on the welding material and its thickness, the tool and the machine, the optimal parameters changes [1]. The friction stir welding outcomes are strongly related to the temperature reached in the nugget zone [2]. If proper temperature ranges are achieved, proper material flow during agitation can be ensured, and if not, a defective joint is likely to be obtained [3]. Similarly, in heat-treatable aluminium alloys, the mechanical properties of friction stir welds are strongly related to the heat affected zones, the weaker zone in joint [4]. The severity and the duration of the thermal cycle undergone by the welding material is responsible for microstructural modifications leading to certain mechanical properties [5]. For all these reasons, the thermal modeling and simulation of the process continue to be an essential tool to predict temperature reached in the various zone that composes the weld. Over time, various more or less complex thermal models have been proposed to predict the temperature generated by a given experimental setup. Some of these have considered contact conditions, only friction [6,7], or both friction and plastic deformation, by introducing the concepts of sticking and slipping [8,9] and rely on temperature-dependent physical quantities generated during the process (yield strength, axial force, etc.). In other cases, punctual heat sources were used in which the heat source was the power and it was tuned to minimize differences between model end experiments [10] or directly the power obtained by the machine, calibrating the model through boundary conditions [11]. In all these cases, the respective models were calibrated on experiments done under specific conditions, and no comparisons to other cases already available in the literature were proposed. Also, many of them are based on complex models and not easily maneuverable by people not trained in modeling. For this reason, in this work, a straightforward thermal model is proposed for the prediction of the temperature in an FSW bead, given the power required for the execution of the joint and the setup characteristics (i.e., welding parameters, tool geometry, workpiece geometry and properties, backing plate). The model is calibrated with a specific experimental setup, validated with additional experiments on the same setup, extended and validated to a different setup, and finally compared to the literature.

2. Model description

2.1. Moving heat source

A finite element framework in ABAQUS™ has been used to model and simulate the thermal fields in FSW. The heat source is the mechanical power (P) obtained from the machine. An efficiency parameter η is used to calibrate the model and express what percentage of this mechanical work is converted into heat (P_{heat}). In order to calculate the surface heat flux (φ), P_{heat} is divided by the tool surface, *considering only the shoulder and the pin lateral surface, which generally represents 90% of the total surface and responsible for the heat generation.* For the sake of simplicity, features such as grooves and threads are not considered. The surface heat flux is applied to the top of the welded plates on a surface equal to the shoulder without considering the pin and moves at a speed equal to the welding speed. In these assumptions, in commercial software to perform thermal analysis, it is sufficient to model the welding plates, assign them thermal properties according to the considered material and create a moving surface heat flux. Regarding the tool contact that is not modeled here, thermal losses through the tool are taken into account through the efficiency parameter. The thermal field obtained at the top of the plates and the FSW joint cross-section running the simulation is shown in Figure 1.

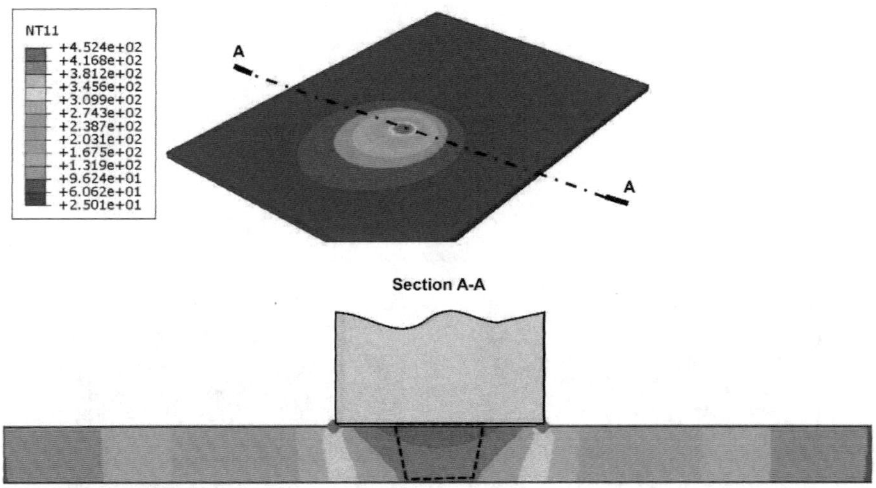

Figure 1. Plot obtained from the simulation. The thermal field in the upper plates and in the cross-section. Temperature in °C.

2.2. Boundary conditions

The boundary conditions are represented by the losses that occur with the air (lateral and upper surface of the plates) and the support system (lower surface here divided into two parts, the central one in contact with steel and the lateral part aluminum). For simplicity, these losses are all incorporated into a contact condition through a convection coefficient. These coefficients have been taken from previous work [11], 15 and 1000 W m^{-2} °C^{-1} for air and steel, respectively. The aluminum one was set to 100 W m^{-2} °C^{-1} to improve the prediction in the heating and cooling cycle of the model, but it does not significantly influence the peak temperature. The interaction with the clamping system was neglected while the initial temperature in the plates was set at 25 °C.

3. Experimental

3.1. CNC lathe machine

The tests were performed on an adapted CNC lathe machine, Somab Genymab 900, in position control (Figure 2). The employed process parameters and the measured mechanical power are listed in Table 1. The experimental campaign consisted of ten bead-on-plate tests performed on 3 mm sheets (200 mm long and 180 mm large), five on AA7075-T6, and five on AA6082-T6. Of these ten tests, two from each material were used to calibrate η (C1 to C4). The validation was made on the remaining six tests (C5 to C10).

Table 1. Process parameters employed in the various experiments. Base material (BM), rotational speed (N), welding speed (v), plunge depth, axial force (F_z) and mechanical power (P)

Config	BM (AA)	N (rpm)	v (mm/min)	d (mm) / F_z (kN)	P (W)
C1	7075	500	180	0.3 mm	945
C2	7075	1500	180	0.2 mm	1060
C3	6082	1500	180	0.2 mm	1240
C4	6082	2000	180	0.1 mm	1200
C5	7075	1000	180	0.2 mm	1002
C6	7075	500	120	0.2 mm	824
C7	7075	1000	240	0.2 mm	999
C8	6082	1000	180	0.3 mm	1253
C9	6082	1000	240	0.1 mm	1140
C10	6082	1500	360	0.1 mm	1255
K1	7075	800	60	6 kN	951
K2	7075	1400	60	5.5 kN	1112
K3	7075	1400	180	7.5 kN	1281

Figure 2. CNC lathe machine setup

Figure 3. KUKA robot setup

3.2. KUKA robot
The runs were carried out on a Kuka KR500-3MT robot with an Actemium BPA-6700 spindle in force control (Figure 3). The employed process parameters and the measured mechanical power are listed in Table 1. The experimental campaign consisted of three bead-on-plate tests performed on 3 mm AA7075-T6 sheets (200 mm long and 180 mm large). Temperature measurements were used to validate the model without recalibrating the parameter for the new configuration.

3.3. Tool and thermocouple
The H13 steel tool was characterized by an 11,5 mm grooved shoulder diameter and a 2.8 mm height frustum shape threaded pin with an upper and lower diameter of 5 and 4 mm. In both experimental configurations, the support system was made of mild steel in the central part surrounded by aluminum. The width of the steel backing plate was different in the two setups, 40 and 60 mm, and this difference was considered when applying the boundary conditions.
In tests performed with the CNC machine, the thermocouples were embedded in the sheets, grooves 2 mm deep, and at a distance of 1 mm from the welding line on the advancing (AS) and retreating side (RS). Post-welding, the thermocouples' final position was determined through computer tomography. They were found about 2.5 mm away from the center due to the pin push action before its arrival. In contrast, the thermocouples were placed between sheets and backing plate in the robot tests, thus measuring the temperature on the bottom surface. The thermocouples distance from the welding line was 3 and 6 mm on the AS and 7 mm on the RS. Transient temperatures were recorded using K-type thermocouples (accuracy ±2 °C). The measurements were acquired by a National Instruments DAQ system at 15 Hz.

4. Results and discussion

4.1. CNC lathe machine

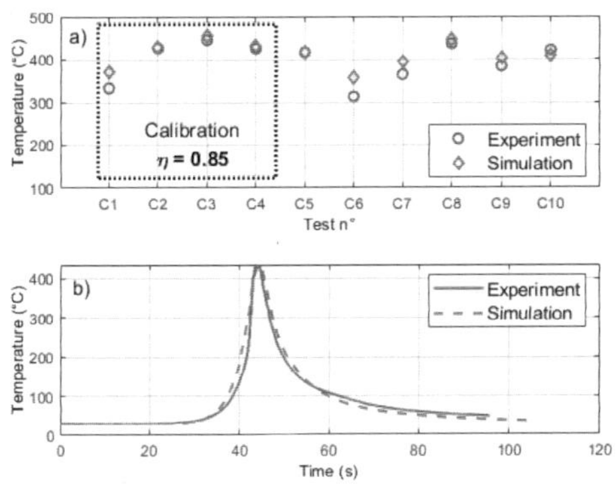

Figure 4. a) comparison between experimental and simulated peak temperatures in test performed on the CNC lathe machine, b) C2 configuration showing the experimental and simulated thermal cycles.

The first four C1-C4 tests were chosen for model calibration. Given the very similar values obtained at both sides, the peak temperature reported in Figure 4a was calculated as the average peak temperature between the two thermocouples. An optimal value of η equal to 0.85 has been chosen in the calibration to minimize the differences between simulated and measured peak temperatures. The same value has been used for all the other cases for validation. In Figure 3a, it can be observed the good approximation offered by the model independently from the material, rotational speed, and welding speed. In eight out of ten tests, the error is less than 5%, confirming a reasonable estimate of the maximum temperature at 2.5 mm from the welding line. Moreover, the two cases in which the prediction error approaches 10% (C1 and C6) are characterized by rotational speeds of 500 rpm (the lowest tested) and the same AA7075-T6 material, thus showing a problem at low rotational speeds. In any case, it is plausible to think that the efficiency parameter is inevitably dependent on both the process parameters and the tool-workpiece materials. However, without considering this dependence, the results are satisfying considering its application to two alloys and ten welding configurations. In

Figure 4b, the comparison of the experimental and simulated thermal cycles of the C2 configuration is displayed. The thermal cycle is predicted to a good approximation.

4.2. KUKA robot

To verify the model's effectiveness, three tests were performed with the same tool and 3 mm-thick AA7075-T6 slabs but in a different configuration (i.e., different machine used to perform the welding, support system, and clamping system). The comparison of measured and simulated peak temperatures is remarkably successful (Figure 5a), with an error on average equal to 5%. The differences due to the system and the different positions of the thermocouples did not reduce the model's prediction ability without any kind of recalibration (i.e., η=0.85). Similarly, the thermal cycle for all three thermocouples is predicted to a good approximation, as shown in Figure 5b.

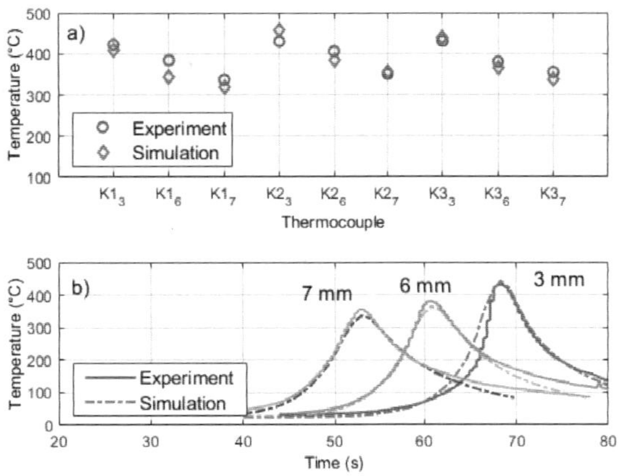

Figure 5. a) comparison between experimental and simulated peak temperatures in test performed on the KUKA robot (Ki_x, i is the test number while x is the thermocouple position), b) K3 configuration showing the experimental and simulated thermal cycles.

4.3. Comparison with literature

Among the few works wherein both temperature measurements and any power or torque measurements are shared, [12] was chosen. Although the most information contained is related to the peak temperatures measured in the nugget zone, through thermocouples embedded in the tool, while the model has been calibrated and validated with temperatures outside the nugget zone, it was decided to evaluate the model without any recalibration. In addition, the process parameter combinations vary widely, not to mention that the plates' thickness has doubled (from 3 mm to more than 6 mm) as well as the tool geometry has also changed. Despite all of these differences, in six out of ten combinations, the prediction error of the peak temperature in the nugget zone is on average 5% (Figure 6a).

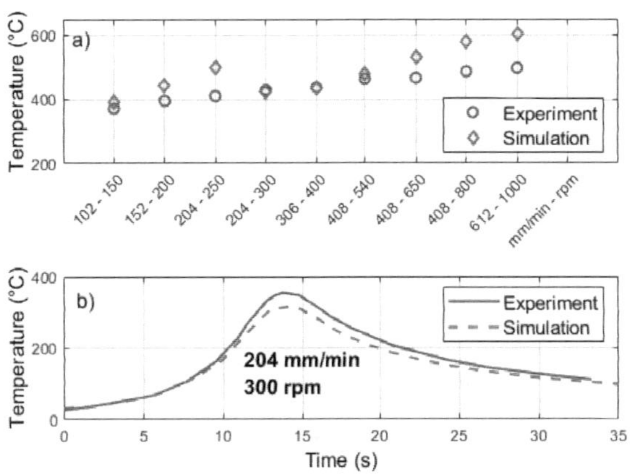

Figure 6. a) comparison between experimental and simulated peak temperatures in [12], b) comparison between experimental and simulated thermal cycles.

However, in other cases, the temperature is too high as the model has no upper limit for heat generation. In the physics of the process, once temperatures above solidus are reached in the nugget zone (around 480 °C for the alloy tested in [12], AA7050), heat generation drops dramatically, and then the temperature stabilizes. In contrast, the model's predictive ability on the heating and cooling cycles is established by comparing the thermal cycle, displayed in Figure 6b, with a measurement taken 6 mm from the welding line.

5. Conclusion

The excellent prediction capability of thermal cycles generated during FSW through a simple thermal FEM model based on a single input has been demonstrated. Its robustness has been verified in various contexts predicting both heat-affected zone and nugget zone temperatures. However, there is a need to incorporate an upper bound into the heat generation to improve the prediction of nugget zone temperature to avoid too high and too far temperatures above the solidus one. Furthermore, this model could be enhanced by studying the efficiency parameter dependence on the process parameters and the tool-workpiece coupling (both material and dimensions). The strength of this macroscopic approach is that it is purely thermal and disregards physical quantities that vary with temperature or are challenging to determine (i.e., yield strength, viscosity, etc.) or the interface tool-workpiece condition without losing prediction quality. In addition, if reliable empirical or analytical models defining the dependence of power on process parameters in a specific configuration are developed, the temperatures achieved with certain process parameters could be easily traced with this thermal model. Hence, in future works, the thermal model can be improved to estimate the mechanical properties of friction stir welding as a function of the thermal cycle induced by the chosen configuration. Also, if optimal temperature ranges to be guaranteed in the nugget zone can be defined, the model could identify suitable process parameters.

6. References

[1] Leitão C, Louro R, Rodrigues D M 2012 *Mater. Des.* **37** 402–409.
[2] Fehrenbacher A., Duffie N A, Ferrier N J, Pfefferkorn F E, Zinn M R 2014 *Int. J. Adv. Manuf. Technol.* **71(1–4)** 165–179.
[3] Ambrosio D, Wagner V, Garnier C, Jacquin D, Tongne A., Fazzini M., Cahuc O, Dessein G 2020 *WELD. WORLD.* **64(5)** 773–784.
[4] Simar A, Bréchet Y, de Meester B, Denquin A, Pardoen T (2008) *Mater. Sci. Eng. A* **486(1–2)**, 85–95
[5] Ambrosio D, Garnier C, Wagner V, Aldanondo E, Dessein G, Cahuc O 2020 *Int. J. Adv. Manuf. Technol.* **111(5–6)** 1333–1350
[6] Song M, Kovacevic R 2003 *Int. J. Mach. Tools Manuf.* **43(6)** 605–615
[7] Zhang X X, Xiao B L, Ma Z Y 2011 *Metall. Mater. Trans. A Phys. Metall. Mater. Sci.* **42(10)** 3218–3228
[8] Schmidt H, Hattel J 2005 *Sci. Technol. Weld. Join.* **10(2)** 176–186
[9] Zhang P, Yan H, Li C, Yu Z, Lu Q 2017 *Sci. Eng. Compos. Mater.* **24(3)** 439–446
[10] Vilaça P, Quintino L, Dos Santos J F 2005 *J. Mater. Process. Technol.* **169(3)** 452–465
[11] Khandkar M Z H, Khan J A, Reynolds A P 2003 *Sci. Technol. Weld. Join.* **8(3)** 165-174
[12] Upadhyay P, Reynolds A P 2010 *Mater. Sci. Eng. A* **527(6)** 1537–1543.

Acknowledgments

This project received funding from the European Union's Marie Skłodowska-Curie Actions (MSCA) Innovative Training Networks (ITN) H2020-MSCA-ITN-2017 under the grant agreement N°764979.

The Structural Design and Analysis of a Symmetrical Twin-tower Conjoined Building with Vibration Absorber

Boyuan Deng[1]

1 School of Civil Engineering, Southwest Jiaotong University, Chengdu, China
Email: dby0119950139168@163.com

Abstract. Throughout the ages, earthquakes have caused numerous instability and destruction of buildings. With the increasing of the height and volume of modern buildings, the seismic force on the buildings gradually increases. To avoid serious damage and ensure the safety, seismic design has become a crucial element in the analysis and design of the structure. In traditional seismic design, increasing stiffness of the building is usually adopted to reduce the structural response during an earthquake. This approach is usually costly, and the effect of practical application is not obvious, which could not fundamentally enhance the bearing capacity of the structure. Basing on existing researches, both theoretical and physical model of symmetrical, conjoined twin tower is made, meanwhile, the reasonable design using damping vibration absorber with the function of energy dissipation, utilizes the movement of additional weight to reduce vibration amplitude of main body structure. The use of shaking table test on the analysis and verification of bearing capacity of the model of symmetrical, conjoined twin tower with vibration absorber, improves the seismic performance of the building. It helps propel innovative and practical design ideas in structural seismic design of symmetrical conjoined twin tower.

1. Introduction

With the continuous enrichment of both theoretical and practical knowledge in the construction industry, human beings present their demands for beauty in complex and beautiful architectural forms. As a special architectural form, the multi-tower conjoined buildings are widely loved and often exist as landmarks because of the aesthetic and visual impact brought by their combination forms and changes. For example, the Petronas Twin Tower, Kuala Lumpur, Malaysia, is the most representative of the world's tallest symmetrical twin tower. It consists of two separate towers, and the most prominent feature of which is a 58.4-meter-long arched, sky corridor, 170 meters above the ground, right between the 41st and 42nd floors of the two main buildings, connecting the two towers with a total of 88 floors, both 451.9 meters high [1]. For a building with a higher height, due to its small natural vibration frequency, the structure is subjected to greater seismic force and overturning moment, causing more damage [2]. Therefore, to ensure the seismic safety of high-rise structures and be economy, it is vital to carry out in-depth research on the aseismic structure and aseismic technology of high-rise buildings.

The twin-tower building connected by a sky corridor is widely adopted because of its multiple strengths such as convenient communication and providing an escape passage during a fire. But a higher internal force in the corridors could occur under an earthquake, and the constructional details are very complex to address if the sky corridor is rigidly connected to the towers. On the other hand, a flexible connection (passive control devices) can overcome these weaknesses to a certain degree. The rational design of flexibly connected element parameters has become a key issue [3]. In the conventional seismic measures, by adjusting the size and strength of the structure to increase structural stiffness, reduce structural seismic response. Such approach is always costly, and due to the increase of dead load when adjusting the structural size, accelerating the collapse during an earthquake. It cannot fundamentally improve the structural seismic performance.

This paper takes the Petronas Twin Tower as the prototype, building a scale model of the symmetrical twin tower with a sky corridor. By making a wooden structure model, mechanical analysis and vibration loading study of the vibration characteristics of the two towers with flexibly connected element, make the high-rise twin-tower conjoined structure meet the requirements of structural and seismic design. In recent years, many scholars at domestic and abroad have introduced energy dissipation and shock absorption equipment like vibration absorber into building structures [4]. Serving as an improvement on the traditional structural design, the vibration absorber can effectively reduce the damage of the main structure in the earthquake. In this study, in consideration of both economy and applicability, vibration absorber is utilized as the flexibly connected element to improve its seismic performance in the earthquake.

The vibration absorber, also called a tuned mass damper, or TMD. Early Ormondroyd used springs and dampers to hang mass blocks on the main structure as substructure attached, laying the foundation for the development of vibration absorber [5]. The principle of vibration absorber is that when the structure is forced to vibrate, the substructure and the main structure vibrate together, and the resulting energy reacts on the main structure through the inertial force, and part of the energy is consumed by the substructure to realize vibration reduction [6]. At present, vibration absorber has become one of the passive energy-dissipation vibration reduction control devices that is widely used in civil engineering and mechanical engineering around the world [7]. In different damping structures with multi-degree of freedom, the performance of dynamic vibration absorption varies. To further expand the application of TMD in civil engineering, researchers have designed, tested and applied vibration absorbers in a variety of complex environments. For example, Makoto Ohsaki from Japan deigned a tetrahedral tuned mass damper (TD-TMD), for three-directional seismic response reduction, which is composed of the viscous damper, the spring and the rigid bar. By utilizing the flexibility of the spring, the three-directional displacement of the viscous damper is increased, and vibration energy is effectively dispersed [8]. JaviDialesaadi A et al. from the United States designed a passive mass damper including physical mass and rotational mass, named rotational inertia double-tuned mass damper (RIDTMD). Compared with the traditional tuned mass damper, RIDTMD is expected to effectively reduce the response of the underlying main system [9]. Chiara Masnata et al. from Italy described an innovative passive control strategy, called the tuned mass damper inerter (TMDI), by working with the base isolated system (BI), to restrain the displacement of the base isolation structure under seismic excitation [10]. Seung-yong OK from Korea installed mass coordination damper on a symmetrical twin-tower structure in an asymmetric way to perturb the symmetry of the twin-tower structure and form an asymmetric coupled system, thus establishing an efficient vibration control system [11] Therefore, an appropriate design of vibration absorber is vital. How to design the shape, size and mass of the vibration absorber reasonably, how to choose the position and number of the vibration absorber appropriately, and so on, all of which need to be studied in depth in this article and in the future to make sure the vibration absorber is functional during an earthquake. In comparison with the above works, by analysing and modelling existing building, this paper will demonstrate the process of designing and application of the vibration absorber, making the research results more convincing and providing others with research ideas.

2. Structural design and demonstration

This structural design is aimed at the symmetrical, conjoined twin-tower building, using the reduced-scale test to simulate the structural response in the actual working state.

2.1. Simplified structural model

To highlight its essence and basic characteristics. First of all, the simplification of actual structure and assumptions, application of AutoCAD software to map the structure calculation diagram under the two-dimensional plane, the structure design model of the overall building height is set to 600 mm,

Journal of Physics: Conference Series **2287** (2022) 012026 doi:10.1088/1742-6596/2287/1/012026

layer number is 3, is divided into two columns, the height is 200 mm, the beam is located in the second layer, the span is 150 mm, caused by seismic horizontal load (Fx) constraints, simplified structure model, as shown in Figure 1.

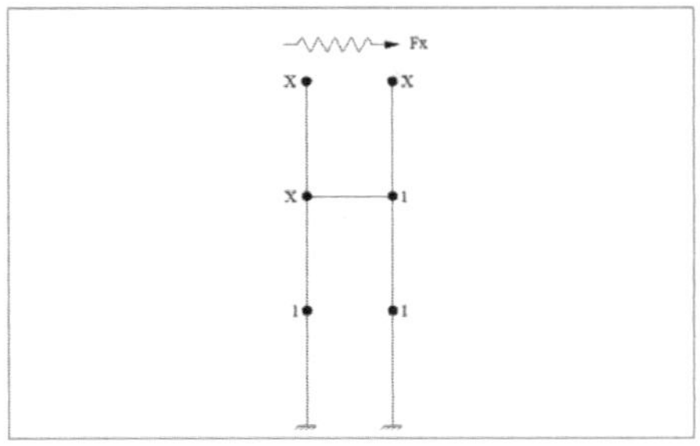

Figure 1. Schematic diagram of structural calculation.

Combined with the structural calculation diagram, the application of Ths-Unibim software to carry out three-dimensional modeling of the symmetrical double-tower joined structure building, the specific steps are as follows, the structural calculation diagram: structural three-dimensional model diagram = 1:10.

The first step is to design the dimensions of each component, including frame column, frame beam, connecting beam, plate, wall, etc.The second step, respectively draw the axis network, column, beam body (including the main beam, secondary beam and connecting beam), plate and wall the base layer, the first layer, the second layer. Taking the first floor as an example, the layout is shown in Figure 2.

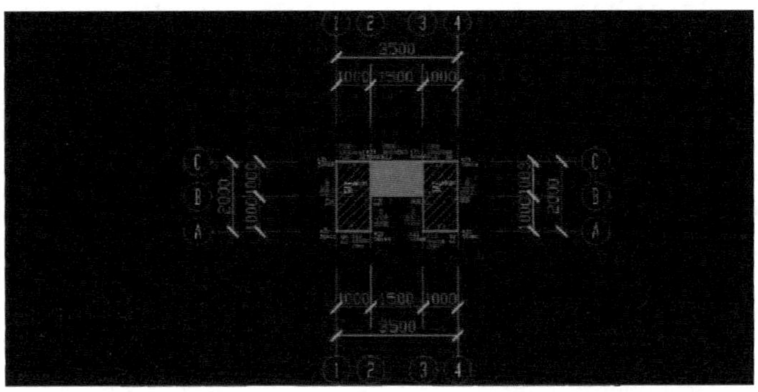

Figure 2. Layout of the first floor.

The third step is to display the layout of every floor in 3D, and carry out the operation of "multi-layer combination", as shown in Figure 3.

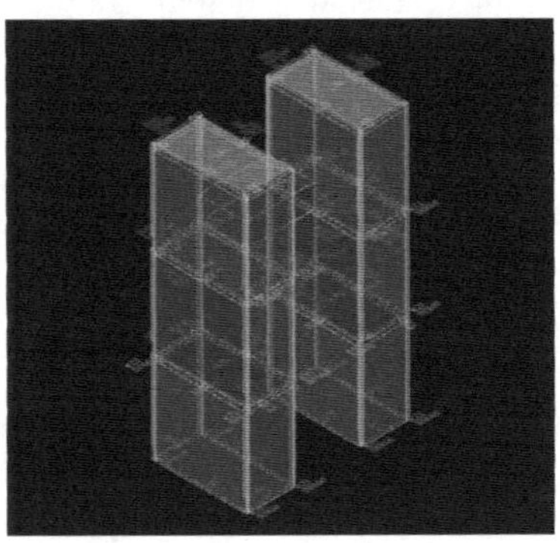

Figure 3. Three-dimensional model of structure

2.2. Structural load test
In order to verify the correctness of the structural design and parameters, it is necessary to carry out load tests on the solid building structure. When loading, it is divided into the following two parts:
The first part is to detect the resistance and deflection of the structure by applying static load to the structure with a heavy object.
The second part is to test the seismic performance of the structure by using the shaking table to apply dynamic load to the structure.
The static load and dynamic load of this test were applied simultaneously. Under the condition of continuous seismic wave loading, the same weight was monotonically added on one side of the second floor and the third floor.
The loading design of seismic simulation shaking table test is very important. If the load is too large, the structure may soon enter the plastic stage or even collapse. If the load is too small, can not reach expectation, produce unnecessary repetition, and multiple loading can produce damage accumulation to the structure. In order to obtain the test data of the system, the design of the test loader must be carefully considered. The specific loading design is as follow:
(1) Static load: weight blocks with asymmetric mass are placed in a total of six floors. The order of weight block placement is: X, X, 1kg from top to bottom on the left side; X, 1,1 kg from top to bottom on the right. X is the added mass of different floors. The static monotone loading of the structure is adopted. In the test process, the static loading starts from X =5kg and gradually increases, with a single loading of X =0.5kg until the structure is destroyed.
(2) Apply dynamic load: in order to improve the reflection performance of the system and reduce the distortion of the acceleration waveform, the acceleration signal input method is used for simulation control by using the electric shaking table. Two kinds of natural seismic waves, Taft and E1-Centro, were input from the table to simulate seismic excitation. The seismic wave was unidirectional, and the vibration only proceeded along a single direction. The acceleration signal of seismic wave is amplified by D/A conversion and analog control system to drive the shaking table about 13 times, that is, the acceleration amplitude is about 0.4g. Computer system of the shaking table analyses and calculates the input seismic waves according to the frequency spectrum characteristics of the shaking table. After processing, D/A conversion and simulation amplification are carried out to make the shaking table reproduce the seismic waves, so as to load the structure, and the seismic waves add A weight to the structure at the end of each cycle until the structure is destroyed.
The overall force analysis of the structure shows that the bearing capacity of the beam is mainly tested when the structure is under static load. Generally speaking, the deflection of the trabecular beam can be effectively reduced by increasing the cross-section area, i.e. increasing the EI of the beam. And because the weight block mass is larger and the distribution is uneven, so the inertia force is larger, higher requirements for the column, which is also the need to focus on the problem to be solved in the design. In general, the vibrating load is more harmful to the structure.

3. Structural design of vibration absorber
In order to effectively analysing the damping effect and the shock absorption of the vibration absorber, the tower structure is modelled by AutoCAD. And the simplified model of the structure without vibration absorber and with a single vibration absorber is shown in Figure 4.
The building is ideal truss structure, which is composed of straight bars, and all the nodes are smooth hinge joints. For the ideal truss, all the bars only bear axial force, so only the axial deformation of the bars is required when calculating the displacement of the truss. To simplify the solving process of axial force, use symmetrical truss and calculate each two layers as one layer.
With the equivalent base shear method, basing on the statics analysis, to simplify the seismic action as an inertial force system act on the object of study. The shear force on the bottom of the structure shares the same horizontal seismic load as equivalent mass point, take the side with larger load for analysis, equivalent inertia force to a periodic transverse load, to reflect the dynamic characteristics of the load.

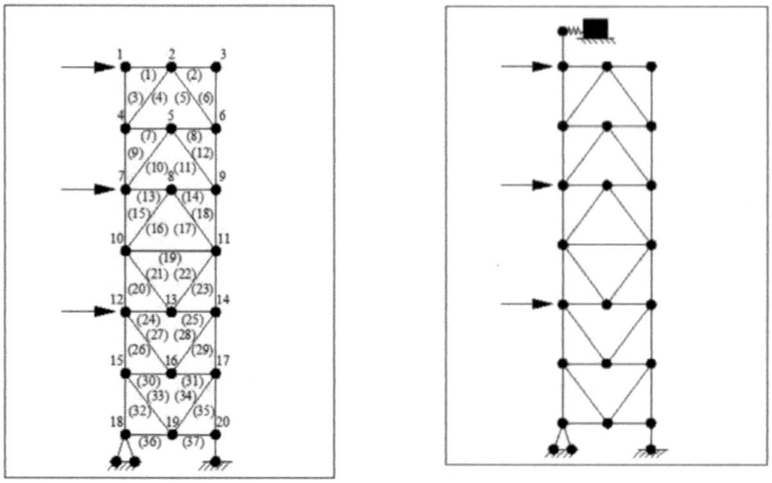

(a) without vibration absorber (b) with vibration absorber

Figure 4. Simplified model of structure with or without vibration absorber.

3.1. Displacement amplitude

MATLAB is used to solve the global stiffness matrix and the mass matrix. Firstly, basic information about the structural components and the magnitude, direction of external load is required to generate the mass matrix and the stiffness matrix. And the vibration absorber is installed on the node under external load. The number of node displacements is increased by one, so is the dimension of the matrix. The displacement is obtained by decoupling the stiffness matrix and mass matrix of increasing order and the external force, and the node displacement diagrams with no vibration absorber, one vibration absorber, two vibration absorbers and three vibration absorbers are shown in Figure 5 (a), (b), (c) and (d). The MATLAB procedure is shown in Appendix A

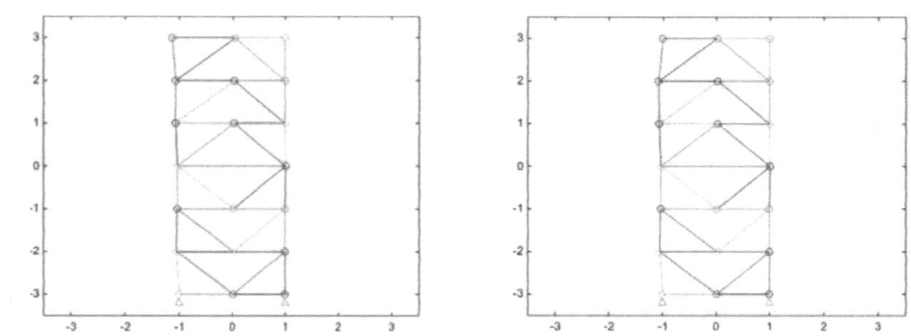

(a) No vibration absorber (b) One vibration absorber

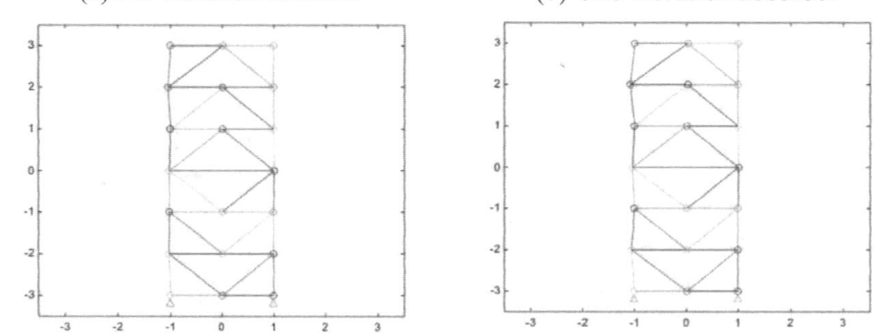

(c) Two vibration absorbers (d) Three vibration absorbers

Figure 5. Amplitude of structural displacement.

Obviously, at 1-kg-weight place and 5-kg-weight middle tier (X = 5, X is the added mass of different floors), the vibration absorber has a good damping effect to bear the vertical load. Damage is caused to the column footings when there is so vibration absorbers, and the bearing capacity is very low, only 3kg. After adding 1 or 2 vibration absorbers to the structure, the bearing capacity has more than doubled to a maximum of 6.5kg. Displacement under dynamic load decreases significantly. Vibration absorber greatly reduces the structural response under the seismic load. The change of displacement of using 3 vibration absorbers is not obvious compared with using 2 vibration absorbers. With comprehensive consideration of both economy and damping effect, 1 or 2 vibration absorbers are recommended.

3.2. Final scheme design

In order to achieve the function of vibration absorber, the study makes use of paulownia strips and other materials to make a physical model of rocking vibration absorber. Vibration absorber design is mainly divided into cable design, platform design and displacement limiting device design.

3.2.1. Cable design of vibration absorber. Vibration absorber cable is made of 10mm wide, 205mm long and 1mm thick wood along the grain direction, the board requires better toughness to ensure a good tensile capacity. Every vibration absorber is designed with 4 cables, and the top segment is attached to the inner side of the beam ends of the second and third floors.

Considering the bearing capacity of loading 1kg weight at the vibration absorber and 5kg weight at the vibration absorber is different, so at the 1kg-weight use a thick layer of wood to make cable, at the 5kg-weight use two layers of thick wood to make cable (use two pieces of 1mm thick wood to stick). After shaking table test, the cables of the first and second floors fully meet the requirements of bearing capacity.

3.2.2. Platform design of vibration absorber. The material of the vibration absorber platform is the wood rod with the closed lap around, and the sectional area of the rod is appropriately increased to ensure a good loading bearing capacity. The end of the cable is clamped with 4 pieces of wood with the size of $B \times H \times L$ =3mm×5mm×100mm in pairs to keep the platform level slightly higher than the nearby beam, and two pieces of wood with the size of $B \times H \times L$ =3mm×5mm×90mm are overlaps above the wood and inside the cable to make up the platform of the vibration absorber.

The platform design of vibration absorber is not universal, so different designs are applied to the platform at the 5kg weight and the platform at the 1kg weight. At 5kg weight, the sectional area of the wood rod of the vibration absorber platform is larger than that used at 1kg weight, which makes the platform stiffness at 5kg weight is larger and the deformation is smaller under the external force. With partial deflections within a controllable range, and the friction due to contact, it enhances shock absorption. Shaking table tests have also been carried out to verify that the first and second floor platforms could fully meet the bearing capacity requirements.

3.2.3. Design of displacement limiter. One of the highlights of the design of the vibration absorber is the utilization of the displacement limiter. In the shaking table test, it is found that the swing amplitude of the vibration absorber is large under seismic wave, which will lower the bearing capacity and generate eccentric load. Therefore, a displacement limiter is made for the vibration absorber.

The displacement limiter is located below the platform and consists of two main parts. One part is glued to the lower side of the two lapped wooden strips on the platform. The part under each lapped wooden strip is composed of two small wooden blocks with the size of $B \times H \times L$ =5mm×6mm×18mm and a piece of wood chips with a width of 9mm, a length of 54mm and a thickness of 1mm. The wood blocks are clamped between the wood chips and the wood strips, and the distance between the two wooden blocks is 18mm, leaving a gap for vibration absorber swing; Another part is with a size of the b * h * l = 3 mm x 5 mm x 200 mm batten, four dimensions of b * h * l = 3 mm * 5 mm * 7 mm small block and two film is 15 mm, 5 mm wide, 1 mm thick, the small board, wood through space and lap between on its vertical beam, wood with four pieces of wood on both ends after two clamping, paste with two pieces of wood on its cover, the general structure of the displacement restrictor.

When the vibration absorber swings beyond a certain limit, the wood blocks at both ends will limit the displacement of the attached wood strips below the platform. When the vibration absorber swings from side to side, the excess displacement will be confined by the wood strips below the platform, thus providing the vibration absorber with restoring force. The limit of displacement is based on the maximum displacement under the seismic wave obtained by many tests and then reduced to a certain extent to get the reasonable displacement limit, which could not only realize the function of vibration absorber, but also could limit the displacement.

After several pre-loading tests, it is obvious that the vibration absorber works, the damage caused by seismic waves of the whole structure is greatly reduced, and the vibration absorber consume a certain amount of energy. The physical model of vibration absorber is shown in Figure 6.

(a) Side view of single tower vibration absorber

(b) Myopia of single vibration absorber

Figure 6. Physical model of vibration absorber structure.

4. Analysis of structural bearing capacity

Considering the safety, durability and practicability of the structure, a further analysis of the bearing capacity of the symmetrical, conjoined twin tower is necessary. The physical model of the symmetrical, conjoined twin tower is made and shown in Figure 7.

Figure 7. Physical model of the symmetrical, conjoined twin tower.

The load-displacement relationship curve cannot be directly obtained so the analysis is mainly basing on the bearing capacity and the experimental phenomenon. According to the initial loading, and further reinforcement of the vibration absorber, it is estimated that the bearing capacity of the structure

simulated by the physical model could reach 6.5kg, high safety factor. The force on the vibration absorber with the bearing capacity of 6kg is rather large, and the safety margin is not big. In the end, there was no overall structural damage after applying seismic force, and the vibration absorber effectively played the role of reducing the overall vibration amplitude. The following conclusions can be drawn from the analysis of structural bearing capacity:
(1) The sky corridor of twin tower could improve the overall stiffness and shorten the natural vibration period of the structure. However, because of the sky corridor, the directional factors of the vibration of the whole structure will be changed and the vibration will be more diverse at the middle and higher level, which makes the vibration of the twin tower more complex than that of the twin-tower structure without sky corridor. However, the sky corridor only has a little influence on the natural vibration period of the structure, and a great influence on the displacement of the top floor and the displacement between floors of the tower.
(2) Vibration absorber could effectively consume energy, so that the bearing capacity of structure is more abundant. While ensuring the bearing capacity, the vibration absorber can also be set to reduce the use of the material of the structure itself, so that the cost of the overall structure is reduced, and the coordination and unification of structural performance and cost can be realized.
(3) Vibration absorber could reduce the amplitude of vibration effectively. As economy factor is considered, the design of the structure with 1-2 vibration absorbers can give full play to its damping function, and to produce a good result.
(4) when the vibration absorber swings beyond a certain limit, the wood block will prevent the displacement of the wood strip connected under the platform. When the swing exceeds the limit, the vibration absorber will be confined by the wood strip below the platform, which provides the vibration absorber with restoring force, limiting the displacement of the vibration absorber. The limit of the displacement is the maximum displacement under the seismic wave obtained from several tests, which is then reduced to a certain extent to obtain the optimal, safe displacement limit, which can not only make full use of vibration absorber, but also limit the displacement effectively.

5. Summary and discussion

This paper develops tall building structure seismic design by respectively designing and analysing the symmetrical, conjoined twin towers and vibration absorber. On the one hand, using the static load test and the shaking table test was carried out on the analysis of physical model structure bearing capacity, on the other hand, using the software for graphical modelling and digital simulation of the structure. And verify the swing type of vibration absorber having good damping effect on high-rise, conjoined towers, and reasonable placement of vibration absorber, so as to achieve the aim of improve seismic performance of structure itself.
The design of vibration absorber is one of the highlights of this research project in seismic design. Through the research and analysis of this paper, different from other seismic means and measures, the optimization of building structure by introducing vibration absorber can be summarized as the following three points:

5.1. Economy

In the conventional structural design, the response of the structure under earthquake is usually reduced by increasing the structural stiffness. By simply increasing the structure stiffness will inevitably lead to the increase of quality, for the high-rise building, it is often expensive, and the practical application is not very effective, could not really improve the seismic performance of the structure, the introduction of vibration absorbing device can effectively reduce the structural weight, reduce the use of building materials, so as to reduce the cost.

5.2. Universality

At present, there are a large number of researches, aiming at one specific building type under the high-rise twin-tower conjoined structure, according to different engineering examples, in terms of material selection and construction techniques, specific seismic design schemes are made, but they are not highly universal for other engineering examples or building types. As a common damping structure, vibration absorber can be placed and applied in the retrofitting stage of existing buildings besides the anti-seismic application of the structure in the design stage and construction stage.

5.3. Fundamental

Vibration absorber effectively resist vibration, radically reduce the vibration amplitude of main structure, to reduce the vibration destruction of the connecting and even the overall structure, Vibration absorber can be effective on the system energy consumption, guaranteeing bearing capacity at the same time, by setting the energy consumption structure to reduce the quality of the buildings themselves, to improve the seismic performance.

It can be seen from the foregoing that the swinging vibration absorber designed in this study is successfully applied to the symmetrical twin-tower conjoined structure. It is always the key and foundation of successful seismic design to do a good job in structural conceptual design. However, taking the Petronas Towers in Kuala Lumpur, Malaysia as an example, the application of dynamic vibration absorber to such a specific high-rise building still needs further research and discussion, mainly from the following two aspects:

- To realize the optimization of shock absorption effect of vibration absorber. Applied in the actual cases, the environment should be considered and the corresponding technical problems, such as the quality and how to implement the vibration absorber which is used to the balance of shock absorption band width, on how to ensure making full use of the vibration absorber when earthquake, how to reasonably select vibration absorber with good quality, appropriate spring stiffness and the effective position so as to realize the optimal damping effect.
- Reduce the occupancy of the vibration absorber inside the building. The Taipei 101 building takes up the ball vibration absorber which occupy 88th-92nd floors. Although achieved good vibration reduction effect, but to a certain extent reduce the efficiency of the space. A deeper study of the structural design is required for ensuring a good damping performance while occupying less space.

6. References

[1] Thornton C H, Hungspruke U, Joseph l. m. the Design of the world 's tallest buildings – Petronas Twin Towers at Kuala City Centre [J]. The Structural Design of Tall buildings, 1997, 6 (4): 245-262.

[2] Code for seismic design of buildings: GB 50011 - 2010[S].2016 edition. Beijing: China Architecture and Building Press, 2016. (in Chinese)

[3] Qing,Lyu, Wensheng,Lu, Weiqiang,Wang, Yue, & Chen.(2020).Mechanism and optimum design of shared tuned mass damper for twin-tower structures connected at the top by an isolated corridor. Structural Design of Tall and Special Buildings,29(8).

[4] Research on seismic design of building structure design [J]. China Residential Facilities, 2017(02):15-16. (in Chinese)

[5] Ormondroyd J, Den Hartog J P. The theory of the dynamic vibration absorber [J].Transactions of the American Society of Mechanical Engineers, 1928, 49, 50: A9-22.

[6] Zhang Fan, Li Chunxiang. Research on the Control of Structural Enhanced Tuned Mass Damper [J]. Structural Engineer,2019,35(05):119-125. (in Chinese)

[7] Jing Ming, Dai Junwu. Earthquake Engineering and Engineering Vibration,2017,37(03):103-110. (in Chinese)

[8] Makoto Ohsaki, Seita Tsuda, Toma Hasegawa. Parameter optimization of tetrahedral tuned mass damper for three-directional seismic response reduction[J]. Elsevier Ltd,2016,126.

[9] Javidialesaadi A, Wierschem N E. Optimal design of rotational inertial double tuned mass dampers under random excitation [J]. Engineering Structures, 2018, 165(jun.15):412-421.

[10] Chiara Masnata, Alberto Di Matteo, Christoph Adam, Antonina Pirrotta. Smart structures through nontraditional design of Tuned Mass Damper Inerter for higher control of base isolated systems [J]. Mechanics Research Communications,2020,105

[11] Seung-Yong Ok. Tuned mass damper asymmetric coupling system for vibration control of adjacent twin buildings [J]. Advances in Structural Engineering, 2020, 23(5).

Journal of Physics: Conference Series **2287** (2022) 012026 doi:10.1088/1742-6596/2287/1/012026

7. **Appendix** A

```
function [ M1,K1,num ] = Slow( M,K )
num=input('Please enter the number of vibration absorbers： \n ');
e=zeros(37,1);
a=zeros(num,1);
s=zeros(num,1);
ks=zeros(num,1);
ms=zeros(num,1);
for i=1:num
   s(i,1)=input('Please enter code number:\n ');
   a(i,1)=2*s(i,1)-1;
   ks(i,1)=input('Please enter the stiffness of the spring connected to the vibration absorber： \n ');
   ms(i,1)=input('Please enter the mass of vibration absorber： \n ');
   e(a(i,1),1)=1;
end
K1=zeros(37+num,37+num);
M1=zeros(37+num,37+num);
for i=1:37
   for j=1:37
      K1(i,j)=K(i,j);
   end
end
for i=1:num
   K1(a(i,1),a(i,1))=K1(a(i,1),a(i,1))+ks(i,1);
   K1(37+i,a(i,1))=-ks(i,1);
   K1(a(i,1),37+i)=-ks(i,1);
   K1(37+i,37+i)=ks(i,1);
end
for i=1:37
   for j=1:37
      M1(i,j)=M(i,j);
   end
end
for i=1:num
   M1(37+i,37+i)=ms(i,1);
end
end
```

A Hybridizable Discontinuous Galerkin and Boundary Element coupling method for electromagnetic simulations

Yan Kong[1] and Mei Yan[2]

1 University of Electronic Science and Technology of China, Chengdu, Sichuan, 611731, PR China
2 Yunnan Normal University, Kunming, Yunnan, 650500, PR China
E-mail: coin030911@gmail.com[1], myanlxy920@hotmail.com[2]

Abstract. In this paper, the hybridizable discontinuous Galerkin (HDG) method is combined with the electric field integral equation (EFIE) for the numerical simulation of time-harmonic Maxwell's equation. The key feature of HDG-EFIE is that the HDG method and the boundary element method (BEM) can be coupled naturally through the extra hybrid variable on the faces of elements, the combined method produces a linear system in term of the degrees of freedom of the extra hybrid variable only. A numerical solution was compared to the analytical solution and found to be in excellent agreement.

1. Introduction

Recently, hybridizable discontinuous Galerkin (HDG) [1] have gained much popularity in the numerical solution of electromagnetic scattering problems. HDG method preserves all of the benefits of the DG method, furthermore, by introducing an extra hybrid term on the boundary of the elements, the number of DOFs is drastically reduced. HDG is well prepared for problems with complex media, however, when the problem is restricted to unbounded domain, artificial boundary condition [2] must be imposed, which yields to a great increase of degrees of freedom (DOFs). In contrast, boundary element methods (BEM) [3] works extremely well for solving open scattering and radiation problems involving conducting surfaces but failed to solve complex structure models particularly those with inhomogeneous dielectric. If we want to take advantages of both HDG and BEM, especially for problems with interior finite subdomains enmeshed in exterior unbounded subdomains, it is a natural idea to find a way to couple HDG with BEM. HDG-BEM has been introduced and analyzed by Cockburn [4, 5], Zhixing Fu [6] and their collaborators, which continued the works in DG-BEM and started In Bustinza, Gatica and Sayas [7–9]. In [10] some work had been done for applying HDG-BEM to waveguide problems. Our purpose here is to study the consequences of the combined method on numerical investigation of the electromagnetic scattering problems.

The rest of this article is organized as follows. In section 2, we introduce the Maxwell's equations and the electric field integral equation (EFIE). In section 3, we propose HDG-EFIE formulation. In Section 4, numerical results are illustrated in details. At last, section 5 in this paper provides some concluding remarks.

2. Probem statement and notations

2.1. Time-harmonic Maxwell's equations

Let Ω be a bounded domain with a Lipschitz boundary Γ (the artificial boundary) and structure area $\Omega_s \subseteq \Omega$ with boundary Γ_m. Then we consider the time-harmonic Maxwell's equations in 2d

$$
\begin{aligned}
& i\omega\epsilon E - \nabla \times \boldsymbol{H} = 0, \\
& i\omega\mu \boldsymbol{H} + \nabla \times E = 0.
\end{aligned} \tag{1}
$$

The electric field is $E = E_z$ and magnetic fields are $\boldsymbol{H} = (H_x, H_y)$, the imaginary unit is i, the angular frequency is ω, ϵ is the relative permittivity and μ is the relative permeability. $E^{inc}(\boldsymbol{r})$ is the incident wave, $\boldsymbol{r}$ and $\boldsymbol{r}'$ denote the observation point and the source point, when they are on the same contour Γ, the exterior branch of electric field $E(\boldsymbol{r})$ satisfy the integral equation as follow

$$
\frac{1}{2}E(\mathbf{r}) = E^{inc}(\mathbf{r}) + \oint_{\Gamma}[E(\mathbf{r}')\frac{\partial G(\mathbf{r},\mathbf{r}')}{\partial \mathbf{n}'} - G(\mathbf{r},\mathbf{r}')\frac{\partial E(\mathbf{r}')}{\partial \mathbf{n}'}]d\Gamma' \tag{2}
$$

with

$$
G(\mathbf{r},\mathbf{r}') = -\frac{i}{4}H_0^{(2)}(k|\mathbf{r}-\mathbf{r}'|)
$$

$H_0^{(2)}$ is the modified Bessel function of order zero and $\boldsymbol{n}$ is the unit normal vector on Γ. To make connection between (1) and (2), we need the following boundary condition on Γ,

$$
\boldsymbol{n} \times (\boldsymbol{E_1} - \boldsymbol{E_2}) = 0, \qquad \boldsymbol{n} \times (\boldsymbol{H_1} - \boldsymbol{H_2}) = 0.
$$

2.2. Notations

We use the notations in [1] to give the discretization for further illustration, A triangulation $\mathcal{T}_h$ of domain Ω is the union of K elements. Γ_h^i is the union of all interior interfaces of $\mathcal{T}_h$ and Γ_h^b is the union of all the boundary interfaces of $\mathcal{T}_h$, therefore $\Gamma_h = \Gamma_h^i \cup \Gamma_h^b$. The polynomial space of degree k or less are restricted to the domain $K \in \mathcal{T}_h$ is denoted by $P_k(K)$. Similarly, $P_k(F)$ is the space for $F \in \Gamma_h$. The discrete spaces for a given $p \geq 0$ are defined as follow

$$
\begin{aligned}
V_h^p &:= \{v \in L^2(\Omega) : v|_K \in V^p(K), \forall K \in \mathcal{T}_h\} \\
\mathbf{V}_h^p &:= \{\mathbf{v} \in (L^2(\Omega))^2 : \mathbf{v}|_K \in \mathbf{V}^p(K), \forall K \in \mathcal{T}_h\} \\
M_h^p &:= \{v \in L^2(\Gamma_h) : v|_F \in M^p(F), \forall F \in \Gamma_h\} \\
Y_h^p &:= \{v \in L^2(\Gamma_h^b) : v|_F \in Y^p(F), \forall F \in \Gamma_h^b\}
\end{aligned}
$$

For $F \in \Gamma_h^i$, we define jump $[\cdot]$ and average $\{\cdot\}$

$$
\begin{cases}
\{\mathbf{v}\}_F = \frac{1}{2}(\mathbf{v}^+ + \mathbf{v}^-), \\
\{v\}_F = \frac{1}{2}(v^+ + v^-), \\
[\mathbf{n} \times \mathbf{v}]_F = \mathbf{n}^+ \times \mathbf{v}^+ + \mathbf{n}^- \times \mathbf{v}^-, \\
[v\mathbf{t}]_F = v^+\mathbf{t}^+ + v^-\mathbf{t}^-.
\end{cases}
$$

The scripts $+, -$ represent the values of the elements on the different side of F, the values from the left side of F is $+$, right side is $-$. $\boldsymbol{n}^{\pm}$ denote the outward unit norm vector and $\boldsymbol{t}^{\pm}$ is the unit tangent vector to ensure that $\boldsymbol{t}^+ \times \boldsymbol{n}^+ = 1$ and $\boldsymbol{t}^- \times \boldsymbol{n}^- = 1$. When applied to boundary faces, these expressions become

$$
\begin{cases}
\{\mathbf{v}\}_F = \mathbf{v}^+, \\
\{v\}_F = v^+, \\
[\mathbf{n} \times \mathbf{v}]_F = \mathbf{n}^+ \times \mathbf{v}^+, \\
[v\mathbf{t}]_F = v^+\mathbf{t}^+.
\end{cases}
$$

3. Hybridization principle

By applying Green's formulas to (1) and take $\widehat{E}_h$ and $\widehat{\boldsymbol{H}}_h$ into the boundary terms, a HDG method seeks an approximate solution for

$$\begin{cases} (i\omega\epsilon E_h, v)_{\mathcal{T}_h} - (\nabla \times \mathbf{H}_h, v)_{\mathcal{T}_h} + \langle \tau(E_h - \lambda_h), v \rangle_{\partial\mathcal{T}_h} = 0 \\ (i\omega\mu\mathbf{H}_h, \mathbf{v})_{\mathcal{T}_h} + (E_h, \nabla \times \mathbf{v})_{\mathcal{T}_h} - \langle \lambda_h, \mathbf{n} \times \mathbf{v} \rangle_{\partial\mathcal{T}_h} = 0 \end{cases} \tag{3}$$

Where we used the formulation in [11] to use a hybrid variable $\lambda_h \in M_h^p$ instead of $(E_h, \boldsymbol{H}_h)$ as

$$\hat{E}_h = \lambda_h, \quad \forall F \in \partial\mathcal{T}_h,$$
$$\hat{\mathbf{H}}_h = \mathbf{H}_h + \tau(E_h - \lambda_h)\mathbf{t} \quad F \in \partial K.$$

The boundary condition for all the interior faces F satisfied in (3), is

$$\langle [\mathbf{n} \times \hat{\mathbf{H}}_h]_F, v \rangle_{\epsilon_h} = 0.$$

To model the exterior equivalent problem, EFIE is employed. To write it into operator formulation, we use the single-layer potential $\mathcal{S}$ and double-layer potential $\mathcal{D}$ below

$$\mathcal{S}\varphi = \int_\Gamma G(\cdot, \mathbf{y})\varphi(\mathbf{y})d\Gamma(\mathbf{y})$$
$$\mathcal{D}\psi = \int_\Gamma \partial_{\mathbf{n}(\mathbf{y})} G(\cdot, \mathbf{y})\psi(\mathbf{y})d\Gamma(\mathbf{y})$$

Using them in (2), then the BEM seeks an approximate solution $E_h^b \in Y_h^p$ for

$$\frac{1}{2}\langle \hat{E}_h^b, v \rangle_F - \langle \mathcal{D} \cdot E_h^b, v \rangle_F + \langle \mathcal{S} \cdot J_h, v \rangle_F = \langle E^{inc}, v \rangle_F \tag{4}$$

while

$$\hat{E}_h^b = \lambda_h, \quad \forall F \in \Gamma_h^b,$$
$$J_h = \frac{\partial E_h^b}{\partial \mathbf{n}_h} = -\mathbf{n}_h \times \nabla \times E_h^b = i\omega\mu(\mathbf{n}_h \times \mathbf{H}_h^b), \quad \forall F \in \Gamma_h^b.$$

Finally, we are supposed to couple (3) with (4) and seek an approximate solution $(E_h, \boldsymbol{H}_h, \lambda_h, E_h^b, J_h)$ for

$$\begin{aligned} &(i\omega\epsilon E_h, v)_{\mathcal{T}_h} - (\nabla \times \mathbf{H}_h, v)_{\mathcal{T}_h} + \langle \tau(E_h - \lambda_h), v \rangle_{\partial\mathcal{T}_h} = 0 \\ &(i\omega\mu\mathbf{H}_h, \mathbf{v})_{\mathcal{T}_h} + (E_h, \nabla \times \mathbf{v})_{\mathcal{T}_h} - \langle \lambda_h, \mathbf{n} \times \mathbf{v} \rangle_{\partial\mathcal{T}_h} = 0 \\ &\langle [\mathbf{n} \times \hat{\mathbf{H}}_h], v \rangle_{\Gamma_h^i} = 0 \\ &i\omega\mu\langle (\mathbf{n} \times \hat{\mathbf{H}}_h), \eta \rangle_{\Gamma_h^b} + \langle J_h, \eta \rangle_{\Gamma_h^b} = 0 \\ &\langle \lambda_h, \eta \rangle_{\Gamma_h^b} - \langle E_h^b, \eta \rangle_{\Gamma_h^b} = 0 \\ &\frac{1}{2}\langle \lambda_h, \eta \rangle_{\Gamma_h^b} - \langle \mathcal{D} \cdot E_h^b, \eta \rangle_{\Gamma_h^b} + \langle \mathcal{S} \cdot J_h, \eta \rangle_{\Gamma_h^b} = \langle E^{inc}, \eta \rangle_{\Gamma_h^b} \end{aligned} \tag{5}$$

4. Numerical experiments

In this section we present some numerical results to demonstrate the performance of the proposed method. Two test cases are considered: scattering plane wave by a dielectric cylinder and a dielectric coated cylinder. The grids for discretization were generated by MATLAB's PDE toolbox.

4.1. Plane wave scattering by a dielectric cylinder

The first model problem is the electromagnetic scattering of a plane wave by an infinite dielectric cylinder with radius $r = 1m$ and the relative permittivity $\epsilon = 2.25$. The formulation of incident field is denoted by

$$E^{inc}(\theta) = e^{-jk_0(xcos\theta + ysin\theta)}.$$

To demonstrate the influence of the angular frequency and mesh sizes to the relative discretization error. we choose the incident plane wave to be $\omega = 2k\pi$, and the mesh sizes is $h = al$ with $l = 0.1r$. The contour lines of E_z and H_x are presented in figure 1.

Table 1. Numerical results for scattering of a plane wave by a dielectric cylinder

	k	1/2		1		2	
	a	Error	Order	Error	Order	Error	Order
	1	1.15e-2	-	3.79e-2	-	1.27e-1	-
E	1/2	2.80e-3	2.04	9.80e-3	1.95	3.58e-2	1.82
	1/4	6.98e-4	2.00	2.50e-2	1.97	1.00e-3	1.84
	1	3.12e-2	-	5.93e-2	-	3.40e-1	-
H	1/2	1.00e-3	1.64	1.85e-2	1.68	1.11e-1	1.62
	1/4	3.40e-4	1.57	6.00e-3	1.62	2.81e-2	1.98

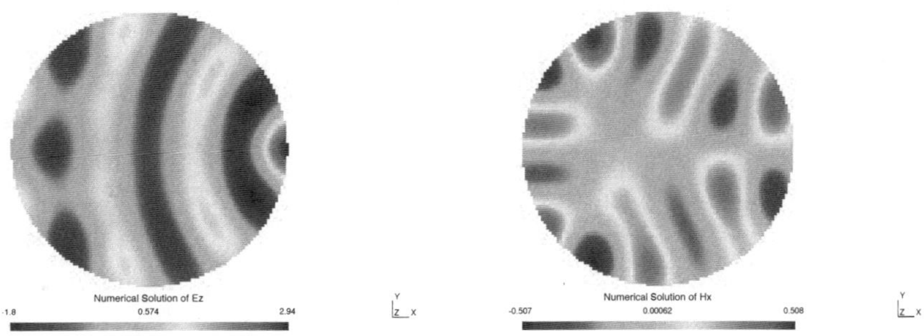

Figure 1. Contour lines of a plane wave scattering by a dielectric cylinder

4.2. Plane wave scattering by a dielectric coated cylinder

Table 2 summarized the relative error of E_z and H_x according to different mesh size and wave length from the numerical results, while figure 2 shows the contour lines. The radius of the cylinder is $r = 1$, the thickness of coating is $d = 0.2m$. The mesh sizes are $h = al$ with $l = 0.1m$. The relative permittivity of the coating material is 2.25, the incident frequency is kf, with $f = 300$ MHz, $k = 1$, $a = 1/2$ in figure 1. RCS of the scattering is presented in figure 3.

Table 2: Numerical results for scattering of a plane wave by a dielectric coated cylinder

	k	1/2		1		2	
	a	Error	Order	Error	Order	Error	Order
	1	2.10e-3	-	9.40e-3	-	7.99e-2	-
E	1/2	5.13e-4	2.01	2.30e-3	2.01	2.37e-2	1.85
	1/4	1.28e-4	2.00	5.96e-4	1.98	6.00e-3	1.97
	1	1.59e-2	-	2.64e-2	-	7.07e-2	-
H	1/2	5.20e-3	1.61	8.10e-3	1.70	1.68e-2	2.07
	1/4	1.80e-4	1.56	2.60e-3	1.64	4.50e-3	1.92

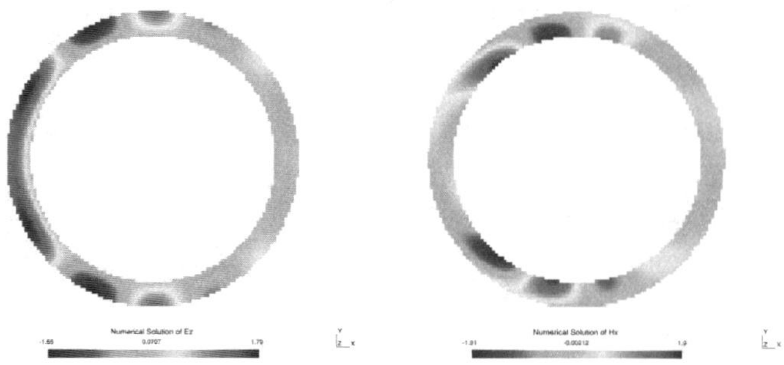

Figure 2. Contour lines of a plane wave scattering by a dielectric coated cylinder

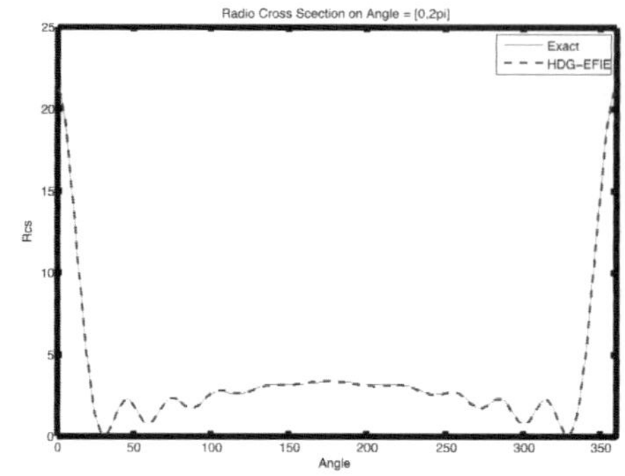

Figure 3: RCS of a plane wave scattering by a dielectric coated cylinder

5. Conclusion

Throughout this paper we conduct a feasibility study of HDG-BEM coupling method and give a numerical investigation of HDG-EFIE. The algorithm is applied to two-dimensional TM-polarized scattering problems with open boundary. We implemented the three test cases and these results yielded similar performance. It is observed in table 1 and 2 that the approximation accuracy of the combined method is controlled by both mesh sizes *a* and wave number *k*, which is similar to the numerical results in [12]. Using HDG-P_2 and HDG-P_3 don't improve convergence order. This may lead to that the performance of the proposed method is determined by the BEM part. Such high accuracy for far field scattering as shown in the presented hybrid method are currently under investigation for complex structure. The experimental results will hopefully serve as useful feedback information for improvements for more complex scattering problems.

6. References

[1] Robert M Kirby, Spencer J Sherwin, and Bernardo Cockburn. *To cg or to hdg: a comparative study.* Journal of Scientific Computing, 51(1):183–212, 2012.

[2] X. Antoine, A. Arnold, C.Besse, M. Ehrhardt and A. Schadle. *A review of transparent and artificial boundary conditions techniques for linear and nonlinear Schrodinger equations*. Commun. Comput. Phys., 4, pp. 729- 796, 2008.

[3] A. F. Peterson, S. L. Ray and R. Mittra, *Computational Methods for Electromagnetics*, IEEE Press and Oxford University Press, New York, 1997.

[4] B. Cockburn, J. Guzman and F. J. Sayas, *Coupling of Raviart-Thomas and Hybridizable Discondinuous Galerkin Methods with BEM*, SIAM J. Numer. Anal., 50(5), pp. 2778-2801, 2012.

[5] B. Cockburn, F. J. Sayas and M. Solano, *Couping at a distance HDG and BEM*, SIAM J. Sci. Comput., 34, pp. A28-A47, 2012.
[6] Z. X. Fu, N. Heuer and F. J. Sayas, *Coupling of HDG with a double-layer potential BEM.*
[7] R. Bustinza, G. N. Gatica and F. J. Sayas, *A LDG-BEM coupling for a class of nonlinear exterior transmission problems, in Numerical Mathematics and Advanced Applications*, Springer, Berlin, 2006, pp. 1129- 1136.
[8] G. N. Gatica, N. Heuer and F. J. Sayas, *A direct coupling of local discontinuous Galerkin and boundary element methods*, Math. Comp. 79 (2010), pp. 1369-1394.
[9] G. N. Gatica and F. J. Sayas, *An a priori error analysis for the coupling of local discontinuous Galerkin and boundary element methods*, Math. Comp., 75 (2006), pp. 1675-1696.
[10] C. Girard, S. Lanteri, R. Perrussel and N. Raveu, *Toward the coupling of a discontinuous Galerkin method with a MoM for analysis of susceptibility of planar circuits*. IEEE Trans. on Magnetics, Institute of Electrical and Electronics Engineers, 2014, 50 (2), pp. 509-512.
[11] L. Li and S. Lanteri, *Numerical investigation of a high order hybridizable discontinuous Galerkin method for 2d time-harmonic Maxwell's equations*, COMPEL, vol. 2, no. 3, pp. 1112-1138, 2013.
[12] Y. F. Jing, T. Z. Huang and Y. Duan, *A novel integration method for weak singularity arising in two-dimensional scattering problems*. IEEE Trans. Antennas Propag. vol.58(8), pp. 2725-2731, 2010.

Acknowledgments
This work was supported by the Doctoral Foundation of Yunnan Normal University (No. 00800205020503127) and the Yunnan Provincial Education Department (No. 2021J0434).

A Short Review of Deep Tissue Imaging Techniques and Applications

J Song[1] and Z Dean[2, 3]

1 Undergraduate Student, University of Massachusetts at Amherst, MA, USA
2 Post-Doctoral Fellow, San Francisco State University, CA, USA
3 Post-Doctoral Fellow, Center for Cellular Construction, CA, USA
Email: ziahdean@sfsu.edu

Abstract. Deep tissue imaging is a fundamental research method in the field of biological imaging. With an intent on obtaining more clear and detailed images deep within tissue, researchers have put much focus on developing deep tissue imaging techniques. In the past few decades, deep tissue imaging techniques have been improved in many aspects. In contrast to increasingly refined techniques, the application of these techniques seems to have not received equally attention. In this review, we will focus on the advanced applications of deep issue imaging that have emerged in recent years.

1. Introduction
In the field of biological imaging, deep issue imaging has always been an essential study tool in peering deep within biological structures and monitoring biological interactions. Deep tissue imaging is made possible with the cooperation of many fields including spectroscopy, advanced imaging, microscopy, optics, and mechanics. In this paper, we focus on three specific deep issue imaging techniques, 2-photon microscopy, adaptive optics microscopy and light sheet microscopy, discussing the benefits and limitations of these approaches. In particular, we will be exploring the biological applications of these techniques and determining whether there is a correspondence between technique and application.

2. Selected deep tissue imaging techniques

2.1. 2-Photon Microscopy
To better understand 2-Photon Microscopy, we need to introduce Confocal Microscopy first. Confocal Microscopy is an imaging technique that takes advantages of a spatial pinhole to eliminate the out-of-focus light signals, which can reduce the influence of unexpected fluorescence light and improve resolution [1]. 2-Photon Microscopy as illustrated in Figure 1, is a technique that is ameliorated from Confocal Microscopy. In traditional imaging techniques, such as Confocal Microscopy, fluorescence dyes are activated by the absorption of single photon. Relatively, in 2-Photon Microscopy, the process of fluorescence excitation is done by 2-photon absorption, which means that two photons arrive at a fluorescence molecule at the same time, usually a time scale within 0.5fs can be recognized as simultaneously, combining their energies to promote the molecule from a ground state to an excited state [2]. 2-Photon Microscopy has some features that make it a stronger tool than Confocal Microscopy for deep tissue imaging purposes. Firstly, 2-Photons of light are capable of penetrating deeper. The absorption wavelength of 2-Photon Microscopy is between about 700-1000nm, which is in the near-infrared range. Light in this range can penetrate deeper and reach more tissues in depth.

Second, 2-Photon Microscopy creates less photo-toxicity. Because near-infrared light has lower energy level and lacks significant endogenous absorbers in most tissues, it causes less phototoxicity in sample tissues, which is crucial for longer time dependent imaging [2]. Third, localized excitation of fluorescence dyes. In 2-Photon Microscopy, the excitation of fluorescence dyes and emission of light signals are confined to the region near focus, whereas in Confocal Microscopy, the emission occurs along the whole laser path. The absence of out-of-focus excitation in 2-Photon Microscopy reduces the influence of unexpected signals and further increases the signal to noise ratio, and improving the resolution of deep tissue imaging [2].

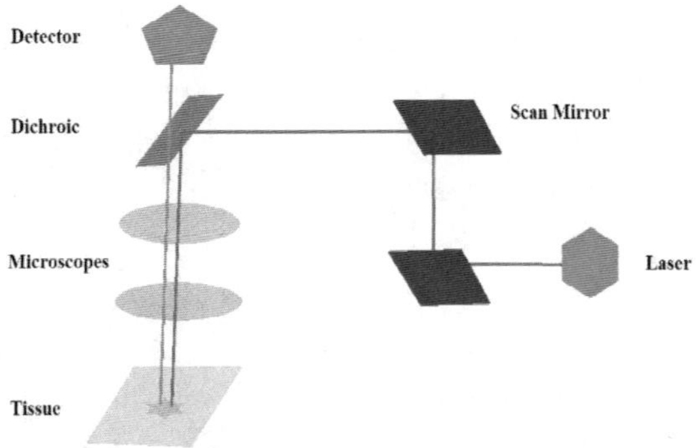

Figure 1. Illustration of a simplified 2-Photon Microscope with Laser transmitter, Scan Mirror, Detector, Dichroic, and Objective.

2.2 Adaptive Optics Microscopy

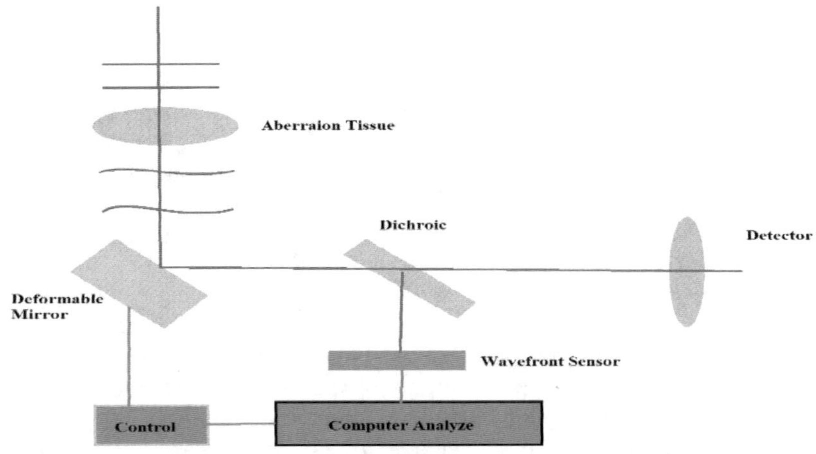

Figure 2. A simplified schematic of an Adaptive Optics system with deformable mirror, wavefront sensor and computer analyser. (red lines indicate light path and blue lines indicate wiring path)

When a fluorescence dye is excited by a laser for fluorescent lamp, the dye emits light which is collected by an objective lens. This is the basic principle of observing in fluorescence microscopy. As the light passes through the biological tissue like cell membrane, aberrations can be introduced that distort the wavefronts, blurring the images. To solve this problem, adaptive optics is introduced into microscopy. Adaptive optics (AO) is a technique that used for improving the performance of optical systems by reducing the effect of incoming wavefront distortions by deforming a mirror in order to compensate for the distortion (Booth 2007). Adaptive optics was first used in telescopes, aiming at reducing the aberration caused by the refraction that occurs when light passes the atmosphere. This situation is similar to the aberrations caused by cellular structures in deep tissue imaging. Thus,

adaptive optics was introduced into microscopy, as an auxiliary to Confocal Microscopy or 2-Photon Microscopy. An Adaptive optics system shown in Figure 2, can be divided into three parts, sensing, computing, and correcting. Most prominent methods of sensing wavefront aberration are the Shack–Hartman wavefront sensor and interferometric sensors. These methods require a point-like reference source, such as the guide star used in astronomical systems, to produce a well-defined wavefront. After getting light signal from the fixed "guide star", a computer will start calculating the aberration data and send it to deformable mirror (DM). Then the DM will deform into certain shapes to neutralize the aberration, recovering the original structure both in 2D and 3D. Adaptive Optics Microscopy is a great tool for enhancing image resolutions, reaching deeper imaging depths, and demonstrating precise biological structures in small scales within cells.
Adaptive Optics has tremendous potential in observing within live biological tissues at the cellular level, this is advantageous for a wide range of biological research fields from neural science to developmental biology.

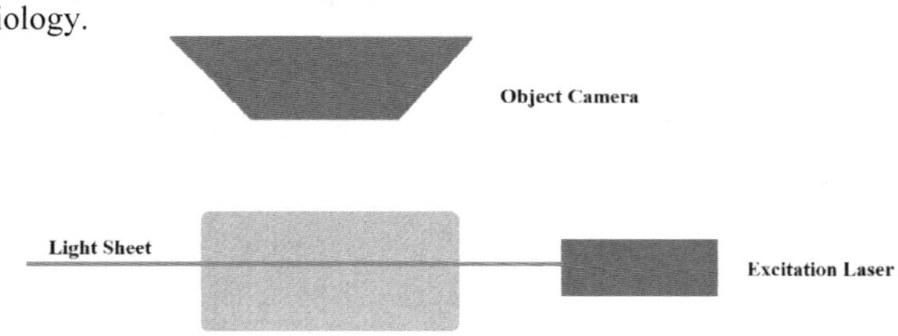

Figure 3. A simplified demonstration of a Light Sheet Microscope, here detection is performed by a camera, and the excitation is accomplished by a laser the illuminates a whole sheet, rather than single point.

2.3. Light Sheet Microscopy
Light Sheet Microscopy is a fluorescence microscopy technique with an intermediate-to-high optical resolution, and fast imaging rate. In contrast to other fluorescence microscopy techniques which excite fluorescence dyes point by point, Light Sheet Microscopy is capable of activating fluorescence by thin slices, so called "light sheet". Light sheet microscopy combines two distinct optical paths, (as shown in Figure 3), one for fast wide-field detection and one for illumination with a thin sheet of light, perpendicular to the detection path. The light sheet is aligned with the focal plane of the detection path, and the waist of the sheet is positioned in the center of the field of view (M, M, and J 2014). During the construction of an image, each slice of specimen is excited once, as a result, the photo-toxicity and photo-bleaching is limited to a minimum. In addition, with Light Sheet Microscopy, cameras can record the whole activated sheet in each frame, making it super-fast to collect images. These features make Light Sheet Microscopy an alternative to confocal microscopy because of some superior features such as high-speed full-frame imaging, extremely low photo-toxicity, and multi-dimensional view of sample rotation, meeting the requirement of long time-lapse imaging (M, M, and J 2014). Such as the time-lapse like recording of embryo development a length biological process. Light Sheet Microscopy and 2-Photon Microscopy are very similar, both have low photo-toxicity and good at long-term imaging. But they still have differences and unique features. Because fluorescence in Light Sheet Microscopy is excited by planes and images are taken by cameras, so it can take much more frames in a period than 2-Photon Microscopy. On the other hand, 2-Photon Microscopy can reach smaller structures and provide higher resolution than Light Sheet Microscopy. So, Light Sheet Microscopy is powerful in extra-long time imaging with large scales, like filming the development process of embryo.

3. Applications
Although the technological advancement of these techniques have received a great amount of attention, here we review several applications of each technique in an effort to demonstrate the breadth of practical impact these techniques have made in our understanding of biological processes. Each technique consists of advantages and disadvantages based on a particular type of application. Thus, we

will be able to answer the critical question of whether each technique is broadly applicable or particularly suited for a precise biological application.

3.1. 2-Photon Microscopy

3.1.1 The study of cerebrovascular dynamics in ultrasound-induced blood-brain barrier.

According to Eunice Cho, Jelena Drazic et al, the blood-brain barrier (BBB) can be disrupted by applying ultrasonic and circulating microbubble contrast agent. In their experiment, a single-element, ring-shaped piezoelectric transducer was attached to male Wister rats' skulls. Within the ring, that part of skull was replaced by a piece of transparent cover glass (as shown in Figure 4), making it possible to directly observe the Texas Red injected into brain vessels through 2-Photon Fluorescence Microscopy. Further, scientists can infer the ultra-sonic disruption rate of BBB by observing the amount and speed of leakage of Texas Red. Finally, they could come up with a conclusion that, by varying acoustic pressure, one can possibly control the opening of blood-brain barrier, causing different kinds of leakage and vessel size.

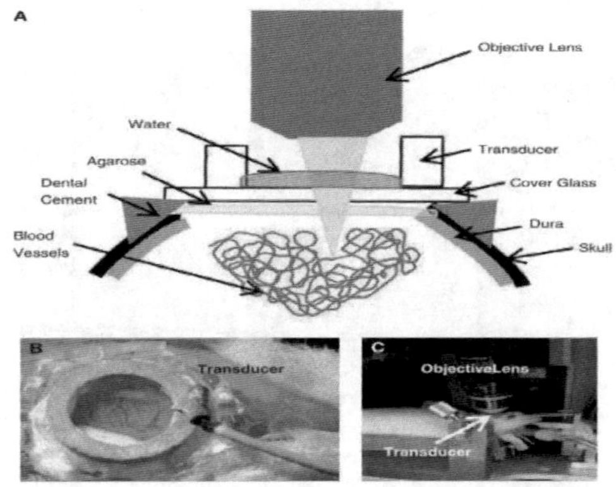

Figure 4. (A) A general structure of the experiment setup. (B) Replace a piece of rat skull with glass cover, ensuring the observation. (C) Real image of experiment setup. [3]

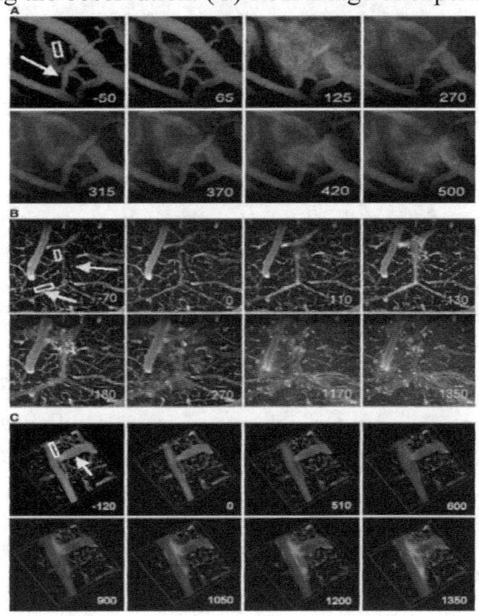

Figure 5. (A) Arrow marks a fast leakage. (B) Arrow marks a sustained leakage. (C) Arrow marks a slow leakage. [3]

In this experiment, 2-Photon Microscopy enables Cho and her colleagues to observe the brain vessels in Wister rats for a long period of time, monitoring the status of leakage of Texas Red, a kind of fluorescence dye. Besides, the high image contrast as well as high resolution of 2-Photon Microscopy provided extreme high quality deep tissue images[3](Figure 5).

3.1.2. Phosphorescence lifetime microscopy of retinal capillary plexus oxygenation in mice.
Another research was operated by Ikbal Sencan, Tatiana Esipova et al, who decided to study retinal capillary plexus oxygenation in mice. The impairment of the delivery and consumption of oxygen can be recognized as a pathophysiology symbol of retinal related diseases. However, in present, there is an absence of effective research tools of measuring oxygen concentration and oxygen transportation in retinal tissues. So, scientists designed this experiment to exam the possibility of using phosphorescence lifetime microscopy to address the absolute partial pressures of oxygen (pO2) in the retinal capillary plexus (results shown in Figure 6).

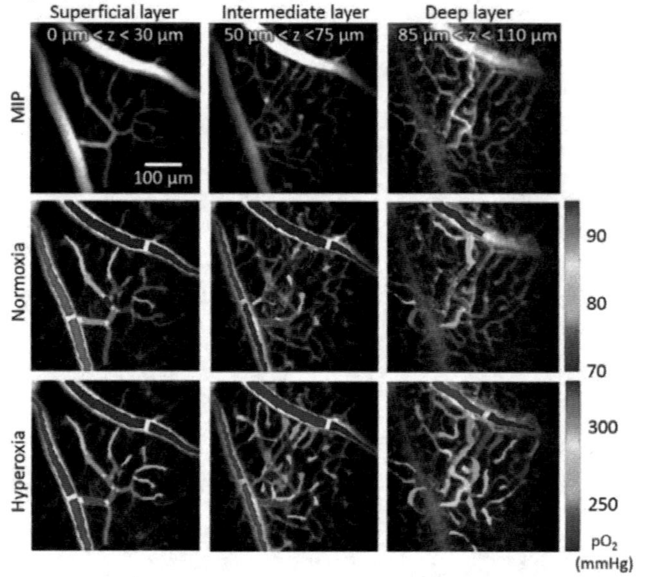

Figure 6. pO2 distribution in three layers of imaging for three depths. [4]

Because retina is sensitive to light, it is crucial to take photo-toxicity into consideration. The low photo-toxicity as well as low photo-bleaching of 2-Photon Microscopy makes it perfectly suitable for this experiment. After the performance of in vivo measurements of this designed system, Ikbal and their team can prove the ability of their setup to detect the change in retinal oxygenation as well as longitudinal gradients from arterioles to capillaries to venules (Sencan et al. 2018).

3.1.3. 2-photon time-lapse microscopy of BODIPY- cholesterol
Cholesterol is an important membrane component, but we do not know much about the transportation of cholesterol in cells. With previous studies, scientists have established a theory that both vesicular transport and non-vesicular transport can be benefit to sterol's transportation from plasma membrane. The significant photo-bleaching of deep tissue imaging methods, however, restricts the possibilities to study sterol in depth or for a long analyze time. In Frederik Lund's publishment, he introduced 2-Photon Microscopy into this research area. 2-Photon Microscopy has relatively low photo-bleaching and photo-toxicity, compared with other microscopy methods, for example, Confocal Microscopy. This feature makes 2-Photon Microscopy meet the requirement of this experiment perfectly. Lund found that BODIPY-cholesterol is surprisingly photostable under 2-Photon excitation, it can endure several hundred frames of photographing without significant photo-bleaching. Lund got a conclusion from long-term tracking of BODIPY-cholesterol that, the movement of sterol-containing vesicles for a

short time scale could reflect dynamic rearrangements of the cytoskeleton, while directed transport of sterol vesicles occurs likely along both, microtubules and actin filaments, and spatially varying anomalous diffusion could contribute to fine-tuning and local regulation of intracellular sterol transport. (Lund et al. 2012).

3.2. Adaptive Optics Microscopy

3.2.1. Fast high-resolution miniature microscopy for brain imaging in freely behaving mice.

A target of neuroscience's study is to understand the principles underlying the neuronal information processed in freely behaving animals in different levels, like subcellular, cellular, circuit, or higher. Developments of microscopy have enabled the observation of brain activities and structural dynamics. However, there still is a challenge to achieve such goals when animals are in self-determined behaviours(Experiment set up shown in Figure 7). Weijian Zong and his team put up with a design and application of a fast high-resolution, miniaturized 2-Photon Microscopy together with Adaptive Optics that accomplishes the challenge of monitoring activities of single dendritic spines in freely behaving animals.

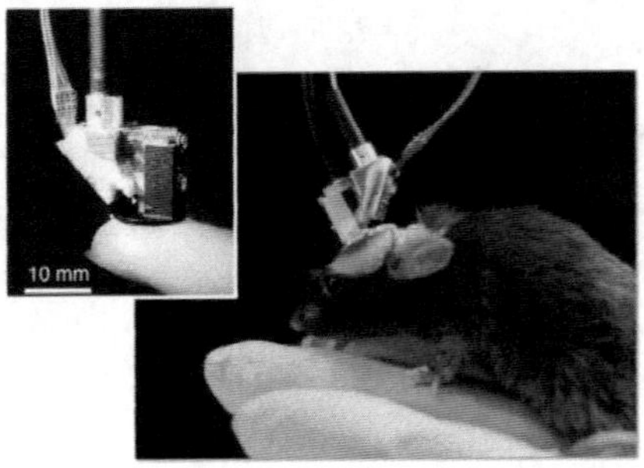

Figure 7. The imaging instrument attached to free behaving rat. [5]

Because the movement of mice is random and chaos, these unpredictable behaviors would largely blur the imaging results of real-time neural system. That is why adaptive optics was introduced into this experiment. With the ability of strongly reducing the influence of aberration and enhancing the resolution of deep tissue images, the unavoidable negative influence originated from animal behaviors could be restrain to a controllable level and get acceptable deep tissue images of single dendritic spines (Zong et al. 2017).

3.2.2. Deep-brain imaging at synaptic resolution over large volumes

Deep-brain high resolution imaging in vivo has always been a great challenge during the past decades. 2-Photon Microscopy is a narrowly approach to this problem, providing a minimally invasive method of imaging deeply buried brain structures. Although the near-infrared light that used in 2-Photon Microscopy has greater penetration depth, its imaging resolution as well as the field of view are limited by aberrations of membranes and lens. Here, a new research performed by Zhongyan Qin and other researchers took advantages of the deep penetration and high resolution of Adaptive Optics to enhance the performance of 2-Photon Microscopy, achieving the challenge of deep-brain imaging at synaptic resolution over large volumes. They developed a new deep tissue imaging system by attaching Adaptive Optics based on direct wavefront sensing to a 2-Photon Microscopy set, which enables recovery of aberration-limited resolution in deep-brain imaging.
After applying Adaptive Optics, the quality of deep-brain images is tremendously increased, both in accuracy and penetration depth (results shown in Figure 8) [6].

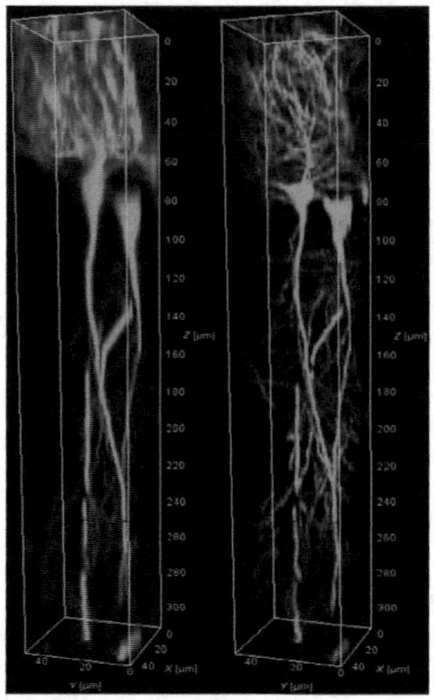

Figure 8. Images of pyramidal neurons, (left) without AO, (right) with full AO. [6]

3.2.3 Super-resolution structured imaging in the living brain

To fully understand a neurobiological processes, it is essential to study them in vivo and within cell level, as cells act as important components of extended neural networks in the brain. To get clear images in such small scale, within cells, Super-resolution Microscopy is needed. Super-resolution Microscopy is a partial resolution to the problem of cell scale imaging. It can provide structural and functional insights that are not easily accessible with other conventional microscopy technologies. However, when apply Super-resolution Microscopy into vivo brain imaging, there are challenges in 3D imaging if a target tissue is constantly moving. So Raphael Turcotte, Yajie Liang and other researchers added Adaptive Optics to Super-resolution Microscopy system, in order to correct the optical aberrations in Super-resolution structured Illumination Microscopy (SIM) in vivo. They used both fish and mammal as experiment animals and successfully imaged the brains of live zebrafish larvae and mice, observing the dynamics of dendrites and dendritic spines at a extremely small nanoscale [7].

3.3 Light Sheet Microscopy

3.3.1. Quantitative imaging of cell dynamics in mouse embryos

Ryan Udan and his team has reported the first 24-hour time-lapse of post-implantation mouse embryo development images. In Ryan Udan's experiment, he used Light-Sheet Microscopy to image cell dynamics in mouse embryos. Light-Sheet Microscopy system has several advantages that make it superior to other fluorescence microscopy methods in live and 3D microscopy. Light-Sheet Microscopy systems are valuable for studying embryonic development. In Light-Sheet Microscopy, the geometry of the light path in a form that requires the sample to be fixed in place as well as accessible from multiple sides, so that it can be rotated around the axis. In present, most used methods of preparing samples include hanging specimen from a pin or embedding it in the 1-2% agarose. These methods are specific to handle certain samples. But for post-implantation mouse embryo, it expands significantly in size and are very delicate and sensitive to mounting. To overcome the sample setting problem and to establish a robust strategy for a long-term imaging of E6.5-8.5 mouse embryos, they developed a method that uses hollow agarose cylinders specially designed to accommodate for embryonic growth and can provide boundaries to minimize tissue drift and enable imaging in multiple axis as well. After the experiment, Ryan successfully demonstrates that Light-Sheet imaging can provide both quantitative data for tracking changes in morphogenesis and reveal new insights into mouse embryogenesis [8].

3.3.2. Reconstruction of Zebrafish Early Embryonic Development

Light-Sheet Microscopy is not only used in mammal study, Philipp Keller and his team uses Light Sheet Microscopy to reconstruction zebrafish's early embryonic development. They got a goal to map all the cell behaviors during the embryogenesis. So, they developed a Light-Sheet Fluorescence Microscopy recording nuclei localization and movement in two kinds of zebrafish embryos, wild-type and the mutant, over the first 24-hour development. They built up a database consist of images filmed at 1.5 billion voxels per minute, which provides the possibility of reconstruction of "digital embryos," in another word, that is a comprehensive database of cell positions, divisions, and migratory tracks(as shown in Figure 9).

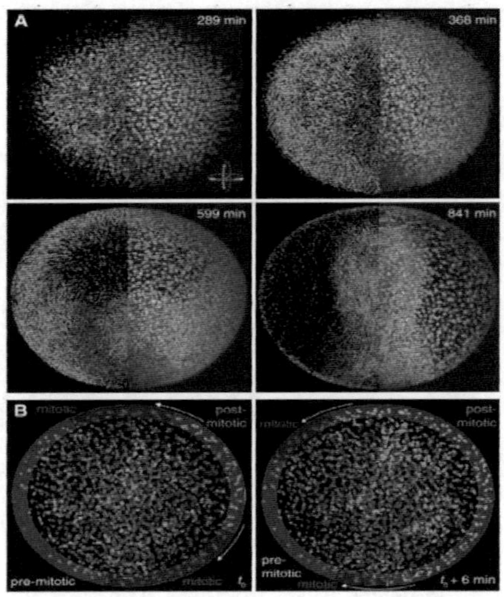

Figure 9. (A) Embryo state tracking images of 289min, 368min, 599min and 841min. (B) Dividing cells are marked with red, daughter cells are marked with blue. [9]

They further put up with a model of germ layer formation and got a conclusion that the endoderm originated from one-third of total embryo's cells at a time. To build their digital embryos, 55 million nucleus entries are provided as a resource. [9].

3.3.3. Imaging adult C.elegans

Observing biological activities in the form of living organisms can provide important insights into the dynamics of these activities. However, the complexity of multi-cellular organisms makes it a challenging environment to implement observations through fluorescence microscopy. Because of the out-of-focus signals, confocal or other widefield fluorescence microscopy suffers from low signal-to-background ratio (SBR), photo-toxicity and photo-bleaching of fluorescent probes. In Light-Sheet Microscopy, only the selected plane of the specimen is illuminated, minimizing the out-of-focus fluorescence lights and photobleaching, thereby enhancing SBR. J van Krugten together with his team present a Light-Sheet Microscope with a 1.0-NA detection objective as well as a fast sample-positioning stage that allows four degrees of freedom. By imaging the sensory cilia and neural system of living adult C. elegans, they demonstrated that this set of microscopes is perfectly suited for relatively fast imaging of large living specimen [10].

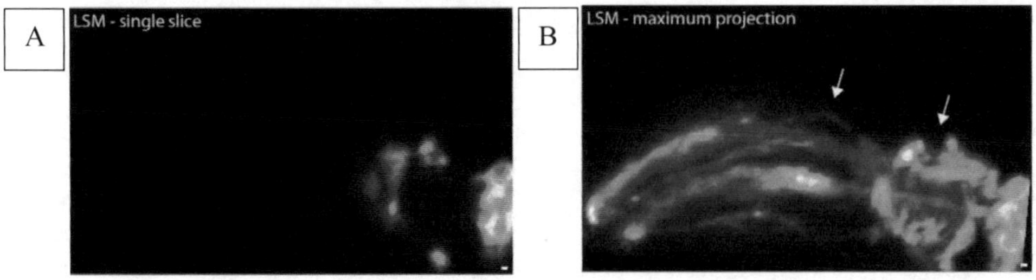

Figure 10. A) One slice of image obtained by LSM. B) A projection of 115 slices of images in the head region of C.elegan. [10]

4. Discussion

The amazing combination of fluorescence and optics provides us with a powerful tool in biophysics research, the fluorescence microscopy. In decades of improvement, fluorescence microscopy technology has evolved from simple Confocal Microscope to complex, more powerful microscopes, such as the 2-Photon Microscope, Adaptive Microscope and Light-Sheet Microscope.

As these advanced fluorescence microscopes are all based on same underlying techniques, fluorescence, and optics for instance, they naturally have commonalities. All three microscopes have a restricted imaging depth, because of the penetration limitation of excitation laser and outgoing fluorescence light. Besides that, the observation time is also constringed by the photo-bleaching phenomenon and fluorescence protein denaturation.

These advanced fluorescence microscopy systems are built to meet different experiment requirements, they all have irreplaceable features that make them different and unique from each other. In the demand for deeper penetrate depth and longer observation time, 2-Photon Microscope is developed from Confocal Microscope and has improvement in photo-damage and penetration. It uses non-linear infrared laser for excitation rather than ultraviolet laser, which is of less energy and larger wavelength, providing less photo-damage and deeper penetration than Confocal Microscope. As a result, 2-Photon Microscope is good at deep tissue imaging like brain structure observation introduced before.

The structure of Light-Sheet Microscope is far more different from original Confocal Microscope than 2-Photon Microscope. It extremely reduces the photo-toxicity and photo-bleaching, offering a possibility of ultra-long tissue observation but sacrifices imaging resolution. This feature makes it a perfect tool for observing embryo development.

Adaptive Optics is an assistant tool to provide resolution upgrade for Confocal Microscope and 2-Photon Microscope. By applying computer and deformable mirror, it can cancel the aberration caused by cell membrane and give us more precise and clearer deep tissue images.

AO can largely enhance the resolution of Confocal or 2-Photon Microscopy, but it cannot be applied to Light-Sheet Microscope. The recording part of LSM is supported by camera, a high-resolution camera is expensive and not easily accessible by most researchers. So, it will be interesting and useful to find out a way to improve the resolution of LSM, making it even a stronger tool for research.

In evaluating different applications of each technique, we can see that a large driving force for advancement of these techniques is the particular application they are set to accomplish. While new biological problems will necessitate further advancements of these deep tissue imaging techniques, the current technology has given way for solving a great deal of problems that were not possible before. It is therefore of utmost importance that we evaluate these applications to allow for the biological community to be able to determine how they might be able to apply it for their outstanding questions.

Reference

[1] Nwaneshiudu A, Kuschal C, Sakamoto F H, Rox Anderson R, Schwarzenberger K and Young R C 2012 Introduction to confocal microscopy *J. Invest. Dermatol.* **132** 1–5

[2] Helmchen F and Denk W 2005 Deep tissue two-photon microscopy.pdf

[3] Cho E E, Drazic J, Ganguly M, Stefanovic B and Hynynen K 2011 Two-photon fluorescence microscopy study of cerebrovascular dynamics in ultrasound-induced blood-brain barrier opening *J. Cereb. Blood Flow Metab.* **31** 1852–62

[4] Sencan I, Esipova T, Yaseen M, Fu B, Boas D, Vinogradov S, Shahidi M and Sakadzic S 2018 Two-photon phosphorescence lifetime microscopy of retinal capillary plexus oxygenation in mice *J. Biomed. Opt.* **23** 1

[5] Zong W, Wu R, Li M, Hu Y, Li Y, Li J, Rong H, Wu H, Xu Y, Lu Y, Jia H, Fan M, Zhou Z, Zhang Y, Wang A, Chen L and Cheng H 2017 Fast high-resolution miniature two-photon microscopy for brain imaging in freely behaving mice *Nat. Methods* **14** 713–9

[6] Qin Z, Chen C, He S, Wang Y, Tam K F, Ip N Y and Qu J Y 2020 Adaptive optics two-photon endomicroscopy enables deep-brain imaging at synaptic resolution over large volumes *Sci. Adv.* **6**

[7] Turcotte R, Liang Y, Tanimoto M, Zhang Q, Li Z, Koyama M, Betzig E and Ji N 2019 Dynamic super-resolution structured illumination imaging in the living brain *Proc. Natl. Acad. Sci. U. S. A.* **116** 9586–91

[8] Udan R S, Piazza V G, Hsu C W, Hadjantonakis A K and Dickinson M E 2014 Quantitative imaging of cell dynamics in mouse embryos using light-sheet microscopy *Dev.* **141** 4406–14

[9] Keller P J, Schmidt A D, Wittbrodt J and Stelzer E H K 2008 Reconstruction of Zebrafish Early Embryonic Development by Scanned Light Sheet Microscopy *Science (80-.).* **322** 1065–9

[10] van Krugten J, Taris K K H and Peterman E J G 2021 Imaging adult C. elegans live using light-sheet microscopy *J. Microsc.* **281** 214–23

Acknowledgments

The authors would like to thank the NSF-Center for Cellular Construction (Z.D) and the Genentech Foundation (Z.D.) for their support. Disclaimer: Any opinions, findings and conclusions or recommendations expressed in this material are those of the authors and do not necessarily reflect the views of Genentech Foundation or National Science Foundation.

The Glauber Model And Flow Analysis with Pb-Pb Collisions at $\sqrt{s_{NN}}$=2.76 TeV

Ceran Hu[1,2,*], Wentao He[3], Chengze Lyu[4], Kangjie Yang[5], Lily Yao[6] and Ruichong Zhao[7]

1 School of Physical Sciences, University of Chinese Academy of Sciences, Beijing, China
2 Department of Physics, University of California, Berkeley, USA
3 WLSA Shanghai Academy, Shanghai, China
4 Shanghai Starriver Bilingual School, Shanghai, China
5 Tsinglan School, Guangdong,China
6 Miss Hall's School, Pittsfield, Massachusetts, USA
7 Shenzhen College of International Education, Guangdong, China
E-mail: huceran18@mails.ucas.ac.cn

Abstract. This work presents data analysis on Pb-Pb collisions with $\sqrt{s_{NN}}$=2.76 TeV in centrality 40%-50%. Introduction and Monte-Carlo simulation results on the Glauber model are presented, which shed light on the initial geometric configuration of heavy ion collisions. Three-dimensional correlation function is plotted, and Fourier decomposition is carried out in order to obtain elliptic flow. Based on the assumption that non-flow effect is less prominent in long-range area, we separate it from the second Fourier decomposition of two-particle correlation function by making polynomial curve fitting.

1. Introduction

Quantum chromodynamics (QCD) is one of the most successful theories to extensively describe elementary particles. To be explicit, it aims at describing the strong interaction widely existing among nuclei and hadrons, where quarks and gluons play a dominant role in deciding physical properties of the system.

One particularly interesting physical system within the framework of QCD is quark-gluon plasma (QGP)[1]. It consists of quarks and gluons with ultrahigh density and temperature that it bears much resemblance to matter existed at the very beginning of the universe. Important as it is, direct insight into such system is, nevertheless, faced with extreme difficulty. The problem is basically twofold. On one hand, extreme density and temperature of QGP can only be facilitated with enormous amount of energy and harsh conditions. On the other hand, quark confinement puts stringent limit on the lifespan of QGP, approximately in the scale of femtosecond. QGP experiences hadronization afterwards, in which quarks and gluons are bounded together and confined within hadrons. Fortunately, both of the questions have been properly addressed nowadays. The production of QGP has been made possible with the help of heavy ion collisions on the Large Hadron Collider (LHC). As for the latter one, though direct study has not yet come into reality, we are indeed capable of analyzing data collected after hadronization to speculate or verify the physical properties of the system.

The evolutionary partterns of QGP has been discovered that, to high extent of accuracy, match the hydrodynamics of perfect liquid, with which the shear viscosity to entropy density is governed by the inequality: $\eta/s < \frac{1}{4\pi}$, as is suggested by anti-de Sitter/conformal field theory (AdS/CFT) conjecture [2]. Moreover, QGP tends to maintain its initial anisotropy, which can be classified as the effects of elliptic flow, triangular flow, etc. Such effects can be extracted from different Fourier decomposition coefficients of the correlation function calculated from the data collected by detectors, in our case, particularly, data from Compact Muon Solenoid (CMS) of Pb-Pb collisions with $\sqrt{s_{NN}}$=2.76 TeV in centrality 40%-50%.

The remaining parts of this paper are organized as follows: Section 2 discusses the Glauber model, with which we simulate the initial geometric figures and distributions in heavy ion collisions. In section 3, we present our plot of the 3-dimensional correlation function against $\Delta\varphi$ and $\Delta\eta$. Several features of the figure are noted and analyzed. Moreover, the Fourier decomposition is carried out and analysis on results is presented. In section 4, we manage to separate non-flow effect from the physical flow effect in the second coefficient of Fourier decomposition, v_2^2. Section 5 aims at providing an conclusions of the text.

2. The Glauber model

It should be clarified that with current experimental apparatus and methods, we are unfortunately not capable of directly probing the primary geometric states of the collision events. However, technological backwardness has in turn stimulated the development of several simulation methods with the help of computers. The Glauber model, has been tested to be one of the most successful methods that shed light on our physical intuition on the initial geometry. It is based on the approximation that collisions are independent, sequential, and is built upon scattering theories in Quantum Field Theory. It aims at tackling high energy scatterings with composite particles, such as nucleon-nucleon scattering on LHC. In this work we present the "Glauber Monte Carlo" Model(GMC), in which nucleus is modeled by a group of uncorrelated nucleons based on a given density distribution. In this model "Optical limit" is utilized to make the integration numerically feasible via computer simulations[3].

The Glauber model is capable of simulating several physical properties for the initial state, including nuclear charge density ρ, number of participating nucleons, N_{part}, number of colliding nucleons, N_{coll}, and the distribution of colliding nucleons. The sole parameter required to be introduced is a randomly generated impact parameter b.

When it comes to nuclear charge density, several empirical distribution functions have been constructed to adapt to different nuclei. Three types of density distribution are pointed out here :

$$\rho(r) = \rho_0 \frac{1+\omega(\frac{r}{R})^2}{1+e^{\frac{r-R}{a}}} \tag{1}$$

$$\rho(r) = \rho_0 \frac{1+\omega(\frac{r}{R})^2}{1+e^{\frac{r^2-R^2}{a}}} \tag{2}$$

$$\rho(r)=4\rho_0 e^{-\alpha r}(\frac{\cosh(-\beta r)}{r})^2 \tag{3}$$

Eq.(1) is a Fermi-like distributions with four parameters. ρ_0 stands for the nucleon density in initial state, ω is a measurement of deviation from a perfect sphere. R is the radius of the nucleus, while a is a characteristic length. The same parameterization works for Eq. (2), in the form of a Gaussian function. As for Eq. (3), the Hulthen formula, $\alpha = 2.807\ \mathrm{fm}^{-1}$and $\beta = 0.9465\ \mathrm{fm}^{-1}$ are two lengths obtained from experimental results. Eq. (1) can be appropriately applied to most nuclei used in collision experiments. Eq. (2) is especially designed for sulfur while Eq. (3) is fitted for deuteron[4].

The rest of this section concentrates on our simulated results in the Glauber model.

On Figure 1, a display of the values of N_{part} and N_{coll} are shown. Both values show a reasonable descend to zero as impact parameter b increases.

After nuclear density ρ, N_{part} and N_{coll} being set up, the rest of work is to input all these parameters to generate the initial geometry in a particular event. A typical example is presented in Figure 2 and Figure 3, in which we present collisions from two rather different views, beam-in view and side view. The green and blue symbols serve as spectators, while red and yellow ones participate in nucleon-nucleon collisions.

One specific parameter matters when it comes to whether the collision could occur or not. The "ball diameter" is defined as [5]:

$$D = \sqrt{\frac{\sigma_{NN}}{\pi}} \quad (4)$$

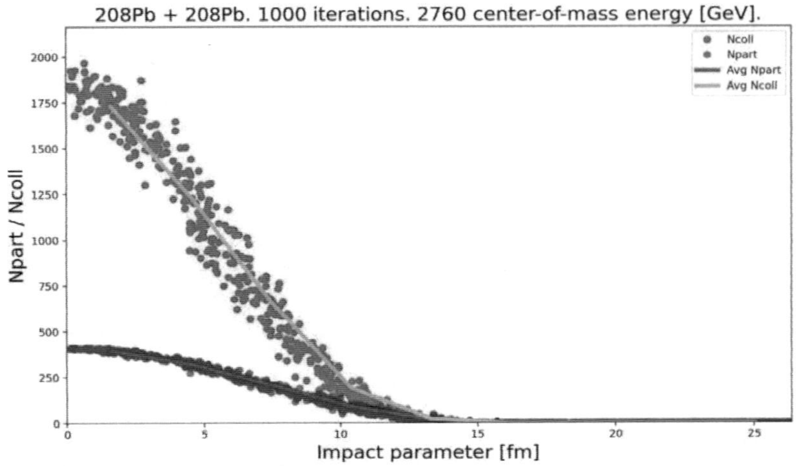

Figure 1. N_{part} and N_{coll} versus impact parameter b with Pb-Pb collision at $\sqrt{s_{NN}}$=2.76 TeV. The scatter points are fitted to the average curves in green and yellow.

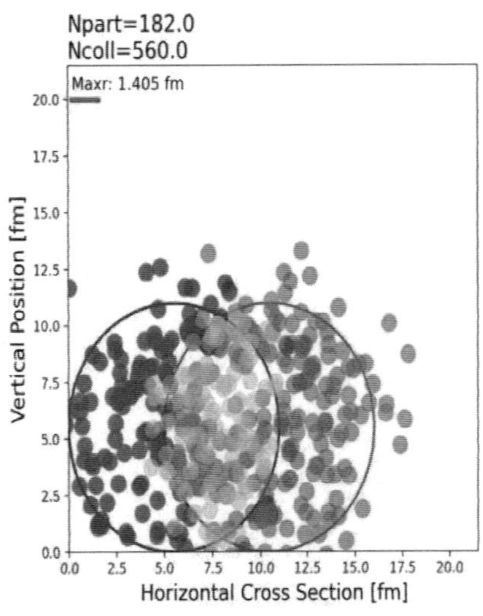

Figure 2. Beam-in View

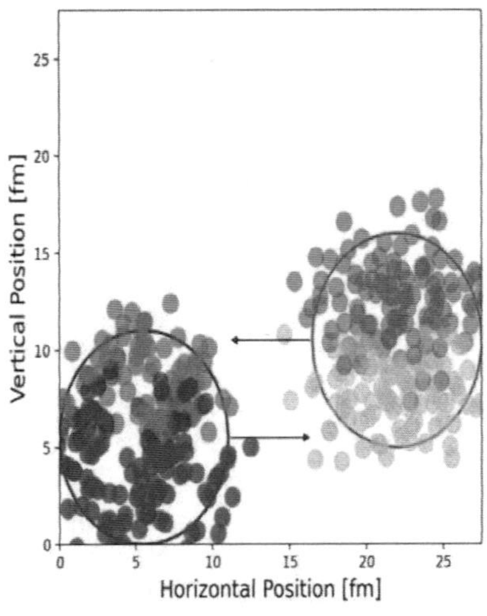

Figure 3. Side View

σ_{NN} refers to the inelastic nucleon-nucleon cross section. Uncertainty arises from the unknown elastic cross section. σ_{NN} solely depends on the collision energy. In this model, it is assumed that the chance for a binary collision between any two nucleons with their relative transverse distance greater than D is ruled out.

3. Correlation and elliptic flow

One piece of interesting information we can extract from data is the correlation function against differences between azimuthal angle,$\Delta\varphi$ and differences between pseudorapidity,$\Delta\eta$. Pseudorapidity η is defined as $\eta = -\ln[\tan(\frac{\theta}{2})]$, where θ is the polar angle. The correlation function aims at providing an insight into the interaction between particles produced in the same collision event from a statistical point of view. Namely, the correlation function is obtained by dividing signal $S(\Delta\varphi, \Delta\eta)$ by background $B(\Delta\varphi, \Delta\eta)$, $\frac{S(\Delta\varphi,\Delta\eta)}{B(\Delta\varphi,\Delta\eta)}$. Signal is defined by making histograms about the $\Delta\varphi$ and $\Delta\eta$ from traces within one event and accumulate over all events, while background is plotted by making histograms with about the $\Delta\varphi$ and $\Delta\eta$ from traces of different events. Background is normalized so that it matches the scale of signal. By definition, correlation function can be verified to be proportional to the true-pair distribution [6]. It is straightforward to test that definition of correlation function is equivalent to $\frac{1}{N}\frac{d^2N}{d\Delta\varphi d\Delta\eta}$. Its plot is shown in Figure 4.

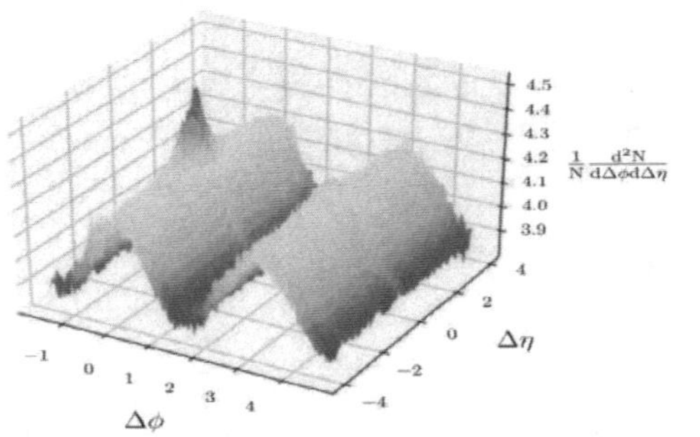

Figure 4. The three-dimensional plot of correlation function against $\Delta\varphi$ and $\Delta\eta$. $\Delta\varphi$ ranges from $-\frac{\pi}{2}$ to $\frac{3\pi}{2}$ in order to match previous work. $\Delta\eta$ varies within -4 to 4, approximately.

Several features of Figure 4 are worth of mentioning.:(1)With centrality 40%-50%, we could discover a symmetry of the configuration with respect to line $\Delta\eta$ =0. (2) Besides, the configuration remains approximately stable along the direction of $\Delta\varphi$ axis (except for the point ($\Delta\varphi, \Delta\eta$)=(0,0)). (3)A spike-like structure locates around the coordinate ($\Delta\varphi, \Delta\eta$)=(0,0), indicating that out-coming particles tend to stay close. Such phenomena can be properly reasoned with Bose-Einstein Correlations. Bose-Einstein Correlations come from the fact that elementary particles are associated with waves, where interference takes place to create correlation. For bosons, they tend to form a compact bunch under such correlation[7]. (4)There are two ridges in the figure. One is along $\Delta\varphi = 0$, it implies the existence of correlations between particles from single jets. Another ridge, on the other hand, locates far from the center along with $\Delta\varphi = \pi$, which corresponds to correlations of back-to-back jets.

As is mentioned in section 1, QGP has the tendency to maintain its initial anisotropy, which can be unfolded by studying the Fourier decomposition coefficient of the azimuthal angle correlation function, $\frac{dN}{d\Delta\varphi}$.

$$\frac{dN}{d\Delta\varphi} = v_0 + \sum v_n \cos[n\Delta\varphi] \tag{5}$$

Here specifically, we study the second Fourier decomposition coefficient, v_2. It is dominated by the effect of the elliptic flow, which mostly makes up the correlation function $\frac{dN}{d\Delta\varphi}$ [8]. However, due to the presence of non-flow effect contributed by secondary decays and emissions, it is inappropriate to make a histogram against $\Delta\varphi$ and decompose directly. Fortunately, non-flow effect is most prominent within short-range area, we instead select traces with $2 < |\Delta\varphi| < 4$ to plot the histogram against $\Delta\varphi$ in order to exclude the interference of non-flow effect. Detailed analysis on separating non-flow effect from elliptic flow will be made in Section 4. The outcome is presented in Figure 5.

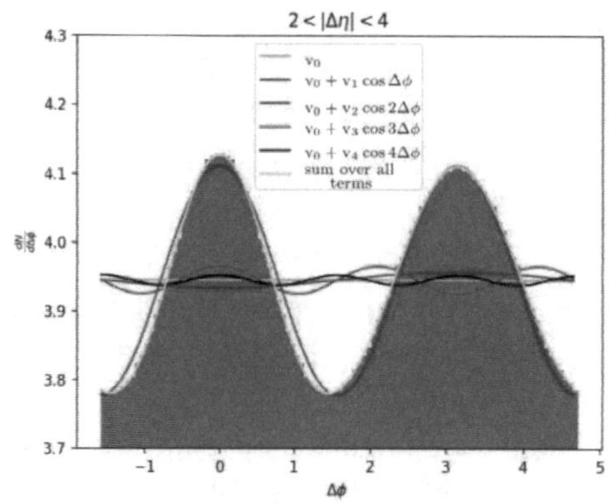

Figure 5: The azimuthal angle correlation function and the components of different orders of Fourier decomposition.

As is shown in Figure 5, the oscillation pattern of $v_0 + v_2\cos 2\Delta\varphi$ term is highly consistent with that of the correlation function. The rest of Fourier component terms, with minor amplitudes, serve as the modifications for $v_0 + v_2\cos 2\Delta\varphi$ term to the correlation function. Such hierarchy of Fourier components implies that elliptic flow predominates over other flows in the data set, including directed flow, triangular flow and rectangular flow, which can be extracted from terms of $v_0 + v_1\cos\Delta\varphi$, $v_0 + v_3\cos 3\Delta\varphi$, $v_0 + v_4\cos 4\Delta\varphi$, respectively.

4. Separation of non-flow

This section is intended for estimating and separating the non-flow contributions from physical flow with two-particle correlation functions. This time, however, instead of presenting the correlation function in the $(\Delta\varphi, \Delta\eta)$ parameter space, we present v_2^2 in (η_1, η_2) parameter space to approach the problem from another point of view.

The Fourier decomposition coefficient, $v_2^2$of correlation function $C(\eta_1, \eta_2, \Delta\varphi)$ in such parameter space [9]:

$$C(\eta_1, \eta_2, \Delta\varphi) = \langle \frac{\rho^{\text{foreground}}(\eta_1,\eta_2,\Delta\varphi)}{\rho^{\text{background}}(\eta_1,\eta_2,\Delta\varphi)} - 1 \rangle \tag{6}$$

$$v_2^2(\eta_1, \eta_2) = \frac{1}{\pi}\int_0^{2\pi} C(\eta_1, \eta_2, \Delta\varphi)\cos 2\Delta\varphi d\Delta\varphi \tag{7}$$

$\rho^{\text{foreground}}(\eta_1, \eta_2, \Delta\varphi)$ stands for foreground pair distribution, which is calculated by making histograms of pair of traces within same events. $\rho^{\text{background}}(\eta_1, \eta_2, \Delta\varphi)$, on the other hand, represents the background pair distribution, which is obtained by histogramming pair of traces from different events. The traingular brackets indicate averages over all events.

It is noteworthy to mention that in this case, such definition for correlation function in (η_1, η_2) space does not take the detector acceptance, secondary decays and scattering into consideration. Further discussions about the influences from such factors on correlation functions are included in [10].

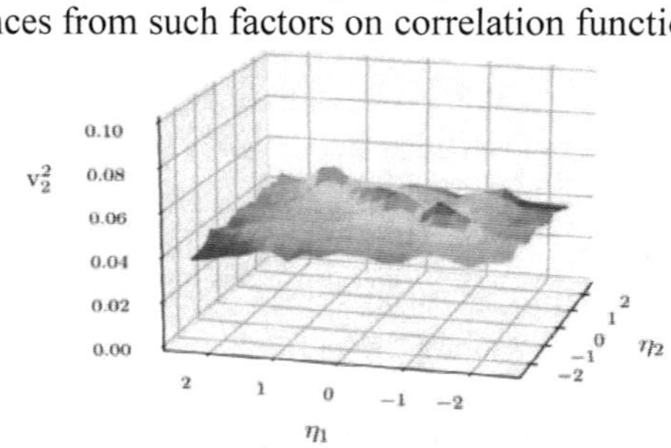

Figure 6. The two-particle correlation function in parameter space (η_1, η_2) with centrality 40%-50% in Pb-Pb collision $\sqrt{s_{NN}}$=2.76 TeV.

The plot of the correlation function is plotted in Figure 6. In detail, we collect traces with η_1 and η_2 in range from -2.5 to 2.5. We divide the parameter space (η_1, η_2) into 20×20 grids. In each grid we perform the Fourier decomposition of correlation function. As is clearly presented in Figure 6, a ridge shows up along the line $\eta_1 = \eta_2$. It corresponds to the Bose-Einstein Correlations mentioned in section 3, or the spike-like configuration in Figure 4. Meanwhile, v_2^2 is less structured in the rest part of the parameter space, approximately maintaining at a constant value. Such differences between two regions can be reasonably attributed to the fact that non-flow effect is most remarkable in short-range area.

The rest of this section mainly focuses on the idea and results of separating non-flow from physical flow.

Without loss of generality, in order to separate non-flow and physical flow, one has to make an assumption that non-flow and physical flow do not couple with each other, otherwise the situation can be much more complicated. As a result, we arrive at:

$$v_2^2(\eta_1, \eta_2) = v_2^{*2}(\eta_1, \eta_2) + \varepsilon(\eta_1, \eta_2) \tag{8}$$

In Eq. (8), v_2^{*2} refers to the contribution of physical flow to v_2^2, while ε stands for component of non-flow term [11].

Furthermore, v_2^{*2} can be divided into $v_2^{*2}(\eta_1) \times v_2^{*2}(\eta_2)$, given the fact that the correlation introduced by physical flow solely depends on pseudorapidity.

Based on these assumptions, the rest of the work is left to make a suitable fit for the physical flow in a specific area in order to accomplish the separation. Keep in mind that non-flow effect is less significant in long-range area. As a consequence, we make a fit for the physical flow in $|\eta_1 - \eta_2| > 2$ in the parameter space and expand the fitting result to the whole parameter space we investigate. In our case, polynomial of fourth order has been chosen to approximate $v_2^{*2}(\eta)$. Indeed, we have to confess that it is a rough fit for physical flow due to the lack of data as well as our fitting method. It does, nonetheless, provide a primary insight into the role physical flow plays.

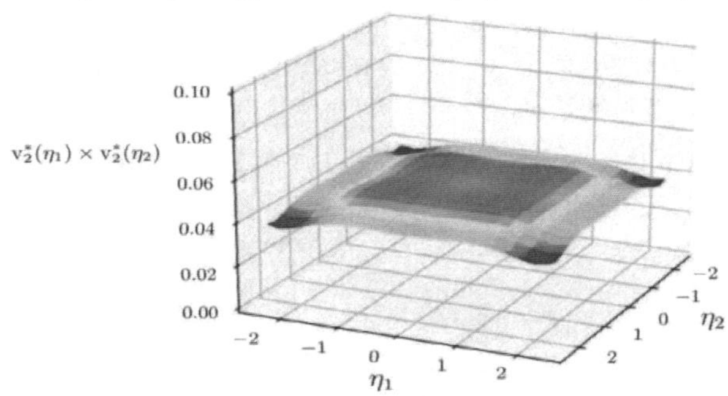

Figure 7. Physical flow's contribution to v_2^2 fitted by degree 4 polynomial in region $|\eta_1 - \eta_2| > 2$ and then expanded to the whole space.

As is presented in Figure 7, the configuration of physical flow term is rather plain, except for four corners in the figure, where slopes are identified because of the polynomial fitting method. Such flat configuration matches our expectation that the effect of physical flow is distributed almost evenly in the parameter plane.

With Eq. (8), we are allowed to obtain the non-flow effect on v_2^2 by subtracting $v_2^{*2}(\eta_1, \eta_2)$ in Figure 6 by the approximated physical term $v_2^{*2}(\eta_1) \times v_2^{*2}(\eta_2)$ in Figure 7. The configuration of $\varepsilon(\eta_1, \eta_2)$ is demonstrated in Figure 8. A few comments are in order. Initially, a ridge-like region lies along $\eta_1 = \eta_2$, indicating the strong non-flow correlation corresponding to the Bose-Einstein Correlation as is discussed earlier. Besides, in the rest of the parameter space, the non-flow correlation fluctuates around 0, which perfectly illustrates that its effect is nearly negligible in long-range area.

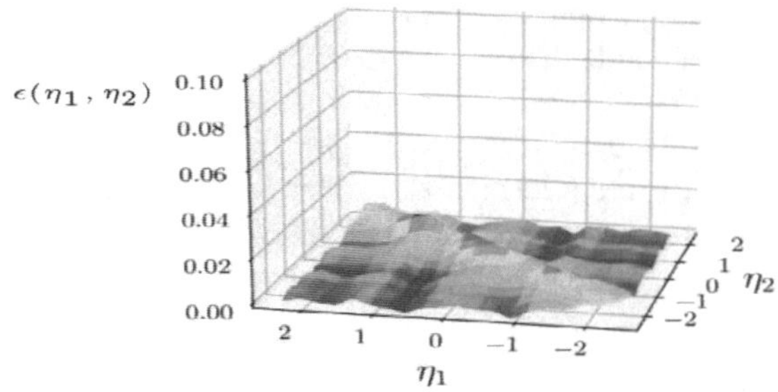

Figure 8. Non-flow correlation $\varepsilon(\eta_1, \eta_2)$ obtained by subtracting the two-particle correlation function $v_2^2(\eta_1, \eta_2)$ by the physical flow's contribution to $v_2^{*2}(\eta_1) \times v_2^{*2}(\eta_2)$.

In this paper, efforts are made to separate non-flow from physical flow. However, we point out that eliminating the effect of non-flow does not complete error analysis. Here we provide a phenomenological formula for error analysis:

$$\sigma^2_{v_2^{obs}} = (f \times \sigma_{\epsilon_2})^2 + (\sigma_{stat})^2 + (\sigma_{non})^2 \tag{9}$$

Eq. (9) is constructed under the assumption that error effects are independent from each other. σ_{ϵ_2} is the fluctuation of initial geometry, f is a factor connecting the initial and final states. σ_{stat} is the statistical fluctuation, and σ_{non} refers to non-flow effect. The separation of the last term has been presented in this paper. One promising future direction can be fulfilling the complete separation of $f \times \sigma_{\epsilon_2}$, σ_{stat} and σ_{non} in order that the capability of extracting physical information from experimental data can be greatly enhanced then.

5. Conclusions

We have presented a brief introduction to the Glauber model and simulated results generated by computer simulations. With open data of Pb-Pb heavy ion collisions at $\sqrt{s_{NN}}$=2.76 TeV from CMS detector with centrality 40%-50%, correlation function $\frac{1}{N}\frac{d^2N}{d\Delta\varphi d\Delta\eta}$ in $(\Delta\varphi, \Delta\eta)$ space is presented, and the Fourier decomposition with traces $2 < |\Delta\varphi| < 4$ is calculated. Important facts include, Bose-Einstein Correlation is obvious on $(\Delta\varphi, \Delta\eta)$ space, which corresponds to the point (0,0). Besides, single jets and back-to-back jets also generate significant effect. Two-particle correlation $C(\eta_1, \eta_2, \Delta\varphi)$ in (η_1, η_2) plane is defined, and the Fourier decomposition coefficient v_2^2 is extracted, with which we perform the separation of non-flow from physical flow. As for the two-particle correlation on the (η_1, η_2) plane, we have made a polynomial fit for the physical flow in long-range area in order to separate it from non-flow. The take-home conclusion is, physical flow is highly plain and flat, with only slopes at the corners because of the polynomial fitting method. Non-flow is most severe in short-range area, or namely, along the line,$\eta_1 = \eta_2$ and oscillating around zero in other areas.
In conclusion, the main focuses of this paper are simulation of initial geometry and interpreting data collected after hadronization. Both of them matter in search for understanding QGP's the characteristic of maintaining initial anisotropy, thus understanding the overall physical properties of QGP.

6. References

[1] L.Van Hove, "Theoretical prediction of a new state of matter, the 'quark-gluon plasma'(also called 'quark-matter',"17th International Symposium on Multiparticle Dynamics,801-818
[2] P. Kovtun, D.T.Son and A.O. Starinets, "Viscosity in strongly interacting quantum field theories from black hole physics," Phys. Rev. Lett. 94 (2005),11601
[3] M.Miller,K. Reygers, S.Sanders and P. Steinberg, "Glauber modeling in in high energy nuclear collisions,"Ann.Rev.Nucl.Part.Sci.57:205-243,2007

[4] B. Alver, M. Baker, C. Loizides and P. Steinberg, "The PHOBOS Glauber Monte Carlo,"arXiv:0805.4411v1
[5] D. d'Enterria and C. Loizides, "Progress in the Glauber model at collider energies,"*Ann.Rev.Nucl.Part.Sci.* 71 (2021) 315-44
[6] A. Adare et al. [PHENIX], "Dihadron azimuthal correlations in Au+Au collisions at $\sqrt{s_{NN}}$=200GeV," Phys. Rev. C 78 (2008), 014901
[7] R.Glauber, "Coherent and incoherent states of the radiation field," *Phys.Rev.* 131 (1963) 2766-2788
[8] B. Alver and G. Roland, "Collision geometry fluctuations and triangular flow in heavy-ion collisions," Phys. Rev. C 81 (2010), 054905
[9] B. Alver et al. [PHOBOS], "Non-flow correlations and elliptic flow fluctuations in gold-gold collisions at $\sqrt{s_{NN}}$=200GeV," Phys. Rev. C 81 (2010), 034913
[10] B. Alver et al. [PHOBOS], "Cluster properties from two-particle angular correlations in p+p collisions at $\sqrt{s_{NN}}$=200GeV and 410 GeV," Phys. Rev. C 75 (2007), 054913
[11] A. M. Poskanzer and S. A. Voloshin, "Methods for analyzing anisotropic flow in relativistic nuclear collisions," Phys. Rev. C 58 (1998), 1671-1678

Acknowledgements
We would like to thank Gunther Roland for delivering lectures on quark-gluon plasma and enlightening us on relevant questions. We also thank Jie-ming Lin, for holding inspiring discussions and sharing interesting ideas with us.

Journal of Physics: Conference Series **2287** (2022) 012030 doi:10.1088/1742-6596/2287/1/012030

Reconstruction, Analysis of the Process ggH Decay to llvv Monte Carlo with MH=125 GeV and Introduction of the Physical Background

Fanli Zeng[1,*], Yiwei Liu[2], Ze Ouyang[3], Bolin Wu[4], Jingyuan Zhang[5]

[1] Physics Department, Department of Nuclear Physics and Quantum Collision Theory, Lomonosov Moscow State University, Moscow 119991, Russia
[2] Taishan College, Shandong University, Jinan 250100, China
[3] School of Physics, Huazhong University of Science and Technology, Wuhan, 430074, China
[4] School of Physical Science and Technology, Lanzhou University, Lanzhou,730100, China
[5] Department of Physics and Astronomy, University of California Irvine, Irvine CA 92697, United States
*Corresponding author. Email: flanriwork@gmail.com

Abstract: This paper outlines analysis the process $gg \to H \to WW \to ll\nu\nu$ Monte Carlo, collision events with MH=125GeV. The discovery of the Higgs boson perfected the standard model and proposed more processes of the production, decay of the Higgs and the Higgs Mechanism. In this analysis, ggH process is reconstructed. Analyzed and calculated the probability of decay WW pairs as productions of decay process. Simulated Monte Carlo signal samples and background are compared to data from Atlas.

1. Introduction

1.1. Summary of The Standard Model

The standard model is a theory that describes the fundamental forces (strong, weak, electromagnetic) between elementary particles and the elementary particles of matter. This model belongs to the category of quantum field theory. At the same time, the Standard Model is also satisfied with relativity (SR) and quantum mechanics. The standard model is not the final answer, it is not perfect yet. So far, the standard model cannot describe gravity. However, experiments on strong, weak, and electromagnetic forces have verified the correctness of the standard model. Each of the three columns on the left constitutes a generation of matter. The next column on the right is the gauge boson, and the rightmost particle is the Higgs boson.

In 1954, Yang and Mills proposed the theory of non-Abelian field gauge groups, which initiated the progress of particle physics across the age. Then Steven Weinberg established the unified theory of weak electricity based on the Young Mills field. The strong interaction was verified afterwards. After adding the Higgs mechanism, the standard model becomes complete. Now, the Standard Model is a necessary theory for studying high-energy physics, particle physics, and nuclear physics.

In the study of the Standard Model, up to 61 common elementary particles are included. These particles are divided into fermions and bosons. The most significant difference between fermions and bosons is the difference in spin. Fermions are semi-odd spins and obey the Pauli exclusion principle.

On the other hand, bosons are just the opposite, their spins are integers and they do not obey the Pauli exclusion principle. In terms of function, fermions are the basic particles that make up matter, while bosons are responsible for transmitting force [1]. The corresponding relationship between each boson and the force is as follows:
Gluons - strong interaction spin = 1. There are 8 known gluons.
Photon - electromagnetic interaction spin = 1.
W, Z bosons - weak interaction spin = 1. There are 3 kinds known.
Higgs - the source of the mass of particles (matter), Higgs is not the gauge boson. The Higgs boson will be discussed in depth in section 3,4.
In particle physics that has emerged, strong interactions are described by quantum chromodynamics (QCD), while weak interactions and electromagnetic interactions are described by the unified theory of electroweakness [2]. The above theory is included in gauge field theory. In short, these theories connect fermions with bosons very well. Since bosons are responsible for transmitting force, the Lagrangian function and gauge transformation related to bosons in each group of particles are the same. Here bosons are also called gauge bosons. The Lagrangian function in QCD is as follows:

$$\mathcal{L}_{QCD} = \sum_{\psi} \overline{\psi_i} \left(i\gamma^{\mu} \left(\partial_{\mu}\delta_{ij} - ig_s G_{\mu}^{a} T_{ij}^{a} \right) \right) \psi_j - \frac{1}{4} G_{\mu\nu}^{a} G_{a}^{\mu\nu} \tag{1}$$

Use theory group to define and describe the standard model: SU(3)SU(2)U(1). This is described using the gauge group, SU(3) is the gauge group describing the strong interaction, and SU(2)U(1) describes the electroweak interaction. It is worth noting that gravity is not described in the standard model. In fact, the concept of 'gravitons' has been proposed, but there is no experiment to prove the existence of 'gravitons'. This is where the standard model is not perfect. The following content will introduce the particles of interest in the Standard Model [3, 4].
Quark is a kind of elementary particles. There are 6 flavors in the standard model, namely up, down, strange, charm, top, and bottom. In addition to electric charge, quarks also possess color charges. It can be imagined that quarks have three color charges, red, green and blue, and their corresponding antiquarks also have corresponding color charges, which are anti-red, anti-green and anti-blue. Color charge does not really mean that quarks and gluons "have colors." In fact, the color charge theory is related to strong interaction, that is, color charge is part of QCD. Due to the mathematical complexity of QCD, we will not discuss the specific principles here.
Quark is the basic unit of matter, and different quarks are combined into hadrons. The protons and neutrons we know all belong to hadrons and are the most stable of hadrons. Quark has a very interesting phenomenon, quark confinement. This makes it difficult to observe quarks. Most of our current understanding of quarks is based on hadron collision experiments. Due to the existence of weak interaction, elementary particles have a phenomenon called decay, and so does quark. The larger mass quarks will eventually decay into smaller masses u and d.
Unlike quarks, *neutrinos* are leptons and common elementary particles. There are currently three kinds of neutrinos, all of them have 1/2 spin, and they are all light and uncharged. The symbol for neutrinos is ν. Neutrinos can be produced in weak interactions between arbitrary particles. Because it is not charged, and its mass and volume are relatively small, it is difficult to find. It was not until 2013 that high-energy neutrinos were first discovered [5].
Neutrinos also have a special phenomenon - neutrino oscillation. We mentioned that there are three kinds of neutrinos, namely electron neutrinos, mu neutrinos, and tau neutrinos. The phenomenon of neutrino oscillation means that the taste of neutrinos will change continuously, that is, the three kinds of neutrinos will switch to each other. This phenomenon is very interesting, and it is also inconsistent with the standard model, because neutrinos have a non-zero static mass. At present, we cannot give the ultimate cause of neutrino oscillation.

1.2. Relativity and Collider Physics
In 1905, Einstein proposed his famous statements about the fast-moving objects. Against the traditional absolute reference system, where the velocity can be simply accumulated or subtracted

when transformation occurs, the special relativity enlightens people with a constant speed invariant for all transformation of reference system -- the light speed c.

$$c = 299792458 \text{ m/s} = \text{constant} \tag{2}$$

This postulate of a constant drastically changes the transformation of two reference system. The transformation of two coordinate system, (x, y, z, t) and (x', y', z', t'), having a relative velocity v along x axis is

$$x' = \frac{x-vt}{\sqrt{1-\frac{v^2}{c^2}}} \tag{3}$$

$$y' = y \tag{4}$$

$$z' = z \tag{5}$$

$$t' = \frac{t-\frac{v}{c^2}x}{\sqrt{1-\frac{v^2}{c^2}}} \tag{6}$$

instead of simple addition of the velocity component in x axis according to Galileo transformation. This form can be further contract into a more compact form, introducing the definition of four-vector

$$a_\mu = (a_0, -a_1, -a_2, -a_3) \tag{7}$$

which is called a *covariant form* of four-vector. For the disposition coordinate mentioned above, the four vector is

$$x_\mu = (t, -x, -y, -z) \tag{8}$$

$$x'_\mu = (t', -x', -y', -z') \tag{9}$$

and a transformation matrix can be attached to the equation array as a concise form

$$\Lambda^\mu_\upsilon = \begin{pmatrix} \gamma & 0 & 0 & -\gamma\beta \\ 0 & 1 & 0 & 0 \\ 0 & 0 & 1 & 0 \\ -\gamma\beta & 0 & 0 & \gamma \end{pmatrix} \tag{10}$$

where the parameter γ and β are defined by $\beta = \frac{v}{c}$, $\gamma = \frac{1}{\sqrt{1-\beta^2}}$, This matrix is called *Lorentz transformation matrix*, which serves as $x'_\mu = \Lambda^\mu_\nu x_\nu$.

Therefore, the displacement vector in four dimension and its transformation between different frames is shown above. Moreover, the four-momentum can be similarly created and performed by the Lorentz Transformation.

$$p_\mu = \left(\frac{E}{c}, -p_x, -p_y, -p_z\right) \tag{11}$$

$$p'_\nu = \Lambda^\mu_\nu p_\mu \tag{12}$$

where the vector is composed of the energy and momentum vector in the Cartesian space.

The applications of Special relativity and Lorentz transformation are of vital importance in the theoretical prediction and authentic detection for particle and high energy physics, as a result of extreme high velocity and transformation between laboratory frame and center of mass (COM) frame. In a typical accelerator, the particles are pumped through the beam pipe, where the detectors are surrounding the pipe; when the collision happens, the particles will be scattered into all directions, going through the detectors and leave information about their velocity. By applying the Lorentz transformation, we can calculate the momentum and energy of the particle just after the collision in COM frame, reconstructing the event.

1.3. LHC and Atlas

The Large Hadron Collider [6, 7], often referred to LHC, is a powerful particle accelerator located in Geneva, Switzerland where the European Center of Nuclear Research lies in. It is composed of a 27-kilometers ring of superconducting electromagnets with a great many of accelerating structures that accelerate hadrons such as protons and ions to approach the speed of light. In the very long and narrow pipe, we assume particles from two high-energy beams, which travel in opposite directions, are made to collide at four sites around the machine. There are four detectors lying in these four sites, respectively named ATLAS, CMS, ALICE, and LHCb.

A Toroidal LHC ApparatuS [8], often referred to ATLAS, is the largest-volume collider detector at present which covers the widest possible range of physics at the LHC. It is designed to precisely measure the properties of Higgs Boson and search for new physics beyond the Standard Model. It is made up of an inner tracking detector, electromagnetic, muon spectrometer, and a hadron calorimeter.

1.4. Monte Carlo Simulation

In the field of particle physics, the Monte Carlo [9] method is often used to model the processes taking place at colliders. Monte Carlo simulations use random sampling and statistical distributions to evaluate the behavior of complex systems [10]. Many simulations are carried out according to this scheme: first, we simulate the processes and obtain the probability density function for the characteristics of the final particles. Further, the statistics necessary for modeling the measurements of the required parameters of the theory are collected. During the simulation, the response of the detector and reconstruction programs are simulated to create a set of events like data recovery [11, 12].

2. Higgs Mechanism

The Higgs Mechanism [13] now provides a convincing demonstration for the mass generation of the bosons, especially the massiveness observed for the particles $W^{\pm}$, Z, which play a part in the weak interaction. This process is now included into the Standard Model (SM) as so-called Electroweak Symmetry Breaking (EWSB) [14].

The original derivation from Higgs comes from the modification of Goldstone's Lagrangian [15] into a gauge-invariant form, containing two Hermitian scalar fields $\phi_1(x), \phi_2(x)$, having a dimension of mass, and one more vector field $A_\mu(x)$.

By maintaining the zero bare mass, which help to maintain the Lagrangian invariant in transformation, the vacuum expectation value for the scalar fields, when choosing a specific gauge [16]

$$\langle 0|\phi_1(0)|0\rangle = 0 \tag{13}$$

$$\langle 0|\phi_2(0)|0\rangle = \eta \tag{14}$$

where the $|0\rangle$ stands for the vacuum state. The $\phi_2(x)$ obviously harvests a non-zero mass η, and the $\phi_1(x)$ can further combined with the vector field $A_\mu(x)$ as

$$B_\mu(x) = A_\mu(x) - \frac{1}{e_0\eta}\frac{\partial}{\partial_\mu}\phi_1(x) \tag{15}$$

which corresponds to a new vector field with mass e_0. This mass induction mechanism, further accomplished by later scientists, is called *Higgs Mechanism*. And the scalar fields in the generation process is called *Higgs Field*. A more intuitive form of Higgs field is in the electroweak dynamics [2], where the scalar fields are contracted into a more concise form of a complex vector

$$\phi = \begin{pmatrix}\phi_1\\ \phi_2\end{pmatrix} \tag{16}$$

and the Higgs potential has the form

$$[\mathcal{V}(\phi^\dagger\phi) = \lambda(\phi^\dagger\phi)^2 - \mu^2(\phi^\dagger\phi)] \tag{17}$$

This form, with $\lambda, \mu^2 > 0$, is a quadratic function with a minimum at $\phi^\dagger\phi > 0$, having a plot shown in the section 4, from which the $\phi^\dagger\phi = 0$ is an unstable local maximum.

3. Reconstructing the $gg \to H \to WW \to ll\nu\nu$ Decay

ATLAS and CMS collaboration studied the use of data sets 7 and 8 TeV Higgs boson decay into a pair of W bosons in the final state leptons are explored [17-19].

In 2015, LHC restarted high luminosity pp collision at $\sqrt{s} = 13$ Tev. The new data is used to further restrict the nature of the Higgs boson: the SM prediction of any significant deviation would be a clear sign of new physics [20]. In the pp collision with $\sqrt{s} = 13$ TeV, the Higgs with 125 GeV is the main production method of gluon fusion (ggH), process is shown in Fig. 1. The branching fraction of the large Higgs boson to the W boson pair makes this channel suitable for accurate measurement of the Higgs boson production cross-section, and allows the study of secondary main production channels, such as through vector boson fusion (VBF) and related Higgs bosons are produced with vector bosons (VH) [20].

3.1. The Decay of $gg \to H \to WW \to ll\nu\nu$

Lepton decay of the W boson is a very clean process, in the final state $W \to l\nu$, where l is an electron or a meson. Select electron or meson pairs with neutrinos as the analysis event. It is also worked to choose two pairs of leptons. The relevant background is shown in Figure 1.

Name	Process	Feature(s)
WW	WW	Irreducible
Top quarks		
$t\bar{t}$	$t\bar{t} \to WbW\bar{b}$	Unidentified b-quarks
t	tW	Unidentified b-quark
	$t\bar{b}, tq\bar{b}$	q or b misidentified as ℓ; unidentified b-quarks
Misidentified leptons (Misid)		
Wj	W + jet(s)	j misidentified as ℓ
jj	Multijet production	jj misidentified as $\ell\ell$; misidentified neutrinos
Other dibosons		
VV	$W\gamma$	γ misidentified as e
	$W\gamma^*$, WZ, $ZZ \to \ell\ell\ell\ell$	Unidentified lepton(s)
	$ZZ \to \ell\ell\nu\nu$	Irreducible
	$Z\gamma$	γ misidentified as e; unidentified lepton
Drell-Yan (DY)		
$ee/\mu\mu$	$Z/\gamma^* \to ee, \mu\mu$	Misidentified neutrinos
$\tau\tau$	$Z/\gamma^* \to \tau\tau \to \ell\nu\nu\ell\nu\nu$	Irreducible

Figure 1. The relevant background to $H \to WW$ [17]

Since there is a large Drell-Yan and top quark background in the event of the same flavor of leptons or

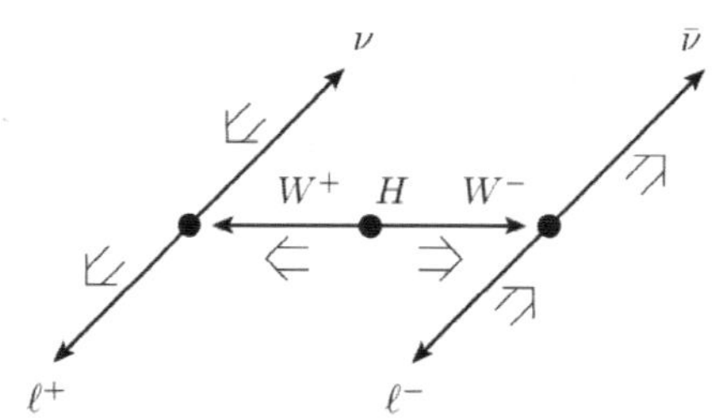

Figure 2. Decay process of $H \to WW$ [17]

ejections, the most sensitive signal area is in the final state of eμ zero ejection. The main background of this category is the production of *WW*, which effectively suppresses the production of *WW* by using the decay characteristics of the W boson and the spin-0 property of the Higgs boson, process is shown in the Figure 2. The next section will process the data provided by Atlas and discuss the details of the decay products, and we will get intuitive conclusions.

3.2. Details of WW Decay

There are many kinds of decay products of W bosons, usually fermion pairs. W bosons can decay to a lepton and anti-lepton or a quark and anti-quark (for example$u, \bar{d}$ and $c, \bar{s}$). Decay channel of W boson is shown in Table 1. The data in the table is obtained in Ref. [21], and the table is rebuilt. It can be calculated that the probability of decay $WW \rightarrow ll\nu\nu$ is 32.57%.

Table 1: Decay channels of W bosons [21]

Decay channel	Probability of Decay
$W^+ \rightarrow e^+ + \upsilon$	10.75%
$W^+ \rightarrow \mu^+ + \upsilon$	10.57%
$W^+ \rightarrow \tau^+ + \upsilon$	11.25%
Hadron	67.60%

4. Data analysis for gluon-gluon Higgs fusion process

As a practice, the gluon-gluon Higgs fusion $\mathrm{gg} \rightarrow \mathrm{H} \rightarrow \mathrm{WW} \rightarrow \mathrm{ll\nu\nu}$ with $\mathrm{M_H} = 125\mathrm{GeV}$ data from the ATLAS open datasets [22] is selected as an example to analyze. The whole process is carried out in Oracle Virtual Machine and Lubuntu OS, which is modified by CERN to optimize the data processing environment, with Python the primary programming language and assistance of ROOT software.

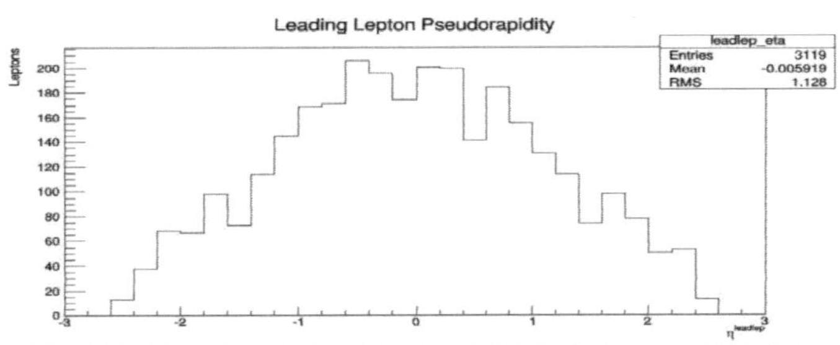

Figure 3. A statistics of the leading leptons falling in the corresponding η area, with the resolution 0.2.

The data package is generated from MC method, which includes $N = 10^5$ events. With such a large data flow, several criteria need to be emphasized in order to locate the Higgs decay process filtering the background noise, which should be appearing in the codes. As we have discussed in the Decay section 4, the characteristic features for the gluon-gluon fusion and double W boson decay will generate 2 leptons visible to the detector, as well as two hardly detectable neutrinos (Fig. 3).

For distinction, the two leptons will be named *leading lepton* and *trailing lepton* separately according to their momentum difference. Once the two spontaneous-generated leptons are detected, or satisfying the criteria in our simulations, the physical properties will be recorded. After a full traverse of data, the statistical module as well as the plotting function goes in, classifying the leptons and generating diagrams, which are available in ROOT file type. A typical statistical diagram is shown in Fig. 3.

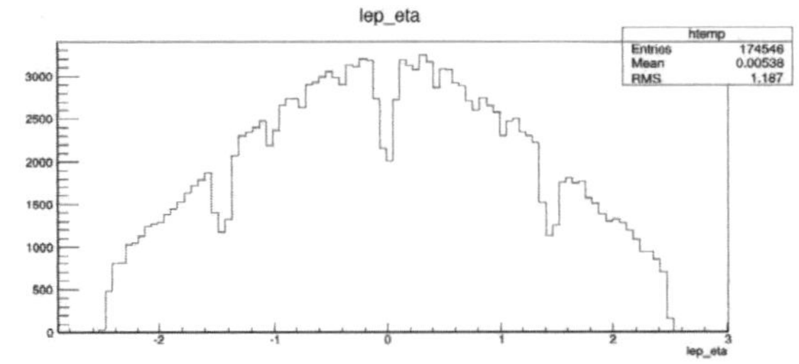

Figure 4. The authentic data for lepton η, noticing the data quantity is nearly 500 times larger than the simulation

In this diagram, the average pseudo-rapidity $\eta = -0.005919 \pm 1.128$, which is indicated in Fig.6. as most the leptons nearly symmetrically distributed at either side of $\eta = 0$ This result is just in correspondence with our intuition, as the leptons should be scattered to all direction equally. The referred authentic data [22] is given in Fig.4. Though rough and inaccurate, the simulated data correctly reflect the shape and average of the realistic ones.
Moreover, from the very same diagram the total events for such Higgs generation and annihilation are counted as n = 3119. Therefore, the probability for getting such a pattern would be about 3%.

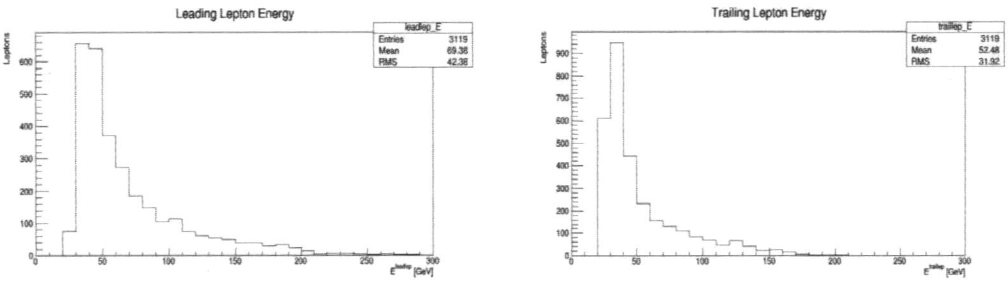

Figure 5. The energy statistics of leading (a, left) and trailing lepton (b, right).

Nevertheless, the pseudo rapidity of the leptons is only one part of its physical features, and the differences are illustrated between leading lepton and trailing lepton since the latter one will generate its own diagram. A discreet comparison of their energy is shown in Fig. 5, showing the leading lepton with $\overline{\mathrm{E}_{\mathrm{lead}}} = 69.38 \pm 42.38\mathrm{GeV}$ and trailing lepton $\overline{\mathrm{E}_{\mathrm{trail}}} = 52.48 \pm 31.92\mathrm{GeV}$, which satisfies our classification that the leading lepton possesses larger momentum, therefore inheriting higher energy. Additionally, the cut-off energy as Fig. 5 shows is about 25GeV, which corresponds to the selection of the events including the Higgs decay pattern.
Finally, the diagrams are converted to *.pdf* type, combining with more information and particles involved in the reaction.
A well-developed, colored diagram considering all the channels are listed in Fig. 6, where the thoroughly predetermined and measured data are inherited in the outputs. Such as in the (a)-1 and (c)-1, the charge of both leptons is separately listed, which is just intuitively correct -- the charge is ± 1. Other distribution along x axis is just as we have predicted and plotted in ROOT -- the energy of the leading one has a climax in about *70GeV*, while for the trailing one the number reduced to about *50*, as shown in Fig. 6. (a)-3 and (c)-3. At the same time, our prerequisites -- the energy limit for choosing a lepton pair must surmount *25GeV* -- is also indicated in the plots, as we have discussed in the context. For all other properties exhibited in Fig. 6, the two series behave nearly the same, since they are just the outcomes from the very same collision.
The overall features and generated plots are listed in Appendix, which are the collisions occurring alongside our main reaction, but have relatively less connection with our discussion. However, the

leptons collected in these data can still have a interference with the Higgs Mechanism, thus making further data excavation necessary.

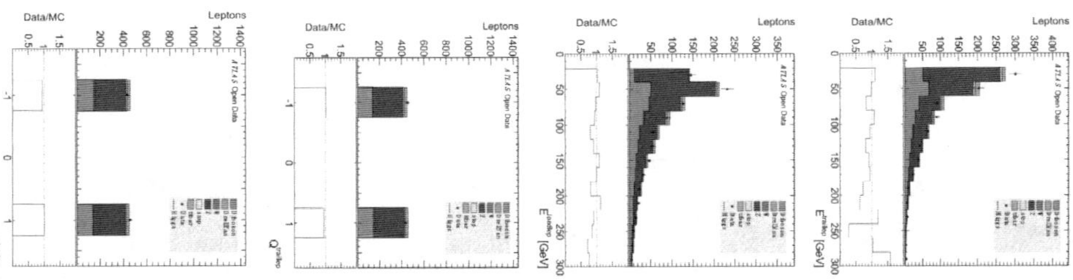

Figure 6. The colored data for leading[(a)-1, (c)-1] and trailing[(a)-3, 3] lepton parameters. The order is from left to right.

5. Conclusions

This article focuses on introducing some physical backgrounds about the $gg \rightarrow H \rightarrow WW \rightarrow ll\nu\nu$ decay process. The Monte Carlo method is used to simulate the decay event and compare it with the given data of the Atlas Open Data.

In the research of high energy physics, it is necessary to make theoretical predictions of differential cross sections and compare them with experimental results. In the theoretical and experimental fields of high energy physics, the Monte Carlo method has been widely applied and developed. Monte Carlo is a very important numerical calculation method guided by probability and statistics theory. Refers to the use of random numbers (or more commonly pseudo-random numbers) to solve many computational problems.

$H \rightarrow WW$ is the cleanest from of Higgs decay because of the productions (leptons and ν). Compared the energy statistics of leading and trailing lepton, which satisfies classification that the leading lepton possesses larger momentum, therefore inheriting higher energy and Higgs decay pattern. Some interesting data analysis results are shown in Appendix A.

6. References

[1] Paul Langacker. The standard model and beyond. In AIP Conference Proceedings, volume 150, pages 142–164. American Institute of Physics, 1986.

[2] RK Ellis, WJ Stirling, and BR Webber. Qcd and collider physics. qcp, page 449, 2003.

[3] Giuliana Davidof, Peter Sarnak, and Alain Valette. Elementary number theory, group theory and Ramanujan graphs, volume 55. Cambridge University Press, 2003.

[4] Edwin Hewitt and Kenneth A Ross. Abstract Harmonic Analysis: Volume I Structure of Topological Groups Integration Theory Group Representations, volume 115. Springer Science & Business Media, 2012.

[5] Masataka Fukugita and Tsutomu Yanagida. Physics of Neutrinos: and Application to Astrophysics . Springer Science & Business Media, 2013.

[6] John M Jowett. The lhc as a nucleus–nucleus collider. Journal of Physics G: Nuclear and Particle Physics, 35(10):104028, 2008.

[7] Lyndon Evans. The large hadron collider. New Journal of Physics, 9(9):335, 2007.

[8] Georges Aad, JM Butterworth, J Thion, U Bratzler, PN Ratof, RB Nickerson, JM Seixas, I Grabowska-Bold, F Meisel, S Lokwitz, et al. The atlas experiment at the cern large hadron collider. Jinst, 3:S08003, 2008.

[9] Robert L Harrison. Introduction to monte carlo simulation. In AIP conference proceedings, volume 1204, pages 17–21. American Institute of Physics, 2010.

[10] Nicholas Metropolis, Arianna W Rosenbluth, Marshall N Rosenbluth, Augusta H Teller, and Edward Teller. Equation of state calculations by fast computing machines. The journal of chemical physics, 21(6):1087–1092, 1953.

[11] Kristen A Fichthorn and W Hh Weinberg. Theoretical foundations of dynamical monte carlo simulations.The Journal of chemical physics , 95(2):1090–1096, 1991.

[12] Kurt Binder. Applications of monte carlo methods to statisticalphysics. Reports on Progress in Physics, 60(5):487, 1997.

[13] Peter W Higgs. Broken symmetries and the masses of gauge bosons. Physical Review Letters, 13(16):508, 1964.

[14] G Bernardi, M Carena, and T Junk. Higgs bosons: theory and searches. Reviews of Particle Data Group: Hypothetical particles and Concepts, pages 10–11, 2007.

[15] Jefrey Goldstone, Abdus Salam, and Steven Weinberg. Broken symmetries. Physical Review, 127(3):965, 1962.

[16] Jeremy Bernstein. Spontaneous symmetry breaking, gauge theories, the higgs mechanism and all that.Reviews of modern physics , 46(1):7, 197

[17] Collaboration ATLAS, Marco Agustoni, Hans Peter Beck, Alberto Cervelli, Antonio Ered- itato, Sigve Haug, Sonja Kabana, Lukas Marti, Federico Meloni, Klaus-Peter Pretzl, et al. Observation and measurement of higgs boson decays to ww* with the atlas detector. Phys- ical review. D-particles, ields, gravitation, and cosmology, 92(1):012006, 2015.

[18] Georges Aad, B Abbott, J Abdallah, R Aben, M Abolins, OS AbouZeid, H Abramowicz, H Abreu, R Abreu, Y Abulaiti, et al. Study of (w/z) h production and higgs boson couplings using hww∗ decays with the atlas detector. Journal of High Energy Physics, 2015(8):137, 2015.

[19] Serguei Chatrchyan, Vardan Khachatryan, Albert M Sirunyan, ArmenTumasyan, Wolfgang Adam, Thomas Bergauer, Marko Dragicevic, Janos Erö, Christian Fabjan, Markus Friedl, et al. Measurement of higgs boson production and properties in the ww decay channel with leptonic inal states. Journal of High Energy Physics, 2014(1):96, 2014

[20] Albert M Sirunyan, Armen Tumasyan, Wolfgang Adam, Federico Ambrogi, Ece Asilar, Thomas Bergauer, Johannes Brandstetter, Marko Dragicevic, Janos Erö, A Escalante Del Valle, et al. Measurements of properties of the higgs boson decaying to a w boson pair in pp collisions at s = 13tev. Physics letters B, 791:96–129, 2019.

[21] Juerg Beringer, JF Arguin, RM Barnett, K Copic, O Dahl, DE Groom, CJ Lin, J Lys, H Murayama, CG Wohl, et al. Review of particle physics. Physical Review D-Particles, Fields, Gravitation and Cosmology, 86(1):010001, 2012.

[22] ATLAS Collaboration. Mc: ggh decays to llvv with m_h =125gev, for 2016 atlas open data release. Technical report, DOI:10.7483/OPENDATA.ATLAS.A43M.KWI7, 2016.

Appendix A. The thorough analyzing plots from the Higgs data package

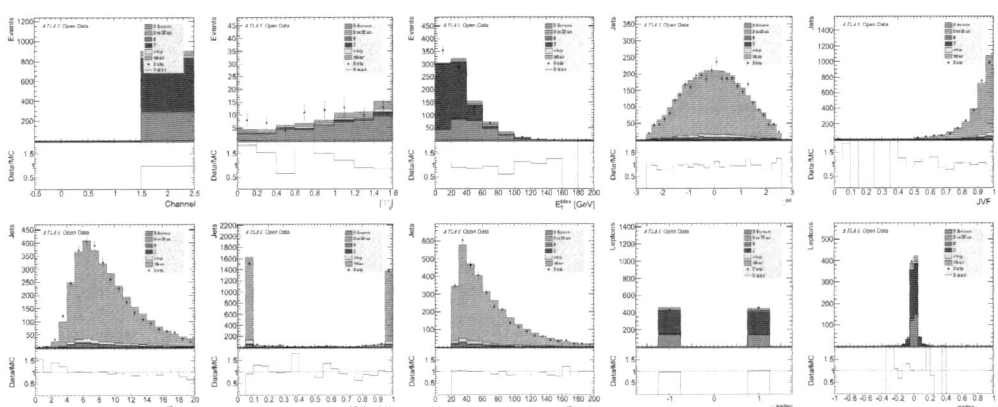

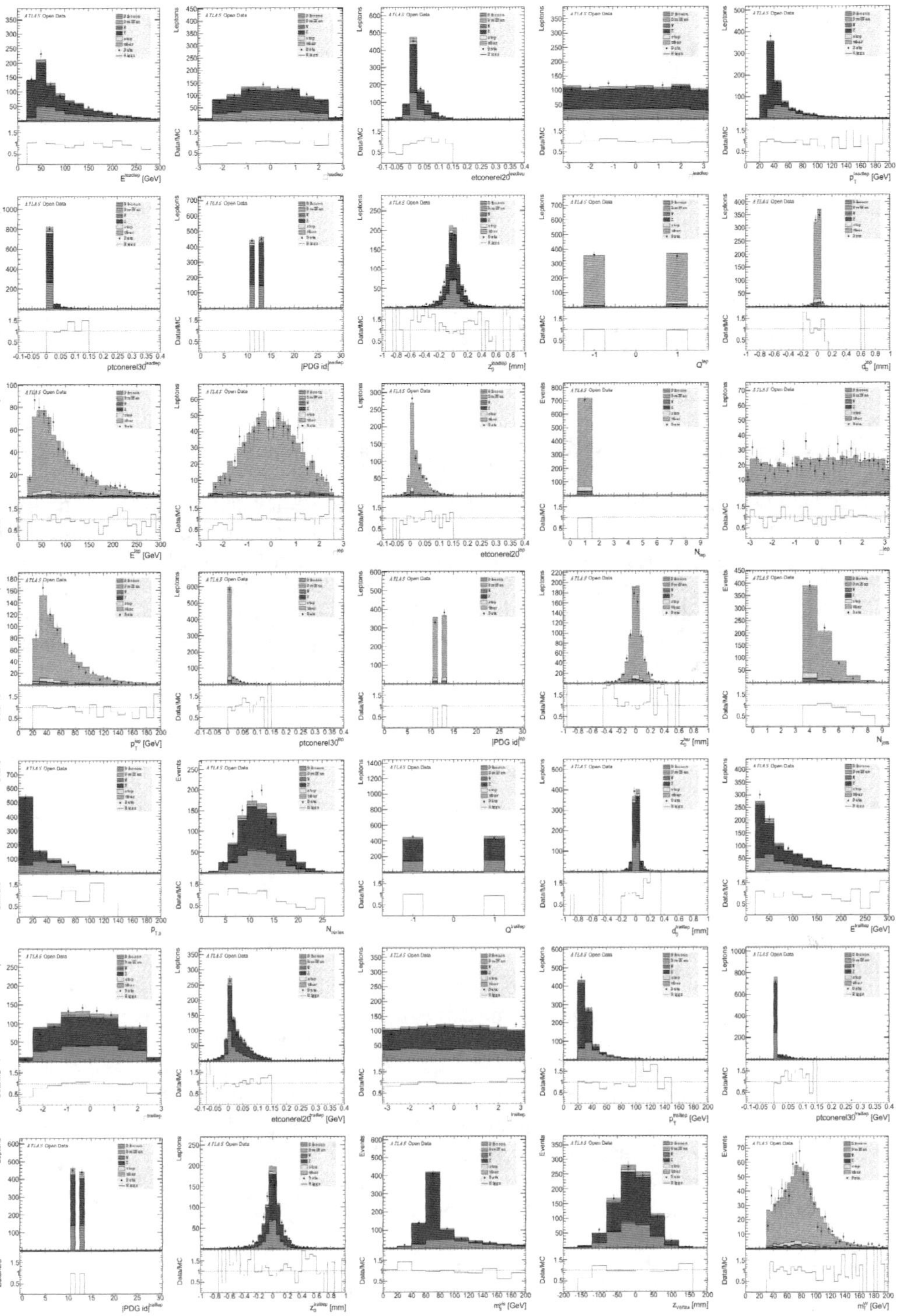

Research on Physical Laws Behind Irreversible Cartesian Diver

Wantong Zhang[1,*] and Zhe Wu[2]

[1] Shenzhen Middle school, Shenzhen, China
[2] Nanjing University of Aeronautics and Astronautics, Nanjing, China
E-mail: zwt2941649765@163.com; 252455858@qq.com

Abstract. The float-sink is an instrument used to demonstrate the pressure transmission of liquids, and it is a vivid proof of Archimedes's principle and Pascal's principle. When pressure is applied to the container in which the sinker is located, the sinker sinks; when the pressure is released, the sinker floats up. This phenomenon qualitatively proves Archimedes's law. But when the sinker sinks to a certain critical depth, even if the pressure is released, the sinker will no longer float up. This phenomenon is the so-called "irreversible sinker". This article mainly analyzes the irreversible principle and influencing factors when the float-sink reaches the critical depth and explores the influence of different influencing factors on the critical depth through mathematical analysis and experimental methods. Finally, it is concluded that the critical depth when the float is irreversible is related to factors such as the mass of the float, the volume of gas in the float and the height of the water in the container, and the critical depth has a linear relationship with the mass of the float.

1. Introduction

1.1. Background

The floatng sink is an instrument used to demonstrate the buoyancy of liquids, the compressibility of gases, and the transfer of pressure to fluids. It was created by French scientist Descartes. There are two types of floats, which are closed floats and open floats. The short float is a closed soft vial, and the available float is a rigid vial with a hole at the bottom. Take the open float sink as an example. According to Pascal's law, the pressure is transferred to the water when the air is compressed. Gravity increases, greater than the buoyancy it receives, it sinks. When the hand leaves the rubber membrane, the volume of air on the water surface in the cylinder increases, and the pressure decreases. The compressed air in the float sink presses the water out. At this time, the gravity of the float sink is less than the buoyancy it receives, so it floats upward.

When the pressure of the float sink reaches a certain depth, after releasing it, the float will no longer automatically float up but will continue to sink. Why doesn't the floater sink? What is the reason? According to specific experiments, there should be a critical depth of sinking. When the sinking position is above this depth, the sinker will automatically float up. When the sinking position is greater than this depth, the sinker will no longer automatically float up, but continue to sink. At the same time, the air volume in the sinker should be related to the critical depth. If so, what should be the relationship? The above phenomenon cannot be quantified through ordinary experiments.

1.2. Research status
At present, scholars at home and abroad have conducted extensive research on the irreversible phenomenon of floats and sinks. Roberto De Luca and Salvatore Ganci mainly studied the pressure and temperature changes of the float sink during the complete sinking experiment and deduced the critical depth formula $h_c = \frac{V_{ext}}{\rho g V_a} p_a$ [1]. Liu Bingsheng mainly studied the factors that caused the sinking and floating of floats and sinks and deduced the formula $x_0 = \frac{p_0 v_0}{G} - H$ [2]. Chang Jian mainly proved the existence of the equilibrium position of the float-sink. He provided a theoretical basis for the lower pressure on the upper part of the float-sink at a deeper level [3-5].

1.3. Research content
In this study, the stress state of the float-sink in the water was analyzed, and the critical depth of the float-sink was determined according to the condition of the force balance of the float in the water. If the float is above the critical depth, the float will automatically float up after the external force is released; if the float is below the critical depth, the float will automatically sink even after the external force is released.

In the study, a specific theoretical analysis of two different types of floats and sinks was carried out to explore which factors (the total mass of the floats and the volume of air in the floats and sinks in the initial state) of the floats and sinks under normal conditions will determine the critical depth of the floats and sinks. And quantitatively study the relationship between the critical depth and parameters of the float-sink so as to get the application conditions of the float-sink. Then through experiments to further discover and demonstrate the floating and sinking phenomenon, through the physical principles to reveal and verify the impact of these influencing factors on the critical depth.

2. Methods

2.1. Floats and sinkers and experiments on their ups and downs
Using simple experimental equipment, make an open float that can hang heavy objects, and verify the irreversible phenomenon of the float through experiments: when it reaches a certain depth, remove the external force, the float will automatically rise; when it reaches a certain depth After the external force is removed, the float will continue to sink and will no longer float.

During the experiment, record the position of the critical depth, and use the scatter plot to summarize the law.

2.2. Floats and sinkers and their physics laws
Determine the factors that affect the critical depth by analyzing the forces at different positions of the floats and sinkers in the water and related theoretical derivations, and analyze how these factors cause the changes in the critical depth, so as to reveal and verify the ups and downs obtained from the floats and sinks experiments law.

Whether an object can float or sink in water is mainly due to the relationship between the buoyancy and gravity of the object. If the buoyancy is greater than the gravity, the object is in a floating state; if the buoyancy is equal to the buoyancy, the object is in a floating or suspended state; if the buoyancy is less than the gravity, the object is in a sinking state. In response to this phenomenon, we will explore the effects of different influencing factors on the irreversible phenomenon and critical depth of floats and sinks through a combination of experiment and theory and derive the essential empirical formula of depth.

3. Shortcomings and prospects
There are two types of floats, which are closed floats and open floats. A fast float is a soft balloon or a plastic bag with a heavy object hanging under it, filled with air, as shown in Figure 1(a); an open float is a hard plastic or glass bottle with a piece hanging under it. Heavy objects contain part of water and air, and there is a small hole at the bottom through which water can enter and drain, as shown in Figure 1(b).

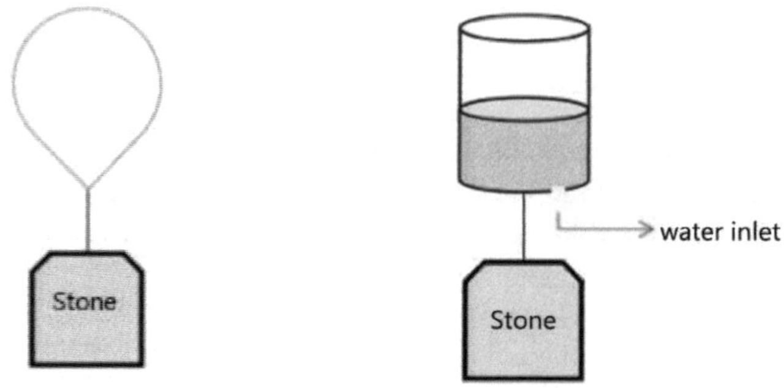

(a) (b)

Figure 1. Float and sink type. (a) Closed float; (b) Open float.

When the closed float sink is fully immersed in water, if the water discharged by the air-bag is heavier than the weight of the rock, the float sink will float up. The float will cause the air-bag to expand and increase, and the speed of floating will increase until it floats to the surface. At this time, the buoyancy is equal to gravity; if the weight of the water discharged by the air-bag is less than the weight of the rock, the float sink will sink, and the sinking will cause the air-bag to shrink and shrink, and the sinking speed will also increase until it sinks to the bottom. Therefore, the enclosed floats and sinks have only two states in the water, namely floating on the water surface and sinking in the bottom.

When the open float sink is placed in the water, water will enter the float sink through the small holes. As the water intake increases, the buoyancy of the float sink becomes smaller and smaller. If the suspended heavy object is too heavy, the interior of the float will be filled with water, and the float will sink until it sinks to the bottom; if the suspended heavy object is light, the water and gas in the float will reach a state of equilibrium. At this time, buoyancy is equal to gravity. Therefore, there are only two states of open floats in the water, namely floating on the water surface and sinking in the bottom.

For the above two different types of floats and sinks, the following numerical analysis is done:

3.1. Theoretical Analysis on the Ups and Downs of Enclosed Floats and Sinks

3.1.1. Conformity and meaning *used in the theoretical analysis process*

The symbol description of the enclosed float is shown in Table 1.

Table 1. Three Scheme comparing.

Symbol	Meaning
m	Total mass of floats and sinks
V_0	The volume of gas in the float (inflation under standard atmospheric pressure)
V_1	Block volume
V	Gas volume after pressure change (dynamic change)
ρ	Density of water ($10^3 \mathrm{kg/m^3}$)
p_0	Standard atmospheric pressure ($1.01 \times 10^5 \mathrm{Pa}$)
h_c	Critical depth

3.1.2. Theoretical analysis process

Assumption: The gas in the float meets ideal gas conditions.

When the float sinks into the water, the resultant force is:

$$F = mg - \rho g(V + V_1) \tag{1}$$

Since there is almost no change in temperature during the process of putting into water, the process of putting into water can be regarded as isothermal compression, according to $pV = nRT$, where T is a constant, so pV is a constant. So when the balloon is placed in the water:

$$p_0 V_0 = (p_0 + \rho g h)V \tag{2}$$

$$V = \frac{p_0 V_0}{p_0 + \rho g h} \tag{3}$$

Substituting into the above formula, we get

$$F = mg - \rho g\left(\frac{p_0 V_0}{p_0 + \rho g h} + V_1\right) = mg - \rho g V_1 - \rho g V_0 \frac{p_0}{p_0 + \rho g h} \tag{4}$$

Let $F = 0$

$$mg - \rho g V_1 - \rho g V_0 \frac{p_0}{p_0 + \rho g h_c} = 0 \tag{5}$$

$$p_0 + \rho g h_c = \frac{\rho g V_0 p_0}{mg - \rho g V_1} \tag{6}$$

The critical depth is:

$$h_c = \frac{p_0 V_0}{mg - \rho V_1 g} - \frac{p_0}{\rho g} = \frac{p_0}{\rho g}\left(\frac{\rho V_0}{m - \rho V_1} - 1\right) \tag{7}$$

The basic condition for the establishment of the above formula is: $\rho V_0 > m - \rho V_1 > 0$, otherwise h_c is a negative value, which does not match the actual situation.

$$F = mg - \rho g V_1 - \rho g V_0 \frac{p_0}{p_0 + \rho g h} \tag{8}$$

It can be seen from the above formula that, except for the variable depth h, the rest are all fixed values. So the curve of F is roughly as shown in Figure 2:

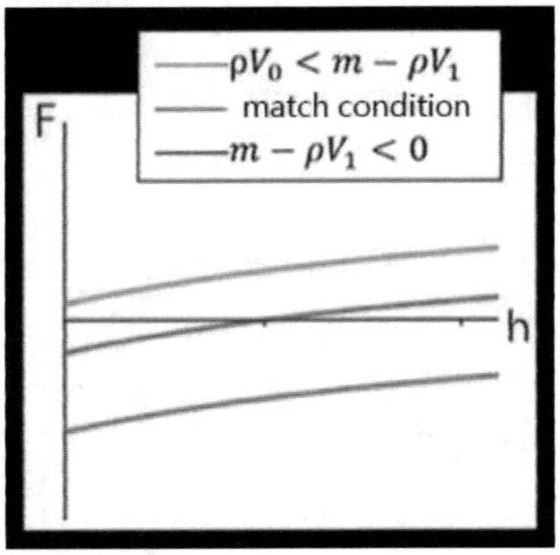

Figure 2. The law of change of the resultant force F.

When $F > 0$, the floater floats up; when $F < 0$, the floater sinks continuously.

$$h_c = \frac{p_0}{\rho g}\left(\frac{\rho V_0}{m - \rho V_1} - 1\right) \tag{9}$$

It can be seen from the above formula that increasing the counterweight m can reduce the critical depth h_c. Therefore, we stipulate the "weight sensitivity" (defined as the partial derivative of the mass of the float-sink to the critical depth) as:

$$\left|\frac{\partial m}{\partial h_c}\right| = \frac{\rho^2 g p_0 V_0}{(p_0+\rho g h_c)^2} \tag{10}$$

It reflects the sensitivity of the critical depth to this change when the counterweight m changes. Subsequent experiments can quantitatively increase the counterweight to study the influence of the counterweight m on the critical depth h_c.

Suppose the mass and volume of the original float and sink are m_s and V_s, the mass and volume of each counterweight block are m_2 and V_2, the number of counterweight blocks is k, and the critical depth expression is:

$$h_c = \frac{p_0}{\rho g}\left(\frac{\rho V_0}{(m_s+km_2)-\rho(V_s+kV_2)} - 1\right) = \frac{p_0}{\rho g}\left(\frac{\rho V_0}{(m_s-\rho V_s)+k(m_2-\rho V_2)} - 1\right) \tag{11}$$

Define the depth coefficient $H = \frac{p_0}{p_0+\rho g h_c}$, so

$$H = \frac{p_0}{p_0+\rho g h_c} = \left(\frac{m_s}{\rho V_0} - \frac{V_s}{V_0}\right) + k\left(\frac{m_2}{\rho V_0} - \frac{V_2}{V_0}\right) \tag{12}$$

It can be seen that the depth coefficient H is proportional to the number of counterweight blocks k, and the resultant force F is:

$$F = mg - \rho g V_1 - \rho g V_0 H \tag{13}$$

3.2. Theoretical Analysis on the Ups and Downs of Open Floats and Sinks

3.2.1. Conformity and meaning used in the theoretical analysis process

The symbol description of the open float is shown in Table 2.

Table 2. Three Scheme comparing.

Symbol	Meaning
m_t	Total mass of floats and sinks
V_t	Total volume of floats and sinks
m_w	The mass of the water in the float
V_w	The volume of water in the float
m_p	The mass of floats
V_p	The volume of the float (ignored as 0)
m_z	The mass of the suspended weight
V_z	The volume of the suspended weight (ignored as 0)
V_{ext}	The volume on the surface of the float
V_s	The volume below the surface of the float
V_0	The volume of gas in the float (initial state)
V_a	Gas volume in floats and sinks (dynamic change)
ρ	Density of water (10^3kg/m^3)
p_0	Standard atmospheric pressure (1.01×10^5Pa)
h_c	Critical depth

3.2.2. Theoretical analysis process

Assumption: The gas in the float meets ideal gas conditions.

Volume equation:

$$V_t = V_{ext} + V_s + V_z = V_0 + V_w + V_p + V_z \tag{14}$$

When the float is floating on the water:

$$F_{Float} = G \tag{15}$$

$$\rho g(V_t - V_{ext}) = (m_p + m_w + m_z)g \tag{16}$$

$$V_{ext} = V_t - \frac{m_p+m_w+m_z}{\rho} \tag{17}$$

When the float is completely immersed in water:

$$F = (m_p + m_z)g - \rho g V_a \tag{18}$$

If there is a critical depth, then $F = 0$, so:

$$(m_p + m_z)g - \rho g V_a = 0 \tag{19}$$

According to $pV = nRT$, T is a constant, so pV is a constant. Ignore the bottle height, so when the balloon sinks into the water:

$$p_0 V_0 = p_1 V_a \tag{20}$$

$$p_0 V_0 = (p_0 + \rho g h)V_a \tag{21}$$

$$V_a = \frac{p_0 V_0}{p_0+\rho g h} \tag{22}$$

So

$$(m_p + m_z)g = \rho g \frac{p_0 V_0}{p_0+\rho g h} \tag{23}$$

Therefore, the critical depth h_c is:

$$h_c = \frac{p_0 V_0}{(m_p+m_z)g} - \frac{p_0}{\rho g} \tag{24}$$

The basic condition for the establishment of the above formula is: $\rho V_0 > m_p + m_z > 0$, otherwise, h_c is a negative value, which is inconsistent with reality.

4. Experimental design

4.1. Experiment equipment
The experimental equipment is as follows:

(1) Measuring cylinder (500ml);
(2) 100ml syringe;
(3) Plastic tube;
(4) High-precision electronic scale;
(5) Float and sinker (lightweight glass medicine bottle);
(6) Glass rod;
(7) Water;
(8) Straightedge;
(9) Dropper.

4.2. Experiment program
The experimental program is as follows:

(1) Use a reagent dropper and two paper clips (plus a part of the other paper clip) together, and the first test is whether it can float in a 200mm high glass. Then put the made float in a graduated cylinder and press it down with a plastic rod to test whether it will automatically sink when it reaches a certain depth. If so, judge the approximate depth and start the experiment; each case needs to be repeated once.
(2) Experimental steps:
1) Place the manufactured float and sink lightly on a high-precision electronic scale to weigh and record the corresponding data.

2) According to the depth of the preliminary test, add an appropriate amount of water to the measuring cylinder to ensure that the float sink can sink to the bottom.
3) Put the suction hose into the water in the graduated cylinder and connect it with the syringe. Use the syringe to slowly draw out the water in the graduated cylinder.
4) Carefully observe the state of the sinker during the slow pumping process until the sinker moves slightly (and then continues to float), and record the corresponding scale.
5) Repeat the test two more times to get three data.
6) Separate the floats and sinks, wipe off the water and weigh them separately.
7) To change the quality of floats and sinks, repeat steps 1-6.

4.3. Improvement of experimental measures

In the float-sink experiment, three methods were used to pressurize the gas in the measuring cylinder, namely: pressing method, drainage method and exhaust method. However, the use of the pressing method will have the problem of greater difficulty in force control, and it is easy to leak air and water when the measuring cylinder is sealed by the drainage method. In the end, we use the syringe exhaust method to pressurize the gas in the graduated cylinder, which is also easy to control.

4.4. Experimental data and its analysis

Since the wall thickness and volume of floats and sinks cannot be ignored, the critical height under test conditions is calculated according to the following formula:

$$V_t = V_w + V_a + V_p \tag{25}$$

$$V_a = \frac{p_0 V_0}{p_0 + \rho g h} \tag{26}$$

$$V_w = V_t - V_p - \frac{p_0 V_0}{p_0 + \rho g h} = V_t - \frac{m_p}{\rho} \tag{27}$$

$$h_c = \frac{p_0 V_0}{m_g g - \rho g V_p} - \frac{p_0}{\rho g} \tag{28}$$

The data directly recorded in the experiment are shown in Table 3.

Table 3. Experimental data record table (including calculation results)

No.	m_p(g)	m_t(g)	m_w(g)	V_p(ml)	V_t(ml)	V_w(ml)	V_0(ml)	h_c(cm) Calculated	h_c(cm) Measured
1	8.75	16.11	7.36	3.74	16.5	7.36	5.40	6.5	5.805
2	8.75	16.05	7.30	3.74	16.5	7.30	5.46	6.8	5.929
3	8.75	16.00	7.25	3.74	16.5	7.25	5.51	7.4	6.032
4	8.75	15.98	7.23	3.74	16.5	7.23	5.53	7.4	6.073
5	8.75	15.94	7.19	3.74	16.5	7.19	5.57	7.9	6.156
6	7.68	14.97	7.29	3.12	15.0	7.29	4.59	5.0	5.068
7	7.68	14.91	7.23	3.12	15.0	7.23	4.65	5.5	5.204
8	7.68	14.85	7.17	3.12	15.0	7.17	4.71	5.9	5.340
9	7.68	14.82	7.14	3.12	15.0	7.14	4.74	5.9	5.408
10	7.68	14.77	7.09	3.12	15.0	7.09	4.79	6.1	5.521

The following scatter plot and its linear regression are only intuitive data relationships that illustrate the correlation between critical depth and influencing factors.

(1) The relationship between h_c and V_0 is shown in Figures 3 and 4. The two have an obvious positive correlation (when $m_p = 8.75g$, $R^2 = 0.976$; when $m_p = 7.68g$, $R^2 = 0.9414$).

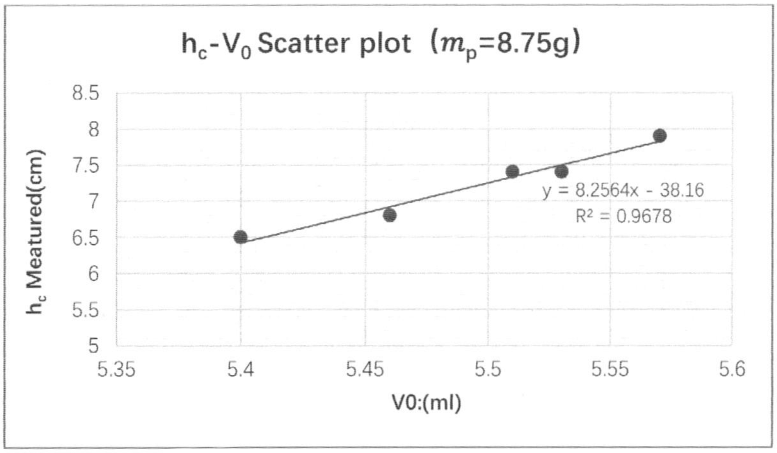

Figure 3. Scatter plot of critical depth when m_p = 8.75g.

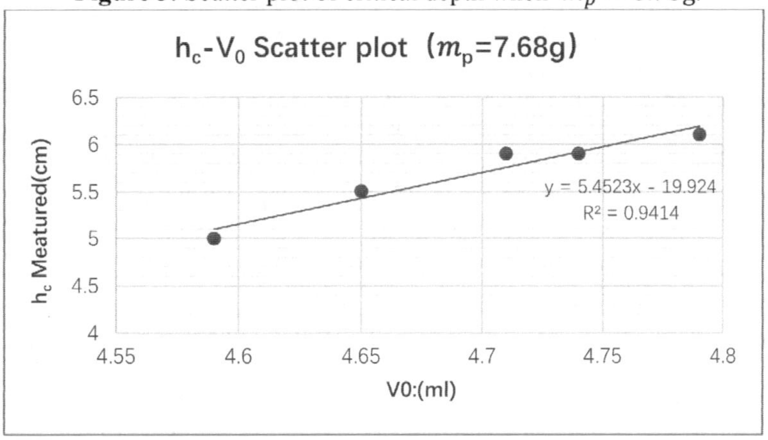

Figure 4. Scatter plot of critical depth when m_p = 7.68g.

(2) The relationship between h_c and m_t is shown in Figures 5 and 6, and the two have an obvious negative correlation (when m_p = 8.75g, R^2 = 0.9768; when m_p = 7.68g, R^2 = 0.9414).

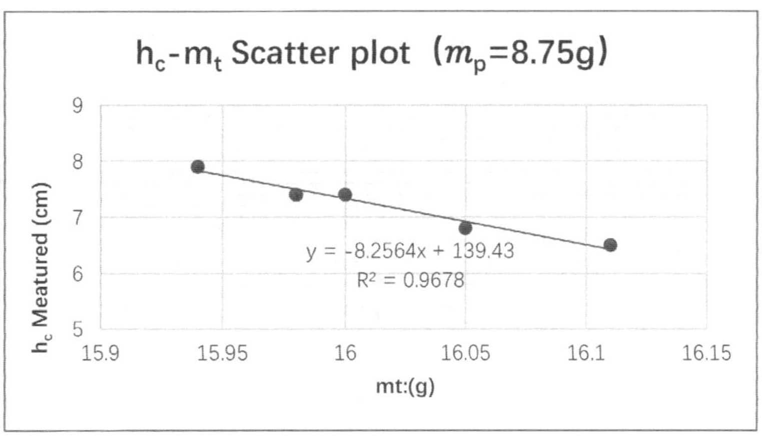

Figure 5. Scatter plot of critical depth when m_p = 8.75g.

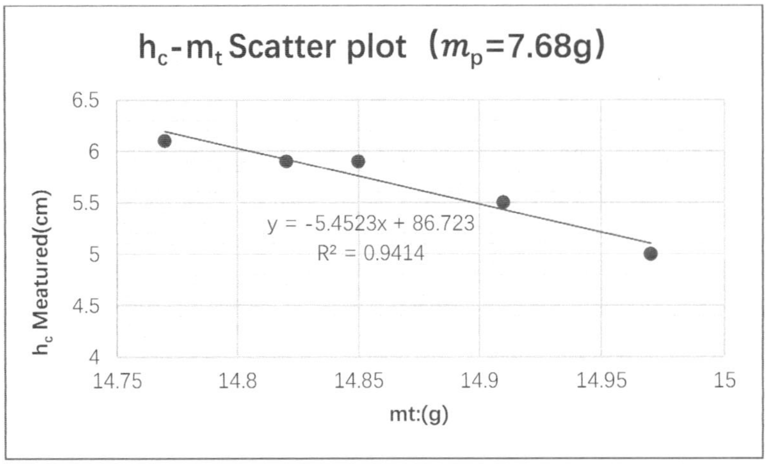

Figure 6. Scatter plot of critical depth when $m_p = 7.68$g

5. Experimental conclusion and summary

It can be found through experiments that the critical depth of the float is mainly related to the height of the water in the container, the mass of the float and the volume of air in the float. When the height of the water in the container is less than the critical depth, the float will return to its original position after decompression, even if the float touches the easy bottom. The critical depth of floats and sinks has a linear relationship with its mass, and the relationship is $h_c = \frac{p_0 V_0}{(m_p+m_z)g} - \frac{p_0}{\rho g}$, which has also been verified during the experiment. The air volume in the float will also affect the irreversible phenomenon of the float. Since the condition of $\rho V_0 > m_p + m_z > 0$ exists, if $\rho V_0 \leq m_p + m_z$, h_c will no longer exist, and the irreversible phenomenon of floats and sinks will not occur.

Through experiments and theoretical analysis, we have a further understanding of the irreversible phenomenon of floats and sinks. Using the conclusion of the critical depth of floats and sinks, practical objects such as vehicles that sail or dive into water, indicator lights floating in the water, floats used for fishing, markers used for water warnings, etc. can be made according to requirements.

6. Shortcomings and prospects

In this study, there are mainly the following shortcomings:

(1) Due to the limitation of the experimental conditions, only the open-type float-sink test has been completed this time, and there is no corresponding closed-type test data.

(2) In the analysis process, the quality of the float and sink itself and the assumption that the water is in a static state are ignored, and the influence of other conditions is not considered.

In future research, we will consider the influence of other conditions in work, life and other scenarios to make the theoretical derivation more suitable for practical application scenarios.

7. References

[1] Roberto De Luca, Salvatore Ganci. A lot of good physics in the Cartesian diver. IOP, 2021.

[2] Liu Bingsheng, Exploring the ups and downs of floats and sinks after decompression, Physics Teaching Volume 33, Issue 2, 2011.

[3] Chang Jian, Quantitative Research on the Underwater Balance of "Floats and Sinks", Physics Teaching Discussion Volume 29, 2011.

[4] Dragia Ivanov and Stefan Nikolov, Once more about the Cartesian diver, Phys. Educ. 55, 2020.

[5] Chen Haishen, Inquiry Teaching of "Principle of Open Floats and Sinks", Physics Teaching Volume 38, Issue 1, 2016.

Observing the Development of Two Solar Flares by Monitoring AR 12882 which Produces a Kp6 Geomagnetic Storm in October 2021

Anis Rossdee[1, 2], N N M Shariff[2,3]

1 Faculty of Applied Sciences, Universiti Teknologi MARA Shah Alam, 40450 Shah Alam, Selangor, Malaysia
2 Institute of Science, Universiti Teknologi MARA Shah Alam, 40450 Shah Alam, Selangor, Malaysia
3 Academy of Contemporary Islamic Studies (ACIS), Universiti Teknologi MARA Shah Alam, 40450 Shah Alam, Selangor, Malaysia
E-mail: anisrossdee@gmail.com

Abstract. In this paper, we study active region 12882 in conjunction with the level of geomagnetic storm based on space weather data. Several flares occurred in this region in October 2021, including M1.6 and C2.1. Data were also collected from the online solar monitor and space weather live database provided by NOAA and SDO/AIA. The study of the solar flare events was done by monitoring the sunspot region for the past four days to observed changes in the earth's atmosphere. The M-class flare event made by AR 12882 produced a CME with a speed exceeding 500 km/sec which directed towards Earth and produced an intense geomagnetic storm with kp6 index.

1. Introduction

The sun produces an abundant amount of energy, and the output energy should be monitored from the surface of the sun because a high energy output could be fatal towards the earth's atmosphere. The sun is now in its 25th solar cycle (SC) and climbing towards solar cycle maximum where we can anticipate in an increase of sunspot numbers. Sunspot number correlate with solar cycle where it can be used to track the progression of a SC [1]. Sunspots are temporary disturbed magnetic field where it develops from a magnetic loop flux that emerges from solar interior of a photosphere. Magnetic field traps gas where hot air are unable to rise to the surface, thus creating areas of dark spot where the temperature is cooler than surrounding areas.

Generally, sunspot groups are classified based on their morphology and size. The magnetic structure, or the relative positions and sizes of concentrations of opposite polarity magnetic flux in the active region (AR), is used by both the Mt. Wilson and McIntosh classifiers in order to define active region class [2]. A large and complex sunspot can produce significant space weather and solar activity [3], including solar flares and coronal mass ejections (CMEs).

In Active Regions (AR), a variety of dynamic process such as plasma heating, solar flares, and coronal mass ejections (CME) would occur. Solar flares are the reconnection of magnetic field on surface of the sun that produces big explosions in the solar system. The energy released could result in electromagnetic emission with different wavelengths of gamma rays to radio waves. Flares are characterized by their brightness in X-rays (X- Ray flux) and can be classified into A-, B-, C-, M- and X- classes which depends on the intensity of electromagnetic energy [4].

CME are large portions of corona blown into space by solar wind and sending huge amount of superheated gas towards outer space known as solar energetic particles [5]. If CME are facing towards earth, its magnetic field would interact with Earth's magnetic field which will result in geomagnetic storm or produce auroras at north and south poles [6].
During the Sun's activity, it releases enormous amounts of energy that can be carried out by the solar wind, disrupting the Earth's thermosphere and ionosphere [7]. It is beneficial to monitor solar activities to avoid potential problems with satellite communications. Based on the latest solar event we can expect cycle 25 to enter a new phase. This paper examines the behaviour of geomagnetic storms after the release of two types of solar flares. Following this, we will monitor the characteristics of AR12882 and discuss the impact of the solar flare towards the earth's atmosphere.

2. Methodology and data collection

In this paper, the solar events, solar flares, and coronal mass ejections during the phase of solar cycle 25, have been selected. The solar events were analysed with its solar wind parameter and geomagnetic storms for three days from 9th October to 12th October 2021. The data of the active region was taken from solar monitor website which can be accessed on the internet. The wavelength used can be focused on the surface of the sun clearly. Image of the active region and solar flare was collected from SDO/HMI and H-alpha website. Data analysis of solar events such as sunspot numbers, sunspot size and solar wind parameter were collected from space weather live forecast. Active region could show the active sunspot that produces solar flare and the occurrence of CME carried out by solar wind that would produce geomagnetic storms.

3. Result and discussion

In this study, selected solar events was taken and observed to understand its impact towards the earth's atmosphere. From Figure 1, there was several flare activities that occurred on October 2021 ranging from X, M and C class flares. The first M-class flare was on 9th October 2021 alongside a C-flare. These flares originated from AR 12882 and projected a CME that contributed to a high Kp index which resulted in a geostorm of G2.

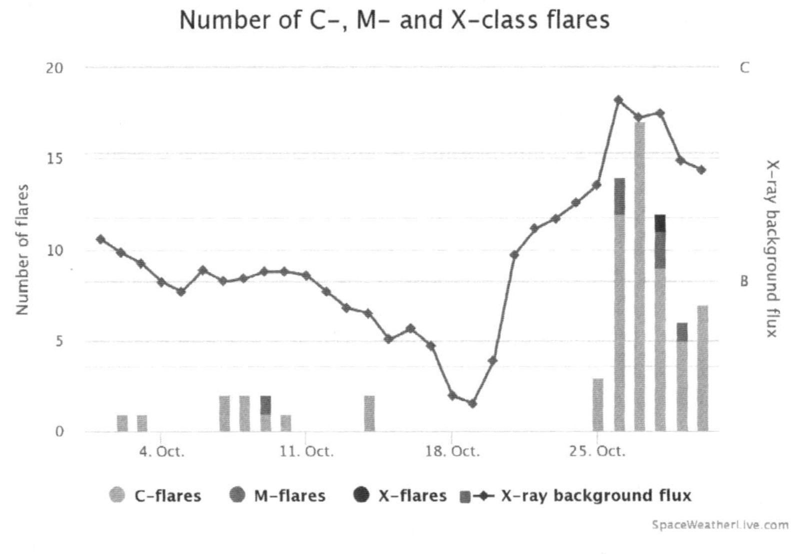

Figure 1. Number of solar flare classes throughout October 2021

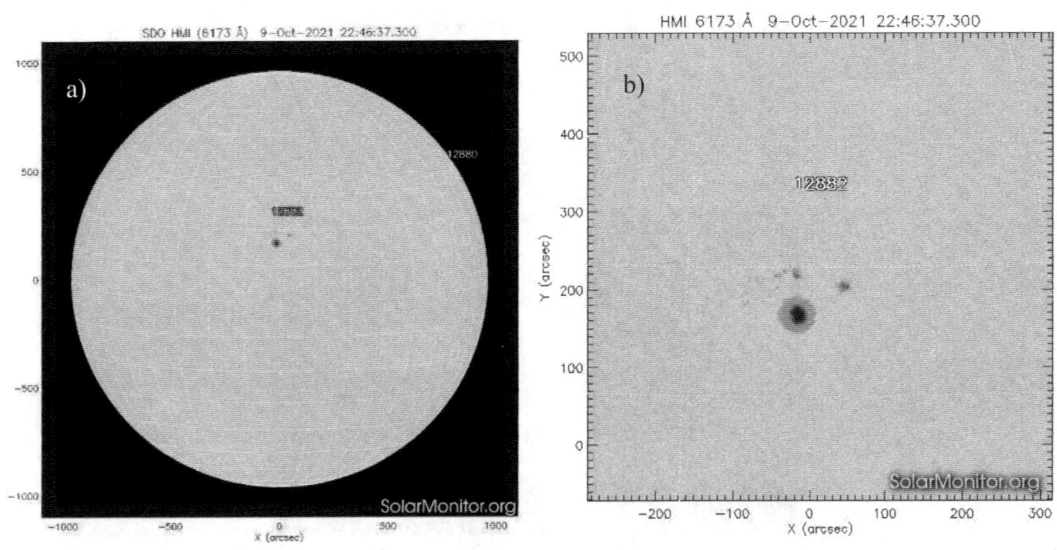

Figure 2. Sunspots on surface of the sun on 9th October 2021. a) two active regions on surface of the sun. b) close-up of AR12882. (Credit to solarmonitor.org)

Figure 2 shows two active regions where AR12882 is directly facing towards earth. This region consists of 4 sunspot numbers with the size of 280 millionths. On 9th October, AR12882 produced a M1.6 and C2.1 class flare at 06:36 UTC and 22:56 UTC, respectively.

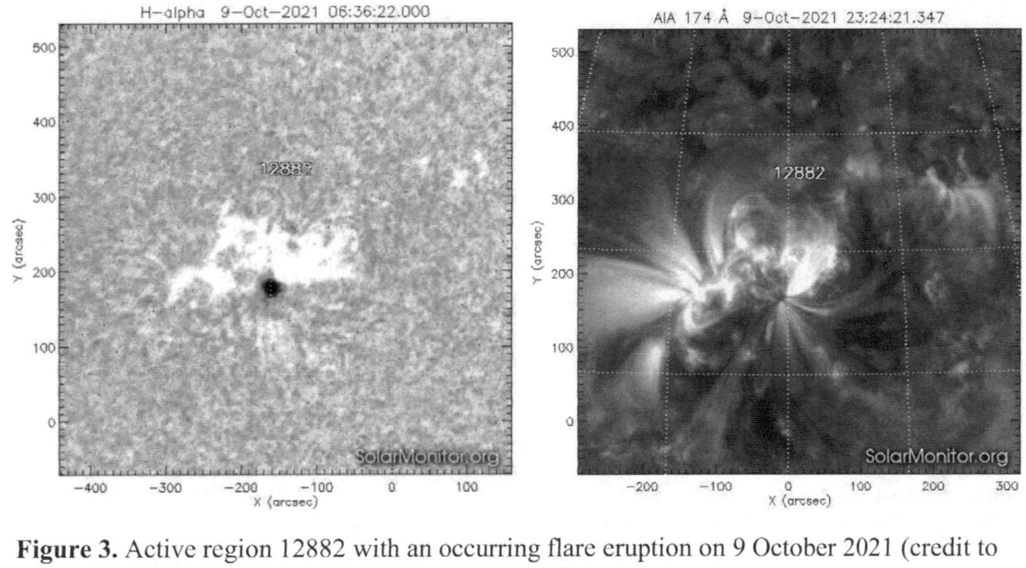

Figure 3. Active region 12882 with an occurring flare eruption on 9 October 2021 (credit to SDO/HMI)

The M1.6 solar flare was released alongside CME from AR12882 as seen in figure 3 and figure 4. Based on the AR position, where the location is facing towards earth, it is expected that the solar energy particles emitted from the solar flare and CME will collide with the earth's atmosphere within 2-3 days depending on the CME speed [8]. The position of a flare on the Sun has an impact on how the earth's atmosphere reacts to the flare [9] .

Figure 4. Coronagraph of the sun showing a halo CME (credit to SOHO)

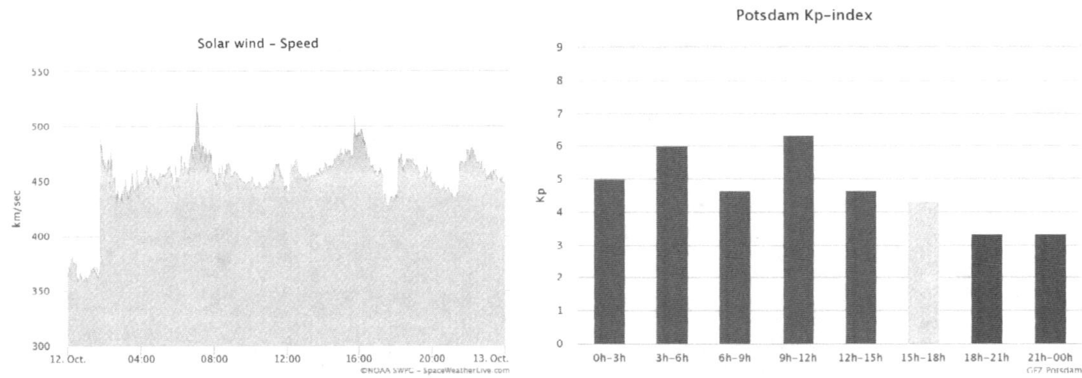

Figure 5. Solar wind speed and Kp-index on 12th October 2021

On the day of the geomagnetic storm on 12th October produced by CME, the solar wind approaches up to 530 kms^{-1} based in Figure 5, alongside a high speed of CME. This event caused an intense geomagnetic storm of G2 on earth and produced auroras with a Kp6.

4. Conclusion

From the results obtained, the strongest flare on 9th October 2021 resulted in a moderate G2 geomagnetic storm on 12th October 2021. AR 12882 occurred in the rising phase of solar cycle 25. It has a high number of sunspots and was at its maximum size which is 280 MH. These characteristics would correspond to the eruption of multiple solar flares on AR12882. To conclude, strong M-class flares could also lead to a medium geomagnetic storm based on the factor of flare location and solar wind speed. High speed wind contributed to a high Kp-index on 12th October 2021 that leads to a minor radio blackout and produced auroras around the poles of the earth.

5. References

[1] Kumar P. Relationship Among Sunspot Numbers , Solar Radio Flux (F-10 . 7 INDEX) And Solar Wind Parameters For Solar Cycle – 24. International Journal of Engineering Research in Current Trends (IJERCT) [Internet]. 2021;3(4):128–32. Available from: https://www.ijerct.com/papers/03-04/relationship-among-sunspot-numbers-solar-radio-flux.pdf

[2] Ireland J, Young CA, McAteer RTJ, Whelan C, Hewett RJ, Gallagher PT. Multiresolution analysis of active region magnetic structure and its correlation with the mount wilson classification and flaring activity. Solar Physics. 2008 Oct;252(1):121–37.

[3] Eren S, Kilcik A, Atay T, Miteva R, Yurchyshyn V, Rozelot JP, et al. Flare-production potential associated with different sunspot groups. Monthly Notices of the Royal Astronomical Society. 2017;465(1):68–75.

[4] Bai T, Sturrock PA. Classification of Solar Flares. Annual Review of Astronomy and Astrophysics. 1989;27(1):421–67.

[5] Norsham NAM, Hamidi ZS. The Relationship between Solar Radio Burst Type III and Geomagnetic Storm on the Earth that Occurred on 4th October 2017. In: Journal of Physics: Conference Series. Institute of Physics Publishing; 2019.

[6] Cho KS, Marubashi K, Kim RS, Park SH, Lim EK, Kim SJ, et al. Impact of the ICME-earth geometry on the strength of the associated geomagnetic storm: The September 2014 and March 2015 events. Journal of the Korean Astronomical Society. 2017;50(2):29–39.

[7] Syed Zafar SNA, Hazmin SN, Jusoh MH, Dagang AN, Adzni MAM, Umar R. Behaviour of Geomagnetic storm, horizontal geomagnetic field and solar wind parameters during solar flare and CMEs event. In: Journal of Physics: Conference Series. IOP Publishing Ltd; 2021.

[8] Gopalswamy N, Yashiro S, Michałek G, Kaiser ML, Howard RA, Reames D v, et al. Interacting coronal mass ejections and solar energetic particles. 2002;(8).

[9] Qian L, Wang W, Burns AG, Chamberlin PC, Coster A, Zhang SR, et al. Solar Flare and Geomagnetic Storm Effects on the Thermosphere and Ionosphere During 6–11 September 2017. Journal of Geophysical Research: Space Physics. 2019 Mar 1;124(3):2298–311.

Acknowledgements
Special gratitude to SDO/AIA, NOAA, SWPC, LASCO, Helioviewer, Solar Monitor, and Space Weather Live websites for making their data available to the public. This project is under the FRGS Grant 600-IRMI/FRGS 5/3 (309/2019) and Kementerian Pengajian Tinggi (KPT). The authors also show their respect to the reviewers and editors for their insightful remarks and recommendations.

Moderate Radio Flux Density of Single and Group III Bursts from AR2887 Active Region on 26th October 2021

Z S Hamidi[1, 3], L. Chen Chen[2], Z. K.D.Nurlisman[2]

1 Faculty of Applied Sciences, Universiti Teknologi MARA Shah Alam, 40450 Shah Alam, Selangor, Malaysia
2 Malaysian Space Agency (MySA), National Space Center, Jalan Turi, Kompleks Teknologi Angkasa, 42700 Banting, Selangor
3 Institute of Science, Universiti Teknologi MARA Shah Alam, 40450 Shah Alam, Selangor, Malaysia
E-mail: zetysh@uitm.edu.my

Abstract. The radio flux density is one of the most important parameters for predicting solar activity, such as solar flares and Coronal Mass Ejections. Compared to other burst types, solar burst type III (SRBT III) occurs most frequently due to solar flare phenomena. In the meter and decimetre wavelength range, gyro synchrotrons dominate, as do localized disturbances in the plasma frequencies caused by SRBT III. In this paper, we examine the most recent solar event that took place on 26th October 2021 in this gradual phase of the 25th solar cycle and can be detected from our site. In order to compare our results to other regions of the electromagnetic spectrum, we also used the data from the Solar Monitor and Space Weather Live website, where the data from NOAA and SDO/AIA were made publicly available. A wavelength of 10.7 cm corresponds to 101 solar flux units. Approximately 4.7 nT of interplanetary magnetic field exists and 1.7 nT in the northward direction. In the past 6 hours and 24 hours, flares of class C2 and M1 have been detected due to active region AR2887 generated by a 'beta-gamma' magnetic field.

1. Introduction

One of the most important parameters for determining solar activity, such as solar flares or Coronal Mass Ejections, is the radio flux density. The flux density is usually measured in Jansky or Decibel units according to the system and the antenna gain. As a result of the structure of the Sun, the solar bursts indicate the Sun is active. Depending on the factors involved, solar bursts can be classified into five types, from type I to type V [1].

Compared to other bursts, solar burst type III is the most common, mostly caused by solar flares. Solar radio burst type III is fascinating because it explains how particles ejected during massive solar activity are emitted [2]. There are three types of solar bursts: single bursts, group bursts, and complex bursts. By studying their correlation, we can improve our understanding of the magnetic reconnection process on the Sun's surface that leads to solar flares and CMEs.

The generation mechanisms in SRBT III mainly consist of two types: nonlinear wave-wave coupling and direct emission mode conversion: the result of density inhomogeneity [3]. The nonlinear wave-wave coupling is the coexistence of two Langmuir waves [4]. SRBT III's localized disturbances in plasma frequencies dominate at meters and decimetre wavelengths, along with gyro synchrotrons. [5]. These are associated with the primary energy-releasing sites in flares. Type III bursts of the Sun are characterized by the very fast movement of disturbance through the solar corona that could exceed 3 x

10^4 to 10^5 km/s. That velocity corresponds to one-third of the velocity of light. These electrons are accelerated to energies between 10^4 and 10^5 eV [6].
Sunspot groups are usually associated with active regions on the Sun. An earth-sized sunspot is similar in size to a moderate-sized sunspot. Sunspots form and disappear over a period of days or weeks when magnetic fields emerge through the solar surface and allow cooling [7]. Penumbra usually develops around the central area or the umbra as the flux accumulates beyond the pores. As the magnetic field rises, it loops back towards the photosphere (North polarity). Sunspot groups with complex magnetic fields are often the source of solar flares. From an astronomical point of view, we can understand and describe the weather and climate conditions of Earth's lower atmosphere by analyzing the magnetic structure of Mount Wilson. [8].

2. Methodology and observation

Activated regions were compiled from the Solar Monitor and Space Weather Live website, which provides publicly available data from NOAA and SDO/AIA. Additionally, we compared the sun's activity with e-Callisto data, which monitors the sun's activity around the world via radio waves. [9]. Solar activity at Earth's surface can be detected by radio telescopes and ground-based antennas [2]. Observations were conducted with the LPDA, low noise amplifier (LNA), CALLISTO spectrometer, and Windows computer. A GPS clock controls the sampling time of the spectrometer in order to standardize the time. An antenna tracking controller controls the direction of the antenna. We will observe from 7 am to 7 pm, 12 hours a day. The CALLISTO network monitors the Sun 24 hours per day within the UNBSSI/IHY and ISWI instrument deployment program [10]. Figure 1 shows the Log Periodic Dipole Antenna that has been installed at National Space Centre, Banting, Malaysia.

Figure 1. Log Periodic Dipole Antenna that has been installed at Banting, Malaysia station

3. Result and discussion

A single and a group of moderate-intensity radio bursts appear in Figure 2, both exploding within less than 14 dB of each other. SRBT III properties have been attributed to a variety of mechanisms. Solar flares of type III are the subject of intense research during the solar cycle's maximum especially. There is usually one flare or flare-like event that propagates electrons through the lower corona emitted incoherently as l-waves (electron plasma waves).

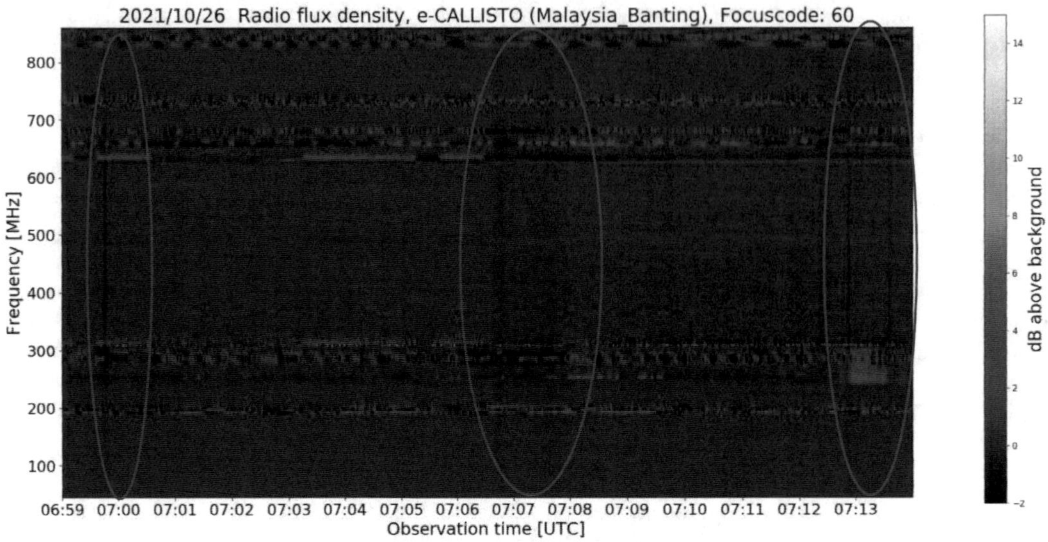

Figure 2. Single and a group solar burst type III at Banting, Malaysia station

Within three hours, the radio flux density reaches a moderate level (Figure 3). The energy release site is believed to be the source of upward- and downward-directed beams of non-thermal electrons that cause the SRBT III. Single bursts at low frequencies lasting a few minutes are quite rare. While isolated single bursts are rare, complex events are usually described as a combination of multiple individual bursts.

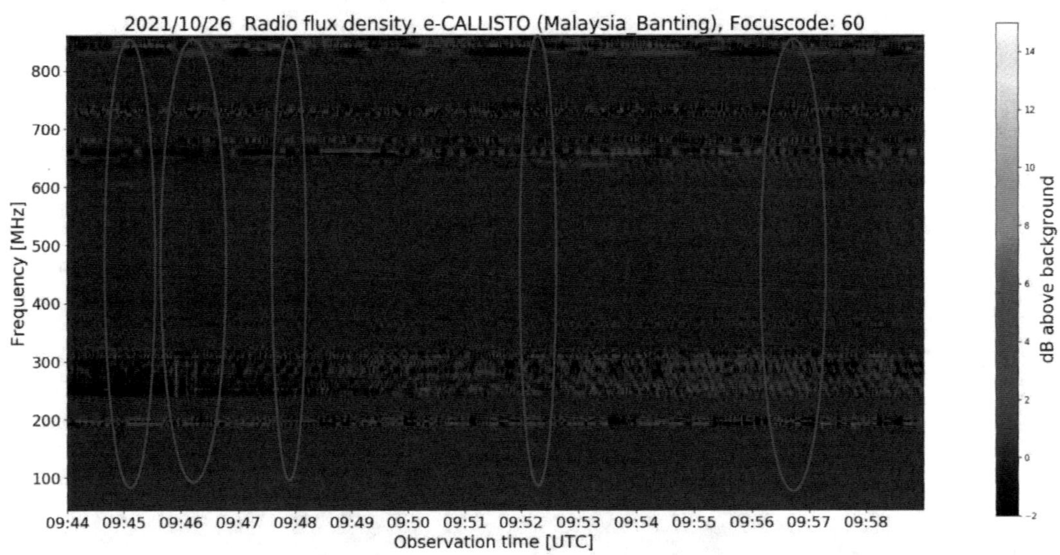

Figure 3. A moderate radio flux density is continuous within few hours on 26th Oct 2021

As shown in Figure 4, the M-class solar flares associated with growing sunspot AR2887 are being generated by a "beta-gamma" magnetic field. Other active regions are AR2886, AR2888, and AR2889, but they are considered less active than AR2887. Due to the stronger magnetic field in active regions with sunspots, solar flares are more likely to occur. At the dividing line between areas of oppositely directed magnetic fields, the magnetic energy in the solar atmosphere is typically released.

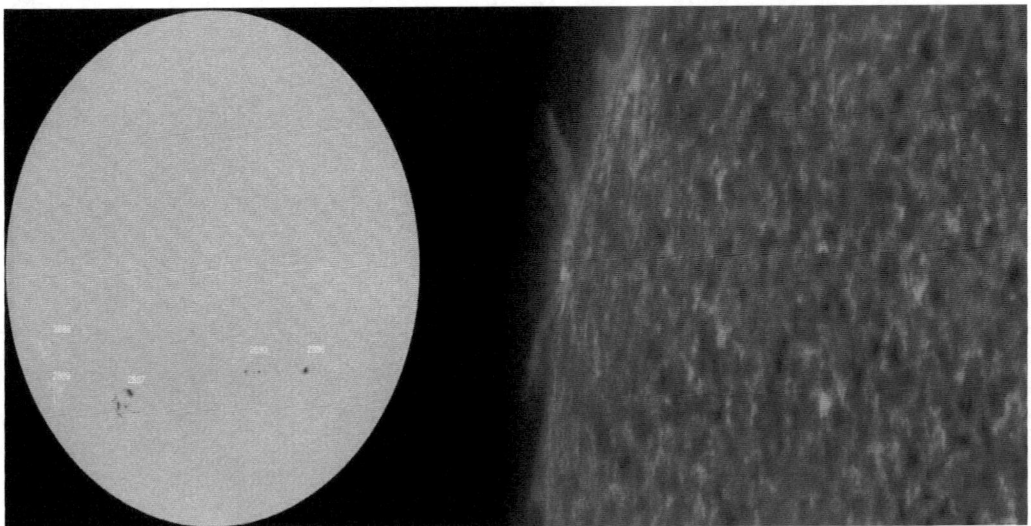

Figure 4. The active regions at the surface of the Sun and the explosion of M-class flare on 26th Oct 2021

Solar magnetic fields are complex and powerful. On this day, 81 sunspots are visible. It has a density of 4.9 protons per cubic centimeter, and it can travel at speeds of up to 378 kms^{-1}. During the past 6 hours and 24 hours, flares of class C2 and M1 have been detected. Temperatures could exceed $49.4x10^{10}$ W in the thermosphere. At a wavelength of 10.7 cm, the flux number shows 101 solar flux units. In the north direction, there is a magnetic field of approximately 1.7 nT and 4.7 nT. Due to magnetic reconnection on the Sun, magnetic energy has been released from the loop top of the reconnection site.

4. Conclusion

Groups of β- γ sunspots that are bipolar, but are sufficiently complex that no continuous line can be drawn between spots of opposite polarity. This is a complex character that results from the larger sunspot. The most intense major solar flares usually originate from the delta configuration active region since its umbrae have mixed magnetic polarity, which originates from AR2887.

5. References

[1] Sabri, S., et al. *The dependence of log periodic dipole antenna (LPDA) and e-CALLISTO software to determine the type of solar radio burst (I-V).* in *2016 International Conference on Industrial Engineering, Management Science and Application (ICIMSA)*. 2016. IEEE.

[2] Husien, N., et al. *Solar radio bursts detected by CALLISTO system and their related events.* in *2016 International Conference on Industrial Engineering, Management Science and Application (ICIMSA)*. 2016. IEEE.

[3] Henri, P., et al., *Evidence for wave coupling in type III emissions.* Journal of Geophysical Research: Space Physics, 2009. **114**(A3).

[4] Ni, S., et al., *Plasma emission induced by electron cyclotron maser instability in solar plasmas with a large ratio of plasma frequency to gyrofrequency.* The Astrophysical Journal Letters, 2020. **891**(1): p. L25.

[5] Kliem, B., et al., *Recent progress in understanding energy conversion and particle acceleration in the solar corona.* Energy conversion and particle acceleration in the solar corona, 2003: p. 263-293.
[6] Sakurai, K., *On the acceleration of relativistic electrons in solar proton flares.* Publications of the Astronomical Society of the Pacific, 1971. **83**(491): p. 23.
[7] Priest, E.R., *Solar magnetohydrodynamics*. Vol. 21. 2012: Springer Science & Business Media.
[8] Shibata, K. and T. Magara, *Solar flares: magnetohydrodynamic processes.* Living Reviews in Solar Physics, 2011. **8**(1): p. 1-99.
[9] Norsham, N.A.M. and Z.S. Hamidi. *A multiwavelength study of the Sunspot of active region which leads to X 17 solar flare*. in *Journal of Physics: Conference Series*. 2019. IOP Publishing.
[10] Ali, M., et al. *e-CALLISTO network system and the observation of structure of solar radio burst type III*. in *2016 International Conference on Industrial Engineering, Management Science and Application (ICIMSA)*. 2016. IEEE.

Acknowledgments
We thank the following organizations for providing their data to the public: SDO/AIA, NOAA, SWPC, LASCO, Helioviewer, Solar Monitor, Space Weather Live and E-CALLISTO. The authors also thank the reviewers and editors for their valuable comments and suggestions. We acknowledge the support provided by grants 600-RMI/FRGS 5/3 (309/2019), 600-RMC/YTR/5/3 (001/2020) and Kementerian Pengajian Tinggi Malaysia.

An Analysis of the X1.59 Flare and Associated Events on 3rd July 2021

S N A Shamsuddin[1], N N M Shariff[2, 3]

1 Faculty of Applied Sciences, Universiti Teknologi MARA Shah Alam, 40450 Shah Alam, Selangor, Malaysia
2 Academy of Contemporary Islamic Studies (ACIS), Universiti Teknologi MARA Shah Alam, 40450 Shah Alam, Selangor, Malaysia
3 Institute of Science, Universiti Teknologi MARA Shah Alam, 40450 Shah Alam, Selangor, Malaysia
E-mail: najihah9881@gmail.com

Abstract. Solar flares are caused by a complex magnetic field on the surface of the Sun, which releases a large amount of radiation. We present an analysis of the X1.5 flare that occurred on 3rd July 2021 at 14:29 UT from the active region AR 2838. It was believed that some other events were related. Prior to satellite communication, the ionosphere was commonly used for radio communication. In case of enhanced excess photoionization due to solar activities, the D layer will not disappear after sunset. This includes the burst of the day, as well as the ionospheric response caused by the flare event. Although numerous studies have been conducted on the sun, its dynamic behavior has continued to draw the interest of astrophysicists to this area to this day.

1. Introduction

Solar flares are one of the major sun's activities described as a sudden, rapid giant explosion that occurs on the sun's photosphere. It included acceleration of erupted charged particles from the built-up magnetic energy to a very high energy of electromagnetic radiation[1]. During the solar maximum, the sun can produce many flares on the same day.

Solar flares classify by A, B, C, M, or X on the peak flux in watt per meter square (Wm^{-2}) along with the logarithmic scale 1-9. A-class is the weakest and X-class is the strongest flare class. As reported by [2], the biggest and strongest flares (X-class flare) can occur almost more than 10 times on average per year during the solar maximum phase. Severe radio blackout may happen when this strong flare directly occurs towards the Earth. The largest X-class flare was recorded on 4th November 2003 with a scale of X28 and this event was associated with a halo coronal mass ejections event with a speed exceed of 2657 kms^{-1} and kinetic energy of 6×10^{32}erg [3]

An important method of detecting different types of solar bursts is the observation of radio region signals from ground-based stations, which are generally responsible for accelerating particles from the sun. Solar burst type III seems to be most dominant with a solar flare. Since it was first introduced by Wild in 1963, solar burst type III is said to be covered the frequency range from 500 Hz to 10 MHz [4, 5]. However, after decades of studies on solar burst characteristics, solar burst types III more commonly to be considered ranges between 10 kHz and 1 GHz. Originating from the interplanetary(IP) medium, there are three- sub-type of solar burst type III which are the one from energy system and small-scale energy (isolated type), the second type which occurs during the CME (complex type), and lastly the storms type of burst. Solar burst type III is a fast frequency drift burst and is often considered to be a pre-flare stage where an indication of electrons acceleration [6]. It can occur from one to three seconds for a single burst, one to five minutes for the group burst, and up to hours for the storm-like burst. Moreover, a high percentage of 60% type III burst was found to be time-synchronized with solar flares [7].

Until satellite communication became common, radio communication was commonly conducted in the ionosphere. It has three layers (D, E, and F), with D disappearing when the sun sets unless there is excessive photoionization from solar activities. As the signal bounces off the region, it travels to the other side of the Earth. Due to the region's ability to absorb and bend signals, it can cause a delay in the signal's transmission. Therefore, monitoring the ionosphere response to solar activities especially solar flares have been implemented to understand and expect the side effect to the communication system. There were a Doppler sounding system and Global Positioning System for the ionosphere monitoring, however sudden fade-out has happened due to abnormal condition within ionosphere region. Then, it said that by using the 24 satellites around Earth's atmosphere, GPS dual frequencies makes the best option in real-time monitoring system with higher precision, higher spatial, and temporal resolution data collected [8, 9].

From solar flare event, the intensity of narrow ribbon of current flowing eastward and electron density in the F-region increase and forming the excess ionization within E-layer and it causing the electron density of the ionosphere regions a significant increment and disrupt the signal transmitting [8, 10]. Signal strength will be reduced, and a poor signal will be received as the wave undergoes the absorption of wave signal or called attenuation. The changes in the region will tend to attenuate the wave signal and higher frequency will happen to attenuates more than the lower frequency [11]. One of the communications that using ionospheric wave propagation is high frequency (HF) radio; where has long coverage, high mobility, persistence, and low cost. This communication tends to be disrupted when there is a disturbance in the ionosphere layer [12]. In this study, we analyse the events that occur after the event of X1.59 flare on 3rd July 2021 includes the solar burst, coronal mass ejections, and ionosphere response.

2. Methods

All data for this study was obtained from the National Oceanic and Atmospheric Administration (NOAA) website, the NASA space weather monitor, and space weather live sites. Solar flare data is available on the NOAA websites with a location of the active region, classification of flares, etc. The magnetogram and the SDO/AIA Fe IX at 131Å band are used to observe the solar sunspot at 10 million Kelvin. As part of the Space Weather Monitor, D-region Absorption Prediction (D-Rap) provides archived data of SID events of HF signal analysis with flares. Space Weather Live is used to identify the Sun's conditions related to the ionospheric disturbance (solar wind speed, solar wind density, Kp index, etc.).

3. Result and discussion

The mechanism of solar flares is currently an area of research that got many astrophysicists' attention these days. There are numerous of intense flare eruption occurred on 3rd July 2021. There were four active regions (AR) that appeared on the sun's surface observed by HMI magnetogram (Figure 1) which were AR2835, AR2836, AR 2837, and AR2838. However, unlike the other AR, AR2838 which is located at the right top north hemisphere side (N24W88) found to be the only active region that erupting numerous flare events on this date with the six high-class flare events on Geostationary Operational Environment Satellite (GOES) scale. Three C-class flares, two M-class flares, and one X-class flare were among the six high-class flares. Our discussion focused on the analysis of the X-Class flare, which is said to be the first X-class flare in four years since September 2017. On the right side of figure 1, there is an image of the sun from the Atmospheric Imaging Assembly (AIA) onboard of Solar Dynamic Observatory (SDO) indicating the flare event in the corona.

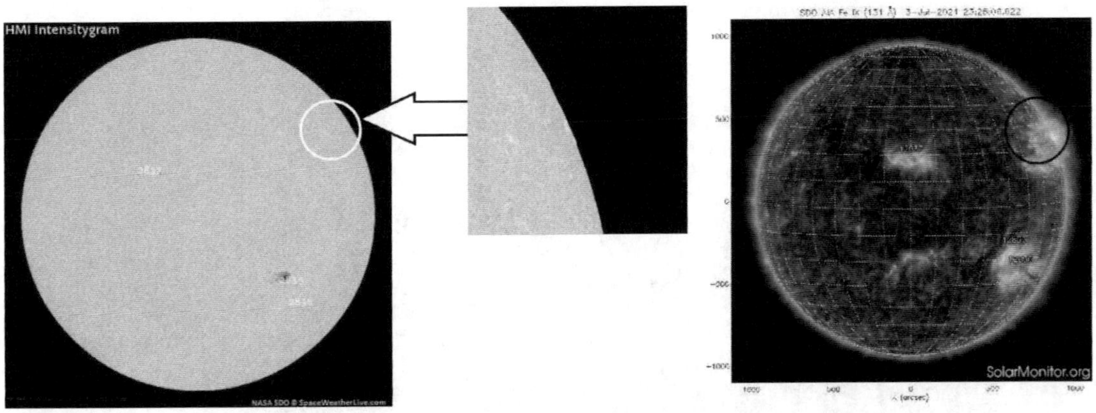

Figure 1. Location of active regions on the day, the AR2838, and the flare events (credited to Solar Dynamic Observation (SDO)).

As observed by GOES Xray Satellite, with a duration of 16 minutes, this X-class flare started at 14:18 UT with $4.81 \times 10^{-7}\ Wm^{-2}$, rose to its peak at 14:29 UT with $1.59 \times 10^{-4}\ Wm^{-2}$. Table 1 is the average value of the parameter of the sun's condition of the day.

Table 1. Parameters of the Sun's condition on 3rd July 2021.

Solar wind		Sunspot Number	Radio Sun (10.7cm Flux)
Speed	Density	72	95 sfu
369.7 km/sec	3.0 proton/cm^3		

Solar wind speed and density appear to be normal. The number of sunspots and the radio sun, however, are significant, increasing flare chances. On the same day, there was a coronal mass ejection event (Figure 2) that was observed on board STEREO as occurring at 105 degrees clockwise from the north and a maximum velocity of 500 km/s (Figure 2).

Figure 2. Coronal mass ejection on the surface of the sun on the same day

In Figure 3, an intermittent burst was detected from Bleien Observatory Switzerland (SWISS-BLEN7M) that can be accessed through eCALLISTO websites. The burst coincides with the flare near the peak time as it is observed with a wide range of radio frequencies from 470MHz to 750MHz. It has a string-like structure that indicates it as the fast drift burst (type III burst). The drift rate of this burst was calculated to be -9.54 MHz/sec which is a fast drift rate.

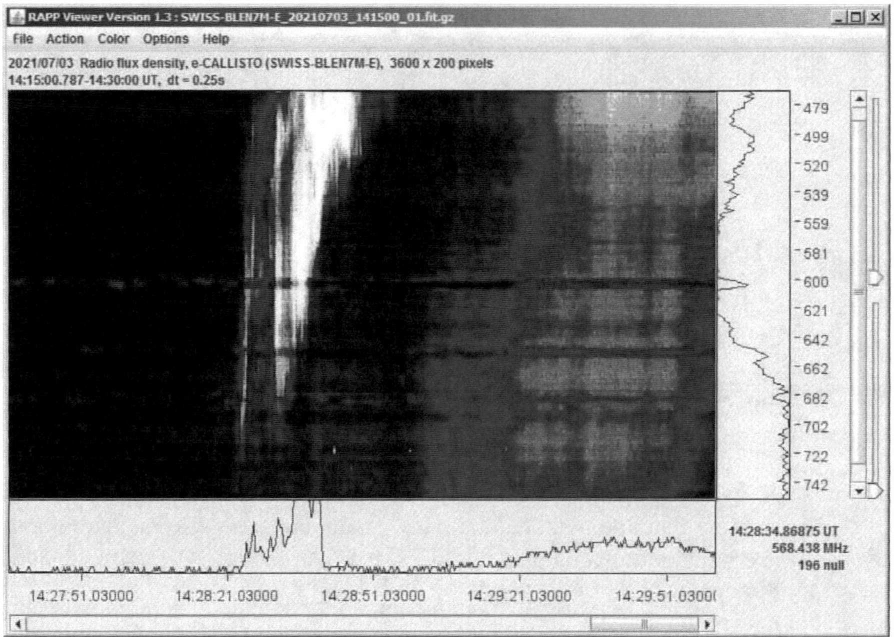

Figure 3. Solar burst type III on the peak time of X1.59 flare.

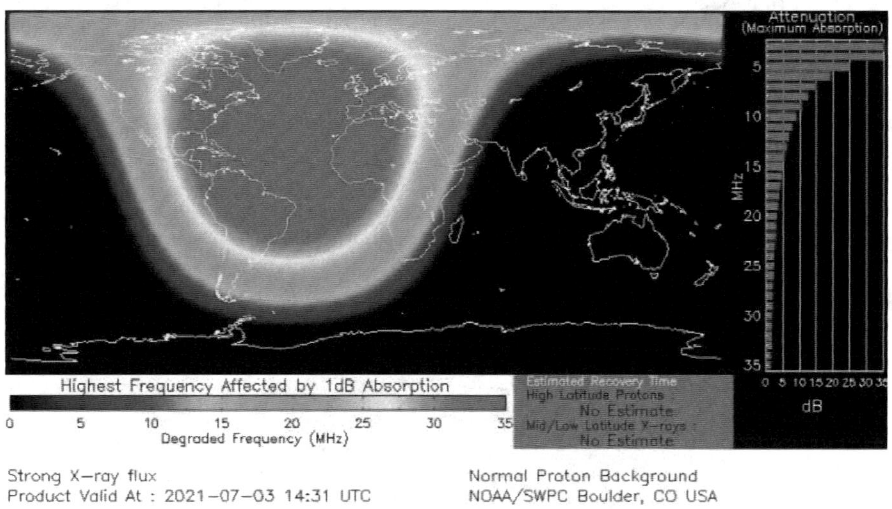

Figure 4. D-region Absorption Prediction (D-Rap) (credited to Space Weather monitor)

There was a shortwave blackout reported at the space weather observatory in Norway on this day. X-ray pulse of the flare believed has been briefly ionized the top part of Earth's atmosphere and gives unusual signals below 30MHz propagations. As a surge of current flow that occurs while the event is in progress, this flare is said to be the one that causing magnetic crochet where the flares produced the radio burst, an ionospheric disturbance, a surge of current in the ground, and the deflection of the observatories' local magnetic field (a rare event). The current flows from 60km to 100-600 km above the Earth's surface due to the ionization and can alter the Earth's polar magnetic field. D-Rap model detected a sudden ionospheric disturbance right after the peak time of the flare at 14:31 UT. The disturbance within the ionospheric region happens when the D-region ionization increases dramatically due to the solar flare event. When the Sun is directly overhead of one point, called a sub-solar point, D-region ionization is at the highest state, and eventually the ionization and absorption

decrease and recovering as it distances away from it. However, the night side of the Earth is unaffected by this excess ionization.
Figure 4 showing that by 1dB absorption, frequencies up to 35 MHz have degraded especially in the red-colored region. At the right side of the figure, there is an Attenuation bar graph that shows the maximum absorption (A_{max}) of frequencies that are attenuated. An example, for a 5MHz signal, the attenuated is high as 25dB. This degraded signal and attenuation event may notice by whom that depends on the high-frequency signal as they faced the unusual propagation signal.

4. Conclusion:
An X-class solar flare event has been observed to have produced a short-lived solar burst type III event. As a result, solar flares can disrupt radio signal transmission and have a significant impact on HF communication like the radio blackout in September 2017. Due to the strong x-ray flux during the solar flare event, high-frequency signals, like those used by aviation, are at risk of attenuation. Based on research done decades ago, solar activity will greatly affect the earth and its surroundings. Hence, it is important to continue studying our sun from time to time in order to understand its dynamic behavior.

References
[1] Heyvaerts J, Priest ER, Rust DM. An emerging flux model for the solar flare phenomenon. The Astrophysical Journal. 1977;216:123-37.
[2] Sharma A, More C. Effect of Solar X-ray Flares on VLF Radio Wave Signal Strength at 19.8 and 24 kHz Received at Khatav (India). Journal of Space Science & Technology. 2017;6(3).
[3] Ciaravella A, Raymond J. The current sheet associated with the 2003 November 4 coronal mass ejection: density, temperature, thickness, and line width. The Astrophysical Journal. 2008;686(2):1372.
[4] Wild JP, Smerd SF, Weiss AA. Solar Bursts. Annual Review of Astronomy and Astrophysics. 1963;1(1):291-366.
[5] Hamidi ZS, Shariff N. The Propagation of an Impulsive Coronal Mass Ejections (CMEs) due to the High Solar Flares and Moreton Waves. International Letters of Chemistry, Physics and Astronomy. 2014;33:118-26.
[6] Dulk GA. Type III solar radio bursts at long wavelengths. Geophysical monograph. 2000;119:115-22.
[7] Swarup G, Stone P, Maxwell A. The Association of Solar Radio Bursts with Flares and Prominences. The Astrophysical Journal. 1960;131:725.
[8] Liu JY. Ionospheric solar flare effects monitored by the ground-based GPS receivers: Theory and observation. Journal of Geophysical Research. 2004;109(A1).
[9] Zhang DH. Study of ionospheric response to the 4B flare on 28 October 2003 using International GPS Service network data. Journal of Geophysical Research. 2005;110(A3).
[10] Liu JY, Lin CH, Chen YI, Lin YC, Fang TW, Chen CH, et al. Solar flare signatures of the ionospheric GPS total electron content. Journal of Geophysical Research. 2006;111(A5).
[11] Hegde S, Bobra MG, Scherrer PH. Classifying Signatures of Sudden Ionospheric Disturbances. Research Notes of the AAS. 2018;2(3).
[12] Uryadov VP, Vybornov FI, Kolchev AA, Vertogradov GG, Sklyarevsky MS, Egoshin IA, et al. Impact of heliogeophysical disturbances on ionospheric HF channels. Advances in Space Research. 2018;61(7):1837-49.

Acknowledgments
We thank the following organizations for providing their data to the public: SDO/AIA, NOAA, SWPC, LASCO, Helioviewer, Solar Monitor, Space Weather Live and E-CALLISTO. The authors also thank the reviewers and editors for their valuable comments and suggestions. We acknowledge the support provided by grants 600-RMI/FRGS 5/3 (309/2019), 600-RMC/YTR/5/3 (001/2020) and Kementerian Pengajian Tinggi Malaysia.

Effects of CME-induced Geomagnetic Storm on Geomagnetic Induced Current at High and Middle Latitudes

N Mohamad Ansor[1, 2], Z S Hamidi[1, 2]

1 Faculty of Applied Sciences, Universiti Teknologi MARA, 40450 Shah Alam, Malaysia.
2 Institute of Science, Universiti Teknologi MARA, 40450 Shah Alam, Malaysia.
E-mail: zetysh@uitm.edu.my

Abstract. Geomagnetic storm has been one of greatest events in space weather studies. Apart from the formation of aurora, geomagnetic induced current (GIC) is also induced during the storm when the storm intensity gets severe. Coronal mass ejection (CME) is a massive eruption created by the Sun and is believed to be the driver of geomagnetic storm. This paper is aimed to examine the effects of CME-induced geomagnetic storm on geomagnetic induced current (GIC) at high and middle latitudes by studying four levels of storm: minor (Kp 5), moderate (Kp 6), strong (Kp 7) and severe (Kp 8). We applied qualitative analysis through descriptive approach to describe the relationship between the storm intensity and GIC activity. Since the exact value of GIC were inaccessible, we adopted horizontal component time derivative magnetic field (dH/dt) as the indicator for GIC activity. At the end of the study, we found that the GIC highly depends on the storm intensity, by which higher levels of storm triggers more actively induced current. Besides, high latitude ionosphere was showing strong reactions to the storm arrival compared to middle latitude ionosphere as the magnitudes of dH/dt recorded by Barrow station at high latitude fluctuated within wider and higher range.

1. Introduction

Coronal Mass Ejection (CME) is an enormous and vigorous explosion of magnetic fields and mass particles coming out from corona. During these powerful eruptions, large number of solar materials are blown out into the interplanetary space in just a few hours, involving $10^{20} - 10^{25}$ J of energy release, travelling at a speed of 400–500 km/s, but can reach up to 2000 km/s [1]. The first 'modern' CME was detected on December 14, 1971, by the white light coronagraph on board NASA's solar observatory [2]. As Earth's magnetosphere encounters the first contact with solar wind plasma, its magnetic field changes with time due to strong shock wave from the CMEs and this disturbance is called solar storm. These storms lead most of space weather processes which may disrupt navigation and communication, long-distance ground transmissions of electric power, oil, and gas [3]. CMEs are believed to be the main driving source of solar storms due to its strong shock wave hitting the Earth. The energy contained by the solar wind is shifted to the magnetospheric current system and is released into the ionosphere [4].

Geomagnetic induced current (GIC) has a strong relationship with geomagnetically induced electric field (GIE) as the GIC travels through electric circuits connecting the power transmission lines with the ground [5]. The Carrington event is marked as the most disastrous event of magnetic storm in history that happened on 1st and 2nd September 1859 [6] that was caused by a strong X class flare on 1st September 1859 and the storm initiated a day after. The studies on GIC have been the centre of interest since a catastrophic power transmission grid failure in Quebec that was caused by an extreme geomagnetic disturbance in March 1989 [7]. The formation of GIC follows the Faraday's law whereby the change in magnetic field causes an induced electric field as explained by Equation (1):

$$\vec{\nabla} \times \vec{E} = -\frac{d\vec{B}}{dt} \tag{1}$$

whereby $\vec{E}$ is induced electric field and $\vec{B}$ is magnetic field.

When the plasma from CMEs approaching the Earth, the magnetized plasma that is in contact with Earth's magnetic field prompts a geoelectric field at the surface and underground [8]. The geoelectric field or in this study we say as GIC gives rise to the destruction of power systems that is brought by the compression of magnetosphere [9]. The estimation of GIC values can be done based on the horizontal component of time derivative magnetic field [10]. According to Faraday's law, the change in magnetic field, specifically in horizontal component (dH/dt) is directly correlated to GIC. As the law represents the pattern of the GIC, the important parameter that should be focused is the dB/dt which is the representativeness of GIC [11]. The power grids would be immensely influenced if the values of dH/dt were greater than 30 nT/min [12]. A sudden impulse (SI) event that manifested prominent fluctuations in magnetic field found that equatorial countries could possibly experience geomagnetic disturbance if dH/dt values go beyond 30 nT/min [13]. Our aim was to analyze the GIC patterns for each storm level (Kp 4, 5, 6, 7 and 8) at high and middle latitudes, and to distinguish the intensity of GIC activity for each latitude. Considering we lack exact GIC values and this study focuses on the overall pattern of the GIC, we adopted dH/dt values to visualize the GIC activity as literature provided good correlations between them.

2. Data acquisition

This study was carried out based on geomagnetic events in 2017 which comprises four level of intensities: Kp 5, Kp 6, Kp 7 and Kp 8. All parameters were obtained and downloaded from free-access databases which are verified by NASA and other space agencies. The 1-min resolution solar wind velocity and IMF Bz parameters measured by ACE/WIND satellites are supplied on OMNIWeb NASA database, despite the presence of some data gaps. Besides, AE index and SYM-H parameters are also considered in this study which can be acquired from the same database. The auroral electrojet (AE) index is a quantitative measurement used to measure the magnetic activity at auroral zone as ionospheric currents circulating in the auroral oval region [14]. Meanwhile, longitudinal symmetric disturbance for horizontal component (SYM-H) is another important geomagnetic index used to measure the intensity of geomagnetic storm. The inclination in SYM-H is due to southward component of IMF Bz which drops below zero value [15]. According to [14], SYM-H must be equal or lower than -50 nT (SYM-H $\leq$ 50) in order to identify an event as a storm [4].

Through a descriptive approach, we investigated the relationship between geomagnetic storm strength and GIC activity at high and middle latitudes by applying qualitative analysis. Values of dH/dt were calculated by substituting values of time derivative magnetic field (dB/dt) in north (X) and east (Y) components obtained from Intermagnet.org into Equation (2).

$$\frac{dH}{dt} = \sqrt{(\frac{dB_x}{dt})^2 + (\frac{dB_y}{dt})^2} \tag{2}$$

where, H is horizontal component magnetic field, B_x is north component magnetic field and B_y is east component magnetic field.

A particular station at all locations was used to study the time derivative magnetic field (dH/dt) of high and middle latitudes during the storm. By measuring dH/dt, we can see how the magnetic field fluctuates at the respective location due to the disturbance. For this study, high latitude is represented by Barrow station and middle latitude is represented by Nurmijarvi station.

3. Results and Discussion

3.1. Minor storm (23rd April 2017)

A minor storm was reported on 23rd April 2017 which commenced at 0600 UT and came to an end at 1800 UT. By referring to geomagnetic and solar wind graphs in Figure 1, all values recorded during minor storm were greater than previous event which explain the higher intensity of the current storm.

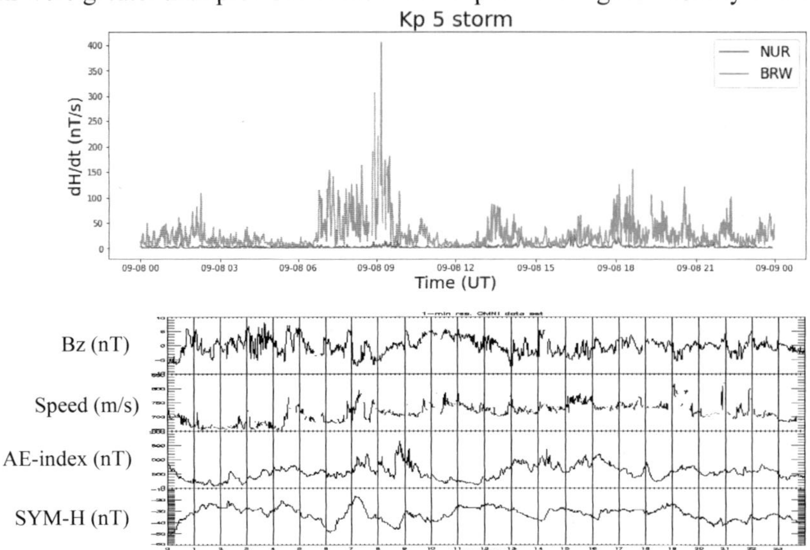

Figure 1. The graph of dH/dt (top), southward IMF (Bz), solar wind speed, AE index and SYM-H

With reference to Bz graph in the first panel, strong and active fluctuations can be observed at the beginning of the day as the graph showed multiple rise and fall phases until it reached the peak point of -7.51 nT at 1259 UT. Besides, the solar wind speed also went through the same phase in the range of 600 km/s to 810 km/s. AE index also recorded significant amplifications with its peak of 1641 nT at 0850 UT as ionospheric currents acquired strength from the contact with Bz. In the intervening period when AE index hit the peak, SYM-H was seen to drop to -46 nT following the third lowest peak of Bz throughout the day. As displayed in the dH/dt graph in Figure 2, the GIC once again was immensely triggered at high latitude while middle latitude detected very weak presence of GIC. There are sudden increases can be observed at 0700 UT onwards with its peak of 405.95 nT/s. The presence of the GIC activity occurred concurrently with the outstanding signals from SYM-H and AE index. However, at middle latitude, dH/dt were recorded at much lower values with similar fluctuations pattern as high latitude data and the peak that Nurmijarvi station obtained was only at 27.88 nT.

3.2. Moderate storm (16th July 2017)

One level higher than the previous event is the moderate storm which was triggered on 16th July 2017 as a consequence from a CME that emerged at 0545 UT. Figure 2 shows dH/dt graph representing the GIC activity, solar wind and geomagnetic data. Based on a rough inspection, the storm was initiated at 0600 UT as all geomagnetic and solar wind data showed sudden fluctuations at the same time when dH/dt graph displayed indications of the presence of GIC activity.

A moderate CME that was believed to be the main cause of this event can be detected from the solar wind data whereby its speed was promptly increased exactly at 0600 UT from 318.5 km/s to 419.1 km/s. Right after the emergence of the CME, all parameters (Bz, AE-index and SYM-H) began to show signals of the disturbance due to the CME. As displayed in Figure 3, Bz slowly dropped to its maximum negative value of -15.49 UT at approximately 4 hours after the launch of CME. This maximum phase of Bz was coinciding with the maximum phase of AE-index as the value has gone beyond 1500 nT. The The moment when the CME was set off at 0600 UT, the dH/dt graph immediately manifested the signs of GIC activity as the magnitudes started to fluctuate until the end of the day. As equivalent to previous storm events, high latitude still gained a stronger disturbance with 225.91 nT/s of its peak and middle latitude's peak at only 47.01 nT/s.

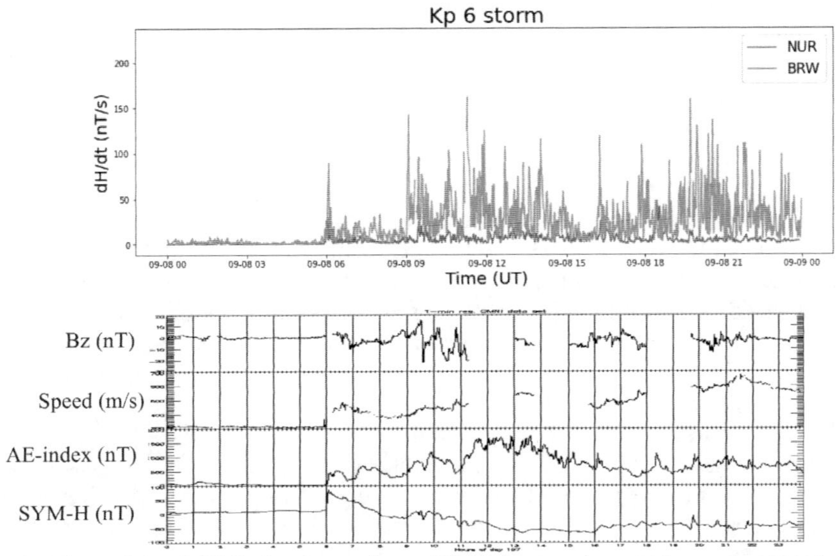

Figure 2. The graph of dH/dt (top), southward IMF (Bz), solar wind speed, AE index and SYM-H

3.3. Strong storm (28th May 2017)

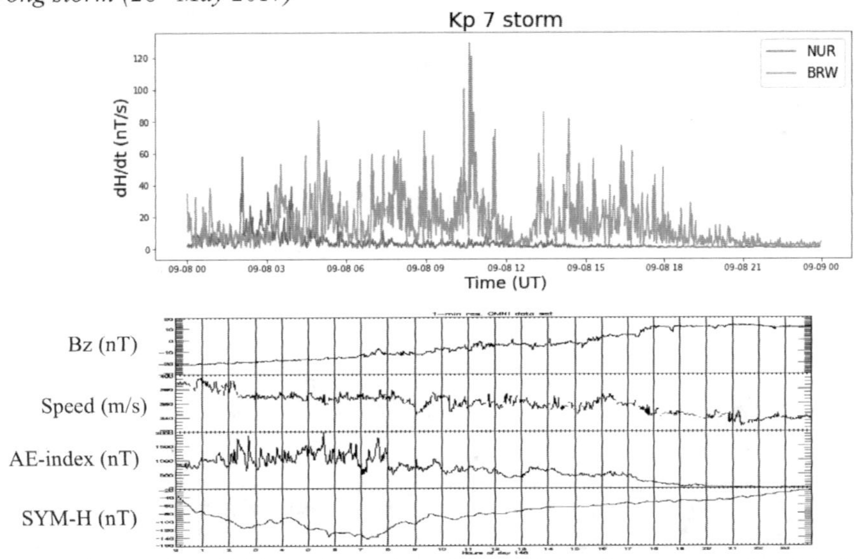

Figure 3. The graph of dH/dt (top), southward IMF (Bz), solar wind speed, AE index and SYM-H

On 28th May 2017, the Earth experienced a strong storm with Kp-index of 7 which can be specifically witnessed at high and middle latitude locations. The CME was travelling at a maximum speed of 629 km/s and it reached into the Earth's atmosphere, inducing the geomagnetic storm. As shown in Figure 3, solar wind speed at the beginning of the day was noticeably active within 0000 UT until 0800 UT, which happened at the same time when southward IMF (Bz) was at its maximum phase with a peak of -21.50 nT. An AE-index recorded vigorous fluctuations when solar wind speed and Bz were at their maximum phases. As solar wind particles making their way into the Earth's magnetic field, the AE-index began to rise until it reached its maximum magnitude at 1960 nT. In the case of SYM-H, it started at -10 nT at 0000 UT and the signal rapidly dropped to lower negative values in the next minute. Over the phase of decrement, the maximum value of SYM-H achieved was -142 nT at 0713 UT and nearly all values were seen to be below -50 nT which demonstrated the condition of strong storm throughout the day. In observing the GIC activity, dH/dt graph gave strong fluctuations at both latitudes in the morning until 1200 UT. For this event, Nurmijarvi station at middle latitude showed quite intense readings which is represented in blue line graph in figure below and the highest point was 60 nT/s. Nevertheless, high latitude still obtained much greater signals with recorded peak of 129.18 nT/s and this peak appeared within the active phase of AE-index.

3.4. Severe storm (8th September 2017)

Severe storm (Kp 8) is the most intense level of geomagnetic storm. During the Kp 8 storm on 8th September 2017, a powerful CME made contact with Earth's atmosphere on 7th September. The event began with a south-directed IMF, and its maximum negative value reached -30.41 nT at 0022 UT. In addition, solar wind speed was seen to undergo major variations in the morning as it went over 700 km/s for several hours. The maximum negative value of Bz took place at the same when AE-index hit its peak of 2337 nT at 0017 UT and simultaneously, SYM-H exhibited a sharp drop to the lowest point of the day which is -146 nT. Within the period, SYM-H had a sudden declination as the reading fell to -115 nT while AE-index was actively evolved. For this level of storm, the GIC was much more significant at both latitudes as dH/dt graphs in Figure 4 display quite many peaks representing the GIC activity. High latitude was seemingly more influenced by the storm with recorded highest dH/dt of 368.27 nT/s, which appears to be the second highest value among all storm events. Meanwhile, Nurmijarvi station at middle latitude obtained 202.59 nT/s as its peak that showed up a couple of hours after Barrow read its peak. As the storm commenced and solar wind speed increased, the southward component of IMF, Bz began to drop to negative values while SYM-H started to decline up to -50 nT which indicated the arrival of the storm. The higher the intensity, the higher the negative values of the SYM-H. At the same time with the down phase of SYM-H, the AE-index also showed multiple inclinations towards higher positive values which implied the intense auroral activities experienced by high and middle latitudes. Therefore, these correlations can be understood as the drop in Bz was due to the increment in solar wind speed.

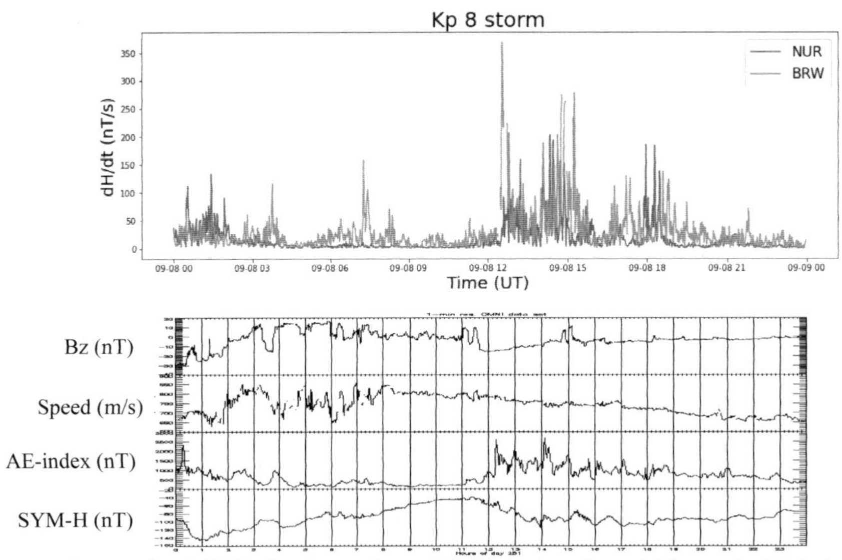

Figure 4. The graph of dH/dt (top), southward IMF (Bz), solar wind speed, AE index and SYM-H

According to all graphs of dH/dt presented in Figure 1 until Figure 4, the GIC activity at high latitude has gone beyond 100 nT/s during all levels of storm. The most intense GIC activity was during Kp 5 storm with the maximum dH/dt hit 405.95 nT/s, followed by the second highest during Kp 8 storm with maximum dH/dt of 368.27 km/s. The magnitudes of dH/dt at middle latitude were much lower compared to high latitude data. Besides, they did not exceed 100 nT/s except for Kp 8 storm that recorded significantly high GIC activity as its peak attained to 202.59 nT/s. For each storm event, the GIC arise phase and duration are not identical as the GIC occurs during the first contact with the southward IMF and the ionospheric disturbance.

4. Conclusion

In all cases, we found a linear relationship between GIC activity and storm intensity. As the storm intensified, the GIC appeared to be greatly provoked. When the geomagnetic parameters deteriorate to higher negative values, the electric currents circulating in the ionosphere intensify as the AE-index climbs and the SYM-H values increase, implying that the storm has intensified. As storm intensity has increased, dH/dt values have risen, indicating a rise in GIC activity and increased storm intensity. It was caused by ionospheric disturbances that altered the magnitudes of magnetic field and, therefore, induced an electric field on the ground and beneath it.

5. References

[1] C. Jacobs and S. Poedts, "Models for coronal mass ejections," *Journal of Atmospheric and Solar-Terrestrial Physics*, vol. 73, no. 10, pp. 1148–1155, 2011, doi: 10.1016/j.jastp.2010.12.002.

[2] R. Tousey, "The solar corona," *Space Res.*, vol. 13, pp. 713–730, 1973.

[3] Q. Bai-han and L. Chuan, "The Study on a Solar Storm and Its Interplanetary and Geomagnetic Effects," *Chinese Astronomy and Astrophysics*, vol. 39, no. 4, pp. 487–496, 2015, doi: 10.1016/j.chinastron.2015.10.006.

[4] W. D. Gonzalez *et al.*, "What is a geomagnetic storm?," *Journal of Geophysical Research*, vol. 99, no. A4, 1994, doi: 10.1029/93JA02867.

[5] Y. Ebihara, S. Watari, and S. Kumar, "Prediction of geomagnetically induced currents (GICs) flowing in Japanese power grid for Carrington-class magnetic storms," *Earth, Planets and Space*, vol. 73, no. 1, p. 163, Dec. 2021, doi: 10.1186/s40623-021-01493-2.

[6] H. Hayakawa *et al.*, "Temporal and Spatial Evolutions of a Large Sunspot Group and Great Auroral Storms Around the Carrington Event in 1859," *Space Weather*, vol. 17, no. 11, pp. 1553–1569, Nov. 2019, doi: 10.1029/2019SW002269.

[7] D. Beamish, T. D. G. Clark, E. Clarke, and A. W. P. Thomson, "Geomagnetically Induced Currents in the UK: Geomagnetic Variations and Surface Electric Fields."

[8] B. Adhikari, N. Sapkota, S. Dahal, B. Bhattarai, K. Khanal, and N. P. Chapagain, "Spectral characteristic of geomagnetically induced current during geomagnetic storms by wavelet techniques," *Journal of Atmospheric and Solar-Terrestrial Physics*, vol. 192, p. 104777, Oct. 2019, doi: 10.1016/j.jastp.2018.01.020.

[9] W. Kai-rang, L. Lian-guang, and L. Yan, "Preliminary Analysis on the Interplanetary Cause of Geomagnetically Induced Current and Its Effect on Power Systems," *Chinese Astronomy and Astrophysics*, vol. 39, no. 1, pp. 78–88, Jan. 2015, doi: 10.1016/j.chinastron.2015.01.003.

[10] W. A. Radasky and J. G. Kappenman, "Impacts of geomagnetic storms on EHV and UHV power grids," in *2010 Asia-Pacific International Symposium on Electromagnetic Compatibility*, 2010, pp. 695–698. doi: 10.1109/APEMC.2010.5475523.

[11] A. Viljanen, H. Nevanlinna, K. Pajunpää, and A. Pulkkinen, "Time derivative of the horizontal geomagnetic field as an activity indicator," *Annales Geophysicae*, vol. 19, no. 9, pp. 1107–1118, Sep. 2001, doi: 10.5194/angeo-19-1107-2001.

[12] F. Eo and N. Beloff, "ESTIMATING GEOMAGNETICALLY INDUCED CURRENTS AT SUBAURORAL AND LOW LATITUDES TO ASSESS THEIR EFFECTS ON POWER SYSTEMS," *SENRA Academic Publishers*, vol. 4, no. 2, pp. 1187–1197, 2010, [Online]. Available: http://swdcwww.kugi.kyoto-

[13] F. A. M. Kasran, M. H. Jusoh, A. Yoshikawa, and Z. Mohd. Radzi, "The Time Derivative of the Horizontal Geomagnetic Field for the Low Latitude MAGDAS Langkawi Station for the Estimation of Geomagnetically Induced Current," in *Space Science and Communication for*

Sustainability, Singapore: Springer Singapore, 2018, pp. 57–71. doi: 10.1007/978-981-10-6574-3_6.

[14] Y. Kamide and G. Rostoker, "What is the physical meaning of the AE index?," *Eos, Transactions American Geophysical Union*, vol. 85, no. 19, p. 188, 2004, doi: 10.1029/2004EO190010.

[15] P. Poudel, S. Simkhada, B. Adhikari, D. Sharma, and J. J. Nakarmi, "Variation of Solar Wind Parameters Along With the Understanding of Energy Dynamics Within the Magnetospheric System During Geomagnetic Disturbances," *Earth and Space Science*, vol. 6, no. 2, pp. 276–293, Feb. 2019, doi: 10.1029/2018EA000495.

Acknowledgement

We would like to thank CALLISTO network; STEREO, LASCO, SDO/AIA, NOAA and SWPC for providing us the free-access online databases. This study was partially supported by the FRGS 2019 grant, 600-IRMI/FRGS 5/3 (309/2019), UiTM grants 600-RMC/YTR/5/3 (001/2020) and Kementerian Pengajian Tinggi Malaysia.

Research Progress Based on *Gaia* DR2 and the Comparison of the *Gaia* Proper Motions to the *Galaxia* Model of the Milky Way

A Ritter[1] and Jiaju Li[2]

1 Laboratory for Space Research, Hong Kong University, Pokfulam Rd, Hong Kong (SAR China)
2 Aidi International School, Louzizhuang No. 7, Chaoyang, Beijing, China
E-mail: azuri.ritter@gmail.com

Abstract. The background of the *Gaia* Data Release 2 (DR2) is introduced and research progress based on Gaia DR2 is reviewed. *Gaia* DR2 coordinates and proper motions are transformed to Galactic coordinates to allow for a comparison of the *Gaia* data to models of the Galaxy. A comparison of the absolute predicted and measured numbers of stars per unit area on the sky as well as the corresponding proper motions in Galactic X, Y, and Z are made. It is found that there is a very big difference between the measurements taken from the Gaia Survey and the predictions made by the *Galaxia* Model of the Galaxy for the proper motions. A further in depth analysis, taking into account the proper motions as a function of distances, is required to fully understand the discovered discrepancies.

1. Introduction

The *Gaia* mission is a space telescope of the European Space Agency (ESA), designed to chart a three-dimensional map of more than one billion stars in our Galaxy, the Milky Way, in the process revealing the composition, formation and evolution of the Galaxy (http://sci.esa.int/gaia/). This massive stellar census will provide the basic observational data to analyze a wide range of important questions related to the origin, structure, and evolutionary history of our Galaxy which has not been possible before. The *Gaia* Data Release 2 [1] contains astrometric measurements (3D positions, radial velocities, proper motions) and key stellar parameters (e.g. spectral type, surface gravity, effective temperature, metallicity) of nearly 1.7 billion stars. The magnitude limit of *Gaia* is $G_{BP} = 25$ mag, which gives a maximum distance to a (bright) star that *Gaia* can measure of 25 kpc. Given that the position of Earth is $\approx$ 8.5 kpc from the Galactic center, and the diameter of the Galaxy is 32.4 kpc, *Gaia* is able to measure at least every bright star in the Galaxy which is not hidden by dust clouds or overcrowding towards the Galactic bulge. During its anticipated lifetime of five years, *Gaia* will observe each of its more than one billion sources about 70 times, resulting in a record of the brightness and position of each source over time, continuously reducing the uncertainties in the measurements.

Galactic models like *Galaxia* [2] or the Besançon Model of the Galaxy [3] are powerful tools to test different scenarios of galaxy formation and evolution, theories of stellar formation and evolution, models of stellar atmospheres as well as dynamical constraints. They are also the key to predictions of future photometric, astrometric, and spectroscopic surveys. Most Galactic models incorporate analytical functions for density distributions, the age/metallicity relation and the Initial Mass Function (IMF) provided for each population of stars the Galaxy consists of. These populations are the thin and thick discs, the bar, the bulge, and the stellar halo with its globular clusters (Fig. 1). More advanced models like *Galaxia* also incorporate N-body models for the known substructures like the Sagittarius stellar stream which may constitute a large fraction of the present halo [4, 5].

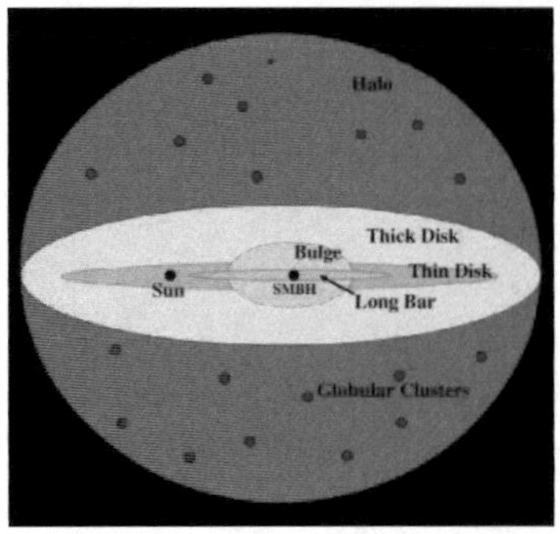

Figure 1. Edge-on schematic of the main components of the Galaxy.

Combining *Gaia* DR2 with the *Galaxia* model of the Galaxy allows for a detailed comparison of the measured data to the predicted parameter distributions. Subtracting the smooth mean model parameters from the mean of the observed values [6] then reveals systematic differences which can be attributed to either new substructures of the Galaxy or real inconsistencies in the model. As an example, Fig. 2 shows the comparison of the mean radial velocity per pixel measured by the RAdial Velocity Experiment (RAVE) Survey [7] to 30 synthetic realisations of the survey predicted by the Besançon Model of the Galaxy. The recently in the RAVE data discovered Stream of Aquarius [8] is marked with a blue circle in the lower left corner.

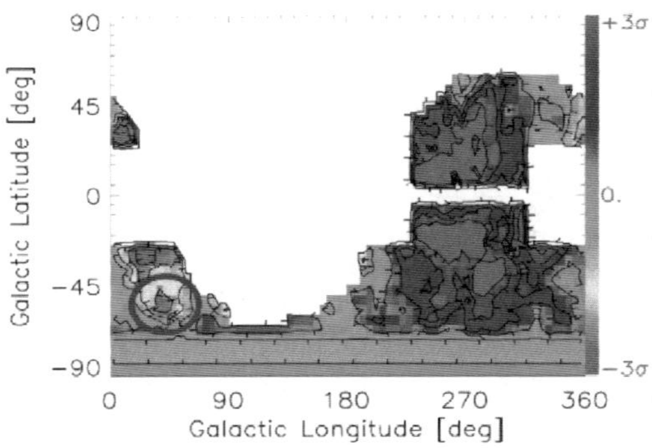

Figure 2. Example of a comparison between model (Besançon Model of the Galaxy) and survey (RAVE).

Shown here is the measured mean radial velocity per pixel compared to 30 realisations of a synthetic survey predicted by the model for giants. The plot has been smoothed once for better visibility. Contour lines show the differences of the mean values in steps of 0.6 standard deviations. Each pixel is 5 degrees wide in l and b, except of the lowest Galactic latitudes, where the swaths get increasingly wider in l due to decreasing numbers of stars. The blue circle in the bottom left marks the position of the recently in the RAVE data discovered Stream of Aquarius.

Gaia provides coordinates and velocities in the ICRS coordinate system. However, for Galactic research it is often desirable to use galactic coordinates instead of ICRS. Also most models of the Galaxy only give Galactic coordinates so in order to compare one data set to the other a coordinate transformation is required. In section 2 we will review the research thas has been based on the *Gaia* DR2. In section 3 we describe in detail the transformation procedure for the *Gaia* coordinates and proper motions to Galactic coordinates. An example application of the transformed *Gaia* proper motions, namely the comparison of the measured parameter distributions for the proper motions to the ones predicted by the *Galaxia* Model of the Galaxy is given in section 4. Shortcomings of this preliminary comparison and future work will be described in section 5.

2. Research progress based on *Gaia* DR2

On the 25th of April 2018, the European Space Agency's *Gaia* satellite task force announced that *Gaia* DR2 contains information on the positions on the sky, parallaxes, proper motions, and brightnesses of more than 1.69 billion stars, three-band photometry of about 1.38 billion stars, and radial velocities of more than 7 million stars.

This unprecendented batch of data not only provides excellent material and opportunity for astronomers studying the Milky Way, but also provides a wealth of information about the Milky Way's star clusters, nearby dwarf galaxies, as well as our neighbouring galaxies, the Large and the Small Magellanic Clouds (LMC and SMC). Based on the *Gaia* DR2, Helmi et al. [9] obtained the proper motions of 75 Galactic globular clusters, 9 dwarf galaxies, 1 ultra-faint system, and the Large and Small Magellanic Clouds. They even derived the rotation curves for 5 globular clusters and the LMC, solely based on proper motions that are now competitive with line-of-sight velocity curves. The unprecedented depth, homogeneity, and precision of the *Gaia* DR2 data in both astrometric and photometric measurements has allowed astronomers to generate the most detailed maps to date.

Using the *Gaia* DR2 sky survey data, Zhang et al. [10] studied the characteristics and structure of the stellar population of the nearby open cluster (OC) Blanco 1. Using unsupervised machine learning they identified a total of 644 members of the cluster. For the first time, the existence of a leading as well as a trailing tidal tail was found in the cluster. The cluster has no obvious mass stratification and is in the early stage of dynamic evolution.

Based on *Gaia* DR2, Yang et al. [11] selected a sample of 24 young (< 3 Myr) pulsars with precise parallax measurements and measured the velocity of their local standard of rest (LSR) and the velocity dispersion among their respective local stellar groups, showing that *Gaia* DR2 is feasible and important to study the velocity of individual systems and the velocity distribution of neutron stars.

Ali & Alharbi [12] investigated the physical and kinematical characteristics of planetary nebulae accompanying PG 1159 stars based on the *Gaia* DR2 parallaxes and proper motions. They found that most of the studied nebulae arise from progenitor stars in the mass range $0.9 - 1.75M_{\odot}$ and tend to live within the Galactic thick disk, moving with an average peculiar velocity of $61.7 \pm 19.2\text{kms}^{-1}$ at a mean vertical height of 469 ± 79 pc. The locations of the PG 1159 stars on the Hertzsprung–Russel diagram indicate that they have an average final stellar mass and evolutionary age of $0.58 \pm 0.08M_{\odot}$ and $25.5 \pm 5.3 \times 10^3$ yr, respectively.

Xu et al. [13] used young massive stars in *Gaia* DR2 to trace the spiral arm structure. Although *Gaia* DR2 can only trace the Perseus, Local, and Sagittarius arms, its data in the fourth quadrant of the Milky Way complement the VLBA's northern results. Lallement et al. [14] built a 3D map of the dust in the Local arm and the surrounding regions using a new hierarchical inversion algorithm. Chen et al. [15] used extinction to calculate the distance and size of the molecular clouds near the Local arm, showing that the molecular gas is extending to the Perseus arm at both ends. They also propose the existence of a spur in the fourth quadrant connecting the Local arm and the Sagittarius arm.

The search for OCs - the fundamental building blocks of the Galaxy - has made new progress with the release of data from the *Gaia* satellite. Stars at different distances can be distinguished by their parallaxes, and the accuracy of the proper motion of stars has also significantly improved. Currently, about 1,500 known clusters have been re-certified in the data. Cluster search and authentication throughout the galactic plane can be performed by many different methods. The most important ones are:

- **Visual examination**

  Using astrometry (proper motion, position, parallax) and photometric measurements from *Gaia* DR2, Sim et al. [16] visually detected 655 (207 of them new) star clusters within 1 kpc of the Sun, taking advantage of the fact that stars in clusters have similar proper motions and are spatially clustered.

- **UPMASK method based on K-means clustering**

  UPMASK is short for Unsupervised Photometric Membership Assignment in Stellar clusters and was developed by Krone-Martins and Moitinho [17]. This method is data driven - apart from the spatial uniformity of field members, UPMASK does not assume a priori parametrisations, such as isochrones or King profiles, of any probability distributions involved. It assumes that cluster members will be clustered in most spaces, including positional space. Field stars, even if clustered in some spaces, are not expected to cluster in positional space. The core principles of this method can be generalized to other types of data. The UPMASK method does not rely on strong physical assumptions about cluster properties, but only requires that cluster members must have common properties and be more closely distributed in the sky than the random distribution between field stars.

- **SHip method based on FoF clustering**
  SHip is short for Star Cluster Hunting Pipeline. In this method developed by Liu & Pang [18], the well established Friend of Friend (FoF) algorithm is first used to cluster stars in a five-dimensional parameter space. The clustering results are then classified and judged by isochrone fitting.
- **Methods based on DBSCAN clustering**
  The density-based clustering algorithm (DBSCAN) is another unsupervised clustering algorithm [19]. This density-based algorithm makes use of the notion of distance between two sources in the data to define a set of nearby points as a cluster; it has the advantage over other methods of being able to find arbitrarily shaped clusters. Castro-Ginard et al. [20] used DBSCAN in combination with a supervised learning method such as an artificial neural network to automatically distinguish between real OCs and statistical clusters.

Already known OCs are easier to re-discover because known clusters usually have more members than new OCs, and the relatively low central star density and fewer member stars in new clusters prevented them from being previously detected. However, not all OCs can be re-certified, and only about half of the known clusters have been identified in *Gaia* DR2 [21]. Combinations of many factors can cause a cluster to be difficult to detect, such as the source density of the background, interstellar extinction, how populated a cluster is, its age, or how its proper motions differ from the field stars.

3. The transformation procedure

For the transformation procedure we followed the *Gaia* Data Release Documentation on the ESA website[1]. The following equations were taken from there: Equation (3.58, here Eq. (1)) to convert Galactic Longitude l and Galactic Latitude b to $[X_{\mathrm{Gal}}, Y_{\mathrm{Gal}}, Z_{\mathrm{Gal}}]$, Equations (3.65, here Eq. (2)) and (3.67, here Eq. (3)) to convert the proper motions to the Galactic Coordinate System, and Equations (3.72) to (3.80) for the propagation of the uncertainties. The program was written in python 3, making extensive use of the *numpy* library.

$$\vec{r}_{\mathrm{Gal}} = \begin{bmatrix} X_{\mathrm{Gal}} \\ Y_{\mathrm{Gal}} \\ Z_{\mathrm{Gal}} \end{bmatrix} = \begin{bmatrix} \cos l \cos b \\ \sin l \cos b \\ \sin b \end{bmatrix} \tag{1}$$

$$\vec{p}_{\mathrm{Gal}} = \begin{bmatrix} -\sin l \\ \cos l \\ 0 \end{bmatrix}, \vec{q}_{\mathrm{Gal}} = \begin{bmatrix} -\cos l \sin b \\ -\sin l \sin b \\ \cos b \end{bmatrix} \tag{2}$$

$$\vec{\mu}_{\mathrm{Gal}} = \vec{p}_{\mathrm{Gal}} \mu_l \cos b + \vec{q}_{\mathrm{Gal}} \mu_b \tag{3}$$

3.1 Comparison of the proper motions in Gaia and Galaxia: Running Galaxia

In order to keep the file sizes manageable we ran the *Galaxia* Model of the Galaxy a total of 650 times for overlapping cones of 157 square degrees, with a V apparent magnitude limit of 21.5 mag. An example parameter file for a run centered on Galactic longitude $l = 175°$ and Galactic latitude $b = 5°$ is shown in Table 1. Using tools provided by the *Galaxia* model and described in http://galaxia.sourceforge.net/Galaxia3pub.html, the absolute magnitudes (in the Johnson–Cousins UBV photometric system) of the synthetic stars were then artificially reddened and converted to apparent magnitudes. The 650 data files were then split up into 40,713 evenly sized pixels in the Hammer projection where each pixel projects the same area on the sky. By doing so we could avoid any projection effects which would have irritated the reader.

---

[1] https://gea.esac.esa.int/archive/documentation/GDR2/Data_processing/chap_cu3ast/sec_cu3ast_intro/ssec_cu3ast_intro_tansforms.html

Example parameter file for a run of the Galaxia model of the Galaxy, centered on Galactic longitude of 175° and Galactic latitude of 5° . For the meaning of each parameter please refer to http://galaxia.sourceforge.net/Galaxia3pub.html.

Table 1. Example parameter file for running *Galaxia*

outputFile	galaxia_175_5
outputDir	/data/galaxia/ubv_Vlt21.5_1.0/175_5
photoSys	UBV
magcolorNames	V,B-V
appMagLimits[0]	-1000
appMagLimits[1]	21.5
absMagLimits[0]	-1000
absMagLimits[1]	1000
colorLimits[0]	-1000
colorLimits[1]	1000
geometryOption	1
longitude	175
latitude	5
surveyArea	157.080000
fSample	1.0
popID	-1
warpFlareOn	1
seed	17
r_max	1000
starType	0
photoError	0

3.2 Comparison procedure

For each pixel in the Hammer projection, we first transformed the *Galaxia* magnitudes from the UBV photometric system to *Gaia* G_{BP} using the transformation described in Ritter & Huang [22]. We then selected all *Gaia* stars in the same pixel with $G_{BP} <= max(G_{BP,Galaxia})$ as the simulated data are not as deep as the *Gaia* data.

We then calculated the mean of the proper motions in X, Y, and Z for the corresponding pixels in the Hammer projection in both data sets and plotted the absolute values as well as the differences as colour–coded maps (Fig.s 6 – 14).

4. Results

The results are visualized in Fig.s 3 to 14. Comparing the absolute numbers of stars in *Galaxia* and in *Gaia* (Fig.s 3 to 5) it is apparent that towards the Galactic plane there are many more stars predicted than measured. This can be explained with the fact that in the Galactic plane stars are too close to each other to be resolved by *Gaia*. In Fig. 5 the Magellanic clouds are easily visible as blue areas in the lower left.

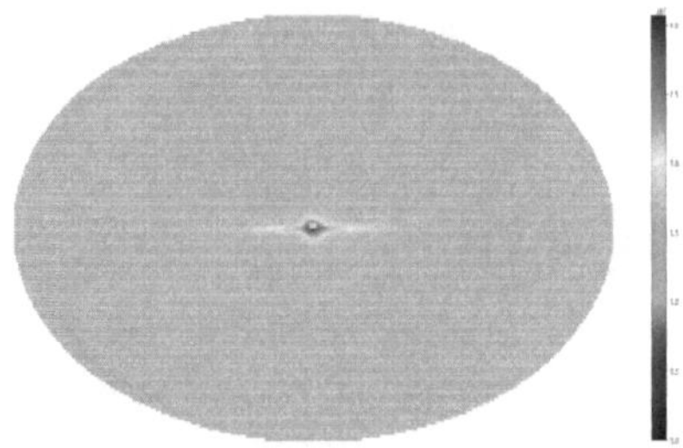

Figure 3. Number of stars predicted by the Galaxia Model of the Galaxy. Note that the missing values in the center are due to a computer glitch.

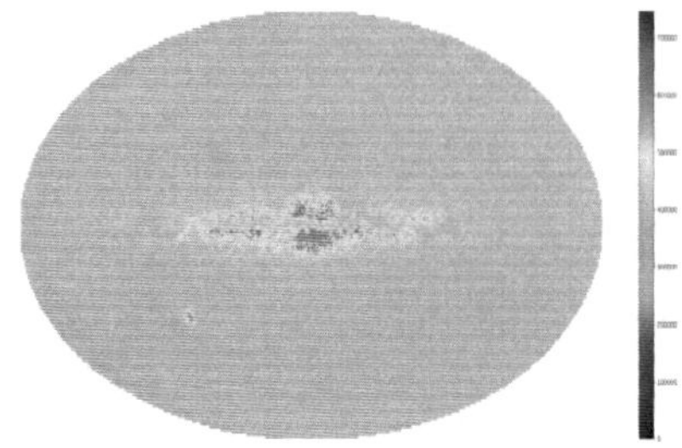

Figure 4. Number of stars measured by Gaia. The Large and Small Magellanic Clouds can be seen in the bottom left.

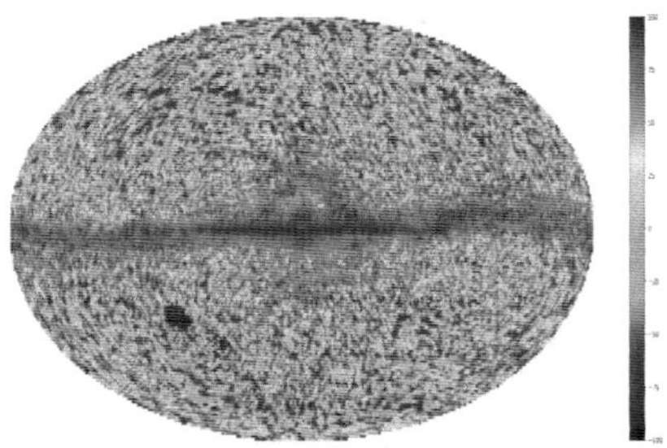

Figure 5. Difference in the number in percent of stars predicted by the Galaxia Model of the Galaxy and measured by Gaia.

Comparing the mean values per pixel of the proper motions in X, Y, and Z we find large systematic discrepancies, which is actually very surprising. Looking at the proper motions in X, both Fig.s 6 and 7 show dipoles, however while in the model the positive values are towards the center of the Galaxy and the negative values are towards the edges of the Galaxy, the measured values by *Gaia* are positive towards the East and negative towards the West. This can also be seen in Fig. 8 where the difference in the proper motions between *Galaxia* and *Gaia* is shown.

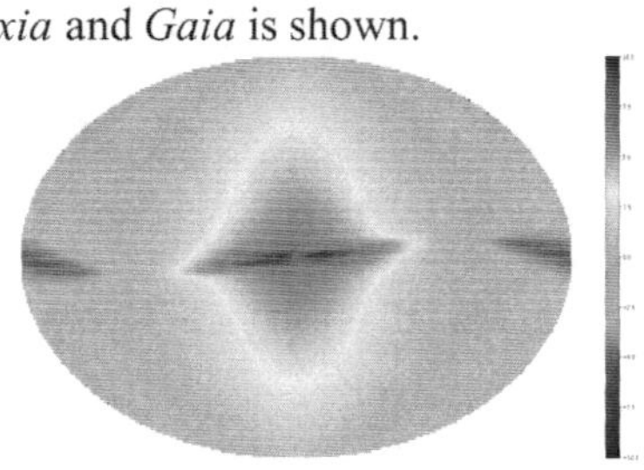

Figure 6. Mean values of the proper motions in X predicted by the Galaxia Model of the Galaxy. Values are in km/s.

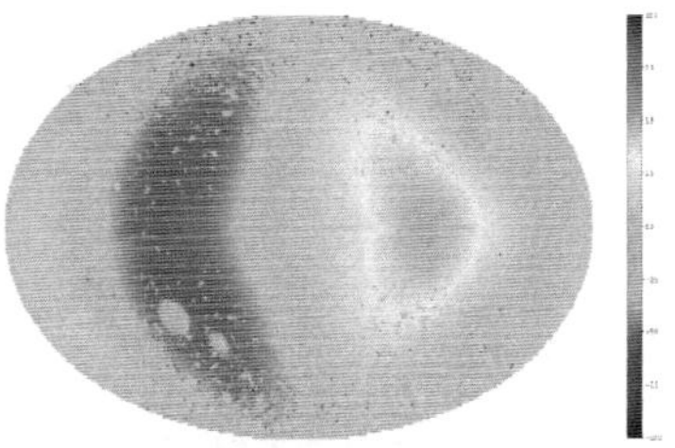

Figure 7. Mean values of the proper motions in X measured by Gaia. Values are in km/s.

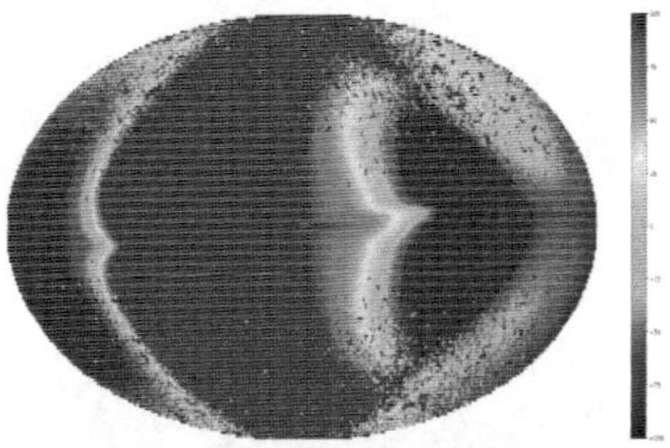

Figure 8. Difference in the proper motions in X predicted by the Galaxia Model of the Galaxy and measured by Gaia. Values are in percent.

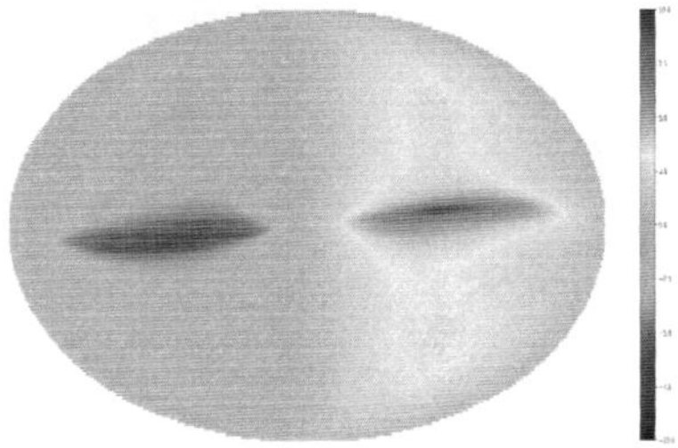

Figure 9. Mean values of the proper motions in Y predicted by the Galaxia Model of the Galaxy. Values are in km/s.

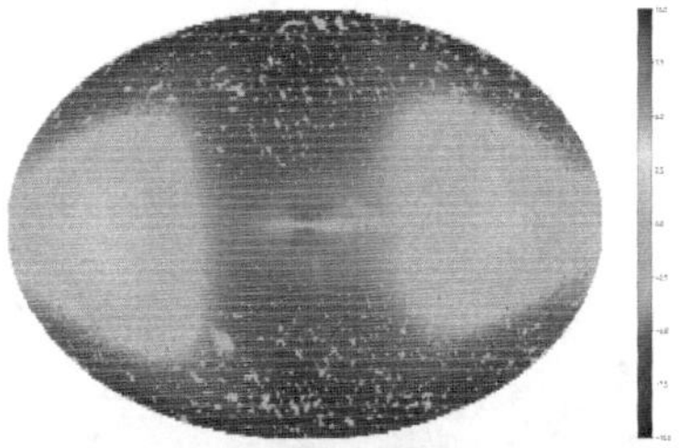

Figure 10. Mean values of the proper motions in Y measured by Gaia. Values are in km/s.

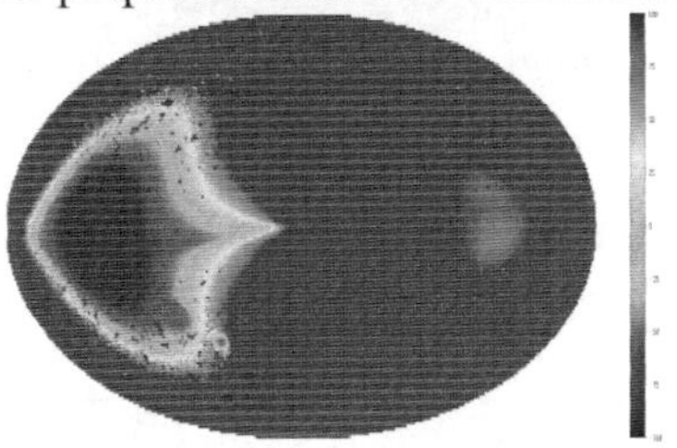

Figure 11. Difference in the proper motions in Y predicted by the Galaxia Model of the Galaxy and measured by Gaia. Values are in percent.

Journal of Physics: Conference Series **2287** (2022) 012036 doi:10.1088/1742-6596/2287/1/012036

For the proper motions in Y (Fig.s 9 to 11) there is a clear dipole in the predicted proper motions with positive values towards the East and negative values with similar amplitude to the West, while the measured proper motions show large negative values everywhere except for the far East and West where the values are slightly positive.
In the Z direction the proper motions predicted by *Galaxia* (Fig. 13) show a clear dipole with positive values towards the North and negative values towards the South. This is in stark contrast to the values measured by *Gaia* (Fig. 12) which show a quadrupole with positive values in the North East and South West and negative values towards the North West and South East. This is also reflected in the difference image shown in Fig. 14.

Figure 12. Mean values of the proper motions in Z predicted by the Galaxia Model of the Galaxy. Values are in km/s.

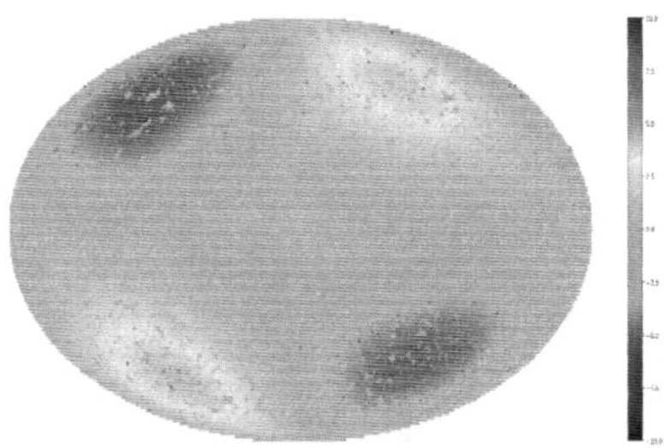

Figure 13. Mean values of the proper motions in Z measured by Gaia. Values are in km/s.

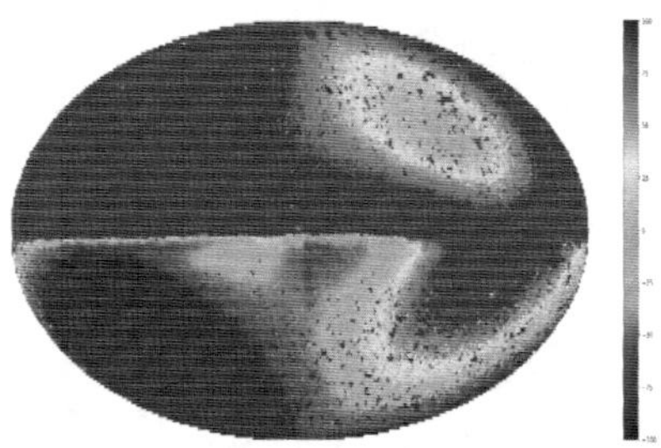

Figure 14. Difference in the proper motions in Z predicted by the Galaxia Model of the Galaxy and measured by Gaia. Values are in percent.

5. Shortcomings and future work

Our preliminary results still only look at the total numbers from all stars predicted in the *Galaxia* model. Given that the uncertainties in the measurements are increasing with the distance to the stars, we should really look at the proper motions as a function of the distances. Another shortcoming of the presented analysis is that we have not convolved the parameters predicted by the model with the uncertainties of the *Gaia* measurements. These shortcomings will be addressed in the subsequent further analysis of the data.

6. References

[1] Gaia Collaboration et al.: 2018, *Astronomy and Astrophysics* **616**, 1
[2] Sharma S., Bland-Hawthorne J., Johnston K. V. and Binney J.: 2011, *Astrophysical Journal* **730**, 3
[3] Robin, A. C., Reyle, C., Derriere, S., and Picaud, S.: 2003, *Astronomy and Astrophysics* **409**, 523
[4] Ibata R. A., Gilmore G., Irwin M. J.: 1995, *Monthly Notices of the Royal Astronomical Society* **277**, 781
[5] Chou M., Cunha K., Majewski S. R., Smith V. V., Patterson R. J., Martinez-Delgado, D., & Geisler, D.: 2010, *Astrophysical Journal* **708**, 1290

[6] Ritter A.: 2012, PhD Thesis: Towards a better model of the Galaxy - Comparison of the RAVE spectroscopic survey to the Besancon Model of the Galaxy
[7] Steinmetz, M.: 2003, *ASP Conference Series* **298**, 381
[8] Williams M. E. K., et al.: 2011, *Astrophysical Journal* **728**, 102

[9] Helmi A., van Leeuwen F., McMillan P. J., Massari D., et al.: 2018, *Astronomy and Astrophysics* **616**, A12
[10] Zhang Y., Tang S.-Y., Chen W. P., Pang X., Liu J. Z.: 2020, *Astrophysical Journal* **889**, 99
[11] Yang, M., Dai S. Li D., Tsai C.-W., et al.: 2021, *Research in Astronomy and Astrophysics* **21**,141
[12] Ali A. and Alharbi W. R.: 2021, *Research in Astronomy and Astrophysics* **21**, 151
[13] Xu Y., Bian S. B., Reid M. J., et al.: 2018, *Astronomy and Astrophysics* **616**, L15
[14] Lallement R., Babusiaux C., Vergely J., Katz D., et al.: 2019, *Astronomy and Astrophysics* **625**, A135
[15] Chen B.-Q., Li G.-X., Yuan H.-B., Huang Y., et al.: 2020, *Monthly Notices of the Royal Astronomical Society* **493**, 351
[16] Sim G., Lee S. H., Ann H. B., and Kim S.: 2019, *Journal of the Korean Astronomical Society* **52**, 145
[17] Krone-Martins A. and Moitinho A.: 2014, *Astronomy and Astrophysics* **561**, A57
[18] Liu L. and Pang X.: 2019, *ApJS* **245**, 32
[19] Ester M., Kriegel H.-P., Sander J., Xu X.: 1996, KDD'96: Proceedings of the Second
[20] Castro-Ginard A., Jordi C., Luri X., Julbe F., et al.: 2018 *Astronomy and Astrophysics* **618**, A59
[21] Cantat-Gaudin T., Jordi C., Vallenari A., Bragaglia A., et al.: 2018, *Astronomy and Astrophysics* **618**, A93
[22] Ritter A. and Huang C.: *Transformations From Standard Photometric Systems To The Gaia Passbands, 2020 Journal of Physics: Conference Series* **1593**, 012039

Acknowledgements

This work has made use of data from the European Space Agency (ESA) mission *Gaia* (https://www.cosmos.esa.int/gaia), processed by the *Gaia* Data Processing and Analysis Consortium (DPAC, https://www.cosmos.esa.int/web/gaia/dpac/consortium). Funding for the DPAC has been provided by national institutions, in particular the institutions participating in the *Gaia* Multilateral Agreement.

A New Method to Reduce Near-Infrared Data of Palomar Observatory

Fred Shao

Pickering College, Toronto, Canada
E-mail: fred.shao88@gmail.com

Abstract. Compared to large optical spectral surveys, near infrared spectra is overwhelmingly outnumbered. Sloan Digital Sky Survey (SDSS), the largest optical spectroscopic survey, has published about 5,789,200 optical spectra in its latest data release DR16, while no large scale near infrared spectra survey has been performed. Typical studies in near infrared are based on limited samples of a few tens to hundreds of near infrared spectra. TripleSpec is one of the workhorses near Infrared (NIR) spectrograph, mounted on the Palomar 200-inch telescope (P200). We have acquired more than 200 new quasar NIR spectra during the past years, compared to the largest sample ever compiled. During the reduction of our latest TripleSpec data, we found that the well accepted data reduction pipeline Spextool (citation: 835) completely ignored one of the five spectral orders (Order 7) of the raw data, probably due to its low throughput. Astronomical observations are always photons starving, we investigate the possibility of incorporation of the photons on Order 7 by developing a new data reduction pipeline. We test the new pipeline on our new quasar NIR data. Compared to the results from Spextool, 1. The pipeline is also featured by its full automaticity, 2. the wavelength coverage extended from 1-2.5 μm to 0.9-2.5 μm, revealing new features such as rest frame Pa ε, [S III] 9531, [C I] 9824, [C I] 9850, [S VIII] 9910, as well as redshifted spectral features for higher redshift targets. 3. The median signal to noise ratio for 0.9-1.1 μm have been elevated by a factor of 2. The pipeline could be easily generalized to accommodate other popular NIR instruments such as IRTF/SPEX, we plan to publish the pipeline and the 200 new quasar NIR spectra to the community. Keck/NIRSPEC.

1. Introduction

We see a totally different sky in infrared than our more-familiar optical band, as well as the distinct physics behind the light (c.f. Orion in Figure 1). In 1800, William Herschel found an imperceptible type of radiation just past the red part of the apparent range. He named this type of radiation infrared ("underneath" red). (Herschel was at that point world-renowned for having found the planet Uranus.) Herschel's revelation was the initial phase in setting up the presence of what we presently call the electromagnetic range. Noticeable light and infrared radiation are only two of the many sorts of electromagnetic energy delivered by objects on Earth and all through the universe. Exclusively by concentrating on this load of sorts of radiation can we completely portray heavenly items and gain a total image of the universe, its set of experiences, and its advancement. In this work, we center around a particular part of infrared space science: date decrease methods for close to infrared spectrograph. We improve a high-citation data reduction pipeline and test our new technique on a large quasar sample observed by using the classic, famous, and still state-of-the-art Palomar 200-inch telescope [1].

1.1. Ground based infrared instruments and NIR spectroscopic surveys of quasars

Only a few NIR spectra samples for AGNs (Active Galactic Nucleus) and Quasars have been published, most of them are with limited sample sizes with a few tens of spectra. Glikman et al. (2006) published an NIR template for AGNs, made from observations of 27 quasars in the redshift range $0.118 < z < 0.418$ [2]. In Riffel et al. (2006), they present a close infrared ghastly chart book of 47 dynamic galactic cores (AGN) of all levels of action in the frequency time frame 2.4 μm, including the transitions of the noticed outflow lines [3]. The sample size is extremely small compared to optical spectral survey such as SDSS, which comprised of 0.75 million quasar optical spectra. A larger sample of NIR spectra of quasars and AGNs are still needed.

Studies based on 1–2.5 μm NIR quasar spectra fall generally into two categories. 1) For low redshift targets, rest frame NIR features are investigated. Continuum shapes, emission, and ab- sorption features are accessed through NIR composite spectrum or individual spectral atlas from small samples with a few tens of targets [3]. The optical-NIR continuum shape could be de- scribed by a broken power law or a single power law plus a blackbody representing hot dust at T~1200 K. Prominent emission features include H I lines Paαλ18756, Paβλ12822, Paγλ10941, Paδλ 10049, He I λ10830, [S III] λ9531, [S III] λ9069, as well as weaker lines from Fe II and O I. Absorption features of He I* λ10830 are seen in some quasars and proposed to be prevalent in at least some sub-class of quasars such as LoBAL [4,5]. 2) For higher redshift quasars, NIR spectra provide coverage for the rest-frame Hα and Hβ line regions, allowing robust black hole mass measurements of broad Balmer lines. A blend of the NIR and optical spectra covers head quasar analytic elements, essentially the C IV λ1549, Mg II λλ2798, 2803, Hβλ4861, and [O III] λλ4959, 5007 emanation lines. These samples have been used to correct black hole mass measurement based on UV broad emission lines such as C IV and Mg II [4,6]. Robust black hole mass measurements also enable studies probing for any difference in masses, Eddington ratios, or rest-frame optical spectroscopic properties of quasar subclasses such as LoBALs as compared to normal quasars, or their evolution over cosmic time. Getting close infrared spectra of satisfactory goal and sign to-commotion proportion (S/N) of even tolerably brilliant quasars remains asset escalated. Generally, these studies utilize NIR samples with limited sizes of a few tens to a few hundreds. The sample sizes are extremely small as compared to optical surveys such as Sloan Digital Sky Survey, with 750,414 optical quasar spectra in its latest data release DR16. Glikman et al. (2006) construct NIR composite from 27 quasars obtained at the NASA IRTF telescope [2]. Riffel et al. (2006) present a rest frame near-infrared spectral atlas of 47 active galactic nuclei (AGN) with $0.0038 < z < 0.549$, including QSO, Seyfert1, Seyfert2 and NLS1 [3]. Zuo et al. (2016) use NIR spectra of 32 luminous quasars with $3.2 < z < 3.9$ to calibrate black hole measurements based on Mg II. Coatman et al. (2017) compiled a sample from heterogenous sources that span wide ranges of source-selection criteria, instrument properties, spectral band and resolution, and signal-to-noise ratio (S/N) [4]. The sample constitutes 230 high-luminosity, redshift $1.5 < z < 4.0$ quasars with both C IV and Balmer line spectra, which is used to correct C IV-Based Viral Black Hole Masses. Matthews et al. (2020) present NIR spectra for a flux-limited sample of 226 quasars with $1.5 < z < 3.5$ from the Gemini Near Infrared Spectrograph - Distant Quasar Survey (GNIRS-DQS), being the largest uniform, homogeneous survey of its kind.

1.2. Aim of this work

We have initiated a near infrared spectroscopic campaign on quasar candidates with He I* absorption line systems. The quasars are selected from SDSS with the presence of optical He I* λ3889 absorption line. The NIR spectroscopic follow-up provide coverage of one important line of He I* multiples, i.e., the He I* λ10830 NIR absorption line [5]. We used Triple Spec spectrograph mounted on the Palomar 200-inch telescope. We are approved 12 nights in the past years, and more than 200 new quasar NIR spectra have been acquired, approx. at the same order of the largest NIR sample ever published.

During the reduction of our latest Triple Spec data acquired on Oct. 2020, we found that the well accepted data reduction pipeline Spextool completely ignored one of the five spectral orders (Order 7) of Triple Spec, probably due to its low throughput. However, as we know that astronomical observations are always photons starving, there is no reason to throw away any photon in priori. In this paper, we investigate the possibility of incorporation of the photons on Order 7 by developing a new pipeline. The paper is organized as follow. In Section 2, we outline the basis procedures of our data reduction pipeline.

In Section 3, the pipeline is applied to Palomar/Triple Spec and compared with result from Spextools. In Section 4, a Triple Spec NIR spectra atlas is compiled and generalization of our pipeline to other NIR instruments is discussed. The paper is organized as follows. In Section 2, we outline the basis procedures of our data reduction pipeline. In Section 3, the pipeline is applied to Palomar/Triple Spec and compared with result from Spextools. In Section 4, a Triple Spec NIR spectra atlas is compiled and generalization of our pipeline to other NIR instruments is discussed.

2. Observation

2.1. TripleSpec NIR spectrograph

TripleSpec is one of the work-horse instruments mounted on Cassegrain focus on the 200-inch Hale Telescope (P200) at Palomar Observatory. The P200/Triplespec is one of the three identical NIR spectrographs built for P200 5.1 m telescope, the Apache Point Observatory 3.5 m telescope, and the Keck 10 m telescope. The entrance slit P200/TripleSpec is 1" × 30". The light is cross-dispersed on 5 orders (order 7, 6, 5, 4, 3) with claimed wavelength range from 1 to 2.45 μm and spectral resolution of 2500–2700 using a 1024×2048 Hawaii-II HgCdTe array. The pixel scale is 0.37"/pixel. Information on the spectrograph design is detailed in Wilson et al. [7]. The spectrograph is also equipped with a K band guider of 4′ × 4′. The throughputs of J, H, K band are measured to be 10%, 20%, 30%, the performance could be found in Herter et al. [8].

Order	Wavelength Coverage (μm)
Order 7	0.80–1.06
Order 6	0.94–1.24
Order 5	1.13–1.48
Order 4	1.41–1.85
Order 3	1.88–2.46

Figure 1. TripleSpec orders and their wavelength coverages.

A typical exposure of TripleSpec is shown in Figure 2 with an exposure time of 100 s. The target is a famous local Active Galactic Neuclei (AGN) NGC 4151, with the order numbers marked. Even with this limited exposure time, we can see the high background for K band (order 3) and vertical OH sky lines, which require dithering the targets during observation (see section 3 for details). The spectrum of the targets is the nearly horizontal bright stripes.

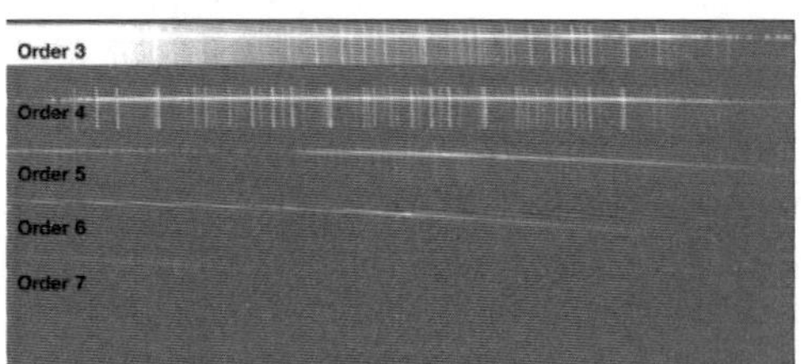

Figure 2. A typical 100 s exposure for NGC 4151, taken on 20130222 by authors using P200/TripleSpec.

2.2. Remote observation

We have carried out a large observational campaign spanning from 2013 to 2021, and still going to. The project is to investigate special absorption line systems in quasars. Our historic observations have accumulated more than 10 nights of TripleSpec data. During our development of the pipeline, we were approved one additional night on Oct. 27, 2020.

The observation log, including targets (with exposure time), seeing, weather, telescopes focus etc., are recorded on standard Triple Spec log sheet as shown in Figure 3. The raw data are downloaded from user1@observer1.palomar.caltech.edu with scp or sftp.

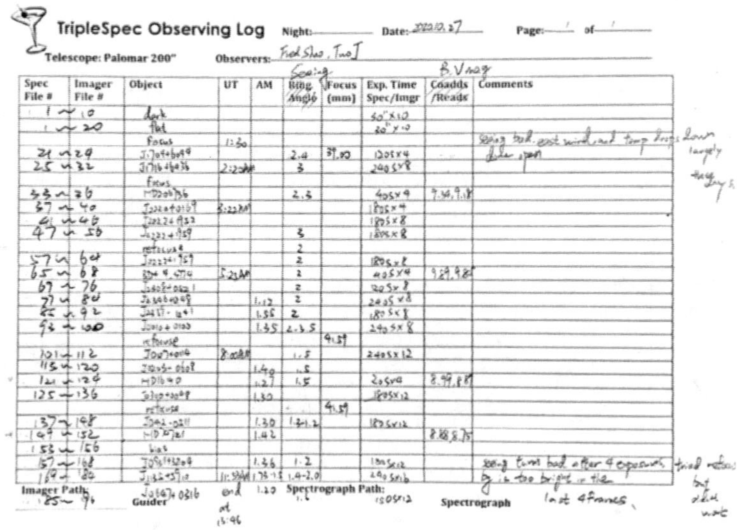
TripleSpec Observing Log Night: Date: 2020.10.27 Page: 1 of 1
Telescope: Palomar 200" Observers:

Spec File #	Imager File #	Object	UT	AM	Seeing Angle	Focus (mm)	Exp. Time Spec/Imgr	Coadds /Reads	Comments

Imager Path: Guider Spectrograph Path: Spectrograph

Figure 3. Logsheet recorded during our Oct. 2020 run. We recorded our observation time, target, focus, seeing, exposure time, and magnitudes. This is the observation the author performed, all other data in the later sections are retrieved from our previous runs taken by collaborators.

3. What's the problem?

The raw data are previously reduced with Spextool [9,10], an Interactive Data Language (IDL)-based data reduction package. Spextool is originally designed for Spex on the NASA Infrared Telescope Facility (IRTF) and subsequently modified to accommodate other NIR spectrographs such as TripleSpec, yet with the limitation (Spextool User's Manual, V4.1) of no optimal extraction [11]. During our recent usage of P200/TripleSpec, we noticed that Spextool ignores order 7 of P200/TripleSpec during processing. In the wavelength map generated by Spextools, there is no coverage of Order 7 (see Figure 4)

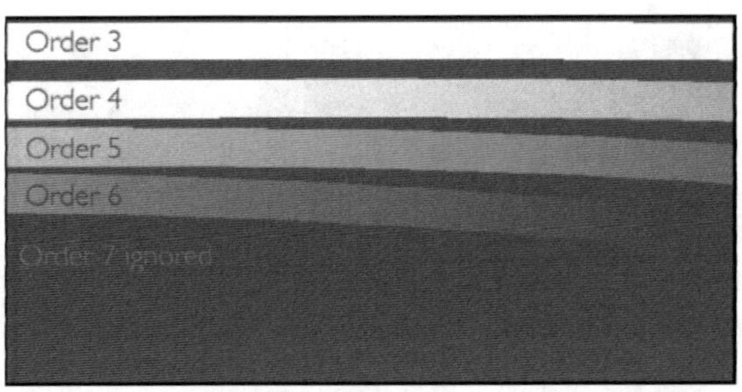

Figure 4. The wavelength map generated by Spextool, i.e., the two-dimensional wavelength as a function of pixel positions on CCD. Please note that the Order 7 is totally ignored.

Although the wavelength coverage of Order 7 (0.8–1.06 μm) is overlapping in large part with Order 6 (0.94–1.24 μm) and the throughput of Order 7 is quite low [8], inclusion of it in data reduction will potentially extend the wavelength coverage (especially for bright targets) and improve the signal noise ratio in the overlapping regions of Orders 6 and 7. The throughput of the five orders is estimated from observation of standard stars. The electron counts per resolution element Δλ are extracted as S(λ) (in e−s−1). The flux intensity I(λ) (in ergs s−1 cm−2 A°−1) is estimated as $I(\lambda) = \frac{S(\lambda)\hbar\nu}{A_T t\Delta\lambda} = \frac{S(\lambda)\hbar\nu R}{A_T t\lambda}$, where R = λ/Δλ = 2700 is the spectral resolution, AT=17.13m^2 is the effective telescope area of P200, t is exposure time and λ is in A°. The throughput is estimated as I(λ)/Ivega(λ), where Ivega(λ) is a model spectrum of Vega [9], scaled to the V magnitude of the telluric star. The FWHM of the telluric star trace is measured and used to correct slit loss as (FWHM/1″)2 with typical values of 2-3. One estimation is shown in Figure 5. The throughput is estimated from HD 77577 observed on 2013-02-22 and includes both the telluric absorption and instrumental response. Depending on different observational conditions such as airmass, seeing, weather etc., the peak throughputs in K band vary from 10% to 30% during our runs.

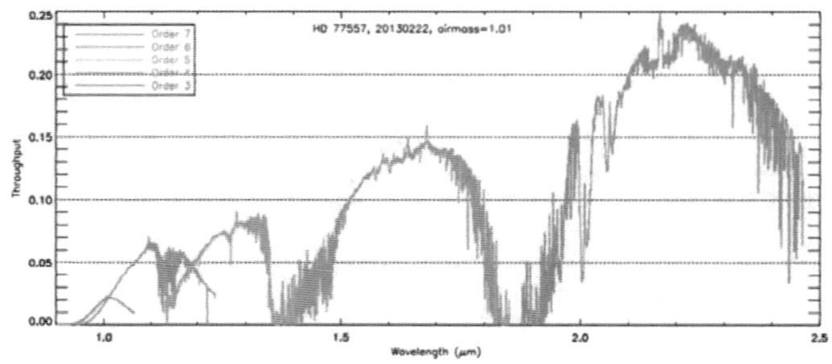

Figure 5. Overall throughput estimated from a standard star HD 77557. The throughput includes telescope, instrumental and atmospheric effects. Note Order 7 contributes significantly in the overlapping part of 0.94–1.06 μm with Order 7.

Astronomical Observations are always photons starving. The throughput of Order 7 is low but not neglectable compared to Order 6, in fact it is clear from the throughput ratio that it is several times higher than Order 6 in 0.94–1.01 μm (Figure 6). We explored the possibility by implementing a new pipeline with IDL. Part of the low-level codes are from the relevant codes distributed with public XIDL package maintained by Prof. Prochaska from UCSC (http://ucolick.org/~xavier/IDL/).

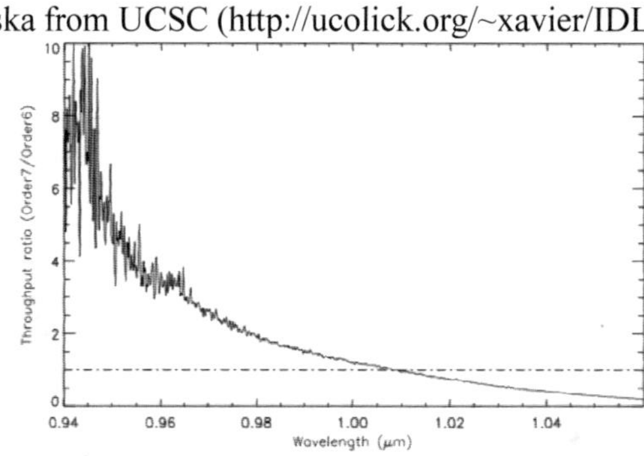

Figure 6. Throughput ratio in the overlapping portion of Order 7 and Order 6. The dot-dashed line indicates the unity ratio. Below 1.01 μm, the throughput of Order 7 is significantly higher than Order 6.

4. Key algorithms of our new pipeline

Our pipeline is similar in spirit to the data reduction pipeline designed for the Magellan Inamori Kyocera Echelle (MIKE) spectrometer and then generalized to the HIRES, UVES, and ESI spectrometers. A comprehensive knowledge of the design, implementation, trade-offs, and limitations of the pipeline is detailed in Bernstein et al. [12]. Our TripleSpec pipeline follows their design with inclusion of the ignored Order 7 by Spextools. Basically, the pipeline performs the standard steps of astronomical spectroscopic data reduction, i.e., the bias subtraction, flat correction, wavelength calibration, telluric correction, and flux calibration automatically. We only highlight our innovative treatments specific to TripleSpec here.

We define X axis as along the 2048 columns, which is the approximate spectral direction with bending orders. The Y axis is along the 1024 rows, i.e., the spatial direction.

4.1. Image processing

First, a super bias frame is generated by averaging the dark frames with 3-sigma clipping. The super bias is subtracted off from each frame in further processing. A flat is selected to trace the order edges and is referred to as a trace flat. A saw tooth filter is applied to the trace flat in the spatial (Y) direction to highlight the slit edges. The edges are then traced and fitted with 4-order Legendre polynomials. The tracing technique is like object tracing and will be detailed in Section 4.3. The averages of the left and right edges will be later used as the initial guess for object tracing. An example of the traced flat edges is shown in Figure 7.

Bias-subtracted flats are averaged with 2-sigma clipping into a super flat, which is then decomposed into an illumination flat and a pixel flat. The average illumination function across the order is fitted.

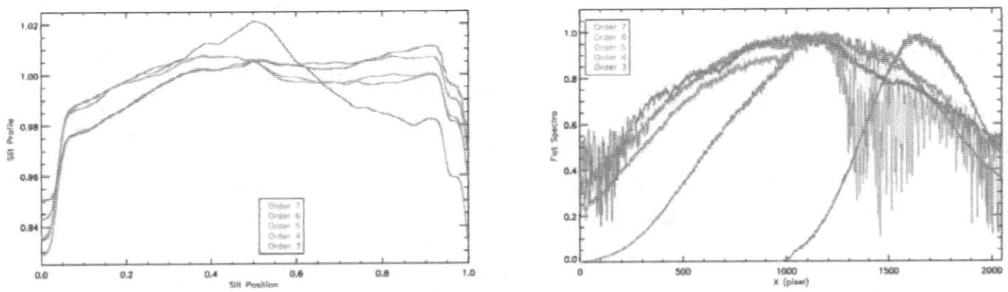

Figure 7. Left: Average Spatial Profile for each order of the flat. The profiles are used to correct Order illumination. Right: Average Spectrum for each order of the flat, used to remove spectral dependence. The two profiles are divided out to generate a pixel flat.

4.2. Wavelength calibration

The sky lines are not strictly aligned to the CCD Y axis. The tilt is corrected using a fitted pixel map. The sky lines are traced for each order of a science frame. Normally a long exposure of 5 minute is used to guarantee enough strong sky lines on Order 7. Examples of Traced lines are shown in Figure 8.

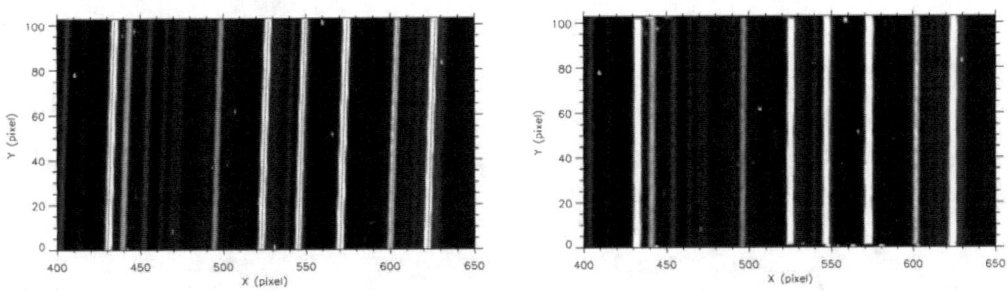

Figure 8. Left: Image cut-out of Order 3 of frame 0189 from Our 2020 run, a five-minute exposure. The curved orders have been rectified to a rectangle image using previously traced order edges. Peaks of sky lines are traced and shown in blue. The lines are not aligned with.

The pixels (Xij, Yij) on the i-th sky line are with the same wavelength, where j is row index. The column index of the center pixel of the i-th skyline Xi is chosen to represent the wavelength. Then we fit traced lines with Xi = L (Xij , Yij) , where L(Xij , Yij) is a 2D Legendre polynomial surface with Order 4 and 3 in X and Y directions. The fitting is evaluated at every pixel and the resulting 2D image is referred to as a pixel map. The pixel map gives the wavelength (recorded in pixels) of each pixel in the original image plane.

A 1-D sky spectrum is extracted along the center of each order and correlated with archived reference sky spectrum with previous wavelength calibration fitting. The shift is calculated, and a new wavelength solution is acquired along the order center. With this 1D wavelength solution and the pixel map, a 2D wavelength solution is calculated and shown in Figure 9. The flux below 0.89 is normally too low and is trimmed in further analysis.

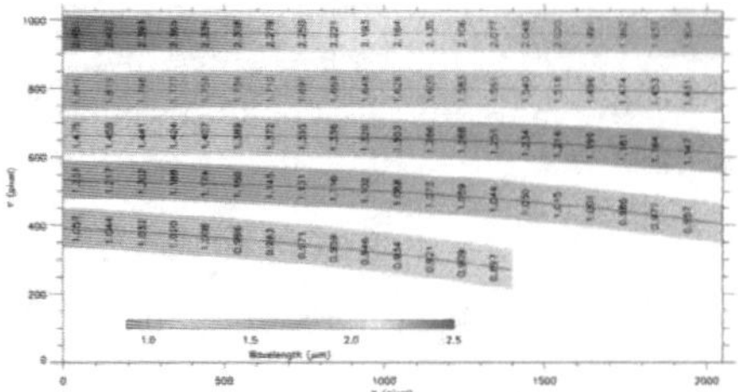

Figure 9. Calibrated 2-D wavelength map. The black lines are averages of traced order edges, indicating order centers. Wavelengths along the order centers are marked. Notice the incorporation of Order 7 (the lowest order) as compared to Spextool result in Figure 12.

4.3. *Traces*

An initial trace of the science target was obtained using IDL code trace crude, then the traces are iteratively refined by weighting the trace positions the target spatial flux distribution. The traces of objects are examined using the data on 2013-02-22. 23 objects were observed. The traces at Position A of Order 3 were plotted in the left panel of Figure 10. The dispersion about the median trace is shown in the right panel of Figure 6 with $\sigma = 1.5$ pixels. The maximum drift is about ± 4 pixels, equivalent to 1.5" with the plate scale of 0.37"/pixel. The traces show that the P200 pointing is extremely stable through the night, we can use one arbitrary trace as the initial guess.

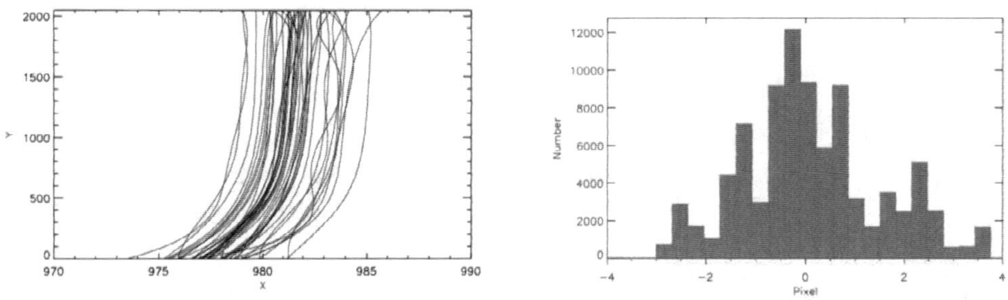

Figure 10. Left: Traces on Order 3 at the same slit position A for all targets on 2013-02-22. The average trace is shown in blue. Right: Distribution of deviations of object traces from mean trace. The sigma of this is distribution is 1.5 pixels.

With the traces at hand, object spatial profiles are evaluated for each order. Optimal extraction is performed for each order. The tracing and extraction are performed for all the science targets and standard stars.

4.4. *Telluric correction and flux calibration*

The pipeline then automatically locates the nearest standard star in position and observational time with the target and choose it for telluric correction and flux calibration. Spectra of the earthly norms were prepared along these lines as targets, trailed by a cautious expulsion of the stars' inherent hydrogen assimilation lines. This process was performed by fitting Lorentzian profiles to the hydrogen absorption lines and interpolating across these features to connect the continuum on each side of the line (Matthews 2020). Following the line cancellation, the sense function of the spectrograph is estimated by dividing the extracted telluric star spectrum with a model stellar spectrum model with same spectral type as the selected telluric standard.

Finally, we divide the extracted spectra of the target by the derived sense function. There are two significant perspectives to creating a last 1D range from a progression of science openings: the coaddition of different openings and the mix of covering echelle orders. The specific ordering of our algorithms is: (1) coadd multiple exposures of each echelle order; (2) flux the individual echelle orders; and (3) combine the echelle orders to produce a final, continuous 1D spectrum.

In parallel, we perform data reduction using Spextool and compare the final 1D spectra with results from Spextool in next Section

5. Result

5.1. Extended wavelength coverage

A detailed comparison is performed for NGC 4151 in Figure 11 (c.f. Figure 19 for Raw data). The result from this work is shown in black and Spextool result in red. NGC 4151 is quite bright in near infrared and we identify near infrared lines such as Paδ, He I 1.0830 μm, [P II] 1.1886 μm, [Fe II] 1.2570 μm, Paβ, [Si X] 1.4300 μm, [Fe II] 1.6436 μm, Paα H2 1.9570 μm, H2 2.1213 μm, and Brγ. By including Order 7, we naturally expect wider wavelength coverage than Spextool, especially for bright targets such as NGC 4151. In the short wavelength side, our pipeline extends down to 0.9 microns for NGC 4151, as appose to 0.98 microns for Spextool. This allows identification of a new feature, [S III] 0.9531 μm.

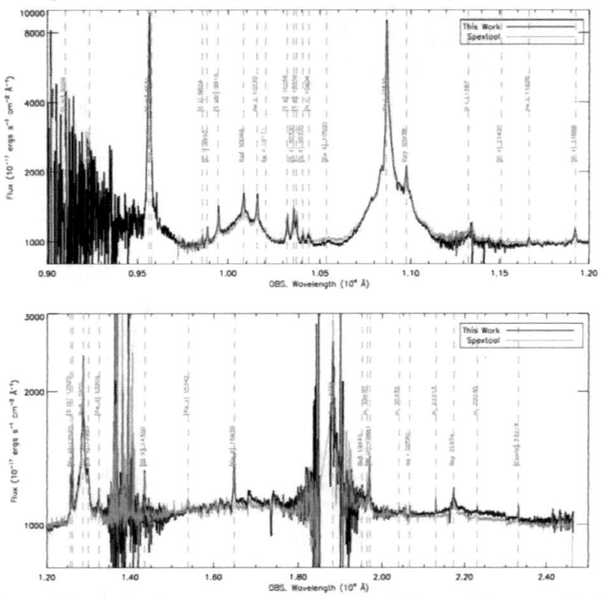

Figure 11. Comparison of results between our new pipeline and Spextool, for NGC 4151. NIR spectrum reduced by Spextool is shown in red and our result is shown in black, with longer wavelength coverage and elevated S/N ratio in overlapping part. Emission features are identified and marked as blue dashed lines.

For higher redshift targets, other important emission features may be redshifted into this addition wavelength coverage between 0.9-1 μm, for example, we see full coverage of [O III] doublets and tentative Hβ in J0352-0711 in Figure 11.

Another benefit from the extended short wavelength coverage is that it overlaps with most optical spectrographs and surveys, such as SDSS and BOSS. For example, SDSS's wavelength coverage is 3800 -9200 Å while 3600-9800 Å for BOSS, which connects with the blue end of our reduction. A continuous wavelength coverage form optical to infrared is important to cross-check the flux calibrations for different instruments involved. Also, it is crucial for sciences involving continuum shape and dust properties.

5.2. Elevated S/N ratio

In this section, we randomly selected quasars from our historic observations to evaluate the signal noise ratio in the overlapping part of Order 7 and Order 6 for our pipeline and Spextool. The results are shown in Figure 12 with an expanded view of 0.9-1.05 μm for five arbitrarily selected quasars and AGNs. Due to the incorporation of the Order 7, all missing photons in Spextool reduction are essentially recovered. Even a visual evaluation clearly reveals the improved signal to noise ratio near 1 μm for our pipeline, for both emission features and continuum region. Statistics based on all 200+ quasars we observed in the past few years shows the S/N ratio from our pipeline is on average higher than from Spextool by a factor of two at this spectral regime. Also, these randomly selected quasars show the benefits of extended wavelength coverage, e.g. the full coverage of interesting emission features and absorption features fall in this regime due to redshift effect.

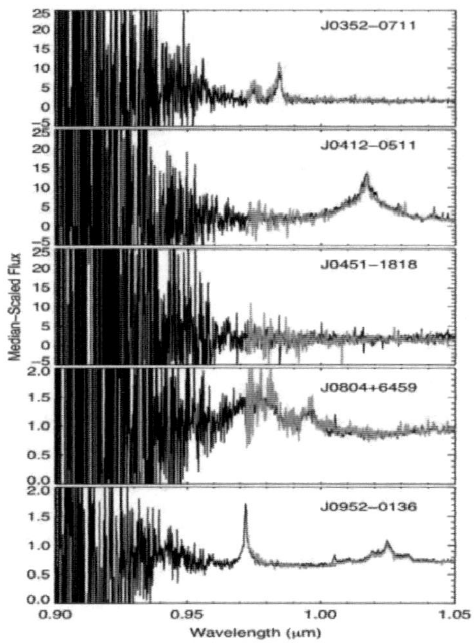

Figure 12. Expanded view of 0.90-1.05 μm for five randomly selected quasars and AGNs in our sample, with our results in black and those from Spextool in red. Even visual inspection reveals the clear elevation in signal to noise ratio in this spectral regime. Also, additional features are recovered by our reduction.

6. Conclusion and future work

We have implemented a new pipeline to reduce Palomar/TripleSpec NIR spectra. Our pipeline recovers the ignored Order 7 by the well accepted NIR spectral reduction pipeline Spextool, 1. The pipeline is also featured by its full automaticity, 2. the wavelength coverage extended from 1-2.5 μm to 0.9-2.5 μm, revealing new features such as rest frame Pa ε, [S III] 9531, [C I] 9824, [C I] 9850, [S VIII] 9910, as well as redshifted spectral features for higher redshift targets. 3. The median signal to noise ratio for 0.9-1.1 μm have been elevated by a factor of two. These new features enable a series of future studies such as optical-infrared continuum shape and spectral features falling on the recovered band (e.g., specific absorption line systems and black hole measurement for specific redshift range).

We only address the importance of incorporating Order 7 in this paper. We have applied the new pipeline to our historic TripleSpec data with 200+ new quasar spectra and plan to release the data set along with the pipeline to the community. The details on the sample, the catalog and the statistics of the sample will be investigated in future work.

The pipeline is also featured by its automaticity and could be easily generalized to accommodate other popular NIR instruments such as IRTF/SPEX, Keck/NIRSPEC. These instruments release their raw data on public archive, which is not ready for scientific analysis. Automatic and systematic reductions using our pipeline will provide a statistically significant IR spectra data base for studies on infrared astronomy.

7. References

[1] Gebhardt, K., Bender, R., Bower, G., et al. 2000, ApJL, 539, L13
[2] Glikman, Helfand, White 2006 ApJ, 640, 579
[3] Leighly, K. M., Dietrich, M., & Barber, S. 2011, ApJ, 728, 94
[4] O'Meara, J. M., Lehner, N., Howk, J. C., et al. 2015, AJ, 150, 111 Sun, L., Zhou, H., Ji, T., et al. 2017, ApJ, 838, 88
[5] Arav, N., de Kool, M., Korista, K. T., et al. 2001, ApJ, 561, 118
[6] Wenwen Zuo, Xue-Bing Wu, Xiaohui Fan et al. 2015 ApJ, 799, 189
[7] Wilson, J. C., Henderson, C. P., Herter, T. L., et al. 2004, in Society of Photo-Optical Instrumentation Engineers (SPIE) Conference Series, Vol. 5492, Ground-based Instrumentation for Astronomy, ed. A. F. M. Moorwood & M. Iye, 1295
[8] Herter, T. L., Henderson, C. P., Wilson, J. C., et al. 2008, in Society of Photo-Optical Instrumentation Engineers (SPIE) Conference Series, Vol. 7014, Ground-based and Airborne Instrumentation for Astronomy II, ed. I. S. McLean & M. M. Casali, 70140X
[9] Vacca, W. D., Cushing, M. C., & Rayner, J. T. 2003, PASP, 115, 389
[10] Cushing, M. C., Vacca, W. D., & Rayner, J. T. 2004, PASP, 116, 362
[11] Horne, K. 1986, PASP, 98, 609
[12] Bernstein, R. M., Burles, S. M., & Prochaska, J. X. 2015, PASP, 127, 911

Journal of Physics: Conference Series **2287** (2022) 012038 doi:10.1088/1742-6596/2287/1/012038

Research on the Galaxy Rotation Model based on 21 cm Hydrogen Spectrum

RUOPENG JIA

Northern Arizona University, College of Engineering, Informatics, and Applied Sciences, 86011 Flagstaff, Arizona, The United States
E-mail: Rj495@nau.edu

Abstract. Glon sky survey is a data acquisition method of 21 cm neutral hydrogen spectrum. Recently, the author has conducted a series of studies on the data of the gLon sky survey. After transforming and analyzing the data, the author found that in the region of the center of the Milky Way galaxy, the linear velocity of galactic rotation increases with the distance away from the center. For the region 15 light-years away, the linear velocity of rotation remains almost unchanged. By transforming the variation of velocity with radius into the chart, it can be found that there are deviations in the early galactic rotation structure model. For the newly proposed combined Exponential and Isothermal model, there is a high degree of agreement. The model confirms that there is a huge dark matter halo at the outer edge of the Galaxy. The weak gravitational interaction between dark matter halo and visible matter affects the current rotational structure of the Galaxy.

1. Introduction

1.1. Comprehensive purpose

It has been more than 5 centuries since Nicolaus Copernicus presented the first study of the solar system in human history [1]. The constituent of each planet, the structure, and rotation model within the solar system are already investigated clearly. Some larger scale constructions in the universe, such as the Local Group [2], the Virgo supercluster [3] and the Laniakea supercluster [4], have been detected. As the intermediate object between star systems and these largest constructions, the Galaxy is a very necessary part of the research in this century. In order to approach the truth of the universe step by step, scientists began to study the larger scale structure of the Milky Way galaxy. In this paper, the author mainly discusses the rotation condition of the Galaxy. By studying the rotation velocity, people can know the position of the solar system in the universe and predict the astronomical events that may be caused by future position changes. Thanks to the rapid development of radio astronomy and pioneers' discoveries about the 21 cm hydrogen spectrum, now people can accurately measure the rotational velocity of the Milky Way galaxy through the Doppler shift of the 21cm neutral hydrogen spectral line.

The main topic of this paper is to ascertain the galactic line velocity changing due to the distance variation. The author is as well as trying to prove the cause of this rotation circumstance. The author summarizes the rotation statuses based on the galactic longitude by transferring the doppler shift of collected 21 cm hydrogen data into seceding/approaching velocity [5]. Then, the author splits the velocity of each longitude and accurately converts it to the distance scale from the silver center. Finally, the rotation model of the Milky Way galaxy is established according to the converted data.

The model of the rotation of the Milky Way galaxy clearly shows the distribution of matter. This provides guidance for astronomers to further study the material composition of each part. At the same time, as a very hot topic in recent years, the discussion of dark matter will also be affected by this research. Through more accurate instruments, it is possible to further determine the mass and radius of dark matter halo. In addition, the galactic rotation model also plays a leading role in the exploration of other galaxies. When people choose to study other nearby galaxies, they can also observe and process through the process analogy in this paper.

1.2. Observation background

During the 1930s, people noticed a periodic extraterrestrial 'hiss' radio signal. First speculation on the signal was to consider it as an emission of the sun. But after a more accurate orientation, people found that it was emitted from the center of the Milky Way galaxy. In 1940, J. Oort published a discovery about this signal [6]. He mentioned the possible detection of the emission lines in the radio part of the spectrum could significantly improve astronomy studies. In 1944, Hendrik C. van de Hulst, then a Dutch graduate student, predicted that it would be possible to observe a 21-cm signal from neutral hydrogen in the galaxy [7]. The 21-cm was first observed in the early 1950's by Ewen and Purcell in the U.S. and by Oort in the Netherlands [8]. Almost at the same time, C.A. Muller [9] of the Netherlands and Christensen [10] of Australia also observed 21 cm spectral line signals. This discovery not only confirmed van der Hulst's prediction, but also initiated the history of radio astronomical spectral line research. After 1952, the first maps of the neutral hydrogen in the Galaxy were made, and revealed for the first time the spiral structure of the Milky Way galaxy [11].

1.3. Hydrogen observation

Before 1963, the 21 cm hydrogen line was the only radio astronomical line observed. It is different from the astronomical optical spectral line. The astronomical optical spectral line is only formed in some extreme conditions. R. Trumpler also verified that the interstellar materials such as dust, dark clouds and black clouds are opaque to visible light [12]. They will block the propagation of visible light. Therefore, it is difficult to study interstellar regions by optical means. However, the hydrogen atoms in these regions can radiate 21 cm spectral lines, and they will not be absorbed by those interstellar materials [13]. Some studies of the Milky Way galaxy through the 21 cm spectral line also show that hydrogen is indeed the most abundant element in interstellar space [14]. Neutral hydrogen atoms form a thin disk centered on the silver center along the galactic plane in the Milky Way galaxy. The thickness of the thin disk is 600 lightyears in the range of 60000 lightyears from the silver center [15]. Its number density has a flat and wide peak in the range of 15000 to 45000 lightyears from the silver center [16]. In 1971, a study by W. Shane estimated that the mass of neutral hydrogen is about 3% of the total mass of the Milky Way galaxy [17]. Then in 2009, C. Francis and E. Anderson published a research of the spiral structure of the Milky Way galaxy by 21 cm spectral line, which successfully depicted the spiral arm structure of the Milky Way galaxy [18]. The study confirmed that the Milky Way galaxy is a spiral galaxy, and measured that the width of the spiral arm on the galactic plane is about 3000 lightyears. After so much research, 21 cm spectral lines have become a significant method of exploring space.

2. Methodology

2.1. Detector

The construction structure of the 18 meters radio telescope is shown in Fig. 1. The feed horn, which is supported by four fiberglass frameworks, is clearly evident. The building to the right of the dish houses the control console. The wing-like structures beneath the dish are counterweights designed to keep the dish in balance independent of its elevation.

Signal processing occurs in two physical locations: in the feed assembly at the focus of the dish and inside the control room located nearby. The raw voltage from the feed will contain two undesirable components in addition to the signal the author wishes to receive. In general, since the interference signals do not fall within the frequency band being used for the radio astronomy observations, they are not a problem. However, sometimes if we take the large amplification factor into consideration, even the signals out of band can make trouble when they are strong enough to overload the electronics [19]. To keep thermal noise at a minimum value, it is recommended to amplify the signals as early as possible. This is done by a low-noise amplifier (LNA) located extremely close to the probe electrode which is used to extract the signal from the feed [20]. As listed above, the author wants the gain (amplification factor) of the LNA not to be too high, since out-of-band signals can drive the LNA into saturation. To avoid this, the voltage gain of the LNA is kept to a modest value and the signal is routed through a band-pass filter (BPF), laid at its output. The BPF removes the out-of-band interference signals, at which further gain can be safely added [21]. At this point, the signal, which has been filtered and heightened well above the noise floor, can be transmitted along a 100-200 m coaxial cable to the control room.

Figure 1. View of the 18-m-diameter dish used in this research.

2.2. Receiver system

Although designs are various, the configuration of a typical dish radio telescope is shown in Fig. 2. The spacial radio waves are incident upon a parabolic reflector that focuses them to a point. At the focal point, a feed horn is placed. It is a form of antenna. The feed horn collects the focused waves and converts them to voltage signals. Then these signals are amplified by a low-noise amplifier and transmitted on a coaxial cable for further processing at the distant end.

Most part of the receiver system is implemented in software, running on a standard computer. A key element in this system is a hardware module that switches the analog signals from the preamplifier to digital bitstreams for the computer to process. In the system that the author uses, this device is called a Universal Software Radio Peripheral (USRP). Technical information on the USRP used in the system is available online. Data from the USRP is transmitted to the computer over a USB3.0, which is now standard on most computers. Radio receivers perform three essential functions: signal amplification, signal filtering and signal demodulation.

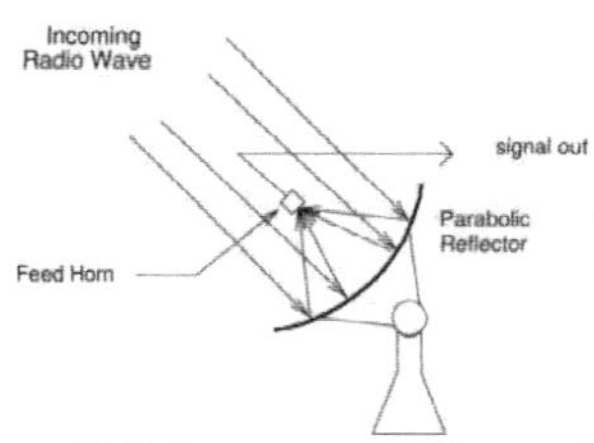

Figure 2. Schematic view of dish type receiving antenna used in the research.

2.3. Detector precision

Angular resolution could be the most important performance parameter of any telescope. The better a telescope's angular resolution, the better its ability to differentiate two closely spaced objects in the sky. This is illustrated for the case of an optical telescope in Fig. 3. The effect of limited angular resolution is shown in the rightmost sketch, where the blurring of what optically should be two independent points is apparent [22]. For a telescope with worse angular resolution, the blurring of the points could increase, probably to a condition where the two stars would appear to merge into one. Various effects affect the angular resolution of ground-based optical telescopes. The most crucial being distortion of the image in the atmosphere, which limits the resolution to about one arc-second [23].

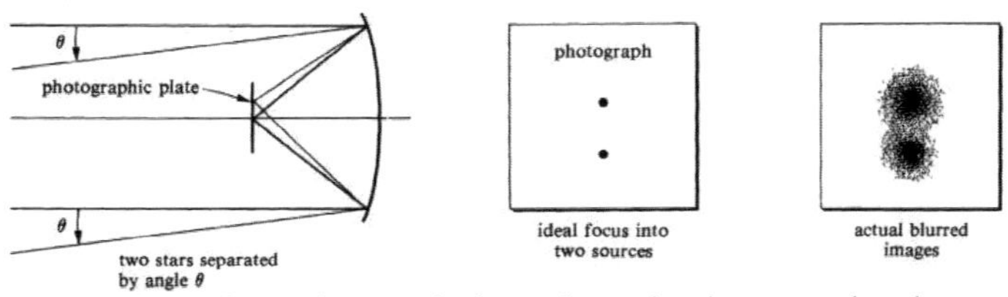

Figure 3. The effect of angular resolution when viewing two closely spaced stars.

3. Results and discussions

3.1. Data

The detector captured a series of hydrogen molecular emission power spectrum data on scale of the Galaxy longitude and frequency. As Fig. 4 shows, there are several peaks indicating higher density regions of the neutral hydrogen cloud at galactic longitude 0° .

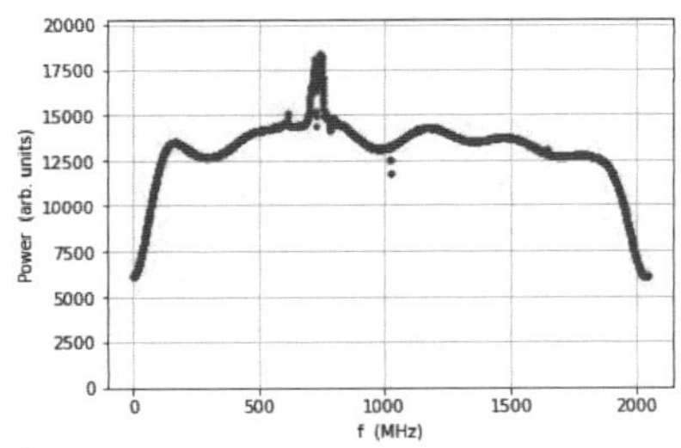

Figure 4. 21 cm neutral hydrogen spectrum at longitude 0° .

In order to establish a model of the Galaxy's rotation, the author transfers the frequency scale into approaching velocities. According to the Doppler effect, the known frequency of emitted signal (f = 1420.41 MHz) can be used as a reference to infer the velocity components of sources from the observed peaks [24]. In the Python environment, the author develops a calculating function to automatically convert scales as well as plot the transferred diagram Fig. 5.
The radial velocities are expressed as

$$v = \frac{c\,(f'^2 - f^2)}{f'^2 + f^2} \quad (1)$$

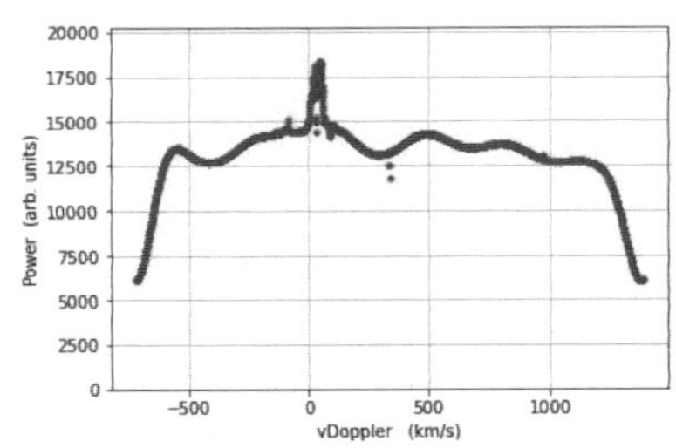

Figure 5. Transferred Doppler spectrum.

In addition to the hydrogen signal, the radio telescope received a lot of white noise. Actually, a lot of this noise is internally generated in the receiver circuitry [25]. The white noise contains approximately equal amounts of energy at all frequencies [26]. So ideally the spectrum of this white noise would be a constant amount of energy in each frequency bin. But in addition to a few clearly distinguishable hydrogen lines, there are even more obvious large-scale numerical differences. These variations are a result of the receiver's response, which depends somewhat on frequency. The author studies properties of the receiver and manually levels the whole spectral band in Fig. 6 to eliminate the variation caused by frequency response difference of the receiver.

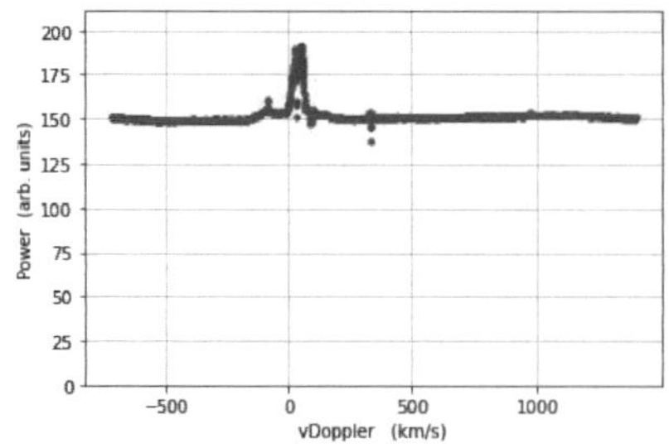

Figure 6. The noise-eliminated spectrum.

Limited in the Galactic metric, the author sets the approaching velocity in a range from -200 km/s to 200 km/s [27]. Then, the author generates a new plot Fig. 7 with this narrower range, in which the peaks became much more significant.

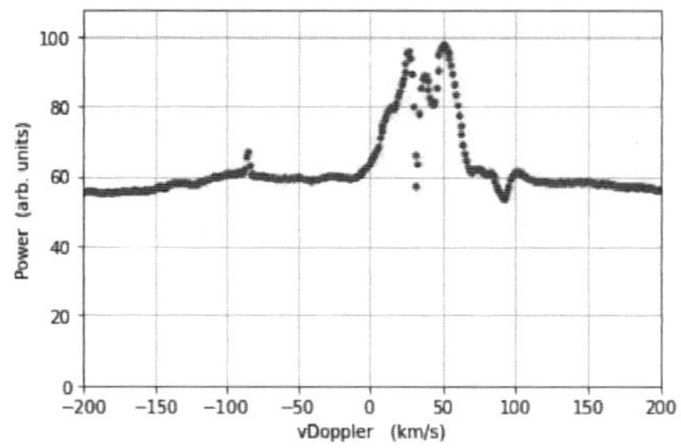

Figure 7. Spectrum on a smaller scale, in which the peaks are clearer.

There are three peaks, two large peaks at $v = +25$ km/s and $v = +50$ km/s, and a smaller sharp peak at -80 km/s. Because of the relatively uniform distribution of neutral hydrogen molecules in the Milky Way galaxy, the radiation emitted by neutral hydrogen molecules mostly appears continuously in the spectrum. Their shift peaks due to the Doppler effect should also be clear and continuous in the spectrum. The sharp peak on the left of Fig. 8 is too narrow for hydrogen regions, so the author considers that this is a false signal caused by a receiver defect [28]. Also, by examining each longitude signal, the author finds that most of them contain this sharp and short peak. This further proves the author's point of view. This unique peak must be artificially smoothed so as not to affect further data analysis.

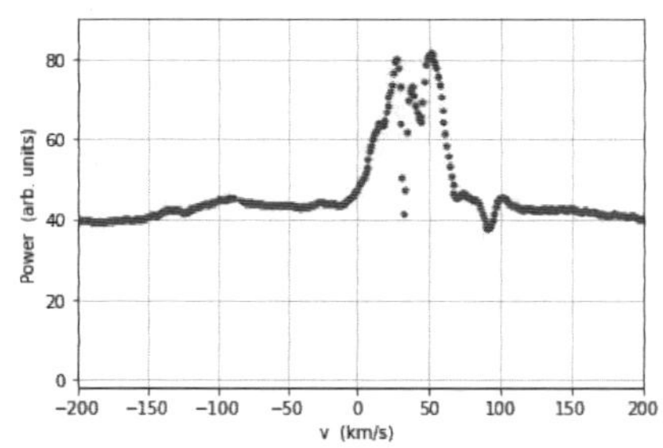

Figure 8. A flat spectrum after eliminating the sharp peak.

All data to be preprocessed can be obtained by performing the same processing on the data of other longitudes. Here is another processed example of the spectrum of longitude 20° in Fig. 9.

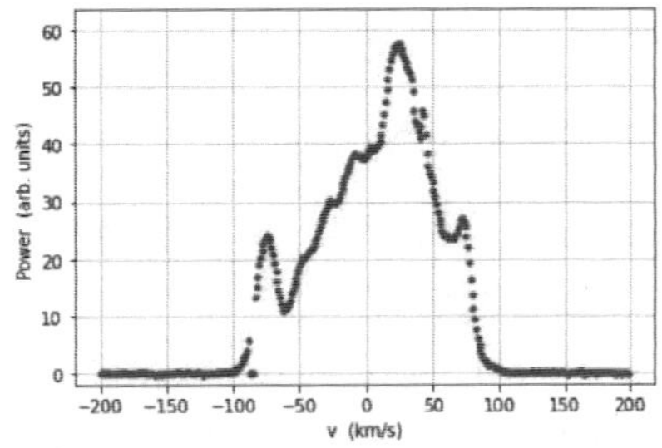

Figure 9. Fully processed hydrogen spectrum at longitude 20° .

At longitude 20 ° , more peaks could be seen. This phenomenon is because of the fact that at this longitude, the gas clouds captured by the detector are more complex. Since our solar system is located at the edge of the Milky way, when the detector is aimed at the middle and low longitudes, the movement directions of hydrogen molecular clouds will no longer be mainly perpendicular to the radial direction of the detector, so the generated spectrum will be more irregular, which can be seen in Fig. 10. The seceding part is because the more central disk is captured, the hydrogen clouds in this region moves at a fast angular velocity and relatively goes away from the Sun. The Galaxy disk on the same radius as the sun is stationary relative to the sun, so there is also a power peak near the velocity of 0. The peak on the right is because the components outside the Sun's radius have a smaller angular velocity relative to the sun, and their distance from the sun is shortening [29]. The data of a single longitude actually contains all of the hydrogen clouds' radiation on the line of sight. All this is a simple analysis based on Kepler's third law. In the later part of the paper, the author will raise a more rigorous model.

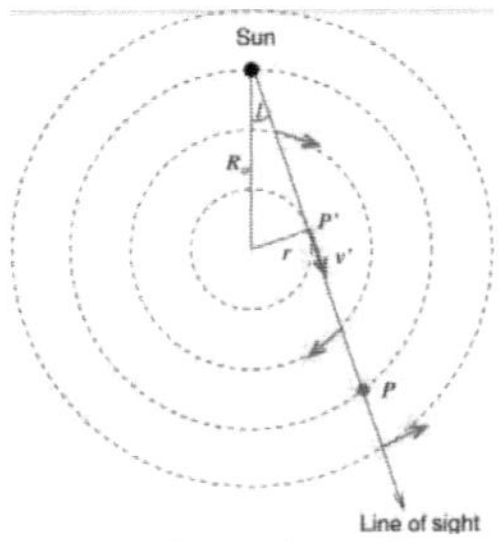

Figure 10. A simple diagram of the predicted motion of parts of the Milky Way galaxy.

The gLon sky survey has observed the disk area over 90 degrees, these longitude-based one-dimensional data need to be displayed on a two-dimensional template to comprehensively describe the hydrogen molecular strength in different longitudes and retrograde velocity dimensions shown in the hot zone diagram.

By transforming dozens of power spectra into a heat map, the structure of the motion of the Milky Way galaxy can be preliminarily displayed. Fig.11 shows that it presents a regular distribution in the longitude range.

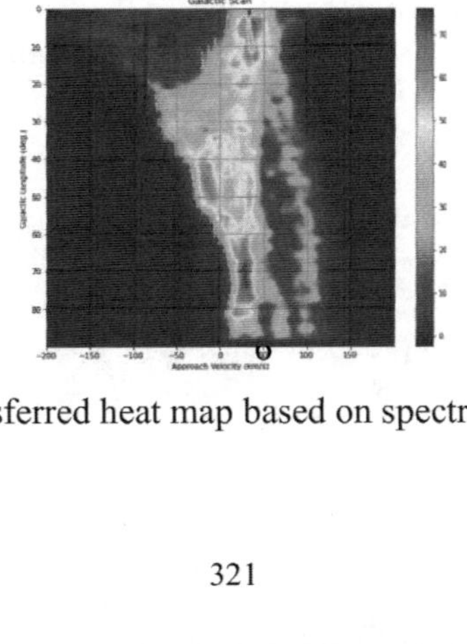

Figure 11. The transferred heat map based on spectra with left thresholds.

Since the Milky way is a rod spiral galaxy [30], it should be more consistent with the rotation characteristics of rod spiral galaxies on the radius scale [31]. In that case, longitude could not be the best reference system to measure the rotation.
First, the author selects the representatives of each longitude, which are shown in Fig. 12. It should be combined with transferring equations below, which requires the value to be the most seceding ones. Then the author derives rotation curves by fitting the velocity fields with a tilted-ring model to obtain initial estimates, which are subsequently refined by building a kinematic disk model [32]. Using a threshold function as in Fig. 12 to mark them as *v'* at the heat map and then plotting the array in another plot.

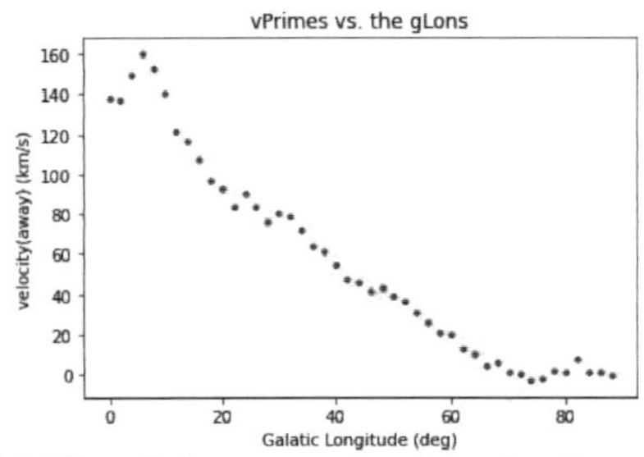

Figure 12. The Galaxy rotation in galactic coordinate.

Then the rotation curve is expressed as

$$v_r = v' + r\frac{2\pi}{T_0} \quad (1)$$

where

$R = R_0 \sin l$

(T_0 is orbital period of the Sun about the center of the galaxy and l is the galactic longitude)

Furthermore,

$R_0 = 2.7*10^{14}$ km

and

$T_0 = 237{,}000$ yr $= 7.47*10^{12}$ s

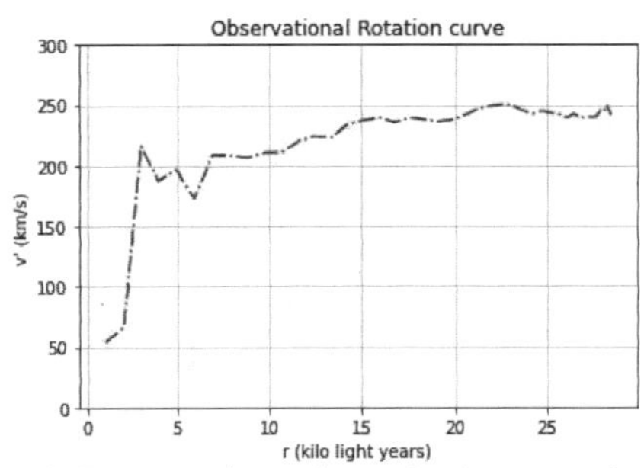

Figure 13. Processed rotating velocity curve in radius.

The author uses thresholds of each longitude as well as the components of the heat map to calculate each radius' rotating velocity as equation (1) and plotted them altogether in the curve in Fig. 13. The transferred plot looks not very likely to the original one. It is because superimposed signals are divided into pieces to be settled at the radius they are at. Each radius has just one rotation velocity.

3.2. Discussions

After J. Oort reported the indications by his research of the stars in the solar neighborhood, a great deal of models about the Galaxy rotation with their own rationality in that era appeared [33]. But with the author's persuasive 21 cm hydrogen spectrum discovered, some of the models could be obsoleted because they have big differences with the gLon data. Here are some representative models in the past decades, the author shows the differences between them and this research's output curve.

3.2.1. The Kepler model. Kepler predicted the motion of the Milky way in the 15th and 16th centuries. He believed that most of the mass of the Milky Way galaxy is concentrated in the central region, which was inconsistent with the current observation facts. The author simply restores the model through Newtonian mechanics.
The Newton's 2nd Law shows

$$F = ma \tag{2}$$

$$\frac{GMM_{\odot}}{r^2} = M_{\odot}\frac{v^2}{r} \tag{3}$$

In the equation 4: M is the mass of the Galaxy, $M\odot$ is the mass of the Sun, r is the distance between the Sun and the center of the galaxy, and v is the orbital speed of the Sun.

$$v(r) = \sqrt{\frac{GM}{r}} \tag{4}$$

Simplifying and solving for vr :

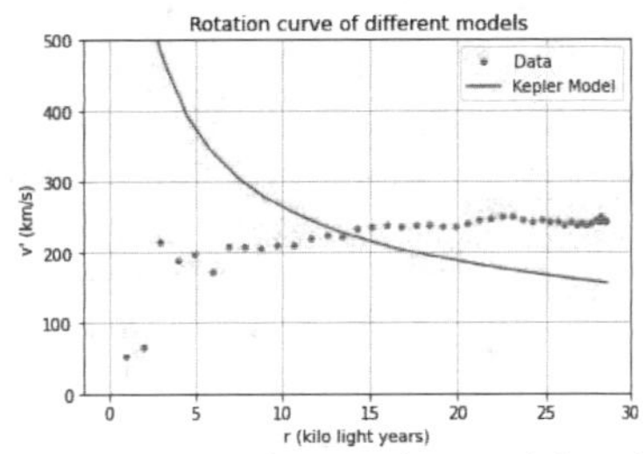

Figure 14. Comparison between the Kepler model and our processed data.

As can be seen from Fig. 14, although tremendous observational data such as *Y.* Sofue's work in 1972 [34] has proved that the Kepler model is significantly contradictory to the real distribution of matter; this model is still a key step for human beings to explore large-scale structure.

3.2.2. Exponential disk model. The Exponential disk model was published by D. Lin and J. Pringle in 1987 [35]. It assumes that the amount of mass is proportional to the amount of light. This is a reasonable assumption and its curve is really close to reality. Even if the mass-to-light ratio varies from star to star, if a large enough collection of stars is taken into consideration, those differences will average out. Given that the Milky Way comprises (way) more than 10 billion stars, it certainly qualifies as a large collection.
In 1999, a research published by A. Misiriotis's team further developed this model [36]. They further improved the fit of the model to the real rotation speed through more accurate measurement. The details of the exponential disk model are a bit involved, but suffice it to say that astronomers have carefully studied the amount of light as a function of r for galaxies similar to the Milky Way and have a pretty good idea of what the mass density of stars is like. We have good reason to expect that the Exponential Disk model will work well.

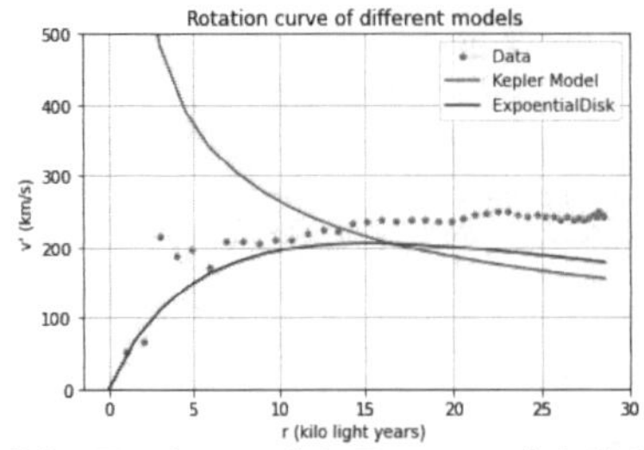

Figure 15. Comparison of the Kepler model, Exponential disk model and processed data.

As can be seen from Fig. 15, even though the Exponential disk model is very close to the real data, there are differences at the center as well as the outermost regions. In the last model, the author takes the dark matter into consideration and chooses the Isothermal sphere model to be blended with the Exponential disk. It is the best one to fit the Python code environment.

3.2.3 Combining the test exponential disk and the isothermal model. When the radius scale goes beyond 15k light years, the Exponential disk model falls a little below the data, suggesting that there are other substances in the Milky way that interact gravitationally. Existing papers have proved that there is a huge dark matter halo at the outer edge of each galaxy. The mass of dark matter in the halo is much larger than the mass of visible matter one can observe [37]. Thus, scientists created an Isothermal sphere (IS) model [38] to describe the movement of visible matter under the influence of dark matter. Dark matter is a kind of matter that cannot emit blackbody radiation. It cannot be found in the spectrum. Therefore, it can only be predicted by the action of gravity. In fact, so far, humans have not been able to know the specific composition of dark matter. The author finds that the curve fitting by combining the Exponential Disk model curve with the existing Isothermal sphere model greatly improves its coincidence with the gLon observation data.

The combined (Ex&Is) model is expressed as

$$v_{tot} = \sqrt{v^2{}_{ED} + v^2{}_{IS}} \tag{5}$$

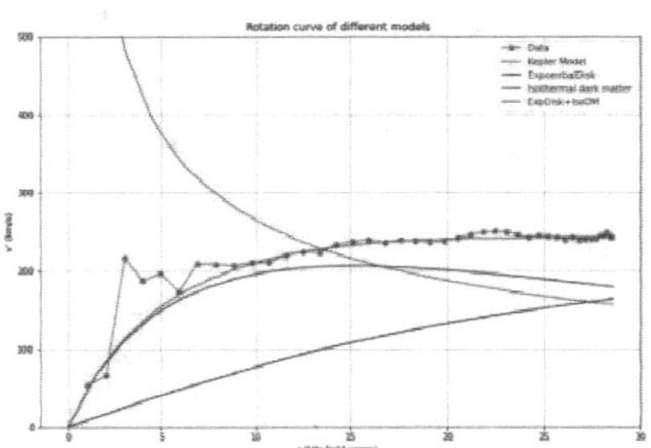

Figure 16. Comparison of all models.

Through comparison, the author finds that in the part beyond the radius of 8 thousand lightyears, Ex&Is model is almost perfectly consistent with the processed gLon data. This similarity could be seen in Fig. 16, which shows that this model has high reliability in predicting the dynamic characteristics of the Milky Way galaxy.

4. Conclusion

Overall, the agreement between data and Ex&Is model is fairly favorable. It is, however, important to notice that the Isothermal sphere model has two free parameters that are adjusted to match the data. They are predicted by a large number of observations and theoretical studies. In the future, there will definitely be more adjustments about the accurate details on the parameters of the Isothermal sphere model. The author might have to add more coefficients to correct the model and even rebuild the whole rotation measurement system. Nevertheless, the failure of the Exponential Disk model alone to describe the data also further confirms the fact that there is a dark matter halo at the outer edge of the Milky way.

In this paper, by comparing the differences between the galactic rotation model and the observation results in different periods, the author establishes a model that is most suitable for the current data. The high coincidence of the velocity of each part of the Milky way by the Ex&Is model shows that this modeling idea is likely to play an important role in the study of the formation and evolution of the Milky way. This model can also be extended to other galaxy research similar to the Milky way, such as the popular research object SPT0418-47 galaxy [39]. At the same time, the model also provides help for the research of dark matter. As one of the hottest topics nowadays, dark matter is not only the key factor for human beings to solve the mystery of the Milky Way galaxy, but also plays an important role in the evolution of the whole universe. Further research on this model is likely to predict the composition of dark matter in theory.

On the other hand, the model still has room for improvement. The velocity in the 5 - 10k light years region of the radius of the Milky way still deviates greatly from the model, which may be caused by some special events in the early evolution of the Milky way. The irregularity of motion speed must be accompanied by specific mass distribution, which requires more accurate detectors and more detailed research. For Ex&Is model, the next thing to do is to determine more factor variables and constantly narrow their scope. In this process, astronomers can also analyze more laws within the Milky way.

The Milky way is the galaxy most studied by human beings for a long time. Only by fully understanding its characteristics can people realistically study the larger scale Local Group, Laniakea Supercluster and the Sloan Great Wall.

5. Reference

[1] N. Copernicus, De revolutionibus orbium coelestium(1543).
[2] O. Marioni et al, Numerical simulations of bar formation in the Local Group(2022).
[3] F. Hu et al, Orientation of Galaxies in the Local Supercluster: A Review, arXiv:0508669(2005).
[4] R. Tully et al, The Laniakea supercluster of galaxies, arXiv: 1409.0880(2014).
[5] J. Zhao and E. B. Ford, FIESTA II. Disentangling stellar and instrumental variability from exoplanetary Doppler shifts in Fourier domain(2022).
[6] J. Oort, Some problems concerning the structure and dynamics of the galactic system and the elliptical nebulae NGC 3115 and 4494(1940).
[7] V. Hulst, Radiogolven uit het wereldruim: II. Herkomst der radiogolven, Nederlandsch Tijdschrift voor Natuurkunde, vol 11 (1945) pages 210-221.
[8] H. Ewen and E. Purcell, Observation of a Line in the Galactic Radio Spectrum: Radiation from Galactic Hydrogen at 1,420 Mc./sec.(1951).
[9] C. Jager et al, Absence of Hydrogen Radiation of Wavelength 21 cm. in the Sun(1951).
[10] W. Christiansen and J. Hindman, A Preliminary Survey of 1420 Mc/s. Line Emission from Galactic Hydrogen, Australian Journal of Scientific Research 5(3) (1952) 437-455.
[11] C. Muller and J. Oort, Observation of a Line in the Galactic Radio Spectrum: The Interstellar Hydrogen Line at 1,420 Mc./sec., and an Estimate of Galactic Rotation(1951).
[12] R. Trumpler, Absorption of light in the galactic system, Publications of the Astronomical Society of the Pacific Vol. 42, No. 248 (1930), pp. 214-227.
[13] D. R. Rybarczyk et al, The role of neutral hydrogen in setting the abundances of molecular species in the Milky Way's diffuse interstellar medium. I. Observational constraints from ALMA and NOEMA(2021).
[14] A. De Cia et al, Large Metallicity Variations in the Galactic Interstellar Medium(2021).
[15] J. Oort, A summary and assessment of current 21-cm results concerning spiral and disk structures in our galaxy(1959).
[16] K. Imamura, Transition between Atomic and Molecular Hydrogen in the Galaxy: Vertical Variation of the Molecular Fraction, arXiv: astro-ph/9607044(1996).
[17] W. Shane, Neutral Hydrogen in an Interior Region of the Galaxy; the Longitude Interval 22° to 42°(1971).
[18] C. Francis and E. Anderson, Galactic spiral structure, Proc. R. Soc. A (2009) 465, 3425–3446.

[19] E. R. Kuhn et al, Design and implementation of a noise temperature measurement system for the Hydrogen Intensity and Real-time Analysis eXperiment (HIRAX) (2021).
[20] A. Perko, The Omniscope: Mapping the Universe in 3D With Neutral Hydrogen(2011).
[21] A. Lalbakhsh et al, A Design of a Dual-Band Bandpass Filter Based on Modal Analysis for Modern Communication Systems(2020).
[22] H. Matsuo et al, Terahertz Atmospheric Windows for High Angular Resolution Terahertz Astronomy from Dome A(2019).
[23] M. W. L. Smith et al, The HASHTAG project: The First Submillimeter Images of the Andromeda Galaxy from the Ground(2021).
[24] C. Doppler, Über das farbige Licht der Doppelsterne und einiger anderer Gestirne des Himmels(1842).
[25] N. Merhav, Optimal Correlators for Detection and Estimation in Optical Receivers(2020).
[26] K. Goswami and R. Chakrabarti, Stochastic resetting and first arrival subjected to Gaussian noise and Poisson white noise(2021).
[27] H. Katz et al, The Tight Empirical Relation between Dark Matter Halo Mass and Flat Rotation Velocity for Late-Type Galaxies(2018).
[28] J. Vinet et al, LISA Telescope: Phase noise due to pointing jitter(2019).
[29] E. M. Di Teodoro et al, Rotation curves and scaling relations of extremely massive spiral galaxies(2021).
[30] A. Kunder et al, The Bulge Radial Velocity Assay (brava). II. complete sample and data release(2011).
[31] A. V. Shelest and F. Lelli, From spirals to lenticulars: evidence from the rotation curves and mass models of three early-type galaxies(2020).
[32] F. Lelli et al, Dynamics of starbursting dwarf galaxies. II. UGC 4483(2012).
[33] J. Jałocha et al, Constraining the vertical structure of the Milky Way rotation by microlensing in a finite-width global disk model(2014).
[34] Y. Sofue, Radio Spurs and Spiral Structure of the Galaxy, I. Optical and Radio Brightness Distributions in the Milky Way, Astronomical Society of Japan(1973).
[35] D. Lin and J. Pringle, The formation of the exponential disk in spiral galaxies, The Astrophysical Journal, 320:L87-L91(1987).
[36] A. Misiriotis et al, Is the exponential distribution a good approximation of dusty galactic disks(1999).
[37] A. Sonnenfeld et al, The dark matter halo masses of elliptical galaxies as a function of observationally robust quantities(2021).
[38] S. Chaudhury et al, Direct detection of WIMPs : Implications of a self-consistent truncated isothermal model of the Milky Way's dark matter halo(2010).
[39] C. Breuck et al, A dense, solar metallicity ISM in the z=4.2 dusty star-forming galaxy SPT 0418-47, Astronomy & Astrophysics manuscript no. SPT0418FSL(2019).

Acknowledgment
Firstly, I would like to show my deepest gratitude to my professor D. Marlow in this research program, who have provided me with valuable guidance in every stage of the writing of this thesis. Further, I would like to thank my research partner Q. Chen and instructor Y. Luo for their encouragement and support. We discussed a lot of possible future developments in the 21 cm spectrum research. Without all their enlightening instruction and impressive kindness, I could not have completed my thesis.

Law of Gravity Blurred by Perturbed Planetary Orbits for Alien Observers

Jessie Wang

Basis Independent Silicon Valley, SAN JOSE, CA 95126, USA
E-mail: wylje6@gmail.com

Abstract. The Fermi paradox is the contradiction between the high probability for aliens and its lack of evidence. This is perhaps the most significant of all questions in astronomy. Nowadays, thousands of exoplanets have been detected, providing us more information directly from alien planets to revisit Fermi paradox. Over recent years, the detection of habitable exoplanets has grown rapidly, and most of them in multi-planet system with fairly compact structures, which is partly due to our observational techniques. In such systems, planets' orbits are significantly perturbed by gravitational interactions from each other. To travel within or out of a planetary system, one must understand the law governing planetary motion, but can extraterrestrials deduce the universal law of gravitation by observing planets in their system, as Johannes Kepler and Isaac Newton had done? Here we show that due to the close planetary orbits in systems hosting habitable exoplanets, planetary orbits significantly deviate from the Keplerian orbit. In our N-body simulation over a period of 300 days, out of the ten planetary systems which all include habitable exoplanets in our sample, the mean anomaly deviates from a Keplerian orbit by more than 5 degrees while the system remains stable as the semi-major axis remains relatively constant. This would significantly complicate what the observers on those planets observe as the motion in the system would not appear to have a pattern periodically. Although planets indeed affect each other in all systems, the compact nature of these systems have magnified the effect and provide no way for the extraterrestrials inhabiting those planets to discover the universal law of gravitation. Extraterrestrials lacking the law of gravity definitely cannot facilitate traveling beyond their planets. That could be one possible resolution of Fermi paradox, reducing the population of high-developed civilization in Universe.

1. Introduction

1.1. Fermi paradox

The Fermi paradox explores the contradiction of the age of the universe, thus the large possibility of extraterrestrials' existence, with the lack of evidence we have collected exhibiting that. With billions of stars in the milky way similar to the sun, there are many exoplanets which are similar to Earth and it is likely intelligent life have developed elsewhere, it is also likely they can facilitate interstellar travel. The age of the universe is 13.8 billion years, which provides plenty of time for Earth-like planets to form and Earth itself formed only just 4.5 billion years ago [1]. From the large amounts of Earth-like planets that should have been produced over billions of years, there must be some which have developed into civilizations capable of interstellar travel [2]. If they have the ability to travel to earth, why is there no evidence of its occurrence? Possible explanations, first analyzed by Michael Hart, can be seen from several perspectives [3].

Journal of Physics: Conference Series **2287** (2022) 012039 doi:10.1088/1742-6596/2287/1/012039

1.2. Discovery of exoplanets

The idea of exoplanets, planets orbiting stars other than the sun, was theorized decades ago, however it was not until 1995 that an exoplanet orbiting a solar-type star was discovered. This discovery of 51 Peg b was awarded the Nobel Prize in Physics 2019 as it contributed to our understanding of the evolution and origin of life in the universe. Ever since then, rapid progress has been made in enhancing the detection of extrasolar planetary systems. As of 2021, the known number of exoplanets exceeds 4000 (Fig 1). The two main methods of detecting exoplanets are radial velocity and transit[4].

The diverse nature of exoplanets was revealed with the discovery of 51 Peg b, as although the 51 peg star was a solar- type star, the planet had a semi major axis of only 0.05 au, an orbital period of about four days and a radius twice of Jupiter's. Prior to the discovery, the perceptions of extrasolar planetary systems were based on the solar system, and it was believed a planet with mass similar to Jupiter's would have a large orbital period as that is the case for planets in the solar system. Although there can be observational bias as the methods of detection favor those with planets closer to the host star, it can be seen the structure of the solar system is not at all a model for extrasolar planetary systems. Through the characterization of exoplanets, the understanding of the perspective of an extraterrestrial civilization can now be furthered, as the diversity of the orbital elements of exoplanets can possibly provide context for the lack of evidence of extraterrestrials in the solar system [5].

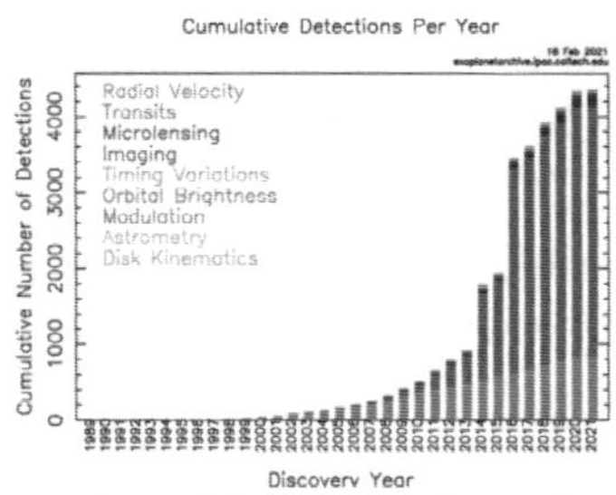

Figure 1. This figure presents the number of detection of exoplanets up until February of 2021, with each color indicating the method used. It can be seen that Transits and Radial Velocity have been used for the majority of the detections and the number of detections has greatly increased in recent years.

1.3. Habitable exoplanets

Out of the over 4000 exoplanets discovered so far, the ones considered habitable only just exceed 60. The habitability of an exoplanet is impacted by factors including the star-planet separation, its atmosphere and its planetary orbit.

An exoplanet's earth similarity index (ESI) considers five parameters: mass, semi major axis, radius, stellar radius and stellar temperature. ESI only characterizes one type of a habitable environment; it ultimately does not determine a planet's habitability as other factors like planetary tidal locking and magnetic field are not considered. The planetary habitability index however, measures habitability based on what requirements are essential to support life, in ways which can differ from Earth. It considers factors including the energy sources, the chemistry composition and the presence of liquids.

The habitable exoplanets that have been discovered so far primarily orbit around M-type stars, whose surface temperatures range from 2000 to 3700K. As a result, the circumstellar habitable zone is much closer to the star compared to the solar system, since the star's radiation falls as the distance from it increases. ($L \propto 1/r^2$)

From the plot (Fig 2) illustrating the mass and radius of those in planetary systems containing habitable exoplanets, it can be seen their size and mass vary but are generally larger than Earth. However, the graph of semi major axis (Fig 3) reflects how compact these systems are, as most measurements seem are only a fraction of Earth's semi major axis, although that maybe a bias from our current detection techniques.

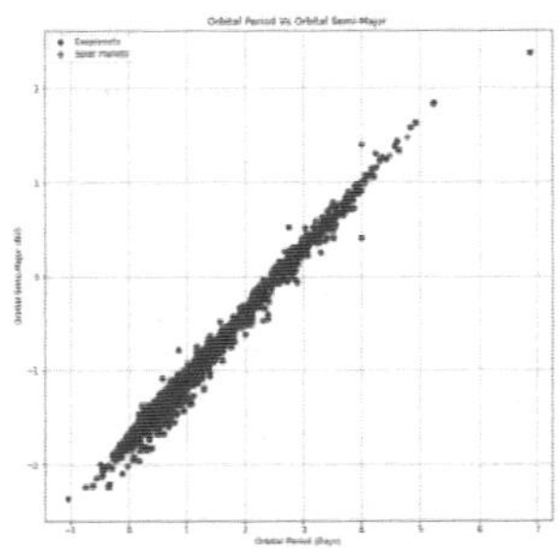

Figure 2. This plots the semi-major axis of the currently known planets against their orbital period the red dots represent those in the solar system.

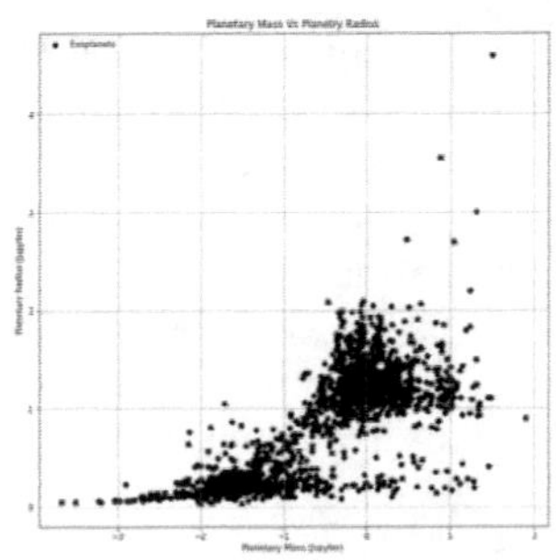

Figure 3. This figure plots the mass of the known planets against their radius, both are represented in Jupiter mass and radius.

2. Theories

2.1 Orbital elements and planetary motions

A planetary body's orbit can be found using Kepler's laws and its motion can be summarized using six orbital elements. The six elements include: a, the semi major axis which is one half of the major axis; *e*, the eccentricity which is how elliptical the orbit is with 0 being a circle and 1 being a parabola and larger values belong to a hyperbola or straight line; *I*, the inclination angle of the orbit which is the angle between the orbital plane and the reference planet; the longitude of pericenter represents the longitude at the point where the body is closest to the star if the inclination angle was 0; Ω, the longitude of the ascending node, represents the angle from the location of the body at a specific time to the ascending node which is when the body crosses from below the reference plane to above it; ϖ, longitude of pericenter is the angle between the ascending node and the pericenter (Fig 4).

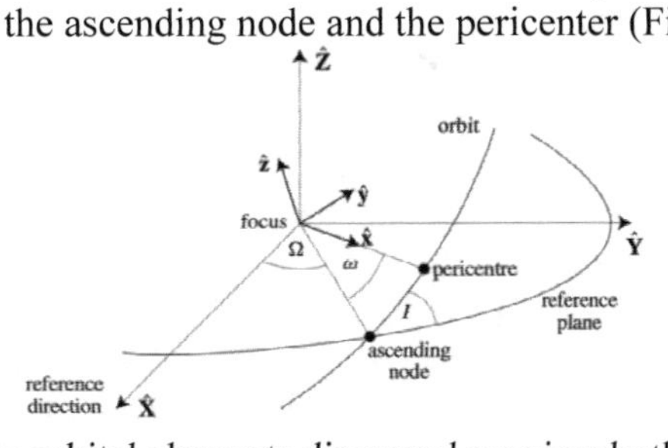

Figure 4. This shows the six orbital elements discussed previously three dimensionally. I is the inclination; Ω is the longitude of the ascending node; and ω is the argument of periapsis. After Carl & Stanley (1999, Fig. 2.13)

2.2 perturbed orbits in three body problem

A solution for the N-body problem has been long sought after and by using simulations, solutions can be found with the given parameters although these are only particular solutions that can not be applied to more than one set of parameters. A system with two bodies interacting through gravity can construct a system with predictable orbits. More importantly, a stable orbit allows the planet's climate and conditions to remain relatively consistent, and eliminates the possibility of collisions, making the planet more habitable.

With the law of universal gravitation, one can deduce the strength of a gravitational force decreases in proportion to the square of the separation. In the case of the solar system, the planet separation is large and as a result perturbations due to other planets is small; for example the force between just Earth and Jupiter, the most massive planet in the solar system, is less than 0.005 percent of the force between Earth and the Sun. However, in compact systems, the small separation and less massive stars produce more impactful perturbations on the planetary orbits as in the TRAPPIST-1 system, the force between b and c is more than 0.01 percent of the force between b and the star, and there are seven exoplanets in the system in total, each exerting a push on planet b. These perturbed orbits produce a more complex system; although the orbits can be predicted using simulations, it is only feasible with prior knowledge of the universal law of gravitation and the relationship it presents between the mass, separation and force between bodies, which has been made much less direct apparent in compact systems like TRAPPIST-1.

A restricted three body system differs from others as it contains one body whose mass is too small to affect the motion of the others for example a satellite between Earth and the Sun. Lagrange points are points at which the third body is able to remain in equilibrium. The Sun-Earth Lagrange points are used for probes, telescopes and satellites although L3 cannot be used because of its placement behind the Sun. Natural objects at Lagrange points include the Trojans and Greeks asteroids which are at the Sun-Jupiter Lagrange points L4 and L5.

3. Simulation and analysis

3.1 Mercury

Mercury is a numerical integrator which was used to create a simulation of effects of the N-body movements which were present in all planetary systems, but especially significant in the compact systems used. The program is written in Fortran 77 by John E. Chambers [6]. Its function includes the ability to calculate the evolution of orbits for objects including comets moving in the gravitational field of a large central body (Fig 5). It allows for the inclusion of general relativity in its calculations although in the cases tested only the Newtonian gravitational forces were used. Mercury6_1.for is one of the drivers for this program and contains the basic integration program which carries out the necessary calculations and produces the desired outputs in a compressed file. Element6.for is used to convert the outputs into files in terms of the keplerian orbital elements for each object of integration, which better demonstrates how exactly the orbit evolves and can be used for the visualization of the simulation.

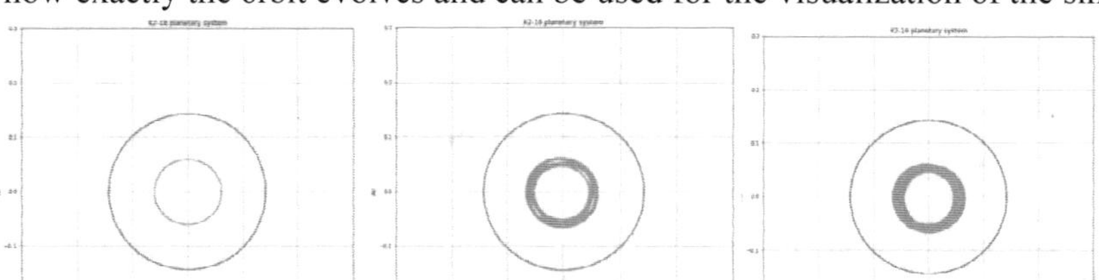

Figure 5. From left to right, the figures present the orbits of two masses over one year, whose mass and semi-major axis have been altered to exaggerate the effects of three body motion to better demonstrate it visually as with the deviations in a real system, although they do exist, are not visually apparent when plotted.

The program takes inputs, the initial data for necessary measurements, as input files like big.in and param.in. Big.in allows the user to enter in the orbital elements for each planetary body involved in the integration. Param.in includes several parameters including those which control how the integration is carried out, like time period and how often an output is recorded, and physical parameters of the central

body. The program produces output files with the orbital elements over the integration for each planet in the input file. For our ten systems, it was necessary for the perturbations to be visible, so two sets of data were taken for each planet in order to contrast its orbital evolution under only the gravitational influence of the central body with if all gravitational influence from other planets were included. The visualization of the integration was achieved using the two sets of data for each planet and the graphing features of python. We also plotted the difference between the two sets of data over the time period of 300 days to analyze these systems more closely.
The simulation used parameters found in the NASA exoplanet archive described in the previous section, and any unknowns were set to zero [7]. We consider the two separate cases of when the planet to planet interactions are included and excluded from the evolution of the orbit over 300 days. The simulation allowed us to consider the change of the mean anomaly over time as it demonstrates change of the planets' orbit distinctly.

3.2 Trappist1
The Trappist-1 star,12 parsecs away, is the host star to seven confirmed planets and has a surface temperature of 2550 K which defines it as an ultracool dwarf, along with its low temperature it also has the mass and radius of 8% and 11.5% those of the sun [8]. Trappist-1b and c were first detected using their transits, which had periods of 1.5 and 2.4 days. Trappist-1d, e, f, g and h were detected later using the 34 transits identified from the light curves constructed from the data collected by the Spitzer space telescope and had periods of 4.04, 6.06, 8.1, 12.3 and 18.9 days. The mass and size of the planets were determined through analyzing transit depths and Transit timing variations, which also reflected the significant planet on planet forces present, influencing their orbits and transit timings. Trappist-1b, c, e, f, g have been found to have sizes close to the Earth and all of them have an equilibrium temperature within the range at which liquid water can exist. The orbital periods of the planets demonstrate multiple examples of orbital resonance (1/2 for 4.04/8.1 and 6.06/12.3) which present the possibility of the planets initially forming further from each other and then migrating inwards (Fig 6).

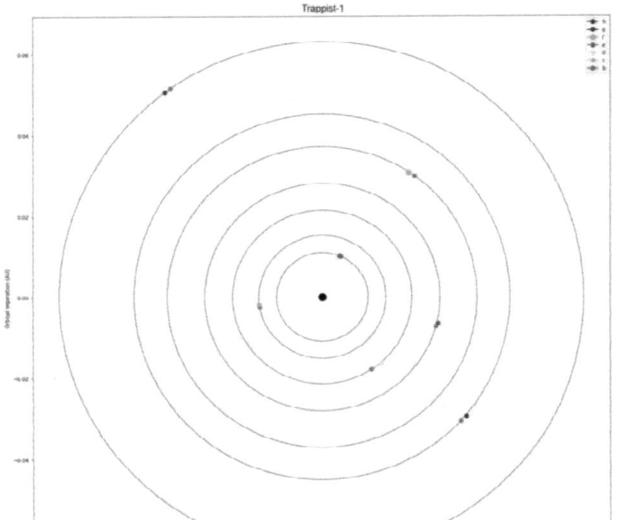

Figure 6. This is a visualization of the simulation at the end of the 300 days, showing multiple aspects of the trappist-1 planets' orbit including their eccentricities, semi-major axis and positions throughout the simulation. It can be observed that the positions of the planets when gravitational influence from other planets is included, represented by the colored dots, deviate significantly from the positions of the planet if it was only engaged in a two-body relationship with the star, represented by the grey dots.

The simulation allowed us to visualize the evolution of the orbits in terms of their mean anomaly and eccentricity in some cases. For the Trappist-1 system, all seven of the planets exhibit visible change in their orbits. Figure 7 shows the difference in mean anomaly which essentially represents the planet's angular position on its orbit, reflects the deviations from a Kepler orbit. Despite the deviations, the system is predicted to be able to remain stable for at least 0.5 million years, as when a simulation is carried out for 1000 years, the semi-major axis's deviations in relation to the semi-major axis is negligible as seen in Figure 8. There are observable patterns in the differences between the mean anomalies, an example would be Trappist-1e, and g, which are due to the orbital resonance between the planets as the ratio between their periods are close to a whole number (Table 1). Although the resonance chains result in periodic gravitational influences as opposed to random pushes, the deviations from a Kepler orbit remains prominent and constructs a complex system that does not directly demonstrate the universal law of gravitation. The most apparent demonstration of the universal law of gravitation in a system would be the gravitational influence exerted by a star on the planet but in compact systems like Trappist-1, the complex planet to planet interactions, despite also being governed by the universal law of gravitation, blur the relationship between the star and the planet.

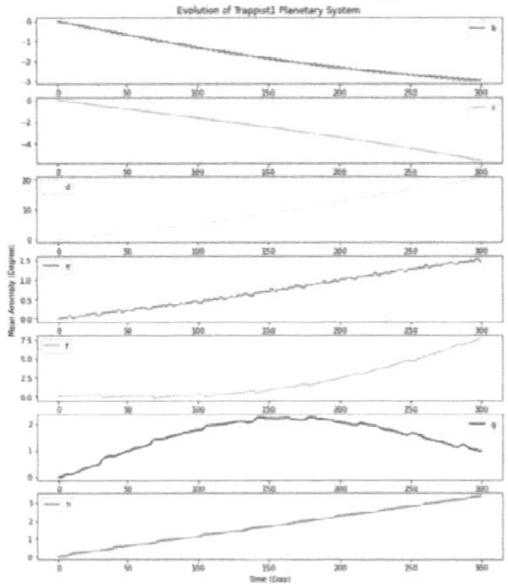

Figure 7. This shows the mean anomaly change of the Trappist-1 planets over the period of 300 days in degrees.

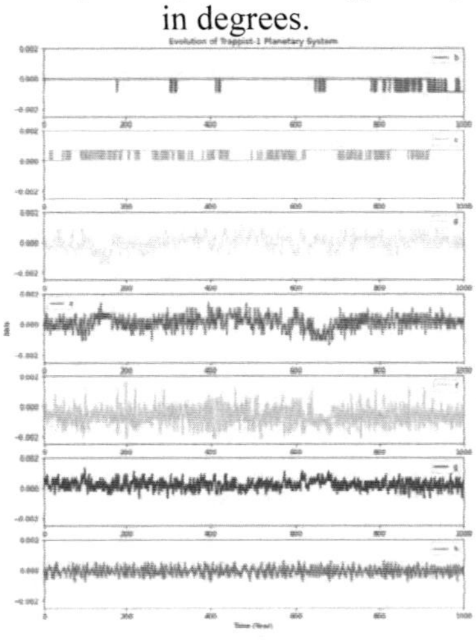

Figure 8. The variations of planetary semi-major axis in a dynamical simulation with a long period of 1000 years are presented. The difference in their semi-major axis when only the gravitational relationship with the star's mass is considered and when the gravitational influence from other planets are also considered, remains within a range of 0.0017 and 0.00033 of the planet's measured semi-major axis. The trivial variations of this parameter demonstrates the stability of the system while they are also evident of deviations from a Keplerian orbit.

Table 1. Trapppist1. Col 1: Planet, each planet in the Trappist-1 system from b to h; Col 2: a, semi-major axis in au; Col 3: e, eccentricity; Col 4: P, the orbital period in unit of day; Col 5: g, the argument of periapsis; Col 6: Inclination in unit of degrees; Col 7: ΔM, change of mean anomaly over 300 days; Col 8: Δa/a, change of semi-major axis over the measured semi-major axis.

Planet	a	e	p	g	I	ΔM	Δa/a
b	0.01	0	1.51	0	89.65	3.011	0.00045
c	0.015	0	2.42	0	89.67	5.580	0.000328
d	0.0214	0	4.049	0	89.75	20.106	0.00139
e	0.02817	0	6.0996	0	89.86	1.539	0.00177
f	0.0371	0	9.206	0	89.68	7.950	0.00229
g	0.0451	0	12.35	0	89.71	2.249	0.0111
h	0.03	0	18.76	0	89.76	3.334	0.000873

3.3 Kepler186

The Kepler 186 system consists of five planets with one being inside the habitable zone of the star. Planet masses are not confirmed for any of the planets however estimates have been made based on their possible compositions. (The distance between the inner four planets (b, c, d, e) are much smaller compared to their distance to planet f whose semi-major axis is around 0.432 au, as shown in figure 9. Being the only known planet in the habitable zone of Kepler-186, planet f has been observed to have the qualities necessary to maintain liquid water. [9] The pattern in the change of mean anomaly, most obviously seen in planets b, c, and d can be attributed to their orbital periods whose ratios with each other are close to those between small integers (p_b:p_c is close to 3:4, p_c:p_d is close to 4:7). The ratios in their orbital periods reflect the periodic pushes to and from each planet, which stack up to produce the resonance pattern seen. Despite the large changes in mean anomaly seen in fig 10, when a simulation over the time period of 1000 years was done, shown in Figure 11, the Kepler 186 system demonstrated its stability over time through the minute changes in its semi-major axis in relation to its measured semi-major axis (Table 2).

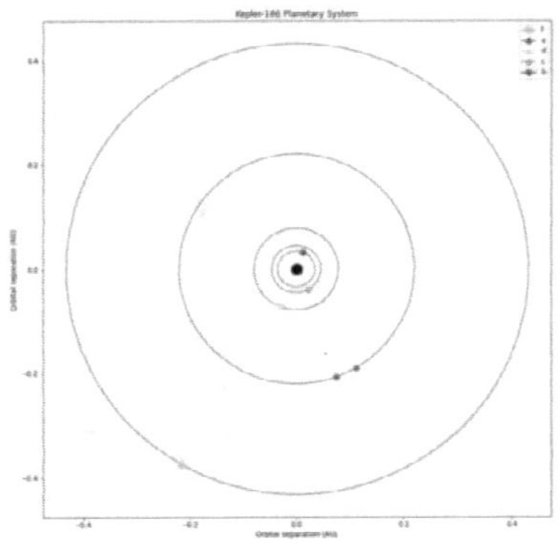

Figure 9. visualization of the simulation of Kepler 186 at 100 days. Colored dots represent the planets' orbit if the gravitational influence of other known planets were included. Grey dots represent the planets' orbits if the only gravitational influence they experience is from the star.

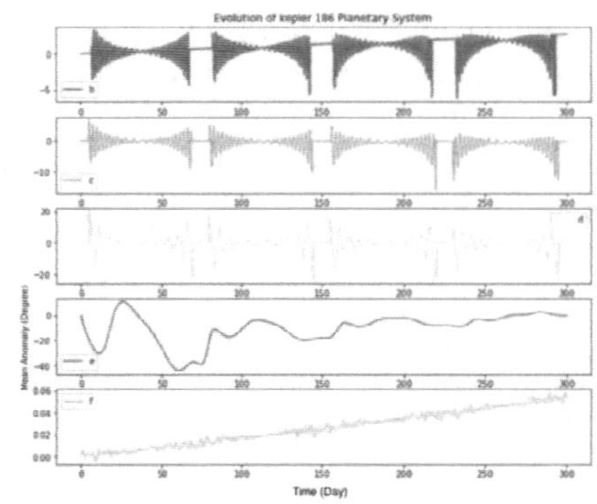

Figure 10. The change in mean anomaly over the time period of three hundred days of Kepler 186 planets is shown in degrees.

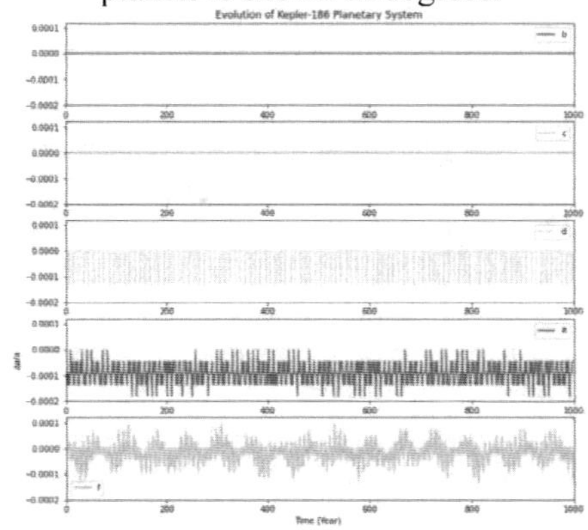

Figure 11. The plots present the ratio between the change in Semi-major axis and the planet's measured semi-major axis for each of the Kepler 186 planets.

Table 2. Kepler 186, see table 1.

Planet	a	e	p	g	I	ΔM	Δa/a
b	0.034	0	3.88	0	0	9.426	0
c	0.045	0	7.26	0	0	22.432	0
d	0.078	0	13.34	0	0	43.331	$6.411*10^{-9}$
e	0.221	0	22.4	0	0	55.272	$9.092*10^{-9}$
f	0.432	0	129.94	0	0	0.0614	0.000127

3.4 GJ 163

The three currently confirmed planets in the system GJ 163 were all found using radial velocity, 150 measurements of GJ 163 were taken by Bonfils et al. over eight years and five significant signals were found. Although two need further interpretation, the other three have been identified as exoplanets. GJ 163 b c and d each have semi-major axes of 0.061, 0.13, and 1.03 au while the other two potential planets e and f would have the semi-major axis of 0.10 and 0.33 au. GJ 163 c, which receives more energy from its star than Earth has been found to potentially be habitable as its equilibrium temperature falls within the range in which liquid water may exist, however its habitable nature needs to be further analyzed [10]. Using our simulations of the mean anomaly of the planets over three hundred days shown in figure 12, we can see that although the ratio between the periods of GJ 163 b and c is 2.97, there is no evidence of resonance in its mean anomaly, which matches with the dynamical analysis performed by Bonfils et al.; also although the plot of GJ 163 d presents a pattern of how the mean anomaly changes, the ΔM over three hundred days is only about 0.043 degrees. When we performed a simulation of the system (specifically looking at the semi-major axis) over 1000 years, the system was able to remain stable as the largest change in a planet's semi-major axis was only around 0.002% of the planet's semi-major axis.

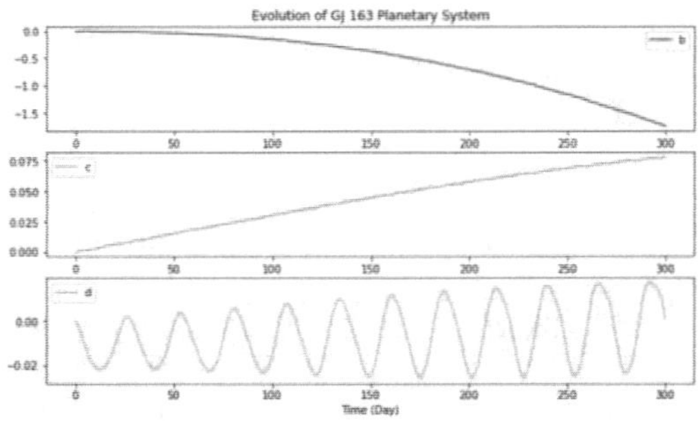

Figure 12. This shows the change in mean anomaly of GJ 163 over the period of 300 days in degrees.

The mean anomaly deviations demonstrated with our simulations are can also be found in transit timing variations, which have also been used to detect exoplanets (Table 3). In certain systems, the period of time between the occurrence of transits does not remain constant, which is observational evidence of the shift of the mean anomaly of a planet, one possible explanation of this may be the existence of an external force interfering with the planet's orbit (Fig 13). This force may be from an unconfirmed planet however it likely needs to have other observational evidence for it to be confirmed.

Table 3. GJ 163. see table 1

Planet	a	e	p	g	I	ΔM	Δa/a
b	0.06	0.073	8.63	70.7	0	1.745	$2.778*10^{-5}$
c	0.1254	0.099	25.631	227.1	0	0.0781	$6.084*10^{-4}$
d	1.03	0.373	603.95	118.3	0	0.0437	$1.394*10^{-11}$

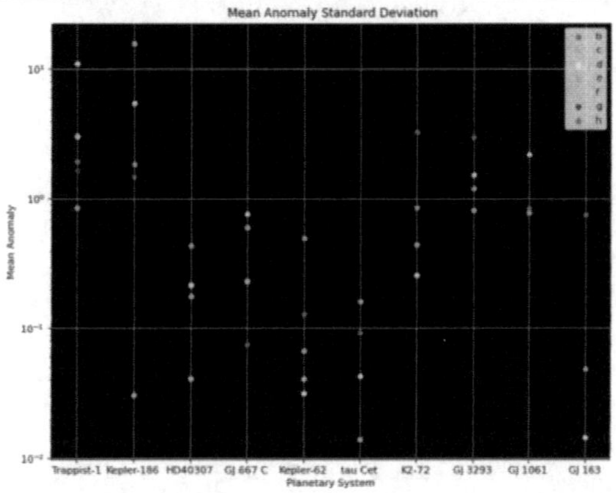

Figure 13. Standard deviation for the mean anomaly for each planet in the ten systems. The color for each lettered planet is the same across all systems.

4. Conclusion

Using the simulations, we specifically examined how in compact planetary systems, the perturbed orbits of certain planets give rise to an unstable environment for their possible inhabitants. In Trappist-1 we found habitable planets d, e, f, and g to deviate by 20.106, 1.539, 7.950 and 2.249 degrees each in terms of their mean anomaly and although the most perturbed are not habitable as they cannot contain liquid water, inhabitants of the habitable planet can often observe those planets' orbits as we observe the other planets in our solar system. In Kepler 186, we found the average of the deviations in terms of mean anomaly for all the known planets in the system was 26.09 degrees. The habitable planet Kepler 186 f, which has the largest orbital period in the system of 129.9 days, is minimally perturbed, partly because of its great distance to the other planets; however as observers view the other planets in the Kepler 186, they would be able to pick up on the irregular orbital periods of the other planets, such as Kepler 186d, which has a ΔM of around 43.33 degrees over 300 days, and to an observer on Kepler 186f, it would appear as a complicated system that doesn't follow something like Newton's universal law of gravity.

In GJ 163, we found the average of the deviations in terms of mean anomaly for the confirmed planets in the system was 0.622 degrees. Although in comparison to other systems, this value appears to be much smaller, the planets are in a compact system, especially GJ 163 b, which has a period of 8.6 days, the seemingly small deviations in mean anomaly will still lead to a change in its orbit. With the analysis of the three systems, out of the ten which we performed the simulation on, we found significant deviations in terms of their mean anomalies. Astronomers on Earth, like Kepler, used observations of other planet's orbits in the solar system to make conclusions which then led Newton to the universal law of gravitation as he attempted to explain the observations; however for the compact systems like Trappist-1, observers from the habitable planets will only be able to record perturbed orbits of the other planets in the system, which do not directly demonstrate the relationship between mass and force as Newton's law of universal gravitation does. An understanding of the law would be the basis of all space exploration attempts, as to launch an object out of Earth, the role played by Gravity must be understood; however in the compact systems seen above, it may be difficult to understand NLUG because of the naturally perturbed orbits, which would then slow the process of exploring the world outside of the planet, providing one of many possible explanations to the Fermi paradox. Further developments in detection of exoplanets can provide a new perspective and understanding of the fermi paradox as they will allow deeper interpretations of the possible habitats for extraterrestrials and whether their absence in the solar system can be attributed to their environment or other factors.

5. References

[1] Mayor, Michel and Queloz, Didier. 1995. "A Jupiter-mass companionto a solar-type star." Nature, 378: 355-359. https://doi.org/10.1038/378355a0

[2] Schulze-Makuch, D., Méndez, A., Fairén, A. G., von Paris, P., Turse, C., Boyer, G., Davila, A. F., António, M. R., Catling, D., & Irwin, L. N. (2011). A two-tiered approach to assessing the habitability of exoplanets. Astrobiology, 11(10), 1041–1052. https://doi.org/10.1089/ast.2010.0592

[3] Hart, Michael H. 1975. "Explanation for the Absence of Extraterrestrials on Earth." Quarterly Journal of the Royal Astronomical Society, 16: 128-135. https://ui.adsabs.harvard.edu/link_gateway/1975QJRAS..16..128H/ADS_PDF

[4] Murray, Carl D. and Dermott, Stanley F. 1999. Solar System Dynamics. Cambridge, UK: Cambridge University Press

[5] Perryman, Michael. 2018. The Exoplanet Handbook by Michael Perryman. Cambridge University Press. Second Edition

[6] Chambers, J. E. 1999. "A hybrid symplectic integrator that permits close encounters between massive bodies." Monthly Notices of the Royal Astronomical Society, 304: 793-799. https://doi.org/10.1046/j.1365- 8711.1999.02379.x

[7] NASA Exoplanets Archive. "NASA Exoplanets Archive.". https://exoplanetarchive.ipac.caltech.edu/

[8] Gillon, M., Triaud, A. H., Demory, B.-O., Jehin, E., Agol, E., Deck, K. M., Lederer, S. M., de Wit, J., Burdanov, A., Ingalls, J. G., Bolmont, E., Leconte, J., Raymond, S. N., Selsis, F., Turbet, M., Barkaoui, K., Burgasser, A., Burleigh, M. R., Carey, S. J., … Queloz, D. (2017). Seven temperate terrestrial planets around the NEARBY ultracool dwarf star trappist-1. Nature, 542(7642), 456–460. https://doi.org/10.1038/nature21360

[9] Quintana, E. V., Barclay, T., Raymond, S. N., Rowe, J. F., Bolmont, E., Caldwell, D. A., Howell, S. B., Kane, S. R., Huber, D., Crepp, J. R., Lissauer, J. J., Ciardi, D. R., Coughlin, J. L., Everett, M. E., Henze, C. E., Horch, E., Isaacson, H., Ford, E. B., Adams, F. C., … Selsis, F. (2014). An earth-sized planet in the habitable zone of a cool star. Science, 344(6181), 277–280. https://doi.org/10.1126/science.1249403

[10] Bonfils, X., Lo Curto, G., Correia, A. C., Laskar, J., Udry, S., Delfosse, X., Forveille, T., Astudillo-Defru, N., Benz, W., Bouchy, F., Gillon, M., Hébrard, G., Lovis, C., Mayor, M., Moutou, C., Naef, D., Neves, V., Pepe, F., Perrier, C., … Ségransan, D. (2013). The harps search for southern extra-solar planets. Astronomy & Astrophysics, 556. https://doi.org/10.1051/0004-6361/201220237

Meshing Simulation of Face-Gear Drives using Multibody Dynamics Software

C K Lee[1], C L Hsieh[1], P C Lin[1] and C Y Wang[2]

1 Department of Industrial Engineering and Management, Cheng Shiu University, 840 Chengcing Rd., Kaohsiung, Taiwan (R.O.C.)
2 Department of Mechanical Engineering, Cheng Shiu University, 840 Chengcing Rd., Kaohsiung, Taiwan (R.O.C.)
E-mail: cklee@gcloud.csu.edu.tw

Abstract. This paper aims to conduct the meshing simulation of two designs of face-gear drives (FGDs) by using the multibody dynamics software, SolidWorks motion. The pinion in the original design of FGD is not provided with any crowning modification, but the pinion in the improved design of FGD is provided with both profile and longitudinal crowning modifications. The results obtained by the meshing simulation show that the edge contact phenomenon greatly reduces the transmission stability of the original design. By giving the pinion a convex dressing in two directions, the transmission stability can be largely improved. The fluctuation in angular velocity and angular acceleration of the improved design of FGD is obviously lower than those of the original design.

1. Introduction

Both bevel gear drives and face-gear drives can transmit the rotation and torque between non-parallel shafts. However, face-gear drives are more suitable for splitting power than bevel gear drives. Many studies on face-gear drives have been reported. In this paper, the term face-gear drive is simply represented by the abbreviation FGD and the term face-gear is simply represented by the abbreviation FG. Litvin et al. [1] proposed an improved version of orthogonal offset FGD. Litvin et al. [2] proposed two new versions FGDs. Litvin et al. [3] studied one FGD in which the FG is generated by a worm. Zanzi and Pedrero [4] studied one FGD in which the pinion is provided with double-crowned teeth. Litvin et al. [5] studied two new versions of FGDs. One FGD is developed based on a screw involute helicoid, and the other FGD is based on two mismatched parabolic racks. Lee [6] designed one FGD to obtain a transmission error function of fourth-degree and contact ellipses with major and minor dimensions controllable. Lee [7] reported the mathematical model and condition of undercutting for a cosine FGD. Lee [8] proposed a cosine FGD having a fourth-degree transmission error function. Shen et al. [9] proposed a method for grinding FG of double crowned tooth geometry on a multi-axis CNC machine. Feng et al. [10] investigated one FGD with an involute helical pinion. Zhou et al. [11] studied a backlash adjustable FGD. Zhou et al. [12] proposed a grinding method of FG mating with a conical spur involute pinion. At present, no literature has been found to perform the meshing simulation of FGDs by using multibody dynamics software.

2. Mathematical modeling

In this section, the mathematical model of the tooth face of one FG is derived by applying the theory of gearing [13]. The FG is generated by a shaper tool. In other words, the tooth face of the shaper tool is the generating surface, and the tooth face of the FG is the generated surface. To mathematically represent the tooth faces, the coordinate system S_s is applied to connect rigidly to the shaper tool and the coordinate system S_F is applied to connect rigidly to the FG, as shown in Figure 1.

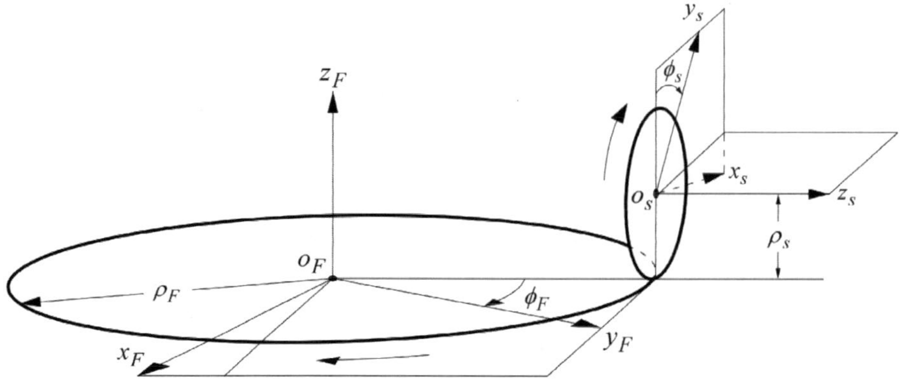

Figure 1. The system S_F connects rigidly to the FG and the system S_s connects rigidly to the shaper tool.

The coordinate transformation matrix from S_s to S_F is denoted by $\mathbf{M}_{Fs}$ and is represented as follows:

$$\mathbf{M}_{Fs} = \begin{bmatrix} -C\phi_s C\phi_F(\phi_s) & -S\phi_s C\phi_F(\phi_s) & -S\phi_F(\phi_s) & -\rho_F S\phi_F(\phi_s) \\ -C\phi_s S\phi_F(\phi_s) & -S\phi_s S\phi_F(\phi_s) & C\phi_F(\phi_s) & \rho_F C\phi_F(\phi_s) \\ -S\phi_s & C\phi_s & 0 & \rho_s \\ 0 & 0 & 0 & 1 \end{bmatrix} \quad (1)$$

Here, symbols $\mathbf{R}_s(u_1,u_2)$ and $\mathbf{n}_s(u_1,u_2)$ are used to denote respectively the position and unit normal vectors in S_s of the tooth face of the shaper tool. Symbols $\mathbf{R}_F^{(s)}(u_1,u_2,\phi_s)$ and $\mathbf{n}_F^{(s)}(u_1,u_2,\phi_s)$ are used to denote respectively the position and unit normal vectors in S_F of the family of surfaces formed by the tooth face of the shaper tool. According to the theory of gearing [13], $\mathbf{R}_F^{(s)}(u_1,u_2,\phi_s)$ and $\mathbf{n}_F^{(s)}(u_1,u_2,\phi_s)$ can be determined by the following equations:

$$\begin{cases} \mathbf{R}_F^{(s)}(u_1,u_2,\phi_s) = \mathbf{M}_{Fs}\mathbf{R}_s(u_1,u_2) \\ \mathbf{n}_F^{(s)}(u,v,\phi_s) = \mathbf{L}_{Fs}\mathbf{n}_s(u,v) \end{cases} \quad (2)$$

The equation of meshing between the tooth face of the shaper tool and the tooth face of the FG is as follows:

$$F = \mathbf{n}_F^{(s)} \cdot \left[\partial \mathbf{R}_F^{(s)} / \partial \phi_s \right] = 0 \quad (3)$$

Simultaneously considering the family of surfaces and the equation of meshing yields the mathematical model of the generated tooth face of the FG as follows:

$$\begin{cases} \mathbf{R}_F(u_1,u_2,\phi_s) = \mathbf{R}_F^{(s)}(u_1,u_2,\phi_s) \\ \mathbf{n}_F(u_1,u_2,\phi_s) = \mathbf{n}_F^{(s)}(u_1,u_2,\phi_s) \\ F = \mathbf{n}_F^{(s)} \cdot \left[\partial \mathbf{R}_F^{(s)} / \partial \phi_s \right] = 0 \end{cases} \quad (4)$$

3. The original design of FGD

Figure 2 shows the original design of FGD, which is consisted of one pinion and one FG. The number of teeth of the pinion-type shaper tool applied to generate the FG is one tooth greater than that of the pinion. The parameters are set as follows: the numbers of teeth of the pinion, pinion-type shaper tool, and FG are 18, 19, and 37, respectively; the normal module is 2 mm; the pressure angle at the normal section is 20 degrees; the helix angle at the reference circle is 0; the axial offset is 2.55 mm; the face-width of the pinion is 10.5 mm; the face-width of the FG is 5.4 mm; the profile shift coefficient of the pinion is 0.28; and the profile shift coefficient of the pinion-type shaper tool is 0.28.

The multibody dynamics software SolidWorks motion is applied herein to perform the meshing simulation. The solver of SolidWorks motion is ADAMS. In the contact property manager, the contact type is set to solid bodies, the materials are set to steel (greasy), the friction is neglected, and the contact force calculation is set to the impact model. As shown in Figure 3, the angular velocity applied to the pinion shaft hole is set to STEP5(time,1,0,2,3) deg/sec, which means the angular velocity is gradually loaded from 0 to 3 deg/sec in the time range from 1 to 2 seconds, and then remains 3 deg/sec unchanged. As shown in Figure 4, the torque applied to the shaft hole of the FG is set to STEP5(time,0,0,1,5000) N-mm, which means the torque is gradually loaded from 0 to 5000 N-mm in the time range from 0 to 1 second, and then keeps 5000 N-mm unchanged. The rotational damping applied on the shaft hole of the FG is set to 500 N-mm/(deg/s). The 3D contact resolution is set to precise contact. The accuracy is set to 0.000001. The integrator type is set to GSTIFF.

Figure 2. The original design of FGD.

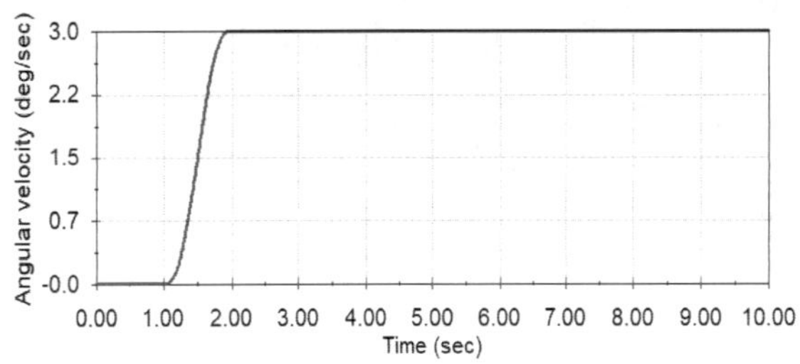

Figure 3. The angular velocity applied to the pinion.

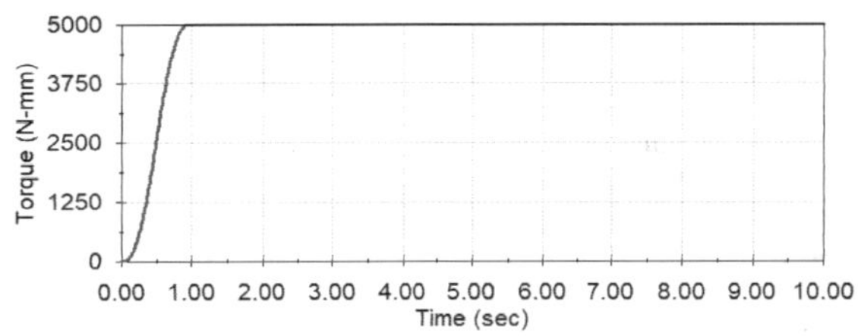

Figure 4. The torque applied to the FG.

The results obtained by the meshing simulation of the original design of FGD are shown in Figures 5 to 8. Figure 5 shows the contact regions when the time is at 7.47 seconds. There are three teeth in contact at the same time. Figure 6 shows the contact regions when the time is at 9 seconds. There are still three teeth in contact at the same time, but the contact region on the far right has been reduced to a very small point. Figure 7 shows the angular velocity of the FG. Figure 8 shows the angular acceleration of the FG. It can be found that the angular velocity and angular acceleration fluctuate violently when one contact region gradually shrinks to a point and reaches the edge of the tooth face of the FG. In other words, the edge contact phenomenon deteriorates the stability of transmission of the FGD.

Figure 5. The contact regions when the time is at 7.47 seconds.

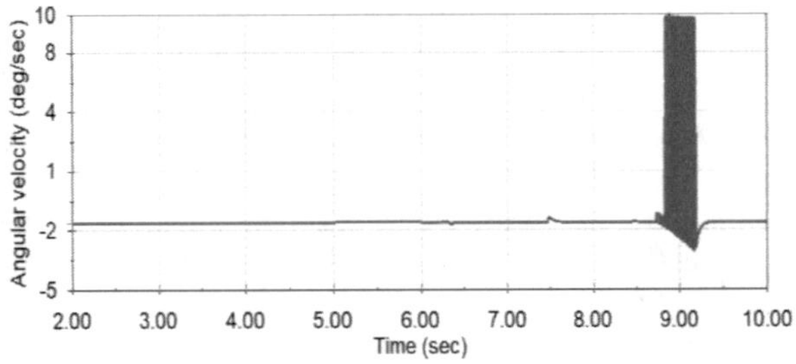

Figure 7. The angular velocity of the FG.

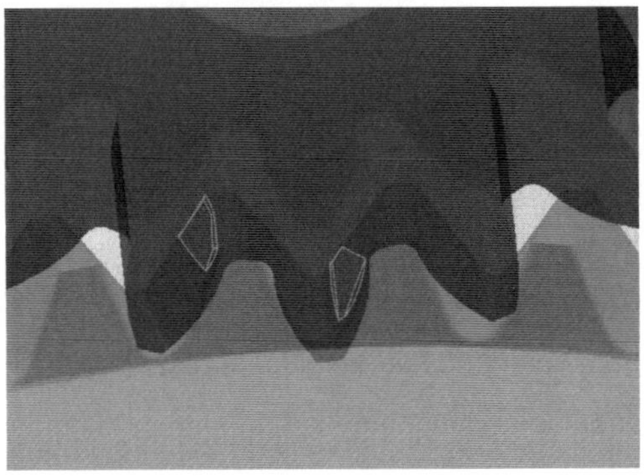

Figure 6. The contact regions when the time is at 9 seconds.

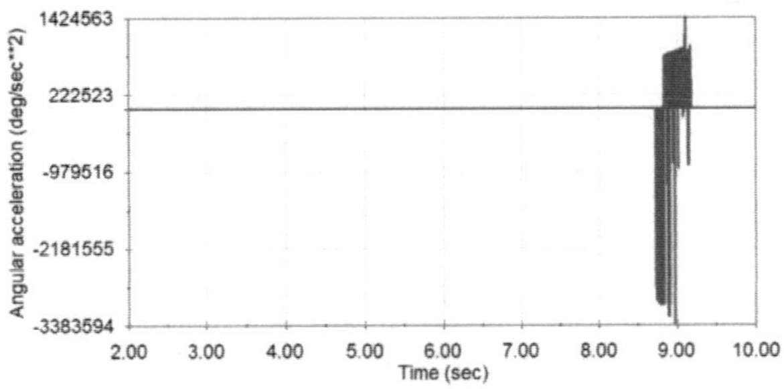

Figure 8. The angular acceleration of the FG.

4. The improved design of FGD

Since the original design of FGD has the defect above, an improved design of FGD is presented as shown in Figure 9. In the improved design, the pinion is provided with both profile and longitudinal crowning modifications as shown in Figure 10. The value of profile crowning is set to 100 μm, and the value of longitudinal crowning is set to 200 μm. To maintain consistency, the settings for the meshing simulation of the improved design of FGD is the same as the settings for the meshing simulation of the original design of FGD. The results obtained by the meshing simulation of the improved design of FGD are shown in Figures 11 to 14. Figure 11 shows the contact regions when the time is at 5.78 seconds. Figure 12 shows the contact regions when the time is at 6.07 seconds. Figure 13 shows the angular velocity of the FG. Figure 14 shows the angular acceleration of the FG. It can be found that the fluctuation of angular velocity and angular acceleration is reduced largely.

Figure 9. The improved design of FGD.

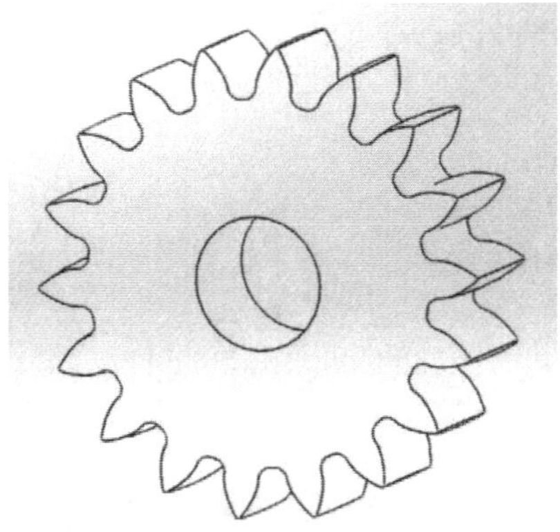

Figure 10. Profile and longitudinal crowned pinion

Figure 11. The contact regions when the time is at 5.78 seconds.

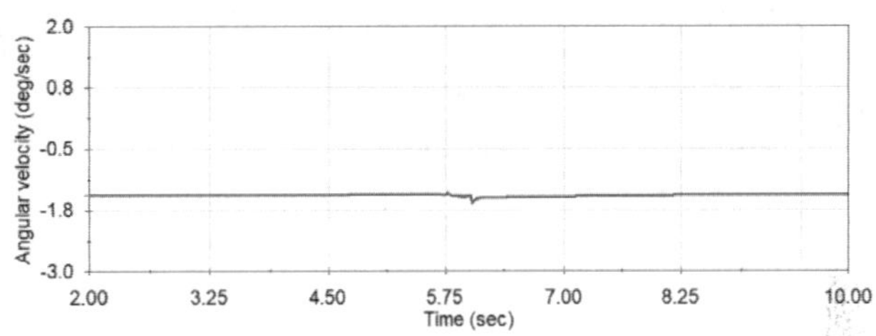

Figure 13. The angular velocity of the FG.

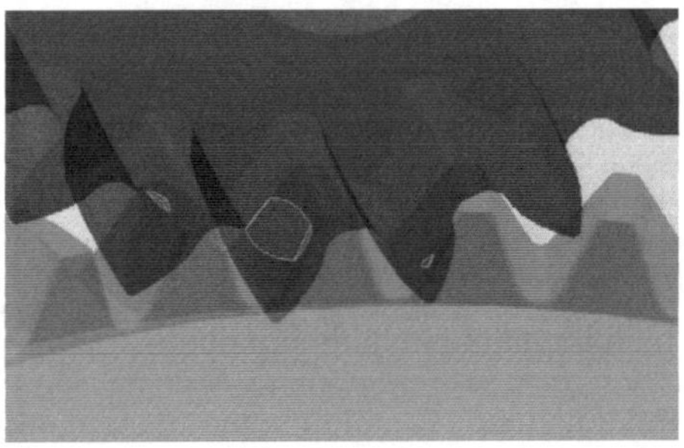

Figure 12. The contact regions when the time is at 6.07 seconds.

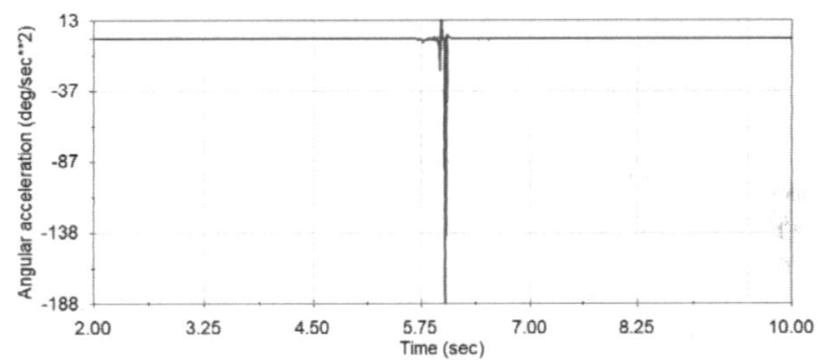

Figure 14. The angular acceleration of the FG.

5. Conclusion

This paper has accomplished the meshing simulation of two designs of FGDs by using the multibody dynamics software, SolidWorks motion. The results obtained by the meshing simulation of the original design of FGD show that the edge contact phenomenon causes serious instability of angular velocity and angular acceleration. The rapid increase of angular acceleration will cause a high level of noise and vibration and shorten the service life of bearing. To improve this drawback, an improved design of FGD is presented, in which the pinion is provided with both profile and longitudinal crowning modifications. The results obtained by the meshing simulation of the improved design of FG show that the stability of angular velocity and angular acceleration can be greatly advanced. Because this improved design has two directional crowning modifications, its low sensitivity to assembly errors can also be expected. It must be noted that two-directional crowning modifications are easy to cause the tip of the pinion to become sharp. Face-width limitation or chamfering modification can be used to avoid the tip sharping problem of pinion.

References

[1] Litvin F L, Egelja A, Tan J and Heath G 1998 *Mech. Mach. Theory* **33** 87-102
[2] Litvin F L, Fuentes A, Zanzi C and Pontiggia M 2002 *Mech. Mach. Theory* **37** 1179-211
[3] Litvin F L, Fuentes A, Zanzi C, Pontiggia M and Handschuh R F 2002 *Comput. Meth. Appl. Mech. Eng.* **191** 2785-813
[4] Zanzi C and Pedrero J I 2005 *Comput. Meth. Appl. Mech. Eng.* **194** 3047-66
[5] Litvin F L, Gonzalez-Perez I, Fuentes A, Vecchiato D, Hansen B D and Binney D 2005 *Comput. Meth. Appl. Mech. Eng.* **194** 3870-901
[6] Lee C K 2018 *J. Chin. Soc. Mech. Eng.* **39** 459-69
[7] Lee C K 2017 The mathematical model and the condition of undercutting for the cosine face gear drive generated by a cosine rack *DEStech Trans. Mater. Sci. Eng.*
[8] Lee C K 2019 *Adv. Mech. Eng.* **11** 1-20
[9] Shen Y, Liu X, Li D and Li Z 2018 *Mech. Mach. Theory* **121** 460-74
[10] Feng G S, Xie Z F and Zhou M 2019 *Mech. Mach. Theory* **134** 169-96
[11] Zhou R, Zhao N, Tan Z, Zhang M and Guo H 2019 *Mech. Mach. Theory* **140** 479-503
[12] Zhou R, Zhao N, Li W, Li R, Guo G and Guo H 2019 *Mech. Mach. Theory* **141**, 226-44
[13] Litvin F L and Fuentes A 2004 *Gear Geometry and Applied Theory (2nd ed.)* (Cambridge: Cambridge University Press)

A Survey of the Influence of Air Distribution on Indoor Environment and Building Energy Efficiency

Longkai Ma[1, *, †] and Zhili Zeng[2, †]

1 School of mechanics and civil engineering, China university of mining and technology, Xuzhou, China;
2 College of Mechanical Engineering, ChongQing University, Chongqing, China
*Corresponding author's e-mail: 26185966@cumt.edu.cn
†These authors contributed equally.

Abstract. With society paying more and more attention to energy conservation, building energy conservation has become a problem that must be considered in the design phase. In addition, modern life makes people spend more time indoors every day. Reasonable indoor air distribution positively impacts a comfortable indoor thermal environment and reduces building energy consumption. This paper discusses the influence of air distribution on indoor thermal comfort and building energy saving. Reasonable air distribution in the air-conditioned room will create a comfortable indoor thermal environment, which is helpful for the physical and mental health of personnel and efficient work. Meanwhile, it can also effectively reduce the equipment's initial investment and operating energy consumption and realize building energy savings. The design of indoor air distribution cannot be a single, isolated project. A comprehensive analysis based on each station's environmental thermal comfort requirements in the work area must be carried out to reduce air-conditioning operation energy consumption and achieve building energy-saving goals. For existing buildings and large space buildings, the corresponding air distribution design scheme should be designed reasonably. This paper also suggests future research directions on indoor air distribution.

1. Introduction

Building energy consumption accounts for a high proportion in the total social energy consumption. In China's total social energy consumption, building energy consumption accounts for about 20.7% [1]. Therefore, one of the most important tasks of energy conservation is to reduce building energy consumption. In the 2014 ranking of world building energy consumption, the energy consumption of American buildings accounted for 17% of the total building energy consumption worldwide, ranking first in the world. China accounts for 16.2%, ranking second globally [2]. Compared with other services, heating ventilation and air conditioning (HVAC) systems are the most energy-consuming services in large buildings, with the energy consumed by HVAC systems accounting for 30% to 50% of the total building energy consumption. Especially, in cold areas where heating is needed, energy consumption will be as high as 80%. The energy consumption of HVAC in large public buildings accounts for more than 60% of the total energy consumption of buildings [3,4]. With the increasingly serious energy crisis, energy saving will have an important impact on the design and operation of the HVAC system because of the huge energy consumption of the air conditioning system itself.

Scholars have studied the airflow distribution of indoor air through computational fluid dynamics (CFD), experiments, and theoretical analysis. The research shows that the main factors affecting the air distribution in the air conditioning room are the air supply parameters, the form and position of the air supply outlet, the position of the return air vent, the position of the heat source, and the shape of the room. According to the different layout positions and forms of air supply outlet and return vent, different air distribution forms will form, mainly including up-supply and down-return, up-return down-supply, sidewall supply, and return, etc.

For the upper supply and lower return vent arrangement, according to the setting of the air supply distance, louver vent, nozzle or swirl vent can be selected. The advantage of this air supply and return mode is that it can reach a uniform temperature gradient for the action area, and it is easy to arrange the air inlet. The disadvantage is that the air volume to be processed is large, and the system's energy consumption is large. Moreover, with many upper duct, the installation is complex. In winter, the vertical temperature gradient is great, so the lower part is cold and the upper part is hot.

The air supply mode of up-return down-supply is mainly aimed at specific situations, such as the high humidity in the space and the autonomous heat source in the room. Down air supply can be achieved through seat vents, floor vents, etc., combined with some composite systems such as radiation cooling. In this air supply mode, the air exchange rate is faster. In addition, the temperature and humidity of the working area can be effectively controlled, and the energy-saving effect of the system is obvious. However, the air inlet is located at the bottom, the speed and temperature of the air supply need to be strictly controlled. Otherwise, it will blow up dust or cause human discomfort. In addition, due to special vents, the cost is high, and the maintenance is inconvenient.

The air supply mode of sidewall supply and return, the action area of the air conditioning, is set in the reflux range of jet. The advantages are that the range of the jet is far and the temperature field is easy to achieve stability. Moreover, the wind speed decays slowly. Therefore, the airflow and temperature in the action area are uniform and more energy-saving. However, in winter, because the buoyancy of hot air is large, it is difficult for heat to reach the action area, so the uniformity of temperature distribution is not as good as other ways [5,6].

This paper discusses the influence of air supply mode on the indoor environment and building energy efficiency. In detail, the influence of air distribution on indoor air conditioning is introduced firstly. Then influence of air distribution on building energy consumption is summarized. After that, the optimization design of indoor air distribution is presented. Finally, conclusions are drawn, and suggestions for future research are given.

2. Influence of air distribution on indoor air conditioning

2.1. Influence of the air outlet layout on the indoor thermal environment

The research on indoor air distribution and indoor thermal comfort was carried out earliest in foreign countries. As early as 1945, the laboratory of the University of Kansas used measuring instruments to study the airflow organization of indoor hot air. This study was an advanced experimental research method of indoor airflow organization at that time. The experimental method concluded that for some buildings, the indoor comfort of the ventilation system with impinging jet was better than that of the displacement ventilation system. The ventilation system should be selected according to the construction conditions [7]. In 1974, P.V.Nielsen [8] from Denmark first used CFD method to simulate indoor airflow state. CFD has been introduced into the design business in Japan, Europe, and The United States, forming an efficient design system. Some famous projects, such as Fukuoka Stadium in Japan, Kansai Airport, Tokyo Metropolitan Stadium, New National Theater, Tokyo International Convention Center, various Winter Olympic Venues and Rose Park Stadium in the United States, have used CFD method to calculate and predict air distribution and temperature distribution to guide the optimization of engineering design [9].

CFD is also widely used in China to investigate indoor air distribution. Xu et al. [10] studied the airflow distribution and air quality in the room under the three modes of up-in and up-out, up-in and down-out, and displacement ventilation. The results showed that the room had good thermal comfort and good air quality with displacement ventilation. In 2015, Wu et al. [11] combined ventilation methods with indoor thermal comfort. With CFD method and zero-equation model to simulate three ventilation modes of up-supply up-return, up-return down-supply, up-supply and down-return in air-conditioning system, they compared the simulation results of temperature field, PMV, PPD and air age.

An [12] uses FLUENT as a platform to combine the custom function UDF. They use the RNG k-ε model combined with the wall function method and the Lagrangian method-based particle random trajectory model to establish the air-conditioned room's turbulence and particle model of the air-conditioned room. In the case of three types of air supply: up-inlet and down-outlet on the same side, up-inlet and down-outlet on the opposite side, bottom air supply and top return air, numerical simulations are performed on the airflow organization with air changes of 5 times/h and 10 times/h. The velocity field, temperature field, and air are obtained. The distribution rule of an age field and PMV-PPD thermal comfort index and the rule of the influence of different temperature and humidity combination conditions on the thermal comfort of personnel are demonstrated. The results show that the airflow organization of up-inlet and down-outlet on the same side is the best. Up-inlet and down-outlet on the opposite side will cause the airflow to not circulate. The temperature stratification of bottom air supply and top return air is obvious, with the highest energy utilization coefficient. The temperature has a great influence on human thermal comfort, while humidity has no obvious influence. Li [13] studied the indoor airflow organization of the nozzle side air supply and the new diffuser top air supply mode in summer. It was found that the temperature field and velocity field of the air conditioning room were greatly affected by the outlet wind speed of the supply air outlet under the condition of the nozzle side air supply. The lateral air distribution has a good mixing effect on the airflow in the room, forming a uniform temperature field and velocity field. The closing or opening of the light plate of the new type of top air distributor has little effect on the temperature. The top air supply attached to the ceiling and wall jet makes the room temperature and velocity distribution uniform. For ultra-low energy buildings that can provide a comfortable indoor environment and minimize energy consumption, Tian et al. [14] used ANSYS to analyze the different forms of fresh air outlets in the room and their position relationship. When the fresh air outlets are arranged near the air conditioning outlets, the form of louver fresh air outlets on the top side is better than that on the bottom diffuser. In addition, the spacing between the top side louver fresh air outlet and the air conditioning outlet should not be too large. If the distance is more than 400 mm, the cold air from the fresh air outlet cannot be heated by the hot air from the air conditioning outlet, making the working area near the fresh air outlet feel blowing.

2.2. Study on the influence of air supply angle on indoor comfort

In addition to the installation location of the air-conditioning air supply outlet, the air supply angle greatly influences indoor comfort [15]. If the angle is reasonable, it will effectively improve indoor comfort and reduce air conditioning energy consumption, which is conducive to building energy efficiency [10,16]. In general, the air supply angle and the installation position of air-conditioning air supply outlet are important factors affecting indoor thermal comfort [17].

In 2014, Li [18] took the atrium of a hotel in Chongqing as the research object, simulating the air conditioning mode combining the nozzle side air supply and the floor air supply on both sides of the atrium through CFD. By analyzing the distribution of temperature, velocity, PMV and PPD, it was found that the optimal air supply angles in winter and summer were + 7°and − 15°, respectively, verifying the effectiveness and reliability of CFD simulation of large-space air conditioning. The research results of Li [19] showed that the air supply angle of the swirl vent had a great influence on the airflow organization, human comfort and energy-saving effect in the multifunctional banquet hall. The results show that under the optimal working conditions of the multifunctional banquet hall, the air supply angle of the cyclone outlet is 45° in summer and 60° in winter. The comprehensive optimal parameters under the design conditions in summer were determined by orthogonal test analysis. Under winter conditions, the simulation results show that under the premise of ensuring fresh air supply, the combination of low-temperature radiant floor heating and convective heat dissipation coil heating has better comfort and lower energy consumption.

Air supply angle is directly related to personnel comfort. With more office buildings using air conditioning for room cooling, due to the long-term direct blowing of air outlets in some stations, indoor personnel have a sense of blowing, overheating or super-cooling, leading to adverse reactions such as headache and dizziness [20]. In response to this phenomenon, Sun [21] simulated the air supply angle of the air conditioning in the used office, and obtained a suitable air supply angle so that most indoor personnel could meet the thermal comfort conditions such as indoor air temperature and velocity. The angle-adjustable side air supply shutters are installed on the west side. In the middle installed angle adjustable card four outlet fan coil. A diffuser with an angle of 45° is installed on the east side. Sun simulated the situation of different air supply angles if each fan coil was opened alone. The simulation results show that if the supply angle of the side supply fan coil is 0°or 30°, the supply distance is far. In the working area with vertical height of 2 meters, the indoor temperature difference and wind speed are small, so it is more comfortable. If the supply angle of the four-outlet fan coil is 45°, the indoor temperature distribution is relatively uniform. With a small wind speed, thermal comfort is ideal. The simulation of three fan coils opening at the same time shows that the ideal thermal comfort can be obtained. The measured results of the simultaneous opening of three fan coils show that indoor the simulation results of indoor temperature are in good agreement with the measured results.

2.3. The influence of air distribution in air-conditioning room on indoor thermal comfort

Unreasonable indoor air distribution of air conditioning will not only make the selection of air conditioning equipment' capacity too large, increasing the energy consumption, which is not conducive to achieve building energy conservation, but also reduce the work efficiency of personnel, affecting the thermal comfort of the indoor environment [22].

The study of air distribution began in the 1920s. Human's poor comfort caused by cold air feeling attracted the attention of researchers in the field of air conditioning [23]. Li [24] studied the air distribution and thermal comfort in the office. He found that in the room with louver vent supplying air at the top, people under the vent felt extremely uncomfortable In the office with side air supply, the area close to the vent had high wind speed and low temperature. In this case, the staff in this area could not work normally. Li used the job satisfaction rate to evaluate the indoor thermal comfort of the office building, analyzing the uncomfortable problems in the office more specifically and putting forward optimization plans. For rooms where air distribution of air conditioning is easily changed, measures such as adjusting air supply angle and additional flow diversion device can improve thermal comfort. Aiming at the room where air distribution is not easy to change, the air distribution is analyzed to adjust the station layout and improve the thermal comfort of the station. Therefore, to improve the comfort of the workstation in the office, two ways can be adopted: improving the distribution form of air distribution in the room and adjusting the layout of the workstation.

Meanwhile, in order to fully consider the requirements of occupants on different levels of thermal comfort environment and air quality, Fanger [25] from Denmark proposed personalized ventilation, which directly sent fresh air into the breathing area of indoor occupants according to their different ventilation requirements resulting from their clothing and mental state. Zhao et al. [26] from Tsinghua University compared the ventilation effects of four rooms by introducing a modified evaluation parameter of air exchange efficiency. They found that different ventilation methods had various effects on different personnel distribution in the room.

Reasonable air distribution can improve the comfort of indoor personnel and effectively discharge indoor pollutants. Jiao et al. [27] from Hunan University conducted numerical simulation of indoor temperature field, velocity field, pollutant concentration field and PMV-PPD distribution under mixed ventilation and displacement ventilation. The results demonstrated that displacement ventilation had a better effect on thermal comfort, while mixed ventilation better affected indoor pollutant removal ability. In 2016, An [12] carried out a numerical simulation of the distribution of two kinds of particulate matter generated by indoor personnel activities under different air supply forms, concluding that down-to-top delivery has better suppression and removal effect on particulate matter.

3. Influence of air distribution on building energy consumption

The form of air distribution is closely related to building energy consumption. Because some designers do not pay enough attention to the indoor airflow organization, resulting in uneven indoor ambient temperature, designers have to increase the capacity of the equipment. In addition, due to unreasonable air flow organization, the operation time of air conditioning increases, so the operation energy consumption of air conditioning increases [28]. Because of large volume and large air conditioning load, large space buildings' air conditioning energy consumption is considerable, generally accounted for 40%-60% of the total building energy consumption. In this case, the energy saving of large space building's air conditioning has attracted more and more attention.

With the increasing attention to the energy-saving of air conditioning systems in large space buildings, Cheng et al. [29] carried out research on stratified air conditioning systems. The study found that the absolute energy-saving potential of the stratified air conditioning system is proportional to the discharge air temperature. Therefore, if the return air inlet and the discharge air outlet are arranged separately, increasing the discharge air temperature and improving the building energy-saving potential is beneficial. Under the test conditions, if the return air inlet and discharge air outlet are arranged separately, the energy-saving potential of stratified air conditioning is greatly improved with the decrease of return air inlet height. In 2017, Peng et al. [30] took Xinjiang International Convention and Exhibition Center as an example to theoretically analyze the composition of stratified air conditioning load and the calculation of airflow organization, and verified the theoretical design calculation through CFD simulation. It was found that stratified air conditioning had obvious advantages in energy saving in summer but no obvious advantages in winter.

Waiting station is also a common large space building. The waiting hall on the ground floor of Xiongan Station is a typical tall space. The design adopts a decentralized air supply unit system and air supply with attached jet combined with the end of side nozzle. By means of computer simulation, Li et al. [31] compared the air distribution in the personnel activity area of the waiting hall under three working conditions in summer (attached air supply, nozzle side air supply, attached air supply combined with nozzle side air supply). They obtained the better air supply mode, checking the design working conditions in winter. The results show that compared with the centralized all-air system, the energy consumption of the distributed system is reduced by about 30% in summer, and the air supply mode with attached jet combined with nozzle is reduced by 15% in summer compared with the traditional nozzle side supply mode.

In view of the "chimney effect" in large stadiums and the uneven temperature distribution in the front and rear seating areas, Hua [32] found that compared with the up-send down-return air distribution form for large stadiums, down-send up-return form was more energy saving. If the occupancy rate is greater than 70%, the lower air supply and radiant cooling plate are used to eliminate the space cooling load. If the occupancy rate is below 70%, down-send up-return form can be directly used to eliminate the cooling load. Gao [33] from South China University of Technology took a university gymnasium in Guangzhou as an example to optimize the original air distribution. He optimized the unreasonable distribution of velocity field and temperature field in the gymnasium. On the basis of satisfying PMV-PPD, the optimized air conditioning system saved 16.9% in average during the whole year, achieving good economic effect. Lai [34] first modified the solution of the inner surface temperature of the building envelope in the stratified air conditioning load calculation, proposing a solution model to correct the simplified calculation results. The theoretical calculation was used to analyze the whole room air conditioning load with upper supply and lower return and the stratified air conditioning with side supply and lower return under various working conditions. The air conditioning load of the two methods and the theoretical load energy saving rate of the stratified air conditioning were obtained. To verify the distribution of temperature and air speed fields during actual operation and find out the optimal air distribution design scheme, Lai used FLUNT to simulate the working conditions of typical rooms in summer and winter, obtaining the applicability of full-room and stratified air conditioning. The energy-saving effect of stratified air conditioning is significant in summer, but the energy-saving rate in winter is lower than that in summer. Due to cold air infiltration and unreasonable design in Chongqing, the heat capacity is much higher than the calculated heat load. This will result the heating effect of air conditioning is insufficient in winter. To verify the accuracy of theoretical calculation, Lai analyzed working conditions in summer and winter for three large space buildings in Chongqing: the hall of inpatient building, the hall of outpatient building and the hall of office building. For a large space building, by comprehensively considering its refrigeration energy consumption and air cabinet transportation energy consumption, the energy consumption of the air conditioning system when the design requirements are met is obtained.

4. Optimization design of indoor air distribution

4.1. Optimization of air distribution in indoor thermal environment of existing buildings

Reasonable air distribution form not only can guide people to carry out the air conditioning design of new houses reasonably, but also plays an important role in guiding people to effectively improve the air distribution of existing buildings. In 2016, the urban public building area in China reached 11.7 billion m2. As the indoor environment design standards and specifications of some existing public buildings were not yet perfect when they were built, the existing indoor environment was more prone to problems, so it has been attracting people's attention [35]. In 2013, Awbi [36] obtained the average indoor PPD and PMV values through experiments. The results showed that the comfort of some buildings using displacement ventilation system was not as good as that using impinging jet ventilation system. In 2014, Di and Wang [37] studied the low-frequency sinusoidal air supply for different air distribution models in a dynamic environment. They concluded that if air supply and return vents are located on both sides of the room, air flow fluctuations have a positive impact on human thermal comfort. Zheng et al. [38] carried out CFD simulation on floor air supply and side air supply in a typical office room. They found that the average temperature of the side air supply's working area and non-working area was lower than that of floor air supply, and the air mixing degree was higher than that of floor air supply. At 21℃, floor air supply creates better indoor thermal comfort, but side air supply has a higher energy utilization factor because of higher air mixing and heat exchange. Zhou et al. [39] carried out numerical simulation of different air supply forms in the conference room through CFD. The influence of different air supply modes on temperature field, pollutant concentration field and velocity field in conference room was analyzed and compared. It was found that the indoor temperature, air speed and pollutant concentration were more uniform under the top air supply mode. It is concluded that the top air supply is more suitable for the conference room.

4.2. Optimization design of air distribution of air conditioning system in large space buildings

Since the 1980s, the scale of large-space architecture has become huge because of the progress of architectural technology and the gradual growth of people's demand for cultural life [40, 41]. Different from traditional small and medium-sized buildings, large space buildings are characterized by large space height, large proportion of the outer wall, small proportion of the use area in the whole space, etc. Reasonable air flow organization design is particularly important for its air conditioning design [42].

For the air distribution design of large space air conditioning system, the main methods for predicting indoor air distribution include jet formula, simulation experiment, regional model and CFD. Zhao et al. [43] showed that jet formula had a small scope of application, short prediction period and poor reliability of results. The regional model is difficult to be applied to mechanically ventilated rooms. The simulation experiment is the most reliable, but the prediction cycle is long and it is not convenient to use. CFD can consider all kinds of possible indoor disturbances, boundary conditions, and initial conditions to comprehensively reflect the indoor air distribution and find the optimal air distribution scheme. Therefore, CFD method is widely used in air distribution design of large space air conditioning system [42].

The function of large space building is diverse, so there are different air vents in the same space. Since the 1970s, many scholars have proposed different vent models to simplify vent inlet boundary conditions. Vent models can be classified into the direct description and indirect description according to different description processes. The direct description class includes the basic model and momentum model [44~45], which is not limited by whether the supply air jet is isothermal or restricted. Therefore, in large-space simulation, this description of vent model has been widely used [46~47]. However, to obtain the vent air supply parameters, more grids are needed. Therefore, Zhao Bin proposed the momentum model of n-point vent on the basis of momentum model. The study showed that this model could obtain satisfactory results to simulate the common inlet boundary conditions of the air conditioning vent [48~49]. However, although the indirect description of vent model simplifies the problem, it has certain limitations and is not commonly used in large space simulation [50].

Generally, for theater, multi-purpose hall and other crowded occasions, if only the flow field and temperature field of the whole space is studied, the heat dissipation of lamps and equipment can be directly dispersed to the plane, without a specific geometric shape to express. However, when the thermal comfort around the human body is concerned, it is necessary to simplify the personnel, lamps and equipment into the appropriate shape for studying. Meanwhile, the accurate setting of convection and radiation heating rate of heat source has an important influence on the prediction of large space airflow and energy consumption evaluation. According to the different environment of the large space and the difference in the vertical temperature distribution of the wall surface, the boundary conditions of the wall surface are set. Special attention should be paid to the treatment of boundary conditions of glass curtain walls [42, 50].

Through numerical simulation, Yu [5] obtained that the vertical temperature gradient in the whole area of large space stratified air conditioning is obvious, especially in the upper part, where the hot air moves upward obviously and the temperature near the roof is high. The jet flow at the stratification only bears the load below the stratification. Fan coils need to be set separately to ensure the local temperature of the upper layer. Meanwhile, the top discharge mode needs to be set. Heat is brought out through air flow, thereby reducing the heat transfer from the top heat storage to the lower part, and reducing the convective heat transfer in summer. But too much discharge can also have adverse effects.

In practical engineering, there are many kinds of air distribution schemes for ventilation and air conditioning of large space buildings. Among them, the more widely used forms are the whole room air conditioning scheme with upper supply and lower return and the stratified air conditioning scheme with central side air supply. Aiming at the air flow organization engineering calculation of full-chamber air conditioner and stratified air conditioner, Wang [51] simplified the calculation formula by dimensionless processing of parameters.The calculation charts of air flow organization of two air conditioners in winter and summer are drawn. Moreover, it is found that the energy saving rate is 12.6-18.9% when the relative stratification height is less than 0.5. Therefore, in the design of stratified air conditioning, the relative stratified height should be controlled below 0.5. Through the analysis of three kinds of large space buildings with typical "physical characteristics of load", it is found that the buildings with "physical characteristics of load" as the block heat source (such as machinery factory) should choose the stratified air conditioning mode of medium air supply first. The "physical characteristics of load" is the horizontal heat source of buildings (such as the lecture hall) suitable for up- supply and down- return room air conditioning; "Physical characteristics of load" for the vertical surface heat source of buildings (such as the atrium) could choose the middle to lower stratified air conditioning for comfort and energy saving. The change of the number of air inlets in the whole room air conditioning mode has little effect on the average temperature of the working area. However, the change of the average velocity in the working area will decrease significantly with the increase of the number of air outlets, and the increase of the number of air outlets will greatly improve the uniformity of the temperature field and velocity field in the working area. Therefore, under the premise that the air supply speed meets the conditions, a certain number of air outlets should be ensured. When the air supply outlets on both sides of the stratified air conditioning are staggered, the collision strength between the air supply jets can be effectively reduced, so as to reduce the disturbance of the air supply jet to the gas in the non-air conditioning area and reduce the waste of air supply cooling.

Through experiments and numerical simulation, Yu [5] found that stratificated air conditioning can effectively reduce the overall air conditioning load of atrium buildings. Air supply speed is an important factor in the formation of the air flow partition surface of the stratificated air conditioning. If air speed is too high, the air flow of the jet collides and interferes with each other. With overall air flow disorganized, the air flow partition surface cannot be formed, which strengthens the convection between the air-conditioned area and the non-air-conditioned area. When the air supply speed is too low, the air flow partition surface cannot be formed. With same total air supply volume, increasing the number of vent can reduce the air supply speed. So stratification effect and the comfort level of people in the action area are better. Cheng et al. [52] simultaneously simulated the influence of vent arrangement on the energy saving potential and environmental thermal comfort of the stratificated air-conditioning system applied to different buildings. The results show that the uniformity of indoor air distribution can be improved by adding reasonable air supply for large buildings. Thermal comfort problems such as excessive temperature difference between head and foot can be avoided. For small and medium-sized building space, if the height of the return air inlet drops, the return air flow induces the cold air of the supply air, and cooling capacity at the height of the human head is insufficient. In 2016, Wang [53] took a atrium in Shenzhen as an example to adjust and improve thermal comfort by changing quantity of air supply in the middle of each floor of the atrium. After simulation analysis, if quantity of air supply in the first and top floors is 1.1 times and 2.31 times of the standard floor, respectively, the discomfort caused by the chimney effect in summer can be eliminated.

Based on simulation calculation, Huang [54] concluded that the recommended range of air supply speed of stratified air conditioners in the atrium of shopping malls in the Pearl River Delta in summer is 4-6m/s. If the temperature difference of air supply is 8℃, the thermal comfort of human is the best. The recommended range of air supply height is 4.6-6m and discharge ratio at the top of the shopping mall atrium is 5%-15%. Multi-factor orthogonal experiment was used to optimize the thermal comfort under different air supply conditions. The optimal thermal comfort was obtained if the air supply temperature difference was 8℃, with the air supply speed of 5.2m/s, the air supply height of 4.8m, and the air discharge ratio of 12%. Huang described the pearl river delta region shopping mall atrium' thermal comfort evaluation parameters in summer, which showed the evaluation reference in summer. Meanwhile, by changing the air supply temperature, air supply speed, air supply height and air exhaust ratio at the top of the mall, the scheme of optimizing the thermal comfort of the atrium was given, which provided a reference for the design of the atrium air conditioning engineering in the mall. Guo et al. [55] established a three-dimensional model of an underground supermarket in Chongqing to study its indoor airflow organization. This supermarket adopts the air supply on the diffuser and air return from the side of the louver vent. The on-site measurement method and CFD numerical simulation were used to obtain the temperature field and velocity field distribution in the supermarket. The numerical results are in good agreement with the actual measurement results.

5. Conclusion and outlook

This paper discusses the importance of air conditioning air distribution, air distribution on indoor thermal comfort, and building energy saving. The following conclusions can be drawn:

(1) The installation position of the air outlet of the air conditioner and the angle of the air supply are important factors affecting indoor thermal comfort. If its design is reasonable, it will effectively improve the indoor comfort and the work efficiency of worker, and affect the indoor thermal environment.

(2) The form of air distribution is closely related to the energy consumption of the building. Because some designers did not pay enough attention to indoor air distribution, resulting in uneven indoor ambient temperature, designers had to increase equipment capacity. In addition, due to the unreasonable airflow organization, the air conditioner's operating time and long-term operation with large air volume have been increased, and the air conditioner's energy consumption has increased. It is not conducive to building energy saving.

(3) A reasonable form of airflow organization can guide people to carry out the air conditioning design of new houses rationally and play an important role in guiding people to effectively improve the airflow organization of existing buildings. Moreover, in view of the large space height, the large proportion of the outer wall of the large space, and the small proportion of the use area in the entire space, the appropriate airflow organization form is designed.

In terms of the future research direction, the following avenues of improvement can be followed:

1) For large space buildings such as airport waiting rooms and atrium, glass curtain walls are often used. The solar radiation through the large glass curtain walls has a significant impact on the thermal comfort of the human body and greatly impacts the temperature field and flow field of the large space. Therefore, further research is needed to select a suitable radiation model to describe this physical process.

2) During the raging period of the new crown pneumonia virus, for special places such as pharmaceutical factories and hospital wards, in addition to ensuring a good indoor thermal

environment and achieving energy-saving operation, it is also necessary to strictly control the distribution of airflow to ensure the cleanliness of the indoor environment.

Reference

[1] Zhao Donglai, Hu Chunyu, Bai Desheng, etc. Current status and development trend of China's building energy efficiency technology[J]. *Building Energy Efficiency*, 2015(3): 116-121. (in Chinese)
[2] China Building Energy Conservation Association. China Building Energy Consumption Research Report (2016) [R]. *China Building Energy Conservation Association Energy Consumption Statistics Professional Committee*. 2016. (in Chinese)
[3] Zhuo Mingsheng, Liang Rongguang, Xu Shisong. Discussion on energy saving of central air-conditioning system in modern building [J]. *Refrigeration*, 2004(03): 78-81.(in Chinese)
[4] Jeníček V, Krepl V. Energy, environment and sustainable development [J]. *Areal Research & Development*,2014,41(3):427-440.
[5] Yu Xin. Numerical simulation study on air conditioning of large atrium space in large office buildings using CFD[D]. *Beijing University of Architecture and Architecture*, 2017.(in Chinese)
[6] Zhu Qiqi. Simulation experiment research on indoor airflow organization and particle distribution under different air supply modes [D]. *Shandong Jianzhu University*, 2019.(in Chinese)
[7] Agha-Hossein M M, El-Jouzi S, Elmualim A A, et al. Post-occupancy studies of an office environment:Energy performance and occupants' satisfaction[J]. *Building and Environment*, 2013,69:121-130
[8] P.V.Nielsen. Flow in air conditioned rooms(English translation):[Ph D Thesis]. Copenhagen: *Technical University of Denmark*, 1976
[9] Zhang Jingqing. Research on the boundary conditions of the numerical simulation of large space air conditioning[D]. Chengdu: *Southwest Jiaotong University*, 2006.(in Chinese)
[10] Xu Li, Weng Peifen, Sun Weimin. Numerical analysis of indoor airflow organization and indoor air quality under three ventilation modes[J]. *Acta Aerodynamics*, 2003, 21(03): 311-319.(in Chinese)
[11] Wu Xianqing, pay equal. Numerical simulation of the influence of ventilation methods on indoor thermal comfort ［J］ *Refrigeration and Air Conditioning*, 2015,29(1): 16-21.(in Chinese)
[12] An Puyan. Numerical simulation study of indoor air distribution and particle distribution[D]. *Hunan University*, 2016.(in Chinese)
[13] Li Hongmei. Experimental and simulation study of indoor air distribution with side air supply from nozzles and new diffuser top air supply[D]. *Chang'an University*, 2017.(in Chinese)
[14] Tian Jing, Liu Shaoliang, Mao Hongzhi, Zhao Yanyan. Simulation study on bedroom interior environment of ultra-low energy consumption residential buildings[J]. *Building Technology*, 2021, 52(04): 425-428.(in Chinese)
[15] Wang Fujun. Computational Fluid Dynamics Analysis: Principle and Application of CFD Software.[M]. *Tsinghua University Press*.2004. (in Chinese)
[16] Shang Shaowen, Xu Ying, Bo Shiqiang, Li Longxin, Bi Xue. Simulation and analysis of the effect of different ventilation methods in the office[J]. *Journal of Shenyang Jianzhu University (Natural Science Edition)*, 2017, 33(01): 143-151.(in Chinese)
[17] European Standard, EN 15251, Indoor Environmental Input Parameters For Design and Assessment of Energy Performance of Buildings-Addressing Indoor Air Quality, Thermal Environment, Lighting and Acoustics, 2007
[18] Li Chao. Computational research and optimization method of air distribution in hotel atrium air conditioning[D]. *Southwest University*, 2014.(in Chinese)
[19] Li Jiarong. Calculation analysis and optimization research on airflow organization of air conditioning in multifunctional banquet hall [D]. *Southwest University*, 2016.(in Chinese)
[20] Zhang Junfu. Indoor air quality testing and air distribution analysis of office buildings [D]. *Xi'an University of Architecture and Technology*, 2012.(in Chinese)

[21] Sun Yongli. Research on the influence of air-conditioning air supply angle on indoor comfort[D]. *Shandong Jianzhu University*, 2019.(in Chinese)
[22] Yu Wenhong, Li Yang, Gao Yan. Air-conditioning indoor air distribution and building energy efficiency[J]. *Building Energy Efficiency*, 2017, 45(05):7-9+16.(in Chinese)
[23] Huang Feng, Qin Xinghong, Qu Yunxia, et al. The influence of different air supply methods on thermal comfort[J]. *Energy Conservation*, 2009, (1): 46-49.(in Chinese)
[24] Li Yang. Research on indoor air distribution and thermal comfort of air conditioning systems in office buildings[D]. *North China University of Technology*, 2017.(in Chinese)
[25] Fanger P O. Human requirement in future air-conditioned environments: a reasearch for excellence[C].*Proceeding of HVAC 99*, 1999:86-92.(in Chinese)
[26] Zhao B, Wu J, Li X, et al. Evaluating the Ventilation Effect of Different Ventilation Types by Occupant Air Exchange Efficiency[J]. *ASHRAE transactions*, 2005, 111(2).
[27] Jiao Junjun, Gong Guangcai, Shao Chunsheng. Numerical simulation of the influence of airflow organization on indoor air quality[C]. *Proceedings of the 2008 Annual Conference on HVAC and Refrigeration.* 2008:284. (in Chinese)
[28] Liu Fei. The influence of building thermal environment on human thermal comfort and building energy consumption[J]. *Popular Science and Technology*, 2010,(9):58-59.(in Chinese)
[29] Cheng Yuanda, Zhang Xinghui, Jing Shenglan, Du Zhenyu. The effect of vent layout on energy saving potential and head-to-foot temperature difference of stratified air conditioning system[J]. *Heating Ventilation and Air Conditioning*, 2016, 46(06): 116-121+127.(in Chinese)
[30] Peng Jianbin, Zhang Congli, Wang Jiang. Design and simulation of layered air conditioning for Xinjiang International Convention and Exhibition Center [J]. *HVAC*, 2017, 47(4): 73-77.(in Chinese)
[31] Li Guiping, Wang Tiancheng, Sun Zhaojun, et al. Airflow organization optimization design of the air conditioning system in the first-floor waiting hall of Xiong'an Station [J]. *Heating Ventilation and Air Conditioning*, 2021, 51(9): 2430.(in Chinese)
[32] Hua Yawei. Airpak-based airflow organization energy-saving design and energy-saving operation research of large-scale stadiums and gymnasiums [D]. *Jilin Jianzhu University*, 2016.(in Chinese)
[33] Gao Zhiming. Research on Numerical Simulation and Energy-saving Optimization of Air-Conditioning Air Distribution in Gymnasium[D]. *South China University of Technology Master's thesis*, 2012: 5~39(in Chinese)
[34] Lai Hejing. Energy consumption analysis of different airflow organization forms of air conditioning systems in tall and large spaces [D]. *Chongqing University*, 2011.(in Chinese)
[35] Research Center for Building Energy Efficiency, Tsinghua University. Annual Development Research Report on Building Energy Efficiency in China [M]. Beijing: *China Building Industry Press*, 2018.(in Chinese)
[36] Awbi H.B.. Ventilation of Buildings [R]. *New York: Spon Press*, 2013.(in Chinese)
[37] Di Yuhui, Wang Shancong. Effects of different airflow organization forms on human thermal comfort under dynamic conditions[J]. *HVAC*, 2014, (08): 106-109+105.(in Chinese)
[38] Chenxiao Zheng,Shijun You,Huan Zhang,Wandong Zheng,Xuejing Zheng,Tianzheng Ye,Zeqin Liu. Comparison of air-conditioning systems with bottom-supply and side-supply modes in a typical office room[J]. *Applied Energy*,2018,227:
[39] Zhou Bo, Song Shengbo, Dong Liguo. Numerical simulation of the environment in a public office building [J]. *Journal of Shenyang University of Industry and Technology*, 2018(02):213-218.(in Chinese)
[40] Huang Chen, Li Meiling. Research on indoor vertical temperature distribution in large space buildings[J]. *HVAC*, 1999(05): 28-33.(in Chinese)
[41] Fan Cunyang. Air conditioning design of large space buildings abroad[J]. *HVAC*, 1996(04): 39-49.(in Chinese)
[42] Pan Dongmei, Xu Xiangguo, Wang Yilin, Wang Limin, Shao Junqiang. Simulation of airflow organization in tall spaces-literature review[J]. *HVAC*, 2018, 48(01): 131-138. (in Chinese)

[43] Zhao Bin, Lin Borong, Li Xianting, et al. Prediction methods and comparison of indoor air distribution [J]. *HVAC*, 2001, 31(4): 82-86.(in Chinese)
[44] Shafqat Hussain,Patrick H. Oosthuizen. Numerical investigations of buoyancy-driven natural ventilation in a simple atrium building and its effect on the thermal comfort conditions [J]. *Applied Thermal Engineering*,2012,40(7):358-372
[45] NIELSEN P V . Description of supply openings in numerical models for room air distributions [G]. ASHARE *Trans*. 1992,98 (1): 963-971
[46] Chen Q Y, Moser A. Simulation of a multiple -nozzle diffuser [C].*Proceedings of the 12th AIVC Conference on Air Movement and Ventilation Control within Buildings* . Ottawa , 1991 : 1-13
[47] Zhang Dongjie, Jin Yuanguo. Numerical simulation of large space air conditioning system [J].*Refrigeration and Air Conditioning*, 2009, 23(6): 118-122.(in Chinese)
[48] Jiang Haibin. CFD numerical simulation and thermal comfort of airflow organization in large space stratified air-conditioning system Analysis and Research [D]. Zhenjiang: *Jiangsu University*, 2008: 34-35.(in Chinese)
[49] Zhao Bin. N-point vent momentum model for numerical simulation of indoor air flow [J]. *Chinese Journal of Computational Mechanics*, 2003, 20(1): 64-69.(in Chinese)
[50] Zhao Bin, Li Xianting, Yan Qisen. Numerical simulation of air supply range of orifice air-conditioning air outlet [J]. *Mechanics and Practice*, 2002, 24(1): 18-21.(in Chinese)
[51] Wang Long Pavilion. Research on the optimization of air distribution design in different tall and large space buildings [D].*Chongqing University*, 2015.(in Chinese)
[52] Cheng Yuanda, Yang Jinming, Zhang Xinghui, Du Zhenyu. The layout of the air outlets of the layered air conditioning system in different buildings, Optimization Research[J].*HVAC*, 2018,48(04):100-107. (in Chinese)
[53] Wang Shuo. Application of CFD technology in thermal environment design of tall atrium [J].*Building thermal ventilation and air conditioning*, 2016, 35(6): 77-78.(in Chinese)
[54] Huang Huaming. Research on optimization of summer thermal comfort in the atrium of the Pearl River Delta shopping mall[D]. *Guangzhou University*, 2018.(in Chinese)
[55] Guo Yujie, Yao Liping, Zhang Yongdong, Lai Hejing, Wang Mengning. Research on measurement and numerical simulation of indoor air distribution in underground supermarkets[J]. *Fluid Machinery*, 2019, 47(12): 62-66.(in Chinese)

A New Risk-based Excavation Inspection Decision-making for Long-distance Pipelines

Duhui Lu [1], Guangpei Cong [2,*], and Bing Li [1]

[1] China Special Equipment Inspection and Research Institute, Beijing, China
[2] Guangdong University of Petrochemical Technology, Guangdong Maoming, China
E-mail: 125414712@qq.com

Abstract. With the number of long-distance pipelines increasing in China,the excavation inspection and maintenance is the most important means to avoid the leakage of pipeline. However, the current excavation and maintenance decision-making models are absent to make satisfying results based on the comprehensive data of pipeline. To address this problem, a new quantitative risk-based model is proposed to guide decision-making on excavation inspection and maintenance. Based on previous failure cases, the model includes data about the surrounding soils as well as about the pipeline's protective layer, cathodic protection and thickness readings. Case of the proposed model on previous failure cases shows that the new model can correctly predict a rational excavation inspection and maintenance span for a long-distance pipeline during its whole service life.

1. Introduction

As a cheap and efficient way of energy transportation, pipelines are widely used all around the world, and the proportion of developed countries make up about 60-70% of the total energy transportation capacity. China started late in using pipelines for energy transportation but has developed rapidly. By the end of 2015, the total length of long-distance pipeline trunks in China had exceeded 10 × 104 km [1, 2], most of which are buried pipelines. Affected by soil corrosion, stray current, transportation medium corrosion and other factors, the pipeline body has suffered from corrosion damage, and its service life has been significantly reduced [3-7]. Approximately 30–40 % of long-distance pipeline accidents are caused by corrosion [8, 9]. This figure compares with more than 18% of long-distance pipeline accidents in the United States (of which 45% are external corrosion accidents), more than 33% (42.7% external corrosion) in Russia, and 24% in Europe. Furthermore, leakage problems are more severe in China, with corrosion accidents accounting for about 70% of all accidents in some enterprises and 47.4% of accidents being caused by the external corrosions [10, 11]. The long-distance pipeline accident types in China are shown in Fig. 1. NACE proposed a corrosion evaluation method allows for potential external corrosion problems due to damage of the protective layer [12]. However, the method is only to guide the selection of excavation inspection locations. So quantitative model is proposed to predict the optimal excavation and maintenance span based on a semi-quantitative assessment model that comprehensively evaluates the influence of various detection data on the external corrosion risk of long-distance pipelines.

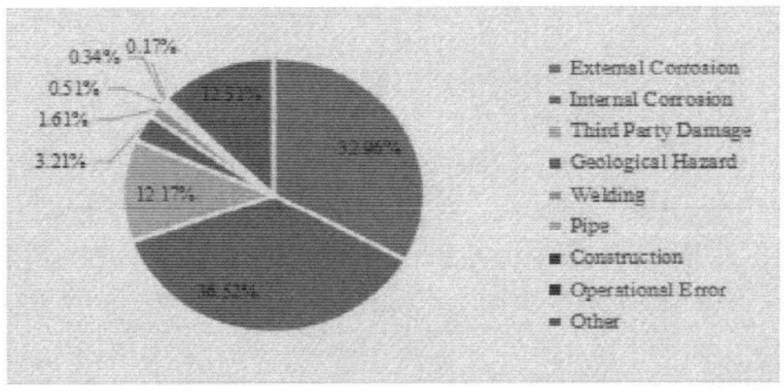

Figure 1. Leakage of long-distance pipelines in China

2. Excavation inspection decision-making model based on risk

2.1. Risk-based decision-making threshold on excavation inspection

Risk (*R*)		Failure likelihood (*S*)				
		Very low <25	Low ≤48	Medium ≤75	Medium-High ≤100	High >100
Failure consequence (*C*)	1 <20	Very low				
	2 ≤50		Low			
	3 ≤80			Medium		
	4 ≤120				Medium-High	
	5 >120					High

Figure 2. The pipe safety risk matrix

Based on the risk matrix shown in Fig. 2 and Table 1, and the total failure risk score,*R*, can be expressed as Eqs (1),(2) and (3) according to Ref [12].

Table 1. Rank standard for pipe safety risk

Risk classes	R score
Very low	[0，1600)
Low	[1600，3600)
Medium	[3600，7800)
Medium-High	[7800，12600)
High	[12600，15000]

$$R = S \cdot C \tag{1}$$

$$S = 100 - (a_1S_1 + a_2S_2 + a_3S_3 + a_4S_4) \tag{2}$$

$$C = \sum_{i=1}^{n} C_i \tag{3}$$

where: S is the total damage score of the pipeline body; a_1 is the correction coefficient of the third-party damage score in the failure probability score correction model during the in-service stage; a_2 is the correction coefficient of the corrosion damage score in the failure possibility score correction model during the in-use stage; a_3 is the correction coefficient of the equipment (device) and personnel operation score in the failure possibility score correction model during the in-service stage; a_4 is the correction coefficient of the body safety score in the failure possibility score correction model during the in-service stage; S_1 is the third-party damage score in the failure probability score correction model during the in-service stage; S_2 is the corrosion damage score in the failure probability score correction model during the in-service stage; S_3 is the equipment (device) and improper operation damage score in the failure probability score correction model during the in-service stage; and S_4 is the body safety score in the failure probability score correction model during the in-service stage;C is the total score of leakage consequence; and C_i is the score of each item leakage consequence.

In Eq.(2),the best choice of correction coefficients a_1 ,a_2, a_3, and a_4 is based on the statistical results of previous failure cases in the pipeline, but in most cases, the statistical results might not be very reliable or even available. In this situation, it can be assumed that the probabilities of the four kinds of pipeline problems are equal; that is, each correction coefficient is 0.25. Then, according to the later statistical results, Bayesian analysis is used to iterate the correction coefficients, as shown in Eq. (4), so as to keep close to the real state of the pipeline body and make full use of the relevant data that can be retrieved from some references.

$$p_i^{(k+1)} = \frac{p_i^{(k)} q_i^{(k+1)}}{\sum_{j=1}^{m} p_j^{(k)} q_j^{(k+1)}} \tag{4}$$

where: $p^{(k)}{}_i$ is the probability of the i-th influencing factor after the k-th Bayesian iteration; and $q^{(k)}{}_i$ is the probability of the i-th influencing factor in the k-th failure statistics.

In the risk model shown in Fig. 2, it can be assumed that when a long-distance pipeline is assessed at being high risk, corrosion perforation may be difficult to control using conventional external corrosion control means according to Ref [13]. Instead, detailed excavation inspection have to be done and its span can be determined for different areas of pipeline. Therefore, the threshold of the prediction model can be obtained according to critical external corrosion thinning rate (*ECT*) according to Ref [12],when the pipeline falls into the high-risk (red) region of the matrix, as shown in Fig. 2. Thus, the threshold of wall thickness reduction, $\Delta Tth(t,r_a)$, is expressed in Eq.(5) .

$$\Delta T_{th}(t,r_a) = ECT_{th} \times T_n \tag{5}$$

where: ECT_{th} is the threshold of *ECT* shown in Fig. 3 as the pipeline fall into high risk (red) region shown in Fig. 2.

2.2. Prediction on excavation inspection and maintenance span

At the same time, the reliability of inspection, local sampling excavation and other uncertain factors need to be considered in a pipeline excavation and maintenance span prediction. Considering the influence of various uncertain factors. the thinning thickness ,$\Delta T(t,r_a)$, based on the expected reliability, r_a ,is now shown in Eq. (6).

$$\Delta T(t,r_a) = v_x\{1 - c_x[0.7797\ln(-\ln r_a) + 0.4501]\}t \tag{6}$$

where: v_x is the equivalent corrosion rate, mm/a, which not only considers the average corrosion rate but also includes the influence of confidence, that is considering the influence of corrosion rate deviation according to the confidence, its calculation is shown in Eq. (7); c_x is the variation coefficient of the corrosion rate, and its calculation is shown in Eq. (8); and r_a is the expected reliability of the protective layer.

$$v_x = \left(x^* + t_{0.90,N-1} \cdot S_x \big/ \sqrt{N-1}\right)\big/ t \tag{7}$$

$$c_x = S_x \sqrt{N / \chi^2_{0.90,N-1}} \big/ x^* \tag{8}$$

where: x^* is the mean value of the maximum corrosion depth, mm, which can be obtained by Eq. (9); S_x is the variance of the maximum corrosion depth, mm, which can be obtained by Eq. (10); N is the amount of corrosion detection pits; $t_{0.90,\ N-1}$ is the t-distribution coefficient under 90% confidence; and $\chi^2_{0.90,N-1}$ is the χ^2 distribution coefficient under 90% confidence.

$$x^* = \frac{\sum_{i=1}^{N} x_i}{N} \tag{9}$$

$$S_x = \sqrt{\frac{1}{N-1}\sum_{i=1}^{N}(x^* - x_i)^2} \tag{10}$$

where: x_i is the measured maximum corrosion pit depth, mm.
And the prediction model on optimal excavation inspection and maintenance span, $t_{optimal}$, can be determined as Eq.(11) according to Eqs.(5)-(10).

$$t_{optimal} = \Delta T(t, r_a)/\{v_x\{1 - c_x[0.7797\ln(-\ln r_a) + 0.4501]\}\} \tag{11}$$

3. Case study

3.1. A long-distance pipeline introduction

A gas transportation pipeline from a natural gas field in the western plateau in China began operating on 31 August 1996. Its route is mainly Yadan landform in the Gobi Desert, where there is less traffic and people flow and cultural activities are underdeveloped. The specifications of the pipeline are presented in Table 2.

Table 2. Specifications of the case study pipeline

Items	Value
Distance	345,295 m
Material	X52
Dimensions	Φ323.9 × 6.0 and Φ323.9 × 7.1 mm
Design pressure	6.4 MPa
Operational pressure	4.5 MPa
Design temperature	atmospheric temperature
Cathodic protection	forced current

3.2. Risk assessment on the long-distance pipeline

The pipeline in this case study was divided into 87 areas for risk assessment. And the 87 areas has been inspected for different key external corrosion factors, such as pipeline buried depth (PBD) (as shown in Fig. 3),

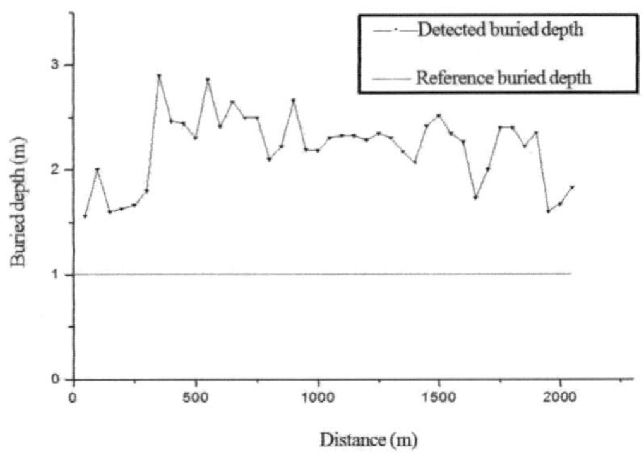

Figure 3. Buried depth inspection along the long-distance pipeline protective layer quality(PLQ) (as shown in Fig. 4),

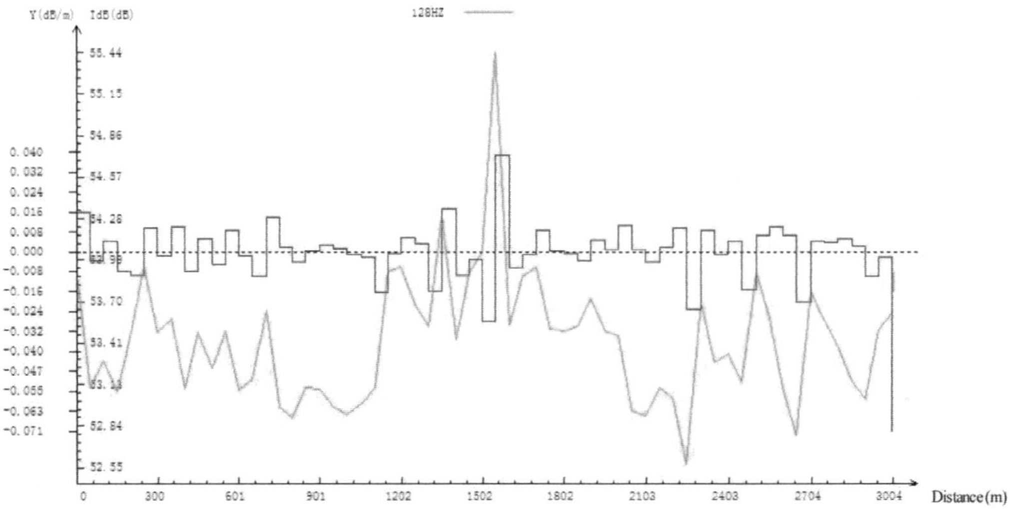

Figure.4. Current decay inspection along long-distance pipeline

protective layer damage(PLD) (applying the same technology with PBD), cathodic protection effectiveness (CPE) (as shown in Fig. 5) and soil corrosivity (as shown in Fig. 6 and Table 3).

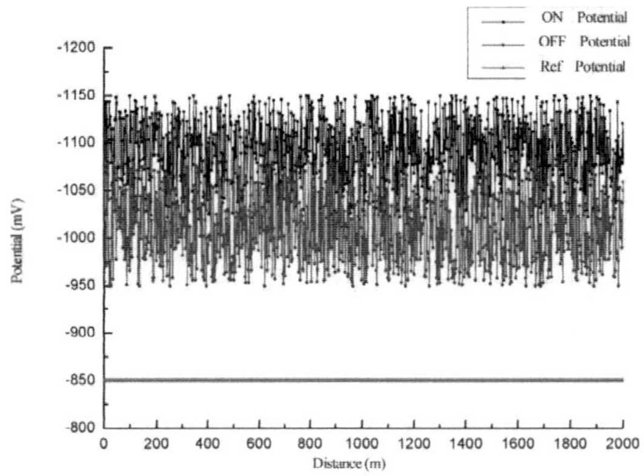

Figure 5. CP - CIPS inspection along the long-distance pipeline

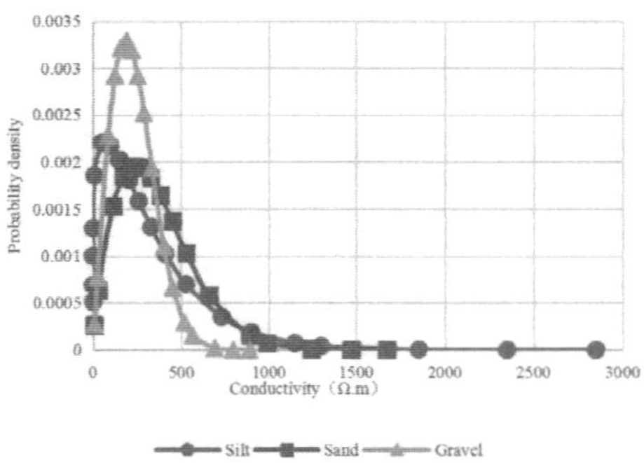

Figure.6. Statistical models of the resistivity of the three soil types

Table 3. Soil corrosive media test data

Samples	PH	Cl^-(mg/kg)	Total salt (g/kg)	Redox potential (mV)	Water content (%)	SO_4^{2-} (mg/kg)
1	8.5	44.08	1.77	398	11.6	49.6
2	8.0	4.2×10^4	186	390	7.89	1.89×10^3

And the inspection data from PBD to PLD are shown in Table 4 with 50m inspection interval, besides, all the critical values are listed in Table 5 to judge a inspection results. And applied inspection techs and their critical values are shown in Table 4. After those inspections, 23 exposed and underburied pipe areas were identified, 9 areas have obvious quality problems in the protective layer, and 209 damage points are across 33 of the 87 risk assessment areas. The combined length of these 23 areas is 1336 m, accounting for 0.39% of the total length of the pipeline, and the most severely damaged areas on their protection layer are shown in Table 4. The soil resistance statistical models, shown in Fig. 6, indicate that almost all the resistances (>90%,as shown in Table 6) are larger than 50Ω, so their corrosivity are all weak in general.

Table 4. Inspection data of the protective layer

Area	Total length (m)	Poor PLQ length (m)	CP potential (V)					PLD (Y/N)
17	2454	10	-1.00	-1.80	-1.35	-0.92	-1.06	N
18	2740	20	-1.39	-1.10	-1.30	-1.08	-1.12	N
56	4452	2	-1.20	-1.40	-1.20	-1.50	-1.10	Y
58	2170	100	-1.10	-1.30	-1.00	-1.30	-0.90	Y
70	4146	48	-1.20	-1.80	-1.26	-1.00	-1.30	Y
72	5630	200	-1.30	-1.60	-1.12	-1.09	-0.96	Y
74	3000	50	-0.90	-1.10	-1.00	-1.35	-1.30	Y
81	2245	60	-1.32	-1.23	-1.23	-1.07	-1.13	Y
87	5500	94	-1.23	-1.19	-1.07	-1.17	-1.03	Y

Table 5. Inspection techs and their critical values

Items	Measurement tech	Data Units	Critical values
PBD	Pipeline current mapper(PCM)	m	1
PLQ	Multi-stage current attenuation	dB/m	0.129
PLD	AC voltage gradient method (ACVG)	dBuV	70
CPE	close-interval potential inspection system (CIPS)	V	-0.85 to -1.2
Soil corrosivity	Soil resistance	Ω	50

Table 6. Soil conductivity assessment scores and their probabilities

Soil types	Conductivity (Ω.m)	Classes	Probability
Silt	<20	Strong	0.04
	≥20–50	Moderate	0.06
	>50	Weak	0.90
Sand	<20	Strong	0.01
	≥20–50	Moderate	0.02
	>50	Weak	0.97
Gravel	<20	Strong	0.01
	≥20–50	Moderate	0.03
	>50	Weak	0.96

Table 7. Weight results of each item influencing pipe safety

Weight	Third party	Corrosio n	Improper operation and maintenance	Defects of pipeline body
$p^{(k)}{}_i$	0.2750	0.4360	0.0270	0.260
$q^{(k)}{}_i$	0.1538	0.6948	0.0017	0.1497
$p^{(k+1)}{}_i$	0.1101	0.7885	0.0001	0.1013

When evaluating the external corrosion risk of a long-distance pipeline, not only should the influence of soil corrosivity be considered, but the damage of the protective layer should also be evaluated. A comprehensive risk,R, can be evaluated according to Eqs (1),(2) and (3), and the weight factor a_i can be calculated according to Eq. (4). it is assumed that the prior value $p^{(k)}{}_i$ in Eq. (4) is based on statistical data from the same industry. The posterior value $p^{(k+1)}{}_i$ and the conditional weight value $q^{(k)}{}_i$ are shown in Table 7. Using $p^{(k+1)}{}_i$ to replace a_i in Eq. (2) to calculate S.

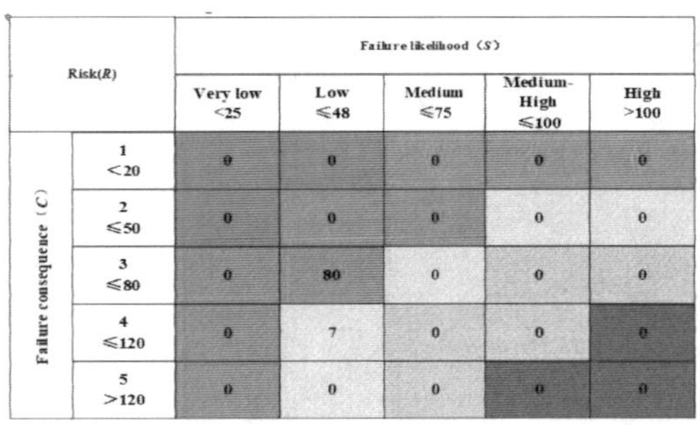

(a)

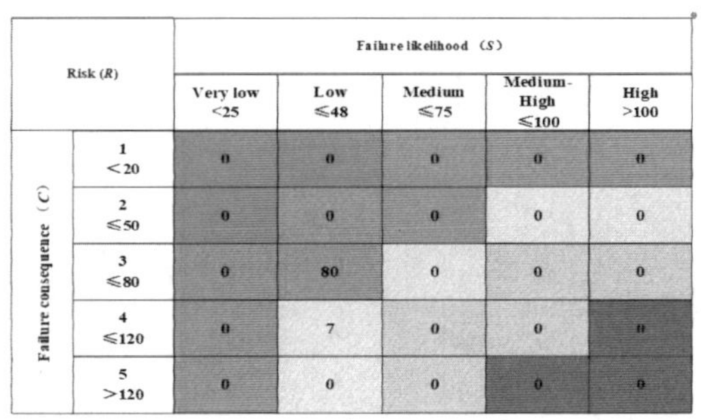

(b)

Risk (R)		Failure likelihood (S)				
		Very low <25	Low ≤48	Medium ≤75	Medium-High ≤100	High >100
Failure consequence (C)	1 <20	0	0	0	0	0
	2 ≤50	0	0	0	0	0
	3 ≤80	0	70	0	0	0
	4 ≤120	0	17	0	0	0
	5 >120	0	0	0	0	0

(c)

Figure 7. Comparing on pipe risk with (a) default mean weight ($a_i = 0.25$), (b) prior weight ($a_i = p^{(k)}{}_i$), and (c) posterior weight ($a_i = p^{(k+1)}{}_i$).

The default mean weight ($a_i = 0.25$), a prior weight ($a_i = p^{(k)}{}_i$) and a posterior weight ($a_i=p^{(k+1)}{}_i$) are separately used to calculate S, and S_i and C_i are calculated according to [13]. Finally, R of all the areas are obtained shown in Fig. 7, where Fig. 7(a) takes the mean weight $a_i = 0.25$, Fig. 7(b) takes the prior weight of $p^{(k)}{}_i$, and Fig. 7(c) takes the posterior weight $p^{(k+1)}{}_i$. Comparing with Figs 7(a), 7(b) and 7(c), the average weight and prior weight do not change the risk level screening results. However, after considering statistical data from similar domestic pipelines in the same industry as the conditional probability and using the posterior value, the risk levels of pipelines changed significantly. Compared with the two earlier calculations using average weight and prior weight, the number of low-risk areas (the yellow regions in Fig. 7) along the pipeline has increased from 7 to 17 when the risk is calculated using posterior weight $p^{(k+1)}$ (see Fig.7 (c)). Thus, because Eq. (4) provides a method to calculate the dynamic evolution of risk that takes into account both historical experience data and current failure statistical data, the model can be used to find any change in current pipeline risk in time.

3.3. Risk-based decision-making model on pipeline excavation inspection span

To research a more effective quantitative excavation decision-making model, some low-risk areas were also excavated and sampled and 185 locations of these 87 areas were selected for excavation. In some locations pipe body are damaged as shown in Fig. 8. And the 39 of the 185 excavated locations revealed areas of the pipeline with a seriously damaged protective layer and the most severely damaged 10 points are shown in Table 8 after ultrasonic thickness reading are done, the rest areas have a less wall thinning than 0.11mm. So except for two areas with 9% and 14%, almost all thinning rate is less than 6%.The maximum thinning rate happens in the location with groove-like damage of the protective layer and the exposed pipe body show in Fig. 8. According to historical statistical data on long-distance pipeline leakage cases, the location with this damage will result in heavy corrosion of the surface of the pipeline. Even if the cathodic protection is perfect, the pipeline will still undergo rapid corrosion perforation.

Figure 8. Damaged pipe body with third-party damage of the protective layer

The location in Fig.8 in this case study with pipeline specification is Φ323.9 × 6.0, the operating pressure is 4.5MPa and atmospheric temperature, transports natural gas and the initial operational date is 1 October 1998, the amount of excavation locations in the excavation inspection was 5. Besides, according to Fig.7(c), its critical values of the thinning rate are between 10% and 30% duo to the different expected reliability. And then all of these data are substituted into Eqs (5)–(11) to obtain the optimal excavation inspection span,$t_{optimal}$, under different reliability expectations, shown in Fig. 9. When the expected reliability of the pipe is higher because the less information about the corrosion of pipe or higher safety expectation to the pipe, then a shorter excavation span should be expected to ensure safer running of the pipe. Due to the uncertainty of pipeline external corrosion information, the $t_{optimal}$ in this prediction model is closely related to variable N, the amount of excavated locations during precious inspections. As shown in Fig. 9, when N increases from 5 to 25, the $t_{optimal}$ under different reliability will increase significantly. Therefore, the model provided in this paper can be applied to rationally extend the excavation maintenance span by selecting a suitable amount of excavation locations, enabling someone to choose an optimal excavation maintenance span according to the expected excavation maintenance cost and safety risk. For example shown in Fig. 9, when the expected reliability of the pipeline is 0.95 and if N were to increase from 5 to 10, effectively doubling the amount of information obtained, then the pipeline excavation and maintenance span can be extended from 6 years to 10 years.

Table 8. Top 10 areas with max wall thinning

Area	Nominal wall thickness (mm)	Wall thinning (mm)				
17	7.1	0.35	0.63	0.31	0.02	0.14
22	6	0.39	0.39	0.11	0.09	0.04
26	6	0.06	0.11	0.14	0.02	0.06
34	6	0.15	0.01	0.09	0.03	0.04
44	7.1	0.38	0.03	0.01	0.04	0.02
62	6	0.1	0.16	0.05	0.02	0.04
74	7.1	0.07	0.13	0.12	0.16	0.05
80	6	0.17	0.19	0.1	0.04	0.82
81	6	0.01	0.02	0.16	0.02	0.01
82	7.1	0.19	0.12	0.12	0.07	0.07

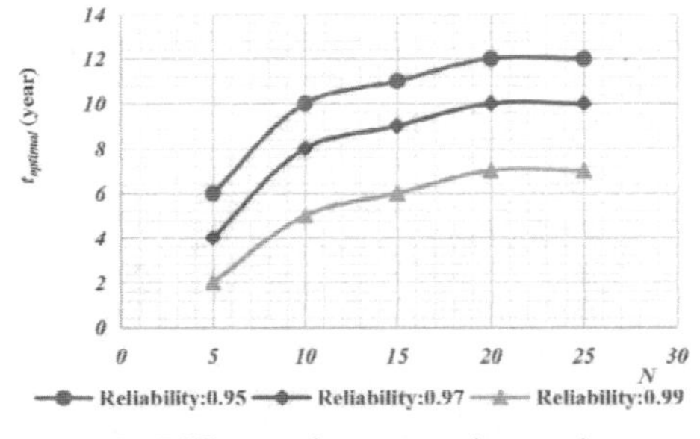

Figure 9. Different the excavation maintenance span the amount of excavated locations

4. Conclusion

A new quantitative excavation inspection span prediction model is proposed to complete a more rational excavation decision-making for long-distance pipeline based on inspection data,such as the protection layer quality, external corrosion thinning and so on. The practical testing results presented in the paper verify that the model can guide the optimal excavation inspection or maintenance span for long-distance pipelines. With the new model, the threshold of the thinning rate of the pipeline can be dynamically determined according to the failure consequence level and the protective layer conditions of the pipeline, and then the optimal excavation and maintenance span of high-risk pipeline can be predicted according to expected reliability and amount of excavation during precious inspections. Besides, The new model can fully consider the impact of uncertainty of pipeline information on pipeline safety and allow 2–5 years time allowance between the early effective medium-high or high-risk identification and the optimal excavation time, it is more enough to draft an optimal pipeline excavation and maintenance plan, consistent with the demands on the inspection and maintenance of long-distance pipelines in China.

5. References

[1] M. Ishwaran, W. King, M. Haigh, et al, *Advances in Oil and Gas Exploration & Production*(Heverlee Belgium, 2017)
[2] J. Wang, H. Jiang, Q. Zhou, et al, Renewable and Sustainable Energy Reviews,84 (2016)
[3] Y. Hou, D. Lei, S. Li, et al, submitted to International Journal of Corrosion(2016).
[4] M. Karami, ASME. J. Pressure Vessel Technol 134(2012)
[5] H. L. Li, X. W. Zhao, and L. K. Ji, Testing - Physical Volume 41(2005)
[6] R. Singh,Pipeline *Integrity Handbook: Risk Management and Evaluation* (Gulf Professional Publishing UK, 2014).
[7] C. R. F. Azevedo, Engineering Failure Analysis, 14(2007).
[8] S. Shin, G. Lee, U. Ahmed, et al, Journal of Hazardous Materials,342(2018)
[9] L. Dai, D. Wang, T. Wang, et al, submitted to Journal of Petroleum Engineering(2017).
[10] K. Siler-Evans, A. Hanson, C. Sunday, et al, International Journal of Critical Infrastructure Protection, **7**(2014)
[11] J. Zhou, X. Wu, Z. Guo: Journal of Safety Science and Technology,10(2014)
[12] NACE, *Pipeline External Corrosion Direct Assessment Methodology* (NACE USA, 2010)
[13] Standardization Administration of China (SAC), *GB/T 27512-2011Risk Assessment for Buried Steel Pipeline* (SAC China, 2011)

Acknowledgements
The authors would like to acknowledge the support provided by the National Key Research and Development Program – Research and Development of Key Technology and Equipment for Typical Petrochemical Process Safety [Grant No. 2018YFC0808600].

Structural Design for Improving the Strength of Flat Wooden Pallets

X Zhu[1], H Toyota[1], D Ito[2] and K Doi[3]

[1] Ehime University, 3 Bunkyo-cho, Matsuyama, Ehime, 790-8577, Japan
[2] Ito Wood Co., Ltd., 2-12-10 Kikumoto-cho, Niihama, Ehime, 792-0801, Japan
[3] Faculty of Engineering, Ehime University, 3 Bunkyo-cho, Matsuyama, Ehime, 790-8577, Japan
E-mail: zhu.xia.mx@ehime-u.ac.jp

Abstract. The purpose of this study is to develop a new pallet structure using the inexpensive hinoki wood, which is produced in the Ehime Prefecture, having the same strength as the conventional Oregon pine wood. A new pallet structure was proposed for achieving high-strength by improving the flexural stiffness and compressive stiffness and achieving high-cost performance by using the inexpensive hinoki wood. The results obtained in this study are summarized as follows: deflection of the new type is reduced compared to the standard type, due to the increased bending rigidity of the edge board and deck board; compressive displacement of the new model decreases, as the compressive stiffness of the pallet legs increases; new mold pattern 6 made is higher in strength and rigidity than the standard mold; cost of the new type #6 is 72% of that of the standard type pallet, and cost can be reduced by 28%.

1. Introduction

Pallets are used in transportation and logistics to facilitate load handling in factories, trucks, containers, and warehouses. During transportation, goods are placed on pallets, and forklift or pallet jacks are inserted into the openings in the pallets for carrying the goods. Pallets are made from various materials, including plastic, steel, aluminum, and paper. However, wooden pallets are characterized by the fact that loads do not slip. Thus, they play a very important role in transportation. Wooden pallets are widely used in transportation and logistics because they have sufficient strength and load carrying capacity. Furthermore, they offer a good overall balance of low-cost production and disposal while being easy to repair when partly damaged.

The purpose of this study is to develop a new pallet structure using the inexpensive hinoki wood, which is produced in the Ehime Prefecture, having the same strength as the conventional Oregon pine wood. A pallet structure was proposed for achieving high strength by improving the flexural stiffness and compressive stiffness.

2. Pallet Fabrication and Strength Experiments

2.1. Production of Flat Pallets

Two types of single-sided square difference (D2 type, JIS Z 0604) flat pallets [1] were manufactured, which are commonly used in distribution sites, as shown in Figure 1. Oregon pine wood was used to manufacture the standard pallet, whereas hinoki wood was used to manufacture the proposed pallet. The external dimensions of both the pallets were 1100 × 1100 mm with a height of 144 mm and

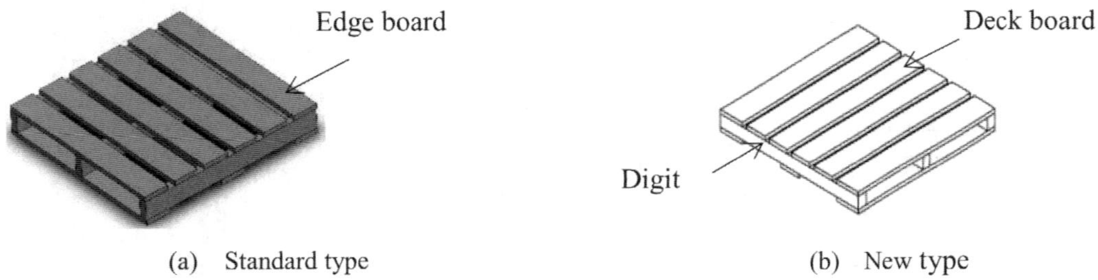

Figure 1. Flat wooden pallets.

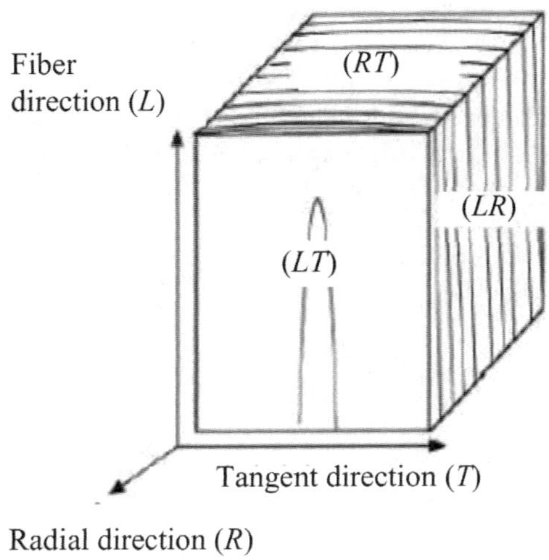

Figure 2. Anisotropy definition.

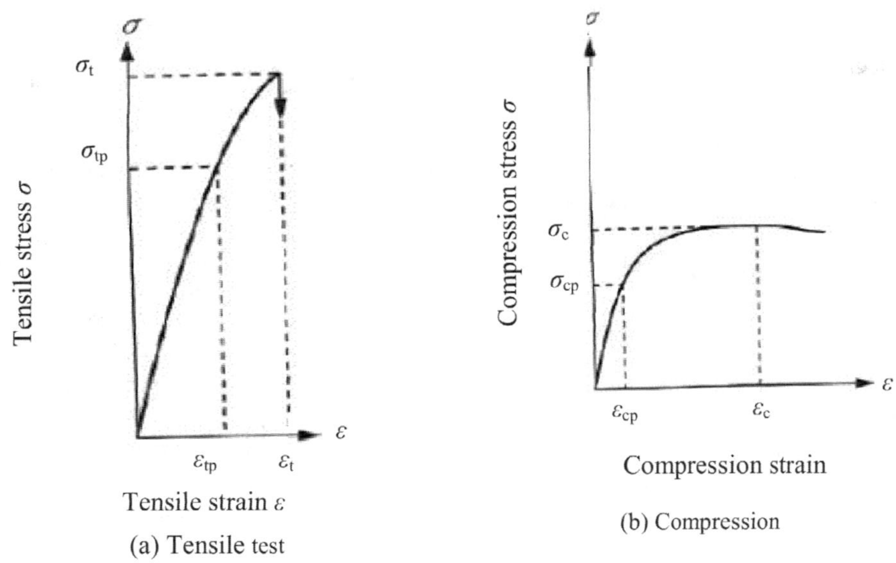

Figure 3. Typical stress-strain curves in L-direction without defects.

Table 1. Average Young's modulus for coniferous tree.

E_R/E_L	E_T/E_L	G_{LR}/E_L	G_{LT}/E_L
0.075	0.040	0.060	0.050
G_{RT}/E_L	ν_{LR}	ν_{LT}	ν_{RT}
0.003	0.40	0.50	0.60

maximum loading mass of 2 t.

2.2. Flat Pallet Strength Tests

Flat pallet bending tests and compression tests were performed based on JIS Z 0602 [2]. In the bending tests, the deflection ratio was calculated by measuring the deflection in the central portion of the lower edge board. In the leg compression tests, the compressive displacement was measured in the central portion of the top surface of the two legs of the pallet and calculated the average compressive strain using the compressive area.

3. Simulation analysis of strength tests

3.1. Wood Anisotropy

The anisotropy of wood is derived from the tissue structure. Wood is generally considered as an orthogonal anisotropic material with axes along three directions: the fiber (L) direction, radial (R) direction, and tangential (T) direction, as shown in Figure 2 [3]. Table 1 shows the Young's modulus E_L, E_R, and E_T, shear modulus G_{RT}, G_{LT}, and G_{LR}, and Poisson's ratio ν_{LR}, ν_{RT}, and ν_{TL}, along the three axes used in the finite element method analysis. In this study, E_L = 12 GPa was used for Oregon pine wood and E_L = 9 GPa was used for hinoki wood [4].

Furthermore, a stress–strain curve along the L direction was used as shown in Figure 3 for analysis [5]. The Young's modulus for compression and tension are almost equal. The ratio of maximum stress of compression to tension is 1:2, whereas the elastic limit stress is approximately 2/3 and 4/5 of the maximum stress during compression and tension, respectively. In this study, the maximum stress was achieved by the bending strength. The strength in the lateral direction (T direction and R direction) for tension and compression was 5% of that in the L direction in the T direction and 10% in the R direction [6].

3.2. Simulation Analysis Model

A bending test model was made, that was a 1/4th of the actual model and accounted for the symmetry of the boundary condition. Additionally, a full-sized model for compression tests was created because of load asymmetry. Elasto-plastic analysis were performed though using three-dimensional isoparametric elements accounting for the anisotropy of wood and different flow stress behaviour between the tension and compression sides. From the perspective of material mechanics, to achieve high structural strength by improving stiffness, we created six types of new pallet models.

Table 2 shows the dimensional differences of each constituent component compared to the standard type. In terms of structural and cost limitations, we modelled a new type of pallet (type #1) having higher bending strength than the standard pallet built using Oregon pine wood. Next, accounting for cost, we designed the type #2 pallet by lowering the bending stiffness while maintaining the same strength as the standard type. Next, to increase the leg compressive strength, we modelled types #3 to #5. Finally, in type #6, all evaluation items exceeded the standard type in terms of bending strength, leg compressive strength, and cost.

Table 2. Dimensional difference between new type and standard type (mm)

Component parts \ New type		# 1		# 2		# 3		# 4		# 5		# 6	
		Width	Height	Width	Height	Width	Height	Width	Height	Width	Height	Width	Height
Top surface	Deck board	-	8	-	4	-	4	-	4	-	4	-	4
	Edge board	-	8	-	4	25	4	27	4	25	4	25	4
Digit	Center digit	-	-16	-	-8	-	-8	-	-8	-	-8	-	-8
	Edge digit	-	-16	-	-8	10	-8	10	-8	12	-8	13	-8
Bottom surface	Deck board	-	8	-	4	-	4	-	4	-	4	-	4
	Edge board	-	8	-	4	25	4	27	4	25	4	25	4

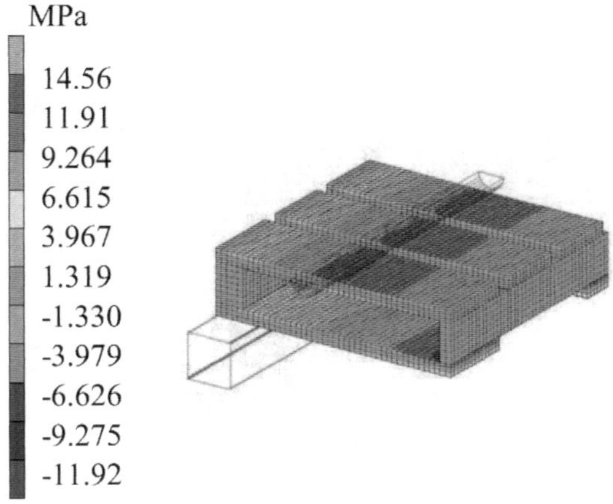

Figure 4. Bending stress contour (# 6).

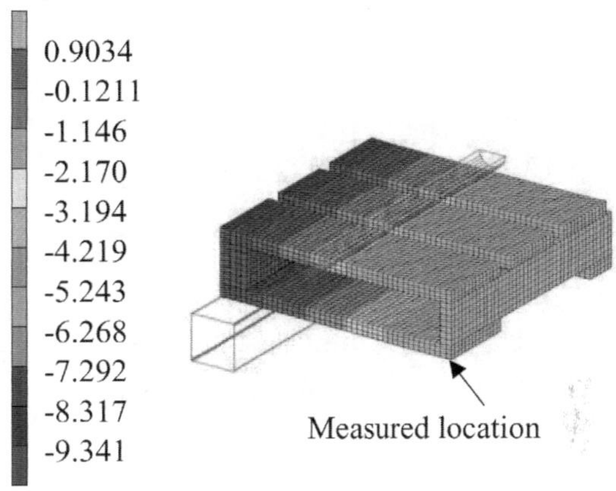

Figure 5. Deflection contour (# 6).

4. Results and Discussion

4.1. Central Plane Bending Stress and Deflection Distribution in Bending Load

Figure 4 and Figure 5 show an example of bending stress and deflection distribution when the bending load is 1.25 times the maximum payload (25 kN), respectively. It was found that in both structural types, the largest bending stress and deflection were produced in the lower central plane of the pallet, excluding an area of concentrated stress near the contact portion with the rigid body support. Incidentally, the bending stress was maximized in the central portion, but the deflection was maximized outward. Figure 6 shows the relationship between flexural stiffness and deflection. It was found that the deflection of all pallets built using hinoki wood was lower than that of the standard type made using Oregon pine wood (excluding the hinoki wood-made standard type #0).

4.2. Distribution of Leg Compressive Stress and Compressive Displacement under Compressive Loading

Figure 7 and Figure 8 show examples of the compressive stress and strain distributions when the compressive load is 1.1 times the maximum payload (22 kN), respectively. The stress distribution was almost even in the cross section of the legs in either structural type. Figure 9 shows the relationship between the compressive stiffness and compressive displacement of the legs. Only the hinoki-made

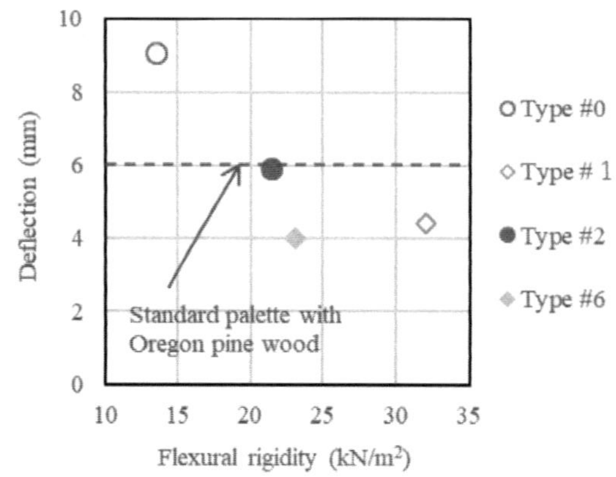

Figure 6. Relation between deflection and flexural rigidity.

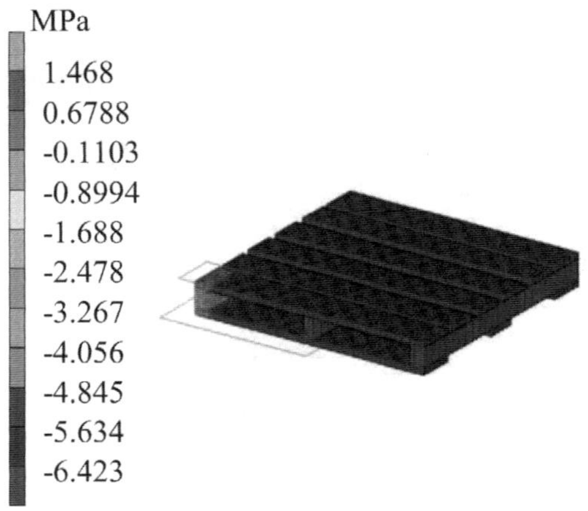

Figure 7. Compression stress contour (#6).

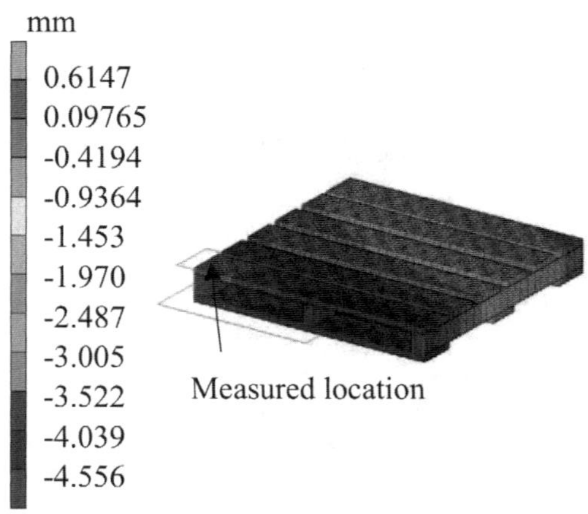

Figure 8. Compressive displacement (#6).

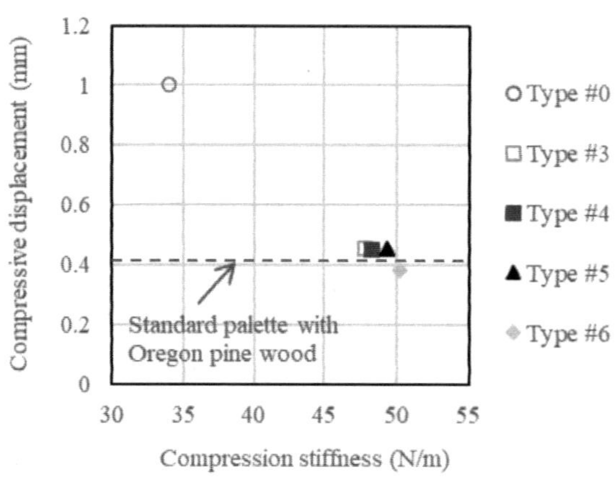

Figure 9. Relation between compressive displacement and compression stiffness.

Table 3. Comparison of new and standard type.

Type \ Items to be evaluated	Deflection (mm)	Compressive displacement (mm)	Cost (Japanese yen)
Standard Type	12.85	2.96	1264
	(6.040)	(0.3952)	
New Type (# 6)	8.83	2.06	906
	(3.994)	(0.3803)	

( ): Analysis value

type #6 showed compressive displacement lower than the standard type created using Oregon pine wood, not use any macros for the figures and tables.

4.3. Comparison between New and Standard Types
Table 3 shows a comparison of the strength test results and cost of the Oregon pine-made standard type and the hinoki-made type #6 pallets. The hinoki-made type #6 pallet was found to be superior to the Oregon pine-made standard type pallet.

5. Conclusion

The purpose of this study is to develop a new pallet structure using the inexpensive hinoki wood, which is produced in the Ehime Prefecture, having the same strength as the conventional Oregon pine wood. A new pallet structure was proposed for achieving high-strength by improving the flexural stiffness and compressive stiffness and achieving high-cost performance by using the inexpensive hinoki wood. The results obtained in this study are summarized as follows.

(1) Deflection of the new type #6 is reduced compared to the standard type, due to the increased bending rigidity of the edge board and deck board.
(2) Compressive displacement of the new type #6 decreases, as the compressive stiffness of the pallet legs increases.
(3) Cost of the new type #6 is 72% of that of the standard type of pallet, and the cost can be reduced by 28%.

6. References

[1] Information on *https://webdesk.jsa.or.jp/books/W11M0070/index*
[2] Information on *https://webdesk.jsa.or.jp/books/W11M0070/index*
[3] Okano T, Sobue N, *Wood Science Handbook (Popular Edition)*, first printing, Asakura Shoten, Tokyo, 2013, pp. 154 –157.
[4] Maekawa J, *Architecture / Environmental Acoustics*, Kyoritsu Publishing, Tokyo, 1990, p. 111, p. 74.
[5] Yamada M, *Science of Woody Environment*, Kaiseisha - press, Otsu, 1987, p.153.
[6] Takahashi T, Nakao T, Inoue M, Tanaka C and Nishino Y, Reduction of floor impact noise of 2x4 construction method by suspended floor construction method and evaluation of L value in 31.5 Hz band, *J. of JWRS*, Vol. 38, No. 3 (1992) 228 – 232.

Development of Particle-fluid Drag Model for Individual Wheat Straw Particle

Yang Yang[1] and Hongguang Zhu[2,*]

[1] College of Mechanical and Energy Engineering, Tongji University, Jiading district, Shanghai, 201804, China
[2] Institute of New Rural Development, Tongji University, Jiading district, Shanghai, 201804, China
*Corresponding author's e-mail address: zhuhg@tongji.edu.cn

Abstract. To deal with the sedimentation of wheat straw particles in the process of conveying and mixing, the drag model of the non-spherical particle is to be developed. In this work the settling velocity of sheared and the defibered particles are acquired by image analysis. The simulating settling velocity of particle is acquired by ANSYS Fluent. As a result, the KG-based drag models with shape factors of Π, AR, BR, and Φ are obtained respectively by fitting the relation between drag coefficient and particle Reynold for 24 sheared particles. The settling velocity of 24 particles are acquired by simulation. In addition, the six drag models are used to simulate settling velocity of 134 defibered particles. At last, the means value of relative error of velocity between the shear and the defibered particles are acquired, they are 20%, 23%, 19%, 59%, respectively, corresponding to the shape factors of Π, AR, BR and Ψ. Considering the pValue of curve fitting, the drag model with shape factor Π is best for simulating the wheat straw particles in particle-liquid flow.

1. Introduction

Approximately 529 million tons' wheat [1] is produced every year over the world, and a nearly equal amount of wheat straw is produced. As the global energy structure changes to clean and low-carbon, the use of crop straw as a traditional fuel has been abandoned and as a new clean energy to convert it to be biogas have been gradually adopted. Unfortunately, the sedimentation of wheat straw during conveying and mixing process is severely to reduce its energy conversion efficiency.

Computation fluid dynamics (CFD) is commonly applied to simulate the particles' motion. Commonly, single phase flow is use to simulate the particle-fluid's motion in anaerobic digesters [2-5]. But the we can't see the sedimentation of the particle from single phase model. Two-phase flow models can reflect the particles' motion in with the drag model included in the CFD code. For instance, Patrícia Metolina et al. [6] studied liquid–solid flow in tapered and cylindrical fluidised beds using the Eulerian–Eulerian two-phase model in which the spherical drag correlation is used.

Infect, wheat straw particle has irregular shape, the spherical drag used in CFD can result in bad accuracy. Pierfrancesco Dellino et al. [7] studied the settlement process of pumice particles using simulated by the spherical and non-spherical drag models and obtained the prediction errors of settling velocity of 65% and 22%.

In this work, the terminal settling velocity of sheared particles and defibered particles are measured through the experiment. The particle shape is acquired and calculated by different shape factors. Then, a drag model of KG-based model was developed with a new shape factor Π . Comparing with relative error of particle settling velocity between measurement and simulation, the best drag model with the appropriate shape factor is acquired for simulating the wheat straw particles in solid-liquid phase flow.

2. Model setup

2.1. Derivation of fluid-particle drag model

For a single particle drag F_D comes from fluid can be expressed by equation(1), F is given by equation(2) [8]. Substitute equation (2) to equation(1), drag coefficient C_D is expressed by equation(3). Because of the drag F_D has relation of particle shape and particle Reynolds number [9], then non-spherical drag model expression can be expressed as equation(4).

$$F_D = m_p F(\vec{u}_f - \vec{u}_p) C_D \frac{Re_p}{24} = S^B(1 + b \cdot Re_p^A S^C) \tag{1}$$

$$F = \frac{18\mu}{\rho_p d_p^2} \frac{C_D Re_p}{24} C_D \frac{Re_p}{24} = \frac{1}{u} \frac{V_p(\rho_p - \rho_f) g d_V}{12 A_p \mu} \tag{2}$$

$$C_D = \frac{24}{Re_p} \frac{\rho_f F_D}{3\pi\mu^2 Re_p} \tag{3}$$

$$C_D = \frac{24}{Re_p} f(S, Re_p) \tag{4}$$

Where, $\vec{u}_f$ is the velocity of flow fluid, $\vec{u}_p$ is the particle velocity, μ is the viscosity of fluid, ρ_f is the density of fluid, ρ_p is density of particle, d_p is diameter of particle, Re_p is particle Reynolds.

2.2. Form of drag model

There are several non-spherical drag models for wheat straw particles, Kishore and Gu [10] studied ellipsoidal particles moving at a moderate Reynolds number and proposed a drag model with the shape factor of particle aspect ratio. Haider and Levenspiel [11] developed a non-spherical drag correlation and an explicit formula of the settling velocity by using the shape factor of sphericity Ψ. Richter and Nikrityuk [12] studied the drag coefficient of particles with shaped of spheres, squares and ellipses at low Reynolds numbers and considered the ratio of the particle length parallel to the flow direction to the diameter of an equivalent-volume sphere as the shape factor. Fabio Dioguardi et al.[13] studied the drag correlation of pumice particles settling in water using the shape factor Φ, which refers to the ratio of the Ψ of a particle to its circularity X.The four drag models are listed in Table 1.

Table 1. Four models with different shape factors for non-spherical particles.

Model	Conditions	Literature
$C_{D,KG} = \frac{24 \cdot AR^{0.49}}{Re_P}(1.05 + 0.152 Re_P{}^{0.687} AR^{0.671})$	$1 \le Re_P \le 200$, $0.25 \le AR \le 2.5$	Kishore and Gu(KG) [10]
$C_{D,HL} = \frac{24}{Re_P}(1 + A \cdot Re^B) + \frac{C}{\left[1 + \frac{D}{Re_P}\right]}$	$Re_P < 2.6 \cdot 10^5, \Psi > 0.67$	Haider and Levenspiel(HL) [11]
$C_{D,RN} = 0.21 + \frac{20}{Re_P} SR^{0.58} + \frac{6.9}{\sqrt{Re_P}} SR^{-1.4}$	$10 \le Re_P \le 250$	Richter and Nikrityuk(RN) [12]

$C_D = \frac{0.9627}{Re^{0.079}\Phi^{1.6}}$	$1 \le Re_P \le 10^5$	Fabio Dioguardi (FD) [13]

2.3. Shape factors
Five shape factors used in this paper, they are Π(Π1), aspect ratio AR, BR, spheric-ity Ψ, and Φ. The calculation of the shape factors is listed in table 2.

Table 2. Five shape factors for non-spherical particle.

Shape factor	Definition	Explanation
Π	$\Pi = ZX$	Heywood shape factor [14] $Z = V_P/(d_P)^3, X = \frac{P_p}{P_{mp}}$
AR	$AR = \frac{2D_{f,max}}{D_{f,min} + d_{f,max}}$	$D_{f,max}$ is circumscribed circle diameter and $D_{f,min}$ is intangential circle diameter of maximum projection area, $d_{f,max}$ is maximum of lateral view
BR	$BR = \frac{L}{B}$	L and B is length and width of the particle
Ψ	$\Psi = \frac{A_{sph}}{A_p}$	A_{sph} is area of the equal-spherical volume, A_p is the real superficial area of the particle
Φ	$\Phi = \Psi/X$	Ψ is sphericity, X is circularity

2.4. Calculation method
According to equation (4) and model KG, a KG-based drag model is acquired by equation (5). $C_D \frac{Re_p}{24}$ is calculated by the balance of the drag force F_{drag} and effective gravity F_g [15]. Particle settling relative velocity , volumeV_p, densityρ_p, particle diameter d_V and projection area A_p, density ρ_f and viscosity μ are also calculated by experiments.

$$C_D \frac{Re_p}{24} = S^B\left(1 + b \cdot Re_p^A S^C\right) \tag{5}$$

$$C_D \frac{Re_p}{24} = \frac{1}{u} \frac{V_p(\rho_p - \rho_f) g d_V}{12 A_p \mu} \tag{6}$$

In order to fit for equation (6), a transformation of equation (5) should be conducted. First of all, let $y1 = \frac{C_D \frac{Re_p}{24}}{S^B} - 1$, then $y1 = ARe^B S^E$, next, Logarithm is used, equation (7) is obtained. let $y = \ln y1$, $x1 = \ln Re$, $x2 = \ln S$, $\beta_0 = \ln A$, $\beta_1 = B$, $\beta_2 = E$, the finale linear regression function is obtained by equation (8). At last, matlab is used to get the coefficients and determination coefficient R^2.

$$\ln y1 = \ln A + B\ln Re + E\ln S C_D \frac{Re_p}{24} = S^B\left(1 + b \cdot Re_p^A S^C\right) \tag{7}$$

$$y = \beta_0 + \beta_1 x1 + \beta_2 x2 C_D \frac{Re_p}{24} = \frac{1}{u} \frac{V_p(\rho_p - \rho_f) g d_V}{12 A_p \mu} \tag{8}$$

3. Experiment

3.1. Particle geometry and density
The wheat straw was obtained from Funan City, Anhui Province, China. 24 sheered wheat straw particles were picked randomly with length of 10mm and width of 1-6mm. Before measuring the particle geometry and density, the particles must be entirely wet.
Outline of a single wheat straw particle is presented in Fig. 1.The maximum projection length L, the maximum projection width B, and the stretched width B' are measured as shown in Fig. 1. The

maximum projection area A_{pmax} is calculated by the equation *LB*. The particle mass m_P is measured by a precision balance with a measurement range of 0–120 g and an accuracy of ±0.1 mg. The density ρ_p of a single particle and the density of water ρ_f are measured by a standard Gay–Lussac pycnometer with the volume of 100 mL. The volume of a particle V_p is calculated using the formula m_p/ρ_p. The particle equivalent thickness T is calculated by V_p/A_{pmax}.

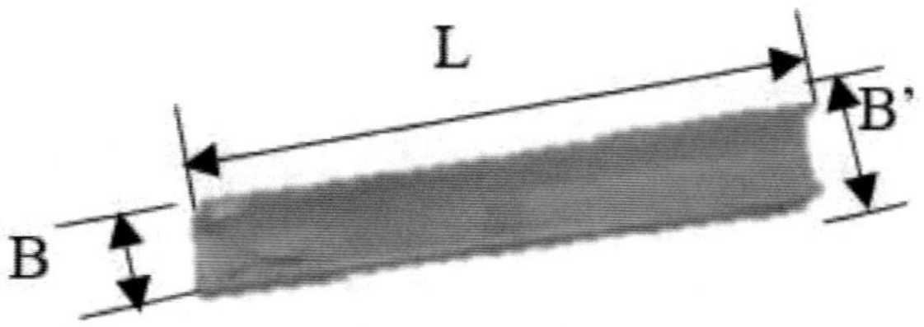

Figure. 1. Outline of a single wheat straw particle.

3.2. Free-settling velocity

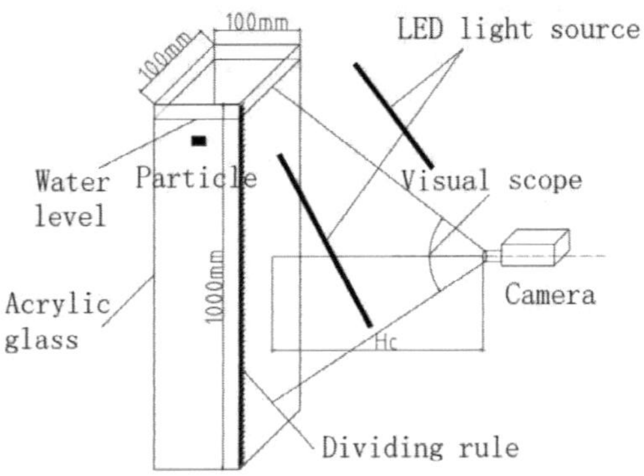

Figure. 2. Diagram of the experimental setup.

The free-settling experiment is conducted in a square tank made of acrylic glass (Figure 2). The length, width, height and thickness of the tank are 100 mm, 100 mm, 1 m and 5 mm, respectively. A white paper is established on one side of the tank, and a V20-3D camera is installed on the opposite side of the paper. The frame rate of the camera is 30 fps and its resolution is 1080*1920 pixel. A 120 W LED light source is used to illuminate the region of interest in the tank. A millimetre-scale ruler is attached to the left side of the water tank and used to assess the falling position of the particles. The distance between the camera and the water tank is denoted as Hc, and the camera focus is adjusted to ensure that the ruler can clearly be seen. The height of the camera's visual scope is 500 mm.

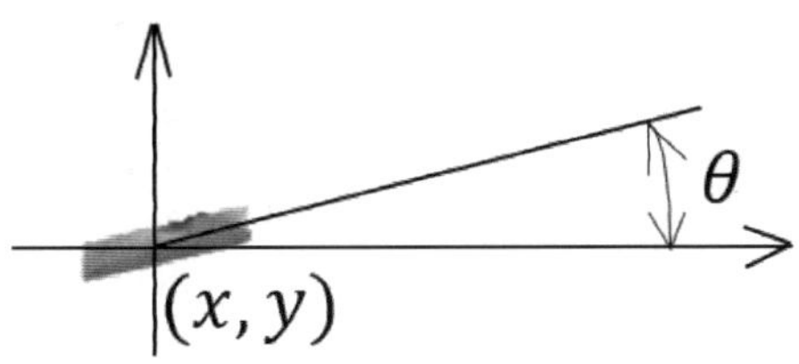

Figure. 3. Particle centroid position and angle θ between the length axis and horizontal plane.

At the beginning of the experiment, the recorder of the camera is switched on. The particles are added to the tank one by one and allowed to fall to the bottom of the vessel. The video recorded by the camera are converted into photographs frame by frame using Photoshop software, and 30 pictures in PNG format are produced in 1s. MATLAB software is used to process the pictures. Firstly, the colour images are converted into grey images. Secondly, the grey images are binarised and denoised. Finally, the centroid y coordinate and the orientation angle θ of the particle in every image is obtained, as shown in Fig. 3. The terminal settling velocity u_{meas} will be obtained by the relationship of y vs t. The maximum projection area A_{pmax} is calculated using the formula $A_{pmax}cos\theta$; d_p, X and Z are also calculated.

4. Numerical setup

4.1. Theory of ANSYS Fluent DPM model

$$\frac{\partial(\varepsilon_f\rho_f)}{\partial t} + \nabla\cdot(\varepsilon_f\rho_f\vec{u}_f) = 0 C_D\frac{Re_p}{24} = S^B\left(1 + b\cdot Re_p^A S^C\right) \tag{9}$$

$$\frac{\partial(\varepsilon_f\rho_f\vec{u}_f)}{\partial t} + \nabla\cdot(\varepsilon_f\rho_f\vec{u}_f\vec{u}_f) = -\varepsilon_f\nabla p + \nabla\cdot(\varepsilon_f\bar{\bar{\tau}}_f) + \varepsilon_f\rho_f\vec{g} + R_{f,p}C_D\frac{Re_p}{24} = \frac{1}{u}\frac{V_p(\rho_p-\rho_f)gd_V}{12A_p\mu} \tag{10}$$

For dispersed phase, the single particle's motion equation is as follows:

$$\frac{d\vec{u}_p}{dt} = F(\vec{u}_f - \vec{u}_p) + \frac{\vec{g}(\rho_p-\rho)}{\rho_p} + \vec{F} \tag{11}$$

Take consideration of the fluid drag force on the particle, number of particles M in the calculation cell, the volume of V_{cell} and the force on the m_{th} particle of $\vec{F}_{D,m}$, Drag $R_{f,p}$ [16]of unit volume of calculation cell is:

$$R_{f,p} = -\frac{\sum_m^M \vec{F}_{mD}}{V_{cell}} \tag{12}$$

4.2. Theory of ANSYS Fluent DPM model
Drag in ANSYS Fluent UDF function is defined by equation (13).

$$DEFINE_{DPM_{DRAG}} = \frac{18C_D Re_p}{24} \tag{13}$$

5. Results and discussion

5.1. Geometry parameters and shape factors
Basic parameters for water and two type of handled particles are listed in table 3. The mass, volume, size and shape parameters of the 24 particles are listed in Table 4.

Table 3. Basic parameters for water and two kinds of treated particles.

	Water	Sheared particles	Defibering particles
ρ_f(g/m^3)	996.62	1138	1050
μ_f(Pa·s)	0.00089	-	-

Table 4. Mass, volume, size and shape parameters of the 24 sheared particles.

	m_p(g)	V_p(mm^3)	L(mm)	B(mm)	T(mm)	A_{pmax}(mm^2)	Dp(mm)	Dv(mm)	Ψ	Φ	Π	AR
1	0.0234	20.56	10.26	4.4	0.46	45.14	7.58	3.40	0.37	0.30	0.038	4.23
2	0.021	18.45	10.42	4.31	0.41	44.91	7.56	3.28	0.39	0.31	0.034	4.41
3	0.0147	12.92	9.72	3.62	0.37	35.19	6.69	2.91	0.38	0.30	0.034	4.88
4	0.0159	13.97	10.16	3.72	0.37	37.80	6.94	2.99	0.43	0.34	0.033	4.97
5	0.0145	12.74	10.1	3.68	0.34	37.17	6.88	2.90	0.42	0.33	0.031	5.02
6	0.0159	13.97	10.2	3.42	0.40	34.88	6.66	2.99	0.40	0.31	0.036	5.34

7	0.0146	12.83	9.76	3.44	0.38	33.57	6.54	2.91	0.44	0.34	0.036	5.11
8	0.0133	11.69	10.96	3.26	0.33	35.73	6.74	2.82	0.39	0.29	0.028	6.11
9	0.0103	9.05	10.52	2.7	0.32	28.40	6.01	2.59	0.46	0.33	0.030	6.97
10	0.0118	10.37	10.62	3.3	0.30	35.05	6.68	2.71	0.41	0.31	0.026	5.91
11	0.0113	9.93	10.32	2.92	0.33	30.13	6.19	2.67	0.46	0.34	0.031	6.35
12	0.0083	7.29	10.36	2.2	0.32	22.79	5.39	2.41	0.49	0.33	0.031	8.22
13	0.0057	5.01	10.6	1.98	0.24	20.99	5.17	2.12	0.42	0.27	0.023	9.56
14	0.0044	3.87	11.1	1.08	0.32	11.99	3.91	1.95	0.58	0.29	0.033	15.83
15	0.0064	5.62	10.48	2.5	0.22	26.20	5.78	2.21	0.36	0.25	0.020	7.72
16	0.0101	8.88	10.76	2.2	0.38	23.67	5.49	2.57	0.54	0.36	0.036	8.36
17	0.0125	10.98	10.62	3.56	0.29	37.81	6.94	2.76	0.39	0.30	0.025	5.52
18	0.0089	7.82	10.3	2.72	0.28	28.02	5.97	2.46	0.42	0.30	0.026	6.87
19	0.0085	7.47	10.46	3.42	0.21	35.77	6.75	2.43	0.32	0.25	0.019	5.77
20	0.0136	11.95	11	4.1	0.27	45.10	7.58	2.84	0.35	0.28	0.022	5.04
21	0.0104	9.14	9.8	3.28	0.28	32.14	6.40	2.59	0.35	0.27	0.027	5.50
22	0.007	6.15	10.28	2.4	0.25	24.67	5.60	2.27	0.41	0.28	0.024	7.76
23	0.0038	3.34	10.1	1.12	0.30	11.31	3.80	1.85	0.57	0.30	0.032	14.27
24	0.0067	5.89	10.64	2.12	0.26	22.56	5.36	2.24	0.43	0.29	0.025	8.94

5.2. Particle settling velocity, drop angle and projection area

Fig. 4 gives the vertical position and velocity of a single particle falling in still water. The trajectory of a particle is a polynomial equation of time t when $t < 2$ s and a linear equation when $t > 2$ s. The drop velocity of a wheat straw particle is calculated by the slope of the trajectory.

Fig. 5 illustrates the horizontal projection position and the falling angle θ of the particle changes with time in every 0.1s. On the left side of Fig. 5, describes the particle falls in the still water, the particles orient themselves with their largest projection area normal to the flow [17]. The maximum area A_{pmax} of a particle is always orient downwards. The vertical projection area is equal to $A_{pmax} \cos \theta_p$, where θ_p is the terminal settling angle when the velocity is in the condition of terminal settling. Due to the falling angle θ changes with times, the mean value in a period of time is calculated as the terminal settling angle θ_p. Settling velocity and related parameters of the 24 particles are listed in Table 5.

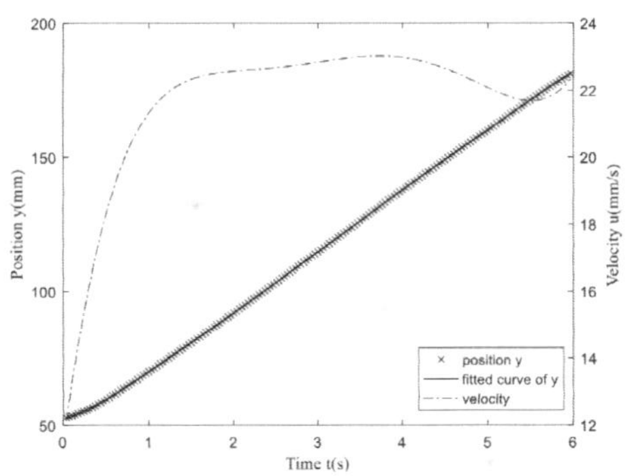

Figure. 4. Falling position and drop velocity of wheat straw particle changing with time.

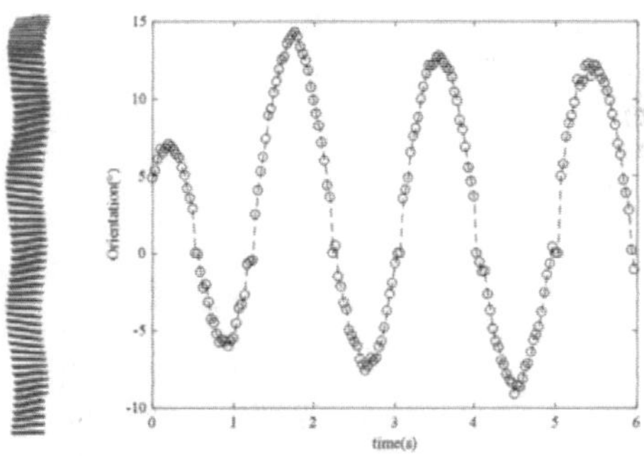

Figure. 5. The horizontal projection and settling angle of wheat straw particle during falling down.

Table 5. Settling velocity and related parameters of the 24 particles.

	u_{meas}(m/s)	θ_p(°)	A_P(mm^2)	$f(Re_P, S)$	Re_p	Re_s
1	0.0227	0	45.144	19.688	191.333	85.77
2	0.0206	1.98	44.883	19.489	173.468	75.25
3	0.0212	3.51	35.120	15.020	157.552	68.60
4	0.0197	0.41	37.794	16.824	152.139	65.52
5	0.0209	2.26	37.139	14.617	159.698	67.28
6	0.0196	−0.37	34.883	17.582	145.586	65.29
7	0.0204	0.98	33.569	15.816	148.606	66.01
8	0.0187	1.3	35.720	15.233	140.560	58.70
9	0.0171	1.442	28.395	14.490	114.429	49.21
10	0.0166	1.79	35.029	15.342	123.816	50.16
11	0.0166	0	30.134	15.916	114.297	60.37
12	0.0166	−0.13	22.792	13.442	99.4019	44.40
13	0.0176	3.03	20.959	9.0780	101.081	41.53
14	0.0148	−0.99	11.986	11.002	64.3783	32.09
15	0.0192	1.8	26.187	8.3556	123.306	47.13
16	0.0203	−0.81	23.670	13.126	123.869	57.96
17	0.0154	0.076	37.807	16.905	119.032	47.33
18	0.0169	−1.06	28.011	12.724	112.604	46.43
19	0.0225	0.95	35.768	8.0761	169.434	60.90
20	0.0196	3.13	45.033	13.214	165.69	62.07
21	0.020	−0.16	32.144	11.764	142.33	57.71
22	0.0149	−3.09	24.636	12.114	93.026	37.77
23	0.0192	−1.01	11.310	7.5534	80.989	39.57
24	0.0195	−0.34	22.556	9.2641	116.43	48.67

5.3. Form selection of drag model

Non-spherical drag model of KG HL RN and FD are used to simulated 24 sheared particles. relative error between simulation velocity and measurement velocity for the 24 particles using four drag models described in Fig. 6. KG model is best for the wheat straw particle with the smallest error of 95%. The KG-based model is used for the drag model development.

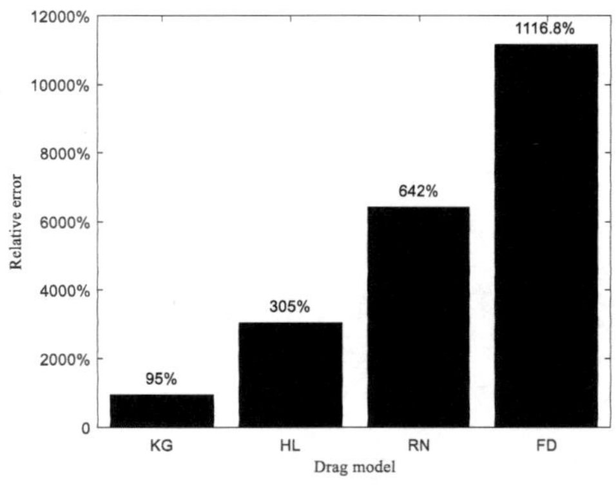

Figure 6. Relative error between simulated velocity and measured velocity for the 24 particles using four KG HL RN and FD models.

5.4. Fitting results of KG-based drag models using sheared particles

Shape factors of Π, AR, BR and Φ are used to fit for the KG-based drag model. Assigns a series of discontinuous values for B, and fit the relationship between $f(Re_p, S)$ and Re_p, corresponding coefficients of β_0, β_1, β_2 are acquired.

According to R2 and R2 (adj)(Adjusted R-Squared) and p-Value, the best B and coefficients are determined. See Fig. 7 and Table 6 and Table 7.

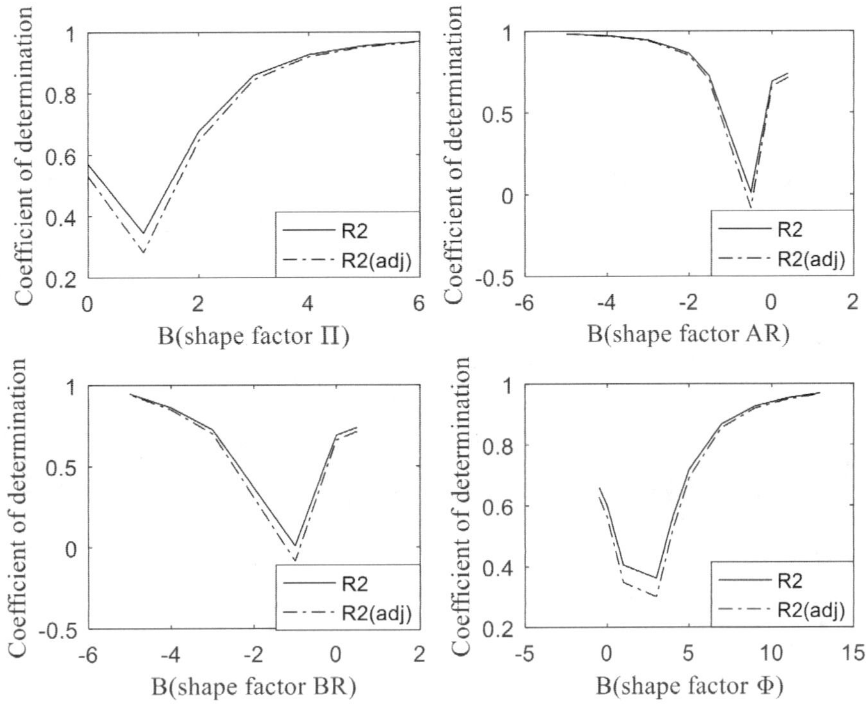

Figure 7. R2 and R2(adj) changes with B by curve fitting with the shape factor Π, AR, BR and Φ used in the drag model.

Table 6. pValue of coefficient from curve fitting of different shape factors.

Coefficient	**Π**	**AR**	**BR**	**Φ**
β_0	0.0082911	0.066814	0.16903	0.0535
β_1	0.0040152	0.99445	0.61977	0.011961
β_2	1.7789e-17	5.3474e-14	1.4163e-15	2.92E-17
overall	3.05E-20	2.06E-19	3.05E-20	1.45E-16

5.5. Prediction effects of KG-based drag models

Drag models with shape factors of Π,AR,BR,and Ψ are used to simulate the 24 sheared particles moving in the still water and Drag models with shape factors of Π, Π1, AR, BR, Φ and Ψ are used to simulate the 134 defibered particles moving in the still water. The relative error of particle velocity between simulation and measurement are listed in Fig. 8 and Fig. 9.

According to Fig. 8 and Fig. 9, drag model with shape factor Π, AR, BR have the lower prediction relative error for sheared wheat straw particles, and drag model with shape factor Π,AR,BR,and Ψ have the lower prediction relative error for sheared wheat straw particles. A comprehensive evaluation is applied by the mean relative error obtained by drag models with shape factors of Π, AR, BR and Ψ, see Table 8. Base on the mean error in Table 8 and pValue in Table 7, KG-based model with shape factor is selected for simulating the settling velocity of wheat straw particle moving in the still water.

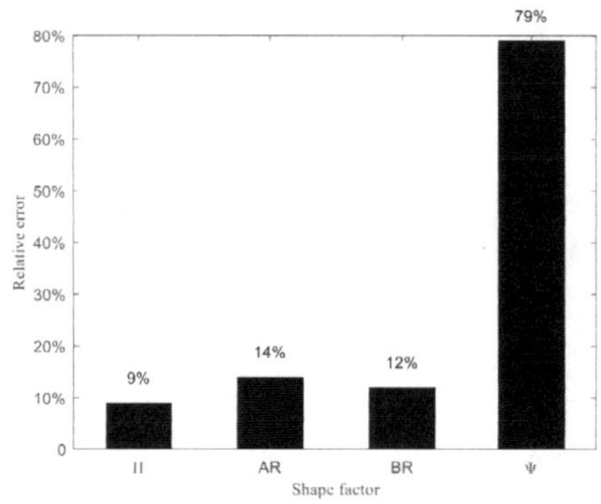

Figure 8. Relative error between simulation velocity and measurement velocity for the 24 particles using shape factors Π,AR,BR,and Ψ .

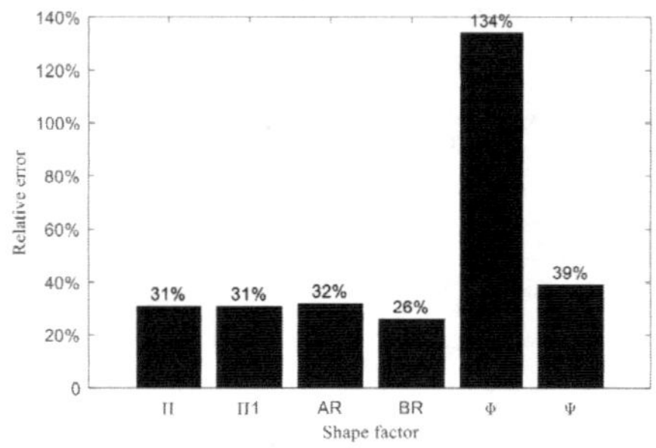

Figure. 9. Relative error between simulation velocity and measurement velocity for the 134 particles using drag models with shape factors of Π, Π1, AR, BR, Φ and Ψ.

Table 7. Relative error prediction by drag models with shape factors of Π, AR, BR and Ψ.

	Π	AR	BR	**Ψ**
Sheered wheat straw particles	9.20%	13.89%	11.94%	78.78%
Defibered wheat straw particles	31%	32%	26%	39%
Mean error	20%	23%	19%	59%

5.6. Error analysis

The error of the drag model to predict the falling velocity mainly come from the experimental measurement. The falling angel and the optical reflection is non-negligible reasons.

6. Conclusions

A KG-based drag model is developed with the form of $C_D = \frac{24\Pi^B}{Re_p}\left(1 + b \cdot Re_p^A \Pi^C\right)$, coefficients of A,B,C and b are 0.48452, -5.1988, 6 and $e^{3.0613}$, respectively.

The drag model with shape factor Π has higher prediction accuracy than other shape factors with the mean relative error of 20% for wheat straw particles.

7. References

[1] Rahman S, et al. 2021 Environmentally friendly superabsorbent fibers based on electrospun cellulose nanofibers extracted from wheat straw Carbohydrate Polymers 251, 1170–1187.

[2] Wu B 2011 CFD investigation of turbulence models for mechanical agitation of non-Newtonian fluids in anaerobic digesters, water research. 45,2082–2094.

[3] Terashima M, et al. 2009 CFD simulation of mixing in anaerobic digesters, Bioresource Technology. 100, 2228–2233.

[4] Wu B 2013 Advances in the use of CFD to characterize, design and optimize bioenergy systems,

[5] Terashima M et al. 2009 CFD simulation of mixing in anaerobic digesters, Bioresource Technology. 100,2228–223

[6] Metolinaa P, Gabriela C 2019 Numerical analysis of liquid-solid flow in tapered and cylindrical fluidized beds for wastewater treatment and biogas production, Energy Conversion and Management. 187, 447–458
[7] Dellino P Mele D 2005 The analysis of the influence of pumice shape on its terminal velocity, Geophys. Res. Lett. 32 L21306
[8] A. Inc 2016 Ansys Fluent Theory Guide 17.0.
[9] MarkkuNikku et al 2014 Characterization method of average gas-solid drag for regular and irregular particle groups. Powder Technology 253, 284–294.
[10] Kishore N Gu S 2011 Momentum and heat transfer phenomena of spheroid particles atmoderate Reynolds and Prandtl numbers, Int. J. Heat Mass Transf. 54, 2595–2601
[11] Haider A Levenspiel O 1989 Drag coefficient and terminal velocity of spherical and nonspherical particles, Powder Technol. 58, 63–70
[12] Andreas Richter, Petr A. 2012 Nikrityuk.Drag forces and heat transfer coefficients for spherical, cuboidal and ellipsoidal. particles in cross flow at sub-critical Reynolds numbers, International Journal of Heat and Mass Transfer. 55, 1343–1354.
[13] Dioguardi F Dellino P Mele D 2014 Integration of a new shape-dependent particle–fluid drag coefficient law in the multiphase Eulerian–Lagrangian code MFIX-DEM, Powder Technology. 260, 68–77.
[14] Carranza F Zhang Y 2017 Study of drag and orientation of regular particles using stereo vision, Schlieren photography and digital image processing, Powder Technol. 311, 185–199.
[15] ROWE P N 1987 A convenient empirical equation for estimation of the Richardson-Zakiexponent. Chem. Engng Sci. 42, 2795–2796.
[16] Mema I 2019 Effect of lift force and hydrodynamic torque on fluidisation of non-spherical particles Chemical Engineering Science 195, 642–656.
[17] Carranza F Zhang Y 2017 Study of drag and orientation of regular particles using stereo vision, Schlieren photography and digital image processing, Powder Technol. 311,185–199.

Acknowledgements
This research was funded by the key technology and demonstration project of the efficient recycling handles of human manure on rural toilets (2018YFC1903204) of the ministry of Science and Technology of China.

Modelling of a Stilling Basins with Sloping Apron in IBER to Improve Efficiency in High-slope Rivers

Y Zabaleta[1,*], O K Espejo[1], J M Fernández[1] and S Santos[1]

[1]Civil Engineering, Peruvian University of Applied Sciences, Prolongación Primavera 2390, Santiago de Surco, Lima, Perú.

u201616464@upc.edu.pe

Abstract. This research shows the influence of stilling basin slopes on energy loss in rivers with a high gradient. This study takes as a case San Pedro water intake (Ayacucho, Peru). The main objective is to improve efficiency of stilling basins in rivers with high slope. Five dissipation pools of different slopes were modelled: 0%, 1.52%, 3.04%, 4.56% and 6.08% to propose the optimum pool among these, for the San Pedro intake. Results were validated by means of a Sensitivity Analysis, trough comparison with the results of previous investigation and results of modelling San Pedro river with HEC-RAS and IBER. It was obtained that the steeper the slope of the stilling basin, the higher the specific energy loss, the higher the output rate, the longer the stilling basin. It can be concluded that the 3.04% slope stilling basin is the most appropriate for the 6.08% slope river since the slope variation is not abrupt as in the case of the horizontal one, that is, 30% more energy loss with respect to the horizontal pool and velocity and Froude results similar to the modelling of the San Pedro river.

1. Introduction

Hydraulic structures placed in natural riverbeds, such as barrages; cause a change in the natural flow, which is why they have an evacuation system that ensures the controlled output of the flow. These systems are called spillways and have a stilling basin to transform supercritical to subcritical flow. In general, the water outlet has a higher kinetic energy and must be dissipated, as it would cause erosion downstream if it were not dissipated. This dissipation occurs by means of the formation of a hydraulic jump in the stilling basin, seeking to join the outlet pipe of the pool with the natural level of the river [1]. These pools are usually horizontal as they ensure a subcritical flow at the outlet, but what happens to rivers with high slopes that does not have subcritical flow?

Flow in the catchment structures located in rivers with high slopes is characterized by a torrential and supercritical regime. The subcritical regime obtained in the stilling basin is once again supercritical due to the difference in slopes between bottom and river natural gradient of the river. In this case, potential energy produced at stilling basin exit, is greater than that of the natural course, generating turbulence and continuous waves. Chanson [1] explains that challenge of a hydraulic jump involves movement of turbulent flow with development of large-scale vortices and trapping of air bubbles at the tip of the jump. He also points to energy transport and dissipation as the key challenges during the weir design process to ensure the life of the structure. Similarly, Legono, Hambali and Krisnayanti [2] show that chaotic jumps of stilling basins downstream of weir cause vibrations in the structure that directly affect it. According to Lempérière et al. [3] the costs of some solutions can be much higher for existing hydraulic structures than for new ones. He also points out that the key to acceptable cost is to reduce probability of failure. Therefore, an optimal solution is sought for designing a stilling basin that guarantees precision and reliability of energy dissipation of supercritical flow at the outlet that reaches the natural flow of the river.

Kumar, M; Kumar, S; Bidhu, S [4] states that with the aim of increasing effectiveness of a sloping channel that maximum energy reduction in relation to the increasing slope of the channel was 45% and according to Pardo's study [5] hydraulic jump in sloping screeds gives high quality efficiency results just like the cases of horizontal screeds. From this, we can model a sloped stilling basin in a CFD program to verify flow behavior, computational fluid dynamics gives reliable results. Thu Hien [6] made a dam spillway physical and numerical model and could prove that the CFD can accurately simulate many real phenomena of the fluids at the stilling basin, since it uses Navier Stokes' equations. By carrying out experimental study by means of numerical modelling of stilling basin with a sloping bottom precision and reliability of energy dissipation of supercritical flow at the pool exit will be guaranteed, reaching the natural supercritical flow of the river. This would provide benefits in improving the performance and safety of the hydraulic structure. The present investigation models flow behaviour in energy dissipation structure located in a high slope river to propose an optimal solution for stilling basin design.

2. Methodology

In order to propose the most efficient stilling basin in high slope rivers, first is necessary to analyse and model main characteristics of flow in river without structures. Based on these data, stilling basins with different slopes will be modelled, results will be compared at pool outlet, so that the one that has the most similarity with the natural river flow characteristics at this point, will be chosen. With the chosen slope, a complementary design accessing the stilling basin will be made for its improvement.

2.1. Hydrological characteristics

Intake is at the San Pedro river, located in Peru, Puquio district, Lucanas province, Ayacucho department. Geographically, in zone 18 L of UTM coordinates approximately 597501.00 m E and 8368463.00 m S. To establish its hydrological and physical characteristics, an investigation carried out by Contreras and Villegas will be used [7]. Table 1 details the main characteristics of the river.

Table 1. Main geomorphological characteristics of the San Pedro river.

San Pedro River					
Length of the river	=	42.10 km	Average slope	=	6.08 %
Average flow	=	6.92 m^3	Design flow rate	=	44.26 m^3/s
Average riverbed width	=	15.40 m	Runoff coefficient	=	0.277
Total annual precipitation average	=	636.43 mm/year	Average annual temperature	=	11.465°C
Average Manning's roughness	=	0.0595	Normal brace	=	0.976 m

2.2. River modelling in HEC-RAS and IBER

The modelling of the river is based on topographic information, in this case NASA's satellite information was used, Alos Palsar, which provides spatial information with a resolution of 12.5 x 12.5 m. From established digital model of elevation, contour lines are generated every 0.50 m by means of Global Mapper program. In the Civil 3D programme river's progressive is created having as a starting point intake, 500m upstream and 100m downstream. The 600 m progressive of the river is created in the opposite direction of the river, as the HEC-RAS programme considers the sections in the opposite direction. Sections are also created at a 10m between them and width of 30m. These values are exported to HEC-RAS. The input parameters are design flow 44,26 m3/s (Q_{50}), average river slope of 6.08% and roughness of 0.0595. The results of the modelling in a three-dimensional view are shown in figure 1.

For the modelling of the surface of San Pedro river in IBER programme, QGIS extension was used. The flow values and the roughness of the river surface were the same as those used in the HEC-RAS modelling; but since this is a 2D software, a 3m unstructured mesh was used and the simulation time was 700s with 10 s intervals. In post-processing of modelling, a longitudinal section is traced along the river to distinguish water level. Profile of this plotted section can be seen in figure 2. To obtain water levels along the river, the results are exported to an Excel spreadsheet.

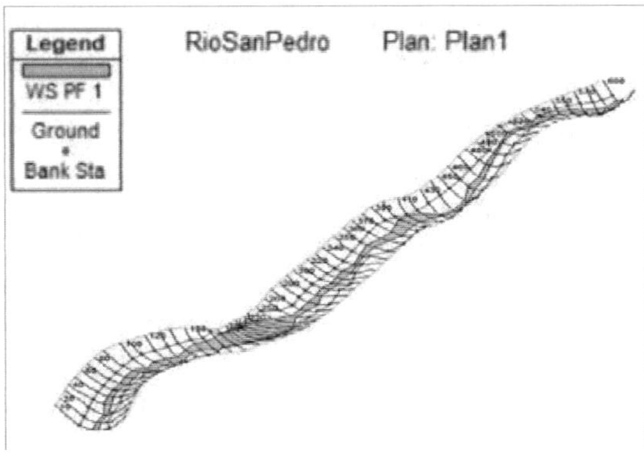

Figure 1. XYZ perspective modelling - HEC-RAS.

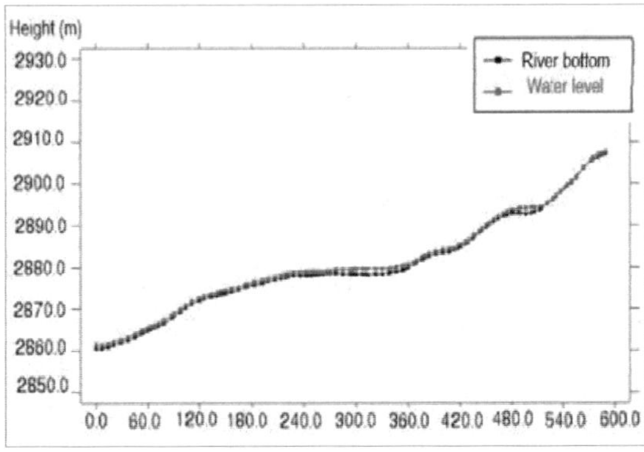

Figure 2. River longitudinal section in IBER.

Hydraulic Characteristics for Q=44.26 m^3/s of San Pedro River were found in area of intake, which corresponds to the progressive 0+500 m of the model. Results were calculated at stilling basin exit. Hydraulic depth of 0.54m was obtained with HEC RAS and 0.52m with IBER. Table 2 shows these results, as well as velocity and Froude number. Velocity values vary in 0.39 m/s. Values of HEC-RAS are higher than IBER's, and with respect of Froude's values, they vary by 0.28. These values confirm that they vary on average by 5.3% between the two softwares. These variations are because in HEC-RAS programme sections have been modelled every 10 m and in IBER programme they have been modelled with the Dem surface with a 3 m mesh.

Table 2. Hydraulic characteristics for Q = 44.26 m^3/s in HEC-RAS and IBER.

Software	Hydraulic depth (m)	Velocity (m/s)	Froude
HEC-RAS	0.54	6.77	4.26
IBER	0.52	6.38	3.98

2.3. Modelling stilling basin

San Pedro's intake is of a conventional type due to topographical, geomorphological characteristics and the area chosen for the location. This structure has a raising water level river function and facilitating entry through the intake window by using a "Creager" weir type profile, to reduce pressure all the points of the weir to almost zero. The model geometry covers only the width of intake and characteristics required are summarized in Table 3.

Table 3. San Pedro intake characteristics.

Components	Measures	Components	Measures
Width	15.00 m	Gate height	1.45 m
Hole Threshold Height	0.60 m	Hydraulic jump depths	y_1= 0.69m
Catchment window height	0.60 m		y_2= 2.63m
Catchment window length	3.00 m	Stilling basin length	13.15 m
Weir height	1.40 m	Stilling basin slope	6.08 %
Horizontal length of barrage profile	2.55 m	Protective rock or breakwater	3.00 m
Barrage load height	1.64 m	Weir width	1.00 m

To simulate the real behaviour of an intake in the modelling, the design flow will not only flow through the weir and the 2 gates, but also part of the flow will be captured by the intake window. Measurements will be taken from Table 3, the stilling basin zone will be modelled with 2 slopes, 6.08% and 0.00%. The length of stilling basin with 6.08% slope is already known, for the horizontal stilling basin it is calculated using the formulas of Pardo [4] and Peterka [8] assuming a hydraulic jump. With a Froude of 2.47 downstream of weir, we found major conjugate depth (2.09 m), which was replaced in the Pardo equation for a stilling basin with a slope of 0%, which gave us a length of 10.20m.

Pre-processing of the 6.08% and 0% slope stilling basin in the IBER software involves modelling the geometry as well as meshing it. Intake is imported as a 3D polyline into IBER. Hydrodynamic conditions such as the design flow of 44.26 m3/s (Q50) and internal conditions such as the location of the structures (2 gates, 1 capture and 1 spillway) are assigned. Similarly, roughness is assigned through Manning's coefficient for concrete material of 0.014 over the entire surface of intake and for the rockfill 0.035 in breakwater area. Subsequently, unstructured mesh of 0.2 m is created and the calculation parameters are set with a total simulation time of 80 seconds with results every 1 second. Calculations are then executed and the results are visualised in the post-process.

Results are validated by sensitivity analysis and comparison with design values. With this validation, 3 stilling basin designs are modelled based on change in slope: 4.56%, 3.04% and 1.52%. These slopes are between 0% and 6.08% so that they vary in same proportion. With exported results of the 5 slopes, hydraulic depth, velocity, Froude number and energy at basically the outlet of the stilling basin are compared. Finally, results of energy dissipation are analysed and efficiencies are compared.

3. Results

3.1. Stilling basins modelling results

Result of modelling in IBER programme for horizontal stilling basin is shown by a plan view in figure 3 and Table 4 details values in 3 important points. In catchment area, which corresponds to the first 5 metres, water levels reach 2.61 m. In weir area the water level gradually decreases until it reaches 0.32 m, which is where stilling basin begins. In this zone, corresponding to basin entrance, the velocity is 7.15 m/s and the Froude 4.04 as shown in Table 4. In the basin zone there is no hydraulic jump, meaning that not enough energy is dissipated for the flow to become subcritical, as can be seen from the values in the table, which indicates that at the outlet of the pool the velocity is 6.79 m/s and Froude number is 3.97. The last zone is 3 m breakwater zone where the material changes to rockfill and the water level increases by 5 cm with respect to the outlet of basin, thus the velocity and Froude decrease.

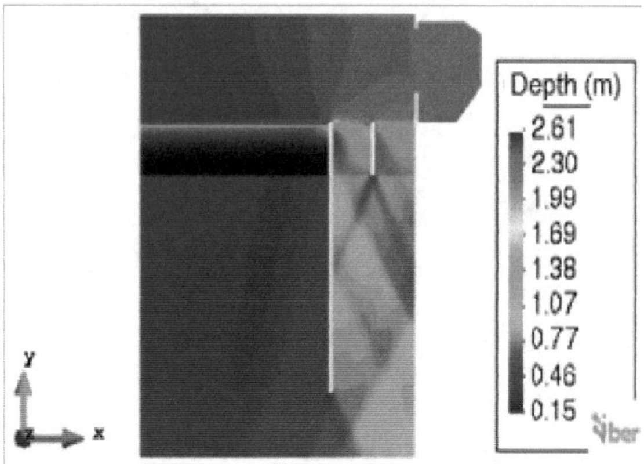

Figure 3. Hydraulic depth plan view - Horizontal stilling basin.

Table 4. Horizontal basin results.

Design Flow Rate (Q_{50}=44.26 m^3/s)			
	Basin entrance	Basin exit	Breakwater exit
Hydraulic depth (m)	0.32	0.30	0.35
Velocity (m/s)	7.15	6.79	5.74
Froude	4.04	3.97	3.11
Energy	2.64	2.57	2.07

Stilling basin with 6.08% slope has 3.48° respect to horizontal. Figure 4 shows plan view of flow through the hydraulic structure and Table 5 shows results for hydraulic depth, velocity, Froude and energy. Catchment area is similar to previous case as flow rate is the same, head reaches 2.62m. The stilling basin area has a slight variation with respect to horizontal modelling as the water level decreases more, in this case by 5cm and in the previous case by 2cm. With respect to values of velocity and Froude in case of this inclined pool, they increase by 0.39 m/s and 1.05. In the breakwater area the values do decrease as in the previous case.

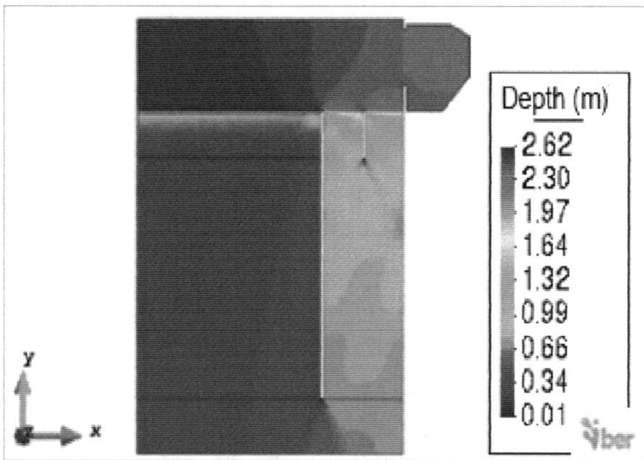

Figure 4. Hydraulic depth plan view - 6.08% stilling basin.

Table 5. Results of 6.08% slope. basin

Design Flow Rate (Q_{50}=44.26 m^3/s)			
	Basin entrance	Basin exit	Breakwater exit
Hydraulic depth (m)	0.31	0.26	0.32
Velocity (m/s)	7.39	7.52	6.40
Froude	3.62	4.67	4.25
Energy	2.65	2.56	1.76

According to both models, the design of the horizontal and inclined pool for a flow of 44,26 m^3/s shows that there is no hydraulic jump in either, which means that there is no energy dissipation. However, according to Fatemeh's study [9], when there is a jump, the factors of turbulence, disturbance and pressure fluctuations increase, which would be more parameters to control.

3.2. Numerical model validation

Information obtained will be validated through a sensitivity analysis [10] in a local approach, that is by changing flow parameter of the numerical model. Likewise, for inclined dissipation pool, the values of model carried out are compared. Furthermore, the results will be compared with the empirically obtained values for the design.

3.3. *Sensitivity Analysis*

It was used to increase model confidence and represent real circumstances. Input parameter taken is water intake flow, this will allow seeing variation it causes in model. To do sensitivity analysis procedure consisted of modelling in IBER, taking stilling basins of slope 6.08% and 0% with flow parameter design flow rate (Q_{50}= 44,261 m^3/s), average maximum flow rate (Q_{MP}= 42.61 m^3/s), and extraordinary flow rate (Q_E= 27.91 m^3/s).

Figures 5 and 6 show hydraulic depths for horizontal basin and stilling basin with a sloping bottom. In both figures similar flow behaviours. It is clearly observed that with a higher flow, water level over the weir increases. Hydraulic depths for discharges 44.261 m3/s and 42.61 m3/s are similar, meaning, this small variation is not very important. It is also highlighted that in stilling basin of both charts water levels are uniform. Tables 6 and 7 also show the values of the graphs of the figures for 5 important points: the entrance to the intake, the beginning of weir, at basin entrance, at basin exit and at breakwater exit.

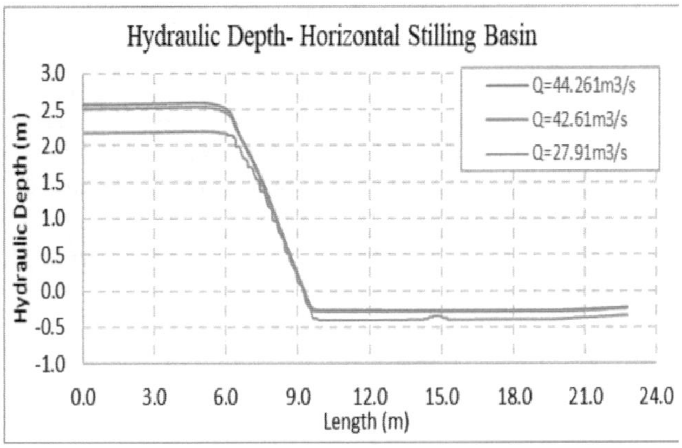

Figure 5. Longitudinal profile- Horizontal basin.

Table 6. Hydraulic depth values - Horizontal basin.

	Hydraulic depth (m)		
Flows (m^3/s)	Q_{50} **44.26**	Q_{MP} **42.61**	Q_E **27.91**
Intake inlet	2.56	2.52	2.16
Weir start	2.58	2.55	2.17
Basin entrance	0.32	0.30	0.19
Basin exit	0.30	0.29	0.18
Breakwater exit	0.35	0.34	0.24

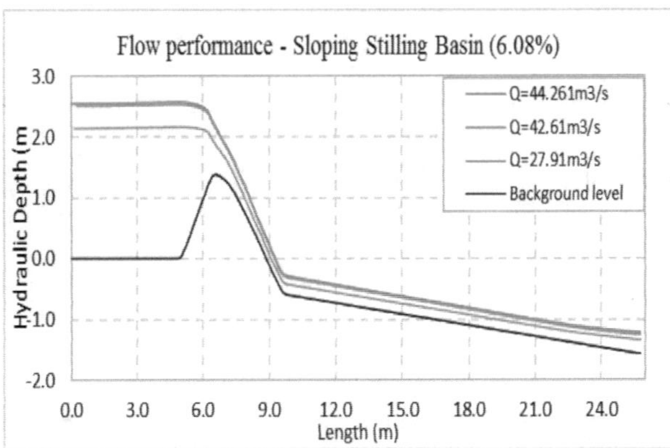

Figure 6. Longitudinal profile- 6.08% slope basin.

Table 7. Hydraulic depth values - 6.08% slope basin.

	Hydraulic depth (m)		
Flows (m^3/s)	Q_{50} **44.26**	Q_{MP} **42.61**	Q_E **27.91**
Intake inlet	2.55	2.51	2.15
Weir start	2.57	2.54	2.16
Basin entrance	0.31	0.30	0.17
Basin exit	0.26	0.25	0.16
Breakwater exit	0.32	0.31	0.21

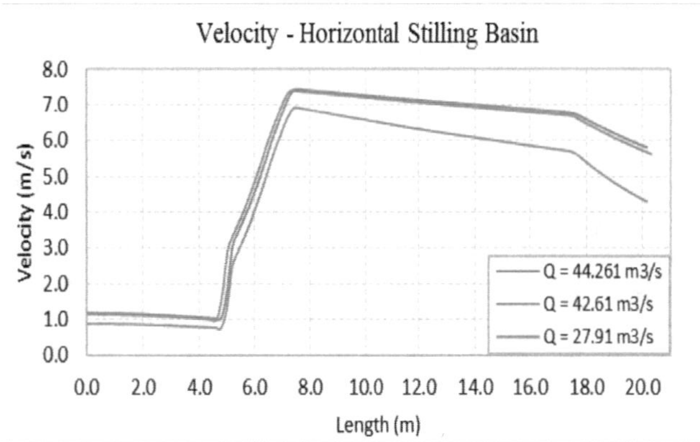

Figure 7. Velocity values in horizontal stilling basin.

Table 8. Velocity values - Horizontal basin.

Flows (m^3/s)	Velocity (m/s)		
	Q_{50} **44.26**	Q_{MP} **42.61**	Q_E **27.91**
Intake inlet	1.18	1.16	0.89
Weir start	1.03	1.00	0.78
Basin entrance	7.15	7.11	6.29
Basin exit	6.79	6.71	5.74
Breakwater exit	5.74	4.23	4.23

Figures 7 and 8 show the velocity behaviours for both models and Tables 8 and 9 show the values at the 5 important points penalised. Behaviour for both the design and extraordinary maximum flow rate are very close, as flow rate only varies by 1.65 m3/s, with extraordinary maximum flow rate being lower than design flow rate. In the case of the average maximum flow rate, velocity values are lower than previous ones, however, it has same behaviour. Velocity gradually decreases until it reaches the breakwater where you see a further decrease.

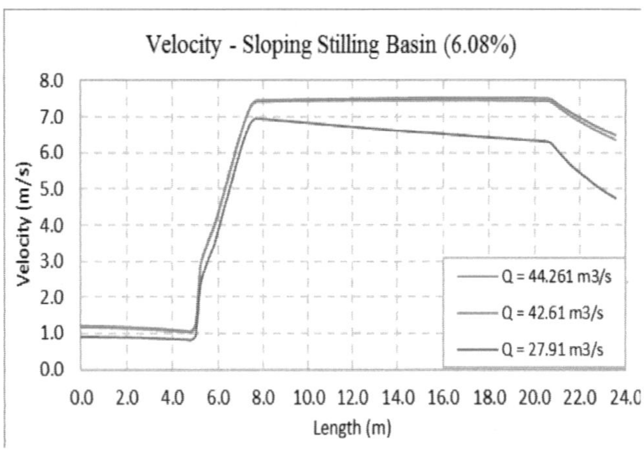

Figure 8. Velocity values in stilling basin with slope of 6.08%.

Table 9. Velocity values - 6.08% slope basin.

Flows (m^3/s)	Velocity (m/s)		
	Q_{50} **44.26**	Q_{MP} **42.61**	Q_E **27.91**
Intake inlet	1.18	1.16	0.89
Weir start	1.03	1.02	0.81
Basin entrance	7.39	7.20	6.90
Basin exit	7.52	7.43	6.31
Breakwater exit	6.40	6.27	4.71

Figures 9 and 10 show values of energy for both stilling basins and the Tables 10 and 11 show the precise values for each point. Initial energy at the inlet is constant for flow rates, presenting a curve in the relief part. The energy values tend to decline along stilling basins, meaning that it loses energy as the section progresses. For stilling basin with a 6.08% slope, it can be observed a greater decrease in energy than in horizontal basin.

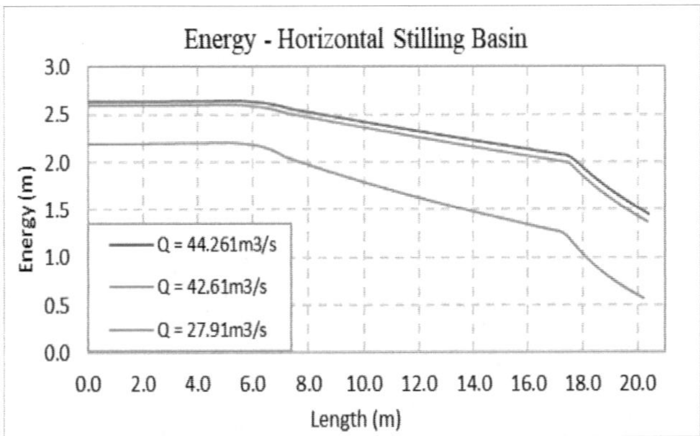

Figure 9. Energy values - Horizontal Stilling basin.

Table 10. Energy values - Horizontal basin.

Energy (m)			
Flows (m^3/s)	**Q_{50} 44.26**	**Q_{MP} 42.61**	**Q_E 27.91**
Intake inlet	2.63	2.59	2.20
Weir start	2.64	2.60	2.21
Basin entrance	2.57	2.53	2.09
Basin exit	2.07	2.00	1.26
Breakwater exit	1.45	1.37	0.57

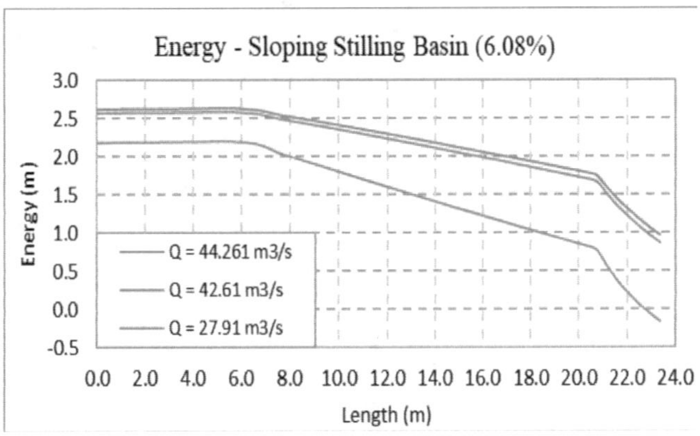

Figure 10. Energy values – 6.08% Stilling basin.

Table 11. Energy values - 6.08% slope basin.

Energy (m)			
Flows (m^3/s)	**Q_{50} 44.26**	**Q_{MP} 42.61**	**Q_E 27.91**
Intake inlet	2.63	2.58	2.19
Weir start	2.64	2.59	2.21
Basin entrance	2.56	2.51	2.07
Basin exit	1.78	1.70	0.81
Breakwater exit	0.98	0.87	-0.16

This sensitivity analysis concludes that values of depth, velocity, and energy differ according to different flow rates with same trend, and the model is approved.

3.4. Comparison with design values

Stilling basin design is based on hydraulic jump behaviour. For these, conjugated depths were calculated y1: Flow depth of pre-jump, y2: Flow depth after jump. Y1 was found based on the data at barrage entrance. Contreras and Villegas [7] values were obtained considering the occurrence of the hydraulic jump applying the Kindsvater equation, with a 6.08 % slope, were modelled in IBER, which works with Saint Venant equations and the results shown in Table 12 were obtained.

Table 12. Comparison between Contreras-Villegas design and by IBER model.

	Contreras and Villegas Design	IBER Model
Flow depth of pre-jump (y_1)	0.69 m	0.31 m
Velocity at the stilling basin inlet (V_1)	6.42 m/s	7.39 m/s
Froude at stilling basin entrance (F_1)	2.47	3.62
Flow depth after jump (y_2)	2.63 m	0.26

A large difference can be seen between calculations and modelling causing uncertainty in the results of the IBER model. However, it must be considered that conditions for both are not the same because in empirical calculation width of the barrage is considered to be 10m and in the case of modelling the entire channel a width of 15 m was considered as well as the basin entrance. As the conditions are not the same it is expected that the results will vary, with this comparison, it is rescued that the design of water intake calculated empirically is not enough to the modelling, since in real conditions result could vary. 10

3.5. Additional proposals for stilling basin

Proposals includes sloping basins for: 1.52%, 3.04%, and 4.56% to see the variation in results as the slope increases. Design is based on Peterka's "Hydraulic Design of Stilling Basins and Energy Dissipators" [8]. Hydraulic jump components are found by using Kindsvater equation. The equation (1) [11] was applied for stilling basin calculations, also used by Contreras and Villegas [7], with this formula y_2 is obtained, coefficient K is used according to Pardo, Wong and Cabrera as in equation (2).

$$\frac{y_2}{y_1} = \frac{1}{2\times\cos\theta}\left(\sqrt{\frac{8.F_{r1}{}^2.(\cos\theta)^3}{1-2.K.\tan\theta}+1}-1\right) \quad (1)$$

$$K = 4.075 - 27.56249.S_0 + 113.0682.S_0{}^2 - 182.9947.S_0{}^3 \quad (2)$$

Y_2 is calculated based on y1 and the Froude number at hydraulic jump beginning. Length of the pool is determined using the U.S. Bureau of Reclamation graph "Length of the jump in terms of conjugate depth, Ds (Basin V, Case D)" [8]. Table 13 shows values obtained for three proposals.

Table 13. Proposal results.

Measures for proposed stilling basins			
Measures	**1.52 %**	**3.04%**	**4.56%**
Tilt angle	0.87°	1.74°	2.61°
Shape coefficient	3.68	3.34	3.04
y_2 (m)	2.24	2.38	2.51
Length (m)	11.10	11.85	12.51

With length found and inclination angle, stilling basin is modelled for each proposal, results and comparison are shown in the following section.

4. Results Analysis

This section compares how slope variation influences flow of the stilling basin for hydraulic depth, velocity, Froude, and energy parameters. Hydraulic depth results are show in Table 14. Differences between 5 slopes vary by 0.01 m and 0.02 m between them, so this parameter is not very significant in variation of the slope of stilling basin.

Table 14. Stilling basin's hydraulic depth (y1) with 5 different slopes.

	0%	1.52%	3.04%	4.56%	6.08%
Basin entrance	0.32	0.31	0.32	0.32	0.31
Basin exit	0.30	0.30	0.28	0.28	0.26
Hydraulic depth variation	0.02	0.01	0.04	0.04	0.05

Concerning velocity values in figure 11 it is observed that higher the slope higher the velocity of the flow at stilling basin ending, the 5 slopes have the same trend. Results at the basin inlet are the same for all of them since the same discharge of 44.261 m3/s for all 5 models, this value is 1.03 m/s. In figure 11, we observe that when slope increases to 1.52%, velocity increases by 0.04 m/s which is not a very significant value: in percentage is 1.17%. When slope increases to 3.04% velocity increases by 5% compared to previous slope value. For slope value of 4.56% velocity increases by 1.2% and finally for 6.08% slope, velocity increases by 3.9%. With these results it is seen that when the slope increases to 1.52% the velocity values do not increase significantly, but when the slope increases to 3.04% the velocity results significantly influence. As the natural flow of the river has a velocity of 6.65m/s approximately the result resembles the horizontal slope.

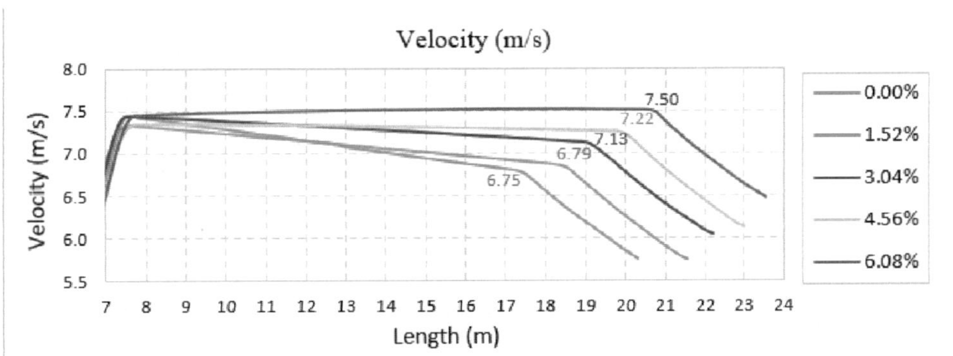

Figure 11. Velocity for 5 different slopes.

Similarly, the graphs for Froude values are displayed in figure 12. For this case, the Froude values at the basin output increase as the slope increases. The results in this case at the output vary from horizontal slope 3.93, 4.01, 4.25, 4.38 and 4.61. In this case it is also seen that the variation of the values when the slope increases to 1.52% and 4.56% the change from the previous slope is 0.08 and 0.13 respectively, the first case does not affect too much and in the second case does, however, when the slope increases to 3.04% and 6.08% the Froude increases by 0.24 and 0.23 compared to its previous slope, it is observed that in these variations that the Froude values does influence. In other words, to have a greater variation in Froude, the slope must increase by 3.04%. Concerning the value most similar to the natural flow of river of the case under study, which has an average of 4.12, is the slope of 3.04%.

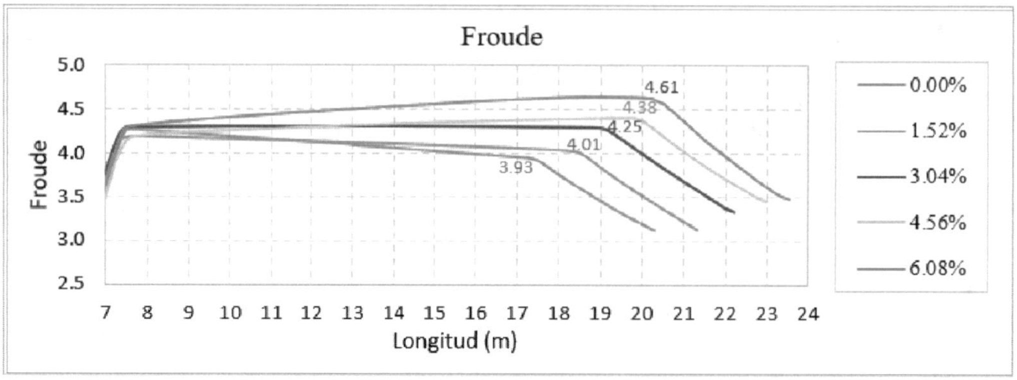

Figure 12. Froude values for the 5 slopes.

Energy variation results for different slopes are shown graphically in figure 13. Efficiency of a stilling basin is measured by amount of energy expelled. More energy its expelled, the more efficient it is. Therefore, energy at the basin end of each stilling basin has been compared. The values in this case are shown in Table 15.

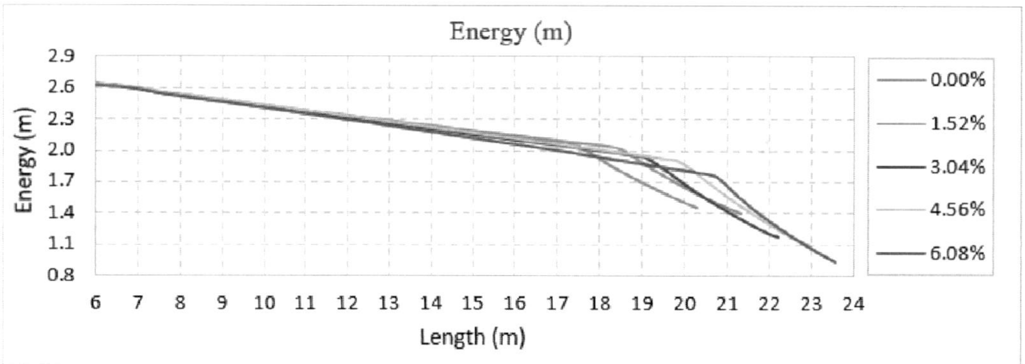

Figure 13. Energy values for the 5 slopes.

Table 15. Energy values for 5 slopes.

	Energy (m)				
	Percentage of slope				
	0%	**1.52%**	**3.04%**	**4.56%**	**6.08%**
Spillway entrance	2.64	2.64	2.64	2.64	2.65
Basin entrance	2.57	2.56	2.57	2.57	2.56
Basin exit	2.07	2.01	1.90	1.87	1.76
Lost energy	0.57	0.63	0.74	0.77	0.89

Concerning the horizontal slope stilling basin, the 6.08% stilling basin is more efficient in 56%. The energy loss of the 4.56% stilling basin relative to the horizontal is 36% more. The energy loss of the 3.04% stilling basin relative to the horizontal is 30% more. The energy loss of the 1.52% stilling basin relative to the horizontal is 11% more. It is observed with the results that the stilling basin with a 6.08% slope equal to the average slope of the San Pedro River is more efficient than the other stilling basins with proposed slopes.

5. Conclusions

From the five different slopes modelled, it was shown that variation in hydraulic depth parameter is not very significant between them as they vary between 1 and 2 cm. These are also low values, so it is necessary to place accessories in the stilling basin to raise the water level in order to connect it with the natural level of the river seen in the case study. Also, according to analysis for results variation of simulated slopes of basin, it was found: with higher slope, higher output velocity, lower y2, higher Froude number, and longer length of the basin.

Stilling basin with a slope of 3.04%, is suitable for San Pedro river. It has an energy loss of 30% more than the horizontal buffer basin. An acceptable and controllable output velocity lower that of the buffer basin of 6.08% and a Froude number that resembles that of the river of 4.12.

IBER software has advantage of being a fast, robust, and free program. It showed that it not only models' geometry on a large scale but can model hydraulic structures 25 meters long like the one presented in this research. On the other hand, IBER software has disadvantages in the fact that some flow conditions cannot be observed as the turbulence that forms and it cannot add three-dimensional elements, since it does not model vertical elements.

It is suggested to complement design of stilling basin with sloping bottom with other dissipation elements such as stilling dice to reduce its length, as well as a modelling of the structure with the river to appreciate the splice of water levels.

6. References

[1] Chanson H 2015 Energy Dissipation in Hydraulic Structures (*Dundee: CRC Press*).

[2] Legono D, Hambali R and Krisnayanti S.D 2019 Experimental study on the side channel spillway and Its impact on the jump, cross flow and energy dissipation *Jurnal Teknologi (Sciences & Engineering)* **81** p 169–178.

[3] Lempériére F, Vigny J P and Deroo L 2012 New Methods and Criteria for Designing Spillways Could Reduce Risks and Costs Significantly (*JOUR)* pp 120-128.

[4] Kumar M, Kumar S and Bidhu 2019 Determination of sequent depth of hydraulic jump over sloping floor with rounded and crushed aggregates using experimental and ANN model *Water Sience and Technology* (*Water Supply*) pp 2240-47.

[5] Pardo G, 2019 «Energy dissipation: part III: Stilling basin with sloping apron and impact – type energy dissipator *Universidad Tecnologica de la Habana,* pp. 73-83.

[6] Thu Hien L T, Agosto 2019 Simulating hydraulic characteristics and enhance design of my lam spillway-stilling basin, Vietnam *International Journal of Innovative Technology and Exploring Engineering* **8** pp 953-957.

[7] Contreras L and Villegas A 2019 Diseño Hidráulico y Estudio de Rentabilidad del Proyecto de la Bocatoma San Pedro para abastecer las zonas agrícolas de San Pedro y Santa Cruz en la región Ayacucho (Lima: *UPC*).

[8] Peterka A 1984 Hydraulic design of stilling basins and energy dissipators (Miami: *Water Resources Technical Publication).*

[9] Kazemi F, Reza Khodashenas S and Sarkardeh H 2016 Experimental Study of Pressure Fluctuation in Stilling Basins *International Journal of Civil Engineering* **14** pp 13-21.

[10] Rincón Ortiz J C 2020 Análisis de Sensibilidad de Párametros Hidrólogicos e Hidráulicos del Modelo SWMM y su Aplicación en Sistemas de Drenaje Urbano (Lara: *UCLA*)

[11] USBR 1987 Design of small dams, fifth ed. (United States).

A TIM-based Method of Automatic Generation of FE Models for Pit Engineering

Lisheng Chen[1], Zhiyi Li[1], Taili Chen[1] , Guoqiang Zhao[1] , Yi Shen[2,*]

[1] Shanghai Urban Construction Municipal Engineering (Group) Co., Ltd., Shanghai 200065, China
[2] Department of Geotechnical Engineering, College of Civil Engineering, Tongji University, Shanghai 200092, China
*E-mail: shenyi@tongji.edu.cn

Abstract. BIM (Building Information Modeling) technology has developed rapidly in many fields of civil engineering. However, in pit engineering, the method of automatically generating FE models by BIM models is not as mature as that in tunnel engineering. There are still some problems in it, such as the lack of stratum data and construction data, the limit of geometric models, and the weak of the integration of numerical results with BIM platform. To solve these problems, this paper presents a TIM (Tunnel Information Modeling)-based on digital solution, which is more specific to underground engineering. Firstly, a normalized numerical pit model is built. TIM models, stratum data and construction data, all these required by numerical model are integrated in iS3 platform. Then a program is developed to obtain the data that has been supplemented or modified from the databases and automatically draw the profiles, then generate the input files to automatic calculation. The integrated TIM/FEM software based on iS3 platform is developed to import the result files to finish the post-processing and schemes assessment and optimization. This method is successfully applied in the pit project of Haikou Wenmingdong Tunnel, shortens the modeling time, and improves the modeling quality and the integration level of TIM/FEM.

1. Introduction

BIM (Building Information Modeling) technology has been widely used in many fields of civil engineering due to its integrated information, unified data standards, and the support of multi-party collaborative work, which can realize the digital management of the whole life cycle of a project. However, compared with architectural engineering, there is still a gap of its application in underground engineering such as tunnels and pits.

Many scholars have carried out relevant researches on the applications of BIM in pit engineering: Hammad et al.[1] proposed a new 4D model for the automatic identification of construction site of pits; Luo et al. [2] provided a method to improve the efficiency and accuracy of checking process of deep pit construction plans; Wu et al.[3] developed a BIM-based monitoring method for deep pits; Li et al.[4] provides a BIM-based automated safety risk recognition mechanism for the pre-construction stage of pits. Many scholars achieved the applications of BIM in pit engineering [5-7]. These researches have used BIM technology into design and construction stage. However, the finite element analysis (FEA), which is important for the safety, has not been linked with BIM, and modeling and result viewing can only be operated in the finite element software. The information in the BIM platform has not been used, and it is valuable to carry out the research on the BIM-based FEA. Based on BIM technology, TIM (Tunnel Information Modeling) technology proposed information solutions to the specific problems of underground engineering. It considers stratum, construction and FEA that are ignored by BIM. Meschk et al. [8], Alsahly et al. [9] and Ninic et al. [10] carried out researches on the link of TIM and FEA of underground projects such as tunnels and presented methods of structural analysis by means of TIM-based numerical modeling. However, most of these researches focused on tunnel projects, and few on pit projects. In addition, their focus of automatically modeling was only on geometrical pre-processing, that is, stratum data or construction data were missed. Therefore, the research is still in the preliminary stage, and there is still a large room for the improvement in modeling efficiency and accuracy.

This paper presents a TIM-based method of automatic generation of FE Models for pit engineering. For its preparation, a normalized numerical pit model is built, and all required data are integrated in iS3 platform [11]. Then, a program is developed to obtain the data from the databases for drawing the profiles and calculating automatically. Finally, the results files are input to iS3 platform again for post-processing and scheme analysis. This method is applied in the pit project of Haikou Wenmingdong and of certain significance for TIM technology to be further integrated into pit engineering.

2. Project profile

2.1. Basic information

Wenmingdong tunnel is located in Haikou city, starting from the west intersection of Bailong Road and Wenmingdong Road, then crossing west Binjiang Road and Nandujiang Road and end at the intersection of Qiongshan Avenue and Second Dongheng Road. The total length of it is 4380 m, of which the tunnel section is 2720 m and the connecting road is 1660 m. In the land section, slope excavation, SMW piles, bored piles, waterproof curtain and underground diaphragm walls are used for support, while bored pile, waterproof curtain and SMW piles are used in the river section. Steel sheet pile and mud pipe bag are used in the cofferdams in the river section. The maximum depth of the pit is 22.6 m, which requires strict control of pit deformation and ground settlement. Adverse geological conditions have a great impact on the bracing system, especially for pit-in-pit, as is shown in figure 1.

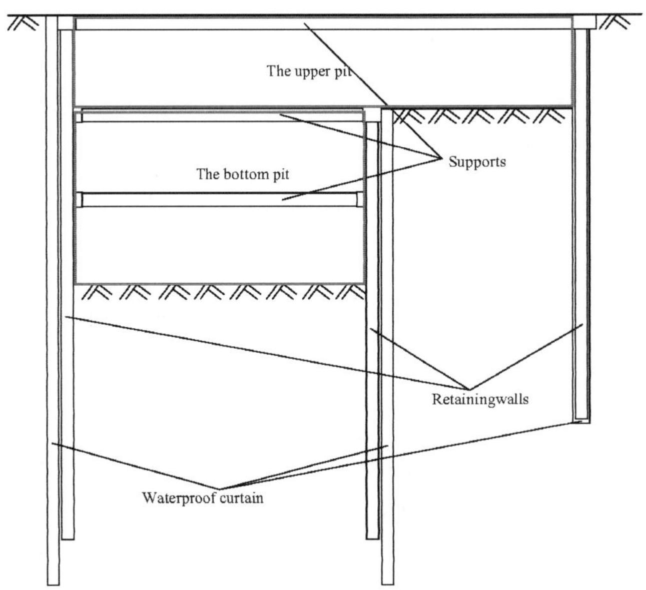

Figure 1. Layout of pit-in-pit.

2.2. Automatic modeling

For automatic modeling, a normalized numerical pit model should be built first, and all the data required by it should be integrated. Then the program should be developed to obtain these data from the databases, and automatically finish the modeling and calculation in order to reduce the workload. The results should also be integrated into the TIM platform, so as to achieve the purpose of information management of whole life cycle of a project in one platform.

Normalized Numerical Pit Model. The normalized numerical pit model is to standardize the data needed for modeling. The data depends on models and finite element software. The normalized numerical model is suitable for the pits with bracing system of retaining walls + supports, and FLAC is selected as the software. According to its modeling steps and characteristics, the required data are divided into four categories: geometric parameters, structural parameters, stratum parameters and construction stages parameters.

The geometric parameters include the width and depth of a pit, which is used to determine the boundary of the model, as shown in figure 2. According to its size and the average distance between nodes, the number of nodes n and m in the horizontal and vertical directions can be determined..

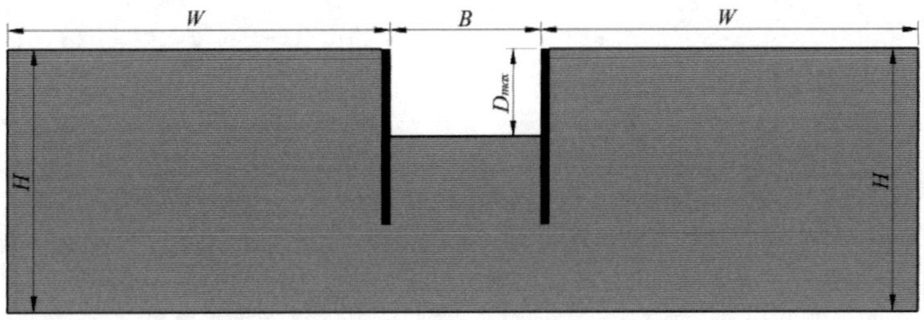

Figure 2. The sketch of a pit model.

2.3. Data integration

Unlike architectural engineering, the normalized numerical pit model requires stratum data and construction data. In this paper, 2D GIS data, 3D BIM data and project databases are integrated by iS3(infrastructure Smart Service System) to build a unified iS3 model, and the iS3-config tool is used to associate these data with the database of DigitalObject, as shown in figure 3.

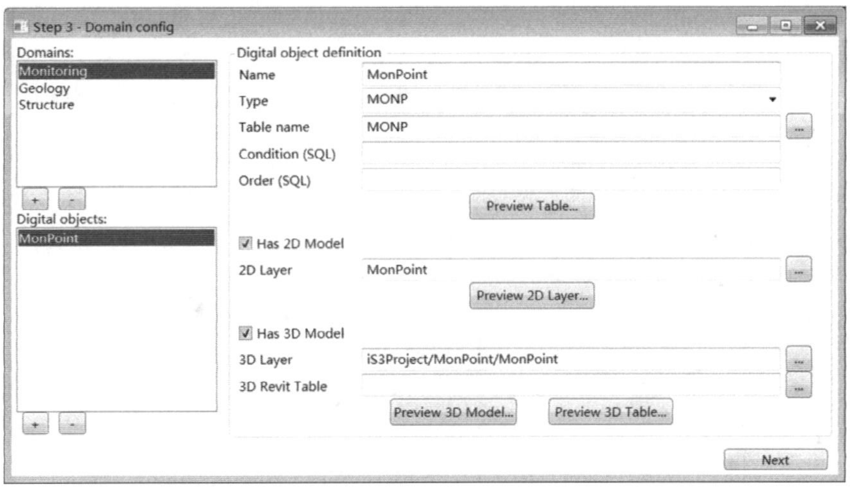

Figure 3. Data integration

2.4. Program design and functions

After finishing the preparatory work, the program based on iS3 platform can be developed. It is divided into four functional modules: Profile data input and integration module, Profile drawing module, Generation of input files and automatic calculation module and Scheme assessment and optimization module. The basic flow is shown in figure 4.

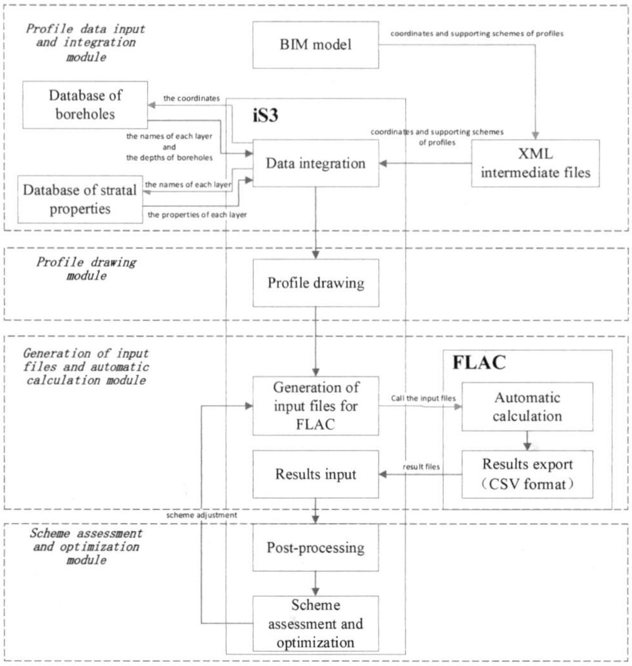

Figure 4. The modules and flow of the program

2.4.1. Profile data input and integration module. XML file, with its simple format, can be written or exported from BIM models through the program, so it is used as the intermediate file. After importing the data of profiles into iS3, data integration of profiles and projects can be completed through the characters such as "ID" and "Name", so as to automatically obtain the data needed for the numerical model. Compared with the complex hand operation of data input, this function greatly improves the efficiency and accuracy of numerical modeling.

2.4.2. Profile drawing. After data integration, DrawProfile, a plug-in iS3, can automatically draws the profiles and shows them visually in iS3 platform, as shown in figure 5. The bracing system, layers and other information can be viewed and checked.

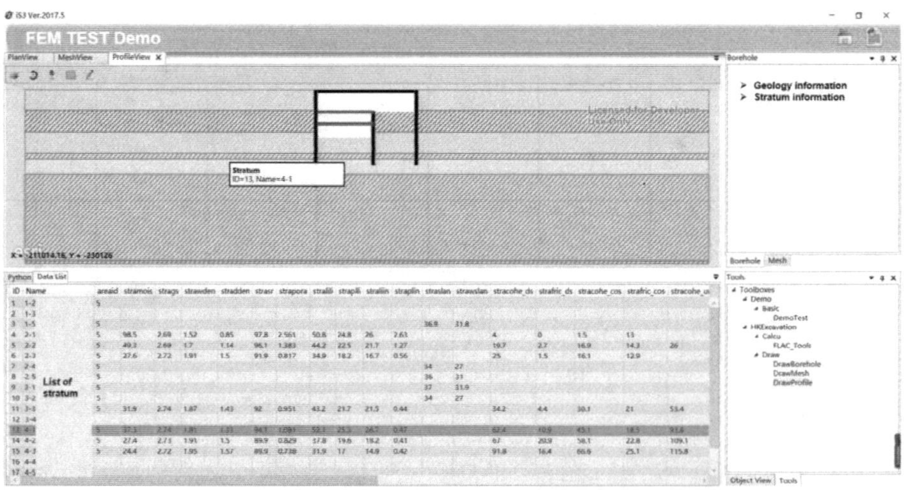

Figure 5. Profile drawing of pit-in-pit.

2.4.3. Generation of input files and automatic calculation module. After checking or modifying the information, all the parameters needed for numerical simulation have been obtained and integrated. Based on the above parametric modeling standard, the program automatically generates the input files for FLAC in order to calculate and obtains the internal force and deformation of bracing system. The result files of each step are output in CSV format and then input in iS3 platform. The results of each step can be viewed not only in the tables, but also shown as auto-drawn diagrams of internal forces and deformation in the profiles that introduced above, as shown in figure 6. This module solves the lack of stratum data and construction data when automatically generating a FE model by a traditional BIM model, and shortens the modeling time, and realizes the integration of TIM/FEM to a certain extent.

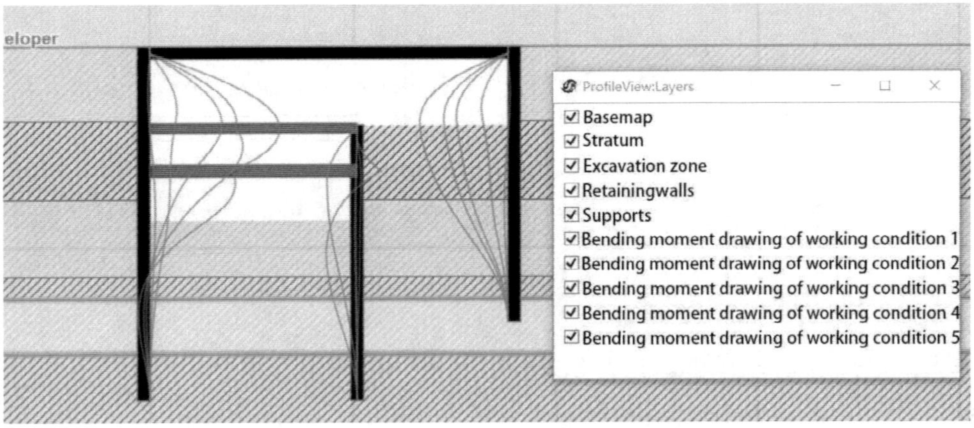

Figure 6. Bending moment drawing.

2.4.4. Scheme *assessment and optimization module.* Based on internal forces and deformations of structures and design codes, the program can assess the scheme from both safety and economics. If the safety factor is lower than the requirements or too large that leads to poor economic benefits, the program will automatically adjust the scheme according to specific situation. It readjusts the relevant parameters of bracing system in the input files, then obtains new results after the calculation, and checks them according to the codes again. When safety is met, the program adjusts the parameters repeatedly until a relatively optimal economic scheme is obtained. The scheme optimization takes a lot of time to modeling and this module not only greatly save the time of repetitive operations and reduce the possibility of errors, but also considers both the safety and the economics of a pit scheme.

3. Conclusions

The method of automatically generating FE models by TIM models of pit engineering now exists many problems, such as the lack of data, the limit of geometric models. Also, the calculation results can only be viewed in FE software. These problems lead to the poor effect of automatic modeling and the low level of TIM/FEM integration. There is little research on the attempt of linking TIM and structural assessment of pit engineering, hence FE technology is difficult to take the advantages of TIM, which widens the gap between pit engineering and tunnel engineering. In order to solve the above problems, this paper presents a method of automatic generation of FE models and TIM/FEM integration post-processing for pit engineering.

For the pits with the brace system of retaining walls + supports and the FE software FLAC, a normalized numerical pit model is built, and the TIM models and the projects data are integrated in iS3 platform, which improves the level of visualization and integration of management and the reliability of calculation results, and lays the foundation for automatic generation and analysis of FE models. The program of automatic modeling is developed. It is a two-dimensional FE models of pit engineering including the data of geometry, structure, stratum and construction. The program can automatically call the input files and calculate, which improves the efficiency and reduce the possibility of errors in the process of modeling. Then according to the results, the schemes assessment and optimization can be carried out automatically to ensure the safety and reduce the cost of the project, which can save a lot of repeated operation time.

Based on iS3 platform, the integrated TIM/FEM software for pit engineering is developed. The results files can be imported in it and the internal force or deformation diagram of profiles can be automatically drawn and viewed without the help of FE software. The platform realizes the visualization of post-processing, and links TIM and structural analysis, and improves the integration level of TIM platform. The method of automatically generating FM models by TIM models in pit engineering and TIM/FEM integrated platform are applied in the pit project of Haikou Wenmingdong Tunnel, which can be used for reference to promote the application of TIM technology in pit engineering.

4. References

[1] Hammad A, Setayeshgar S, Zhang C, Asen Y 2012 *Winter Simulation Conf.*, p 1-10
[2] Luo H, Gong P 2015 *J. Intell. Robot. Syst.*, 79 (3-4), p 549-576
[3] Wu I C, Lu S R, Hsiung B C 2015 *Visualization Eng.* Vol 3, p 1-11
[4] Li M, Yu H, Liu P 2018 *Autom. Constr.* **91**, p 284-292
[5] He J J, Wang S, Yao S Y 2016 *J. Info. Technol. Civ. Eng. Arch.* **8**, p 55-59
[6] Peng S G 2012 *J. Chongqing U. Sci. Technol.(Nat. Sci. Ed.)*, **14**, p 129-131
[7] Zhou K 2019 *Val. Eng.*, **38**, p 186-188
[8] Meschke G, Freitag S, Alsahly A, Ninic J, Schindler S, Koch C 2014 *Bauingenieur*, **89** (11), p 457–466.
[9] Alsahly A, Gall V, Marwan A, Ninic J, Meschke G, Vonthron K M A 2016 *World Tunn. Cong*, p 3178-3186 .
[10] Ninic J, Freitag S., Meschke G 2017 *Tunn. Under. Space Technol*, **63**, p 12–28
[11] Zhu H H, Li X J, Lin X.D 2018 *China Civil Engineering J.*, **51**(01), p 1-12

Acknowledgments
The research is funded by the research project of Shanghai Urban Construction Municipal Engineering (Group) Co., Ltd and Shanghai Sailing Program (20YF1451400).

Journal of Physics: Conference Series **2287** (2022) 012047 doi:10.1088/1742-6596/2287/1/012047

Influence of the P-delta Effect and Stiffness Irregularity on the Structural Behavior of Reinforced Concrete Buildings

D J Zavala*, **C J Torres, and J D Moreno**

Faculty of Civil Engineering, Peruvian University of Applied Sciences (UPC), Surco, Lima, Peru
*Corresponding author's e-mail: U201403828@upc.edu.pe

Abstract. In this research, the influence of the stiffness irregularity and the p-delta effect on the structural behavior of a reinforced concrete building is analyzed. The main objective is to determine the impact of the stiffness irregularity and the p-delta effect on the structural behavior in regular and irregular buildings. First, the linear dynamic analysis procedure is performed in order to determine the structural response in terms of drifts, shear force and moments per floor. Subsequently, we proceed with the nonlinear static analysis procedure to obtain the capacity curve of the structure. The post elastic stiffness and the overall ductility of the structure are determined from the capacity curve. Finally, a comparative analysis of the responses from the linear and nonlinear analysis is carried out to determine the percentage variation of the results. When analyzing the structures that consider the stiffness irregularity and the p-delta effect, variations of up to 16.50%, 11.00% and 14.00% have been obtained in drifts, shear force and moments per floor respectively, which are considerable values. When the p-delta effect is considered in structures with the presence of stiffness irregularity, there is a variation in stiffness of up to 59.85%. With this result it is explained that the p-delta effect produces a greater degradation of the overall stiffness of the structure.

1. Introduction

Buildings are prone to deform laterally from the original position with an eccentricity during an earthquake, when the construction of structures is subjected to seismic loads causing the structure to deform, the eccentricity resulting from the total gravity load due to inclined axes of the structure makes the extra moments exercise at the base [1]. The action of vertical loads acting through structural lateral deformations is well known as the second-order effect or P-Δ effect [2].

P-Δ effect reduces lateral resistance of a system, which under strong excitation may cause partial or total loss of load carrying capacity [3]. This effect can start a pernicious circle against the structural system because the influence of gravity loads increases as the lateral displacement grows while, at the same time, the lateral displacement is magnified as a consequence of gravity loads acting on them [4].

In very flexible buildings the destabilizing effect of gravity loads may lead to a negative post-yield stiffness, and thus, the structural collapse capacity is exhausted at a rapid rate when the earthquake drives the structure into its inelastic range of deformation even for stable hysteretic component behavior [5].

The P-Delta effect varies over the height of the structure as a function of the axial force demand and the interstory drift [6]. Generally, P-Δ effects in low-rise buildings are negligible since the total lateral deflections are kept relatively small by the story drift limitation. However, in taller, high-rise buildings, where lateral deflections may be much larger, satisfying the maximum drift requirement does not ensure that the P- Δ effects will be negligibly small [7]. The importance of this effect may be negligible

when the structure responds elastically, but is very important when the structure responds inelastically. The P-Δ effect generally increases the displacement response of structures. It can even cause dynamic instability when the structure is subjected to severe seismic movements [8].
Regarding the irregularity of stiffness; Zhao, Hu y Long [9] mention that the vertical irregularity or height of a structure is introduced by abrupt changes or discontinuities in the configuration (such as dimension, soil mass, rigidity, resistance or their combination) along its vertical direction. As a consequence, large stresses or concentrations of plastic deformation can occur in the local weakness and spread to global damage or even collapse. In relation to the P-delta effect, Shehu, Angjeliu y Bilgin [10] in their research perform the Push Over analysis to estimate the ductility of a metallic structure, which simultaneously considers the effects of P-Δ and plastic deformations.
In this research, reinforced concrete structures are analyzed that consider stiffness irregularity and geometric non-linearity (P-Delta Effect) in order to determine their influence on the structural response.

2. Method of analysis

In this research it is intended that the influence of the irregularity in stiffness with the P-delta effect to medium-rise buildings on their structural behavior is estimated.
For this reason, 9 models corresponding to a 20-story structure are made, which consider in their configuration resistant to lateral loads a dual-type structural system, in the X and Y directions of analysis. The structural elements that make up the lateral load resistant system are made of reinforced concrete.
It begins with the proposal of 20-story reinforced concrete structures. This approach consists of the development of models or case studies that are organized and ordered into 3 groups based on the irregularity of their structural configuration and the consideration of the P-delta effect in the structure.
In the first group, it is a regular structure that does not present any type of structural irregularity and does not consider geometric non-linearity. The second group, made up of 4 cases or structural models, present stiffness irregularity in a defined floor. Finally, in the last group formed by 4 cases or structural models, they take into account the stiffness irregularity and the geometric non-linearity.
Next, the stability index (Q) of each case is estimated and analyzed, this index is presented in equation (1) according to the Eurocode / ASCE. Subsequently, in the cases described, linear and non-linear analysis procedures are developed.
First, the linear analysis is carried out, which consists of performing the linear dynamic analysis procedure to determine the structural response in terms of drifts, shear force and moments per floor.

$$0.10 < Q < 0.25 \tag{1}$$

Next, we proceed with the non-linear analysis of the research, which implies performing a non-linear static analysis in order to obtain the capacity curve. Previously, the plastic hinges must be defined and assigned to the structural elements.
For the definition of the shape and behavior of the plastic hinges, the modeling parameters established by the ASCE 41-13 standard are used. Figure 1 shows a moment-rotation diagram that is constructed using the modeling parameters.

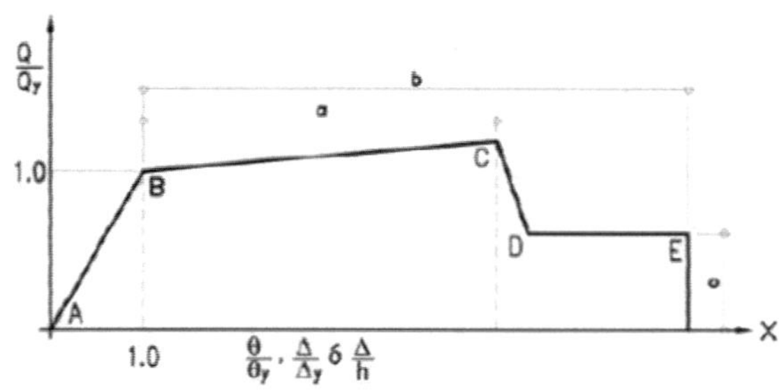

Figure 1. Moment – rotation diagram.

Subsequently, a bilinear representation of the capacity curve is developed according to the following procedure. A line is drawn upward from the origin with a slope equal to the initial stiffness of the structure; then a second line is drawn back from the last point Vu, du, so that when the first line intersects at point Vy, dy, the area designated as A_1 is equal to the area A_2. Figure 2 shows the bilinear representation of a capacity curve.

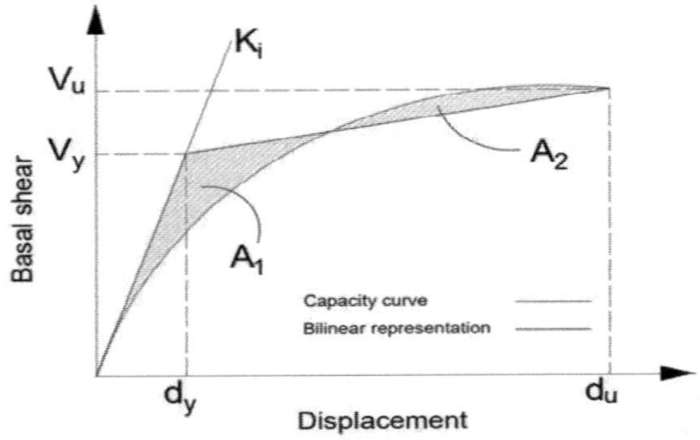

Figure 2. Bilinear representation of the capacity curve.

Based on this curve, both general ductility and post-elastic stiffness are determined and analyzed. The post elastic stiffness of the structure is determined by equation (2). Equation (3) is proposed to estimate the overall ductility of the structure (μδ), which relates the ultimate displacement (δu) and the yield displacement (δy).

$$Kp = (Vu - Vy)\,(Du - Dy)^{-1} \quad (2)$$

$$\mu\delta = \delta u\,(\delta y)^{-1} \quad (3)$$

Finally, a comparative analysis of the responses from the linear and nonlinear analysis of the models or case studies is developed in order to determine the percentage variation of the results.

3. Results

Our study will be valid for structures of irregularity in stiffness, according to the ASCE / SEI 41-13 standard where this type of configuration is identified by equation (4).

$$K_i < 0.70\,K_{i+1} \quad (4)$$

For the confirmation of our method, five buildings were chosen, one regular building and four buildings with the presence of stiffness-type irregularity in height, considering the type of office use. (See Figure 3, 4, 5)

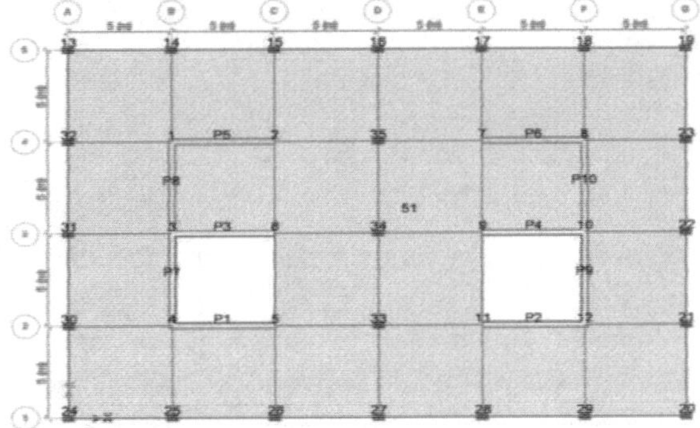

Figure 3. Analysis plant in ETABS.

Floors	Use	Height
1	archive rooms	3
2	archive rooms	3
3	archive rooms	4.5
4	archive rooms	3
5	archive rooms	4.5
6	archive rooms	3
7	archive rooms	4.5
8	archive rooms	3
9	archive rooms	4.5
10	archive rooms	3
11	archive rooms	4.5
12	archive rooms	3
13	archive rooms	4.5
14	archive rooms	3
15	archive rooms	4.5
16	archive rooms	3
17	archive rooms	4.5
18	archive rooms	3
19	archive rooms	3
20	archive rooms	3

CASE 1
CASE 2
CASE 3
CASE 4

Figure 4. Structures cases.

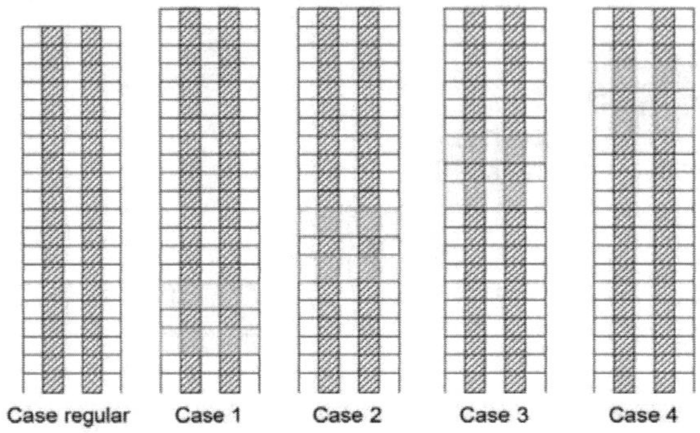

Figure 5. Elevation of study models.

Table 1 shows the verification of the stiffness irregularity for case 2, in the x-x direction.

Table 1. Irregularity verification.

Story	*SIDE RIGIDITY*	*CHECK N°1*		*CHECK N°2*	
		70%		*80%*	
20	*105559.150*	*73891.405*	*REGULAR*		
19	*213993.963*	*149795.774*	*REGULAR*		
18	*309265.732*	*216486.012*	*REGULAR*	*167685.025*	*REGULAR*
17	*408320.758*	*285824.531*	*REGULAR*	*248421.454*	*REGULAR*
16	*539674.638*	*377772.247*	*REGULAR*	*335269.634*	*REGULAR*
15	*765438.544*	*535806.981*	*REGULAR*	*456915.718*	*REGULAR*
14	*7976182.000*	*5583327.400*	*REGULAR*	*2475012.049*	*REGULAR*
13	*8784487.000*	*6149140. 900*	*REGULAR*	*4673628.679*	*REGULAR*
12	*9528250.000*	*6669775.000*	*REGULAR*	*7010378.400*	*REGULAR*
11	*10215007.000*	*7150504.900*	*REGULAR*	*7607398.400*	*REGULAR*
10	*10854193.000*	*7597935.100*	*REGULAR*	*8159320.000*	*REGULAR*
9	*11451919.500*	*8016343 650*	*IRREGULAR*	*8672298.533*	*IRREGULAR*
8	*1200798.750*	*840559.125*	*REGULAR*	*6268509.667*	*IRREGULAR*
7	*1251793.300*	*876255.310*	*REGULAR*	*3707869.747*	*IRREGULAR*
6	*1297735.450*	*908414.815*	*REGULAR*	*1000087.333*	*REGULAR*
5	*1341977.950*	*939384.565*	*REGULAR*	*1037735.120*	*REGULAR*
4	*1374920.050*	*962444 035*	*REGULAR*	*1070568.920*	*REGULAR*
3	*1401011.850*	*980708.295*	*REGULAR*	*1098109.293*	*REGULAR*
2	*1413622.450*	*989535.715*	*REGULAR*	*1117214.493*	*REGULAR*
1	*1418016.650*	*992611.655*	*REGULAR*	*1128706.920*	*REGULAR*

Journal of Physics: Conference Series **2287** (2022) 012047 doi:10.1088/1742-6596/2287/1/012047

Stability indices were calculated for all cases; where it is shown that for all cases of irregularity in stiffness, the application of the P-Delta effect is necessary, because its stability index at one level in each case, exceeds 0.1, established by the ASCE.

3.1. Linear dynamic analysis

A comparison of the drifts, shear force and moments per floor was made; with each case applying the P-Delta Effect.

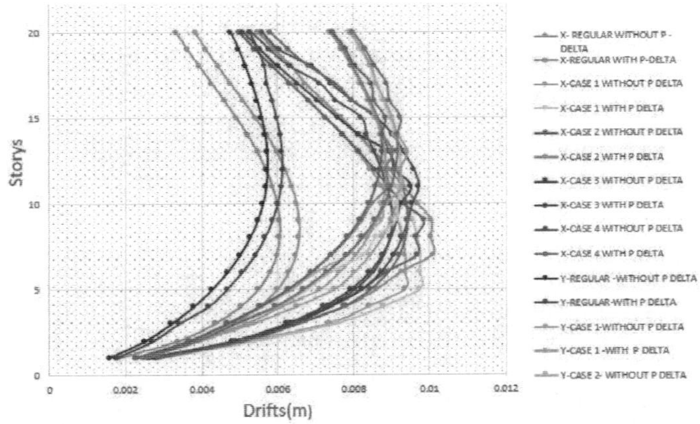

Figure 6. Drifts per floor.

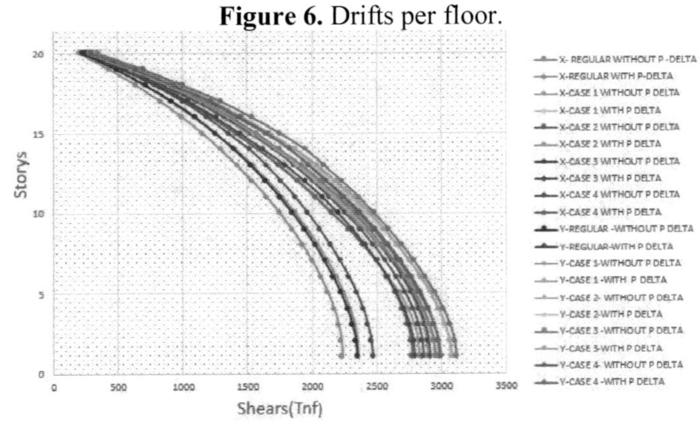

Figure 7. Shear per floor.

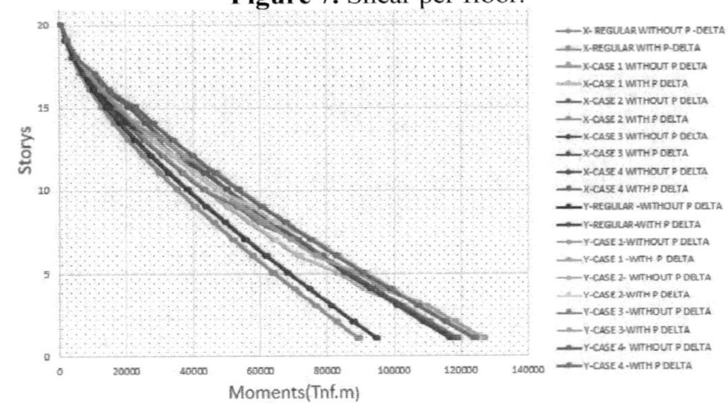

Figure 8. Moments per floor.

The percentages of variation of drifts of the Regular are 4.40%, of Case 1, they are 16.50%; of Case 2, 10.4%; Case 3 is 14.4%; Case 4 is 13.60%; as evidenced in Figure 6; therefore, it can be said that in the different cases, the most critical case is floors 3 and 5, due to the fact that it generates a higher percentage of drifts; in addition, a considerable increase is seen with respect to the regular model, when adding the irregularity in stiffness in different floors.
The percentages of variation of shear per floor, of Regular are 3.50%, of Case 1, they are 11.0%; of Case 2, 9.4%; Case 3 is 9.8%, Case 4 is 8.60%, as evidenced in Figure 7; therefore, it can be said that in different cases, the most critical case is floors 3 and 5 together, because it generates a higher percentage of shear force at the base; in addition, a considerable increase is seen with respect to the regular model, when adding the irregularity in stiffness in different floors.
The percentages of variation of moment per floor, of the Regular are 6.60%, of the Case 1, they are 14.0%; of Case 2, 11.5%; Case 3 is 11.0%, Case 4 is 12.86%, as evidenced in Figure 8; therefore, it can be said that in different cases, the most critical case is floor 9, because it generates a higher percentage of moments; in addition, a considerable increase is seen with respect to the regular model, when adding the irregularity in stiffness in different floors.

3.2. Nonlinear static analysis

Capacity curve. With the nonlinear static analysis procedure, the capacity curves of the study models are obtained in the directions of analysis, this is shown in Figure 9 and Figure 10.

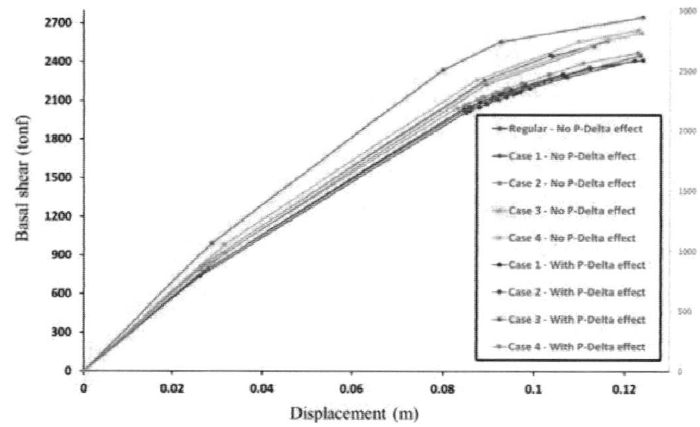

Figure 9. Capacity curve in the x direction.

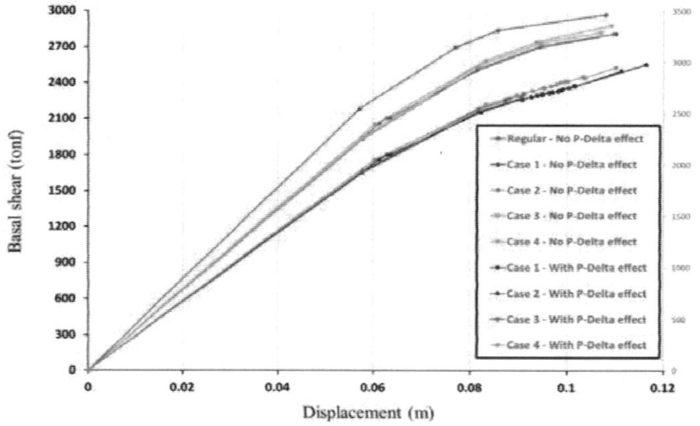

Figure 10. Capacity curve in the y direction.

Ductility. In Figure 11 the ductility values are shown for the x-x direction as a function of the consideration of the P-delta effect. In cases 2, 3 and 4, an increase in ductility is observed when the P-delta effect is considered in the structural models in the non-linear analysis. In case 1 its ductility is not affected by the P-delta effect.

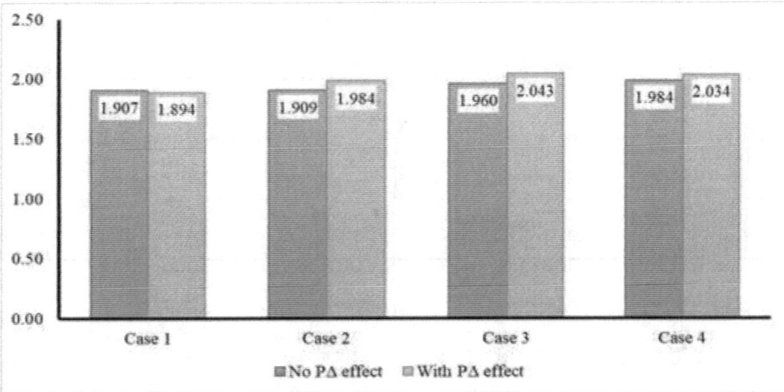

Figure 11. Ductility of cases without P Δ effect vs with P Δ effect – x-x direction.

In Figure 12 the ductility values are shown for the y-y direction as a function of the consideration of the P-delta effect. In case 1 there is a slight increase in ductility caused by the P-delta effect, while in cases 2 and 4 there is a moderate increase in ductility.

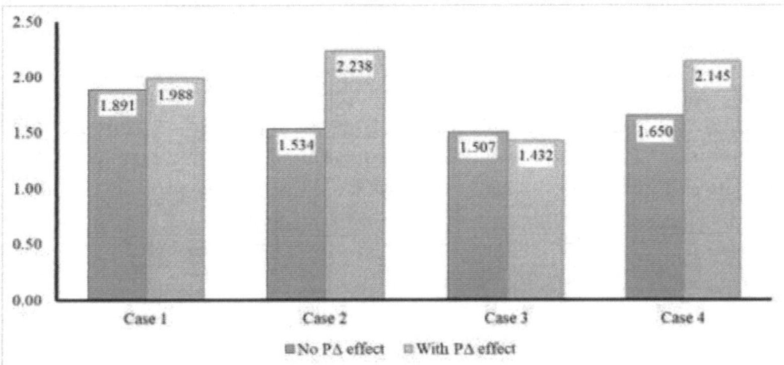

Figure 12. Ductility of cases without P Δ effect vs with P Δ effect – y-y direction.

Stiffness. Table 2 shows the results of the post elastic stiffness that varies according to the incorporation of the stiffness irregularity in a specific floor. In the regular model it is observed that in the x-x direction there is post-elastic stiffness of 20068.05 ton / m and in the y-y direction there is a post-elastic stiffness of 21846.64 ton / m.
In case 1, the post-elastic stiffness in the x-x direction of 13178.28 ton / m is less than the post-elastic stiffness in the other cases. In the y-y direction, case 2 has a post-elastic stiffness of 18626.71 ton / m, which is lower than the other cases.

Table 2. Post-elastic stiffness - no P Δ effect.

Direction	Case 1	Case 2	Case 3	Case 4	Case regular
X-X	13178.28	14477.22	14057.88	15473.52	20068.05
Y-Y	19043.00	18626.71	18975.01	19504.17	21846.64

Table 3 shows the results of the post elastic stiffness that varies according to the incorporation of the stiffness irregularity in a specific floor and the P-Delta effect.
In case 2, the post-elastic stiffness in the x-x direction of 12641.97 ton / m is less than the post-elastic stiffness in the other cases. In the y-y direction, case 1 has a post-elastic stiffness of 14903.95 ton / m, which is lower than the other cases.

Table 3. Post-elastic stiffness - with P ∆ effect.

Direction	Case 1	Case 2	Case 3	Case 4	Case regular
X-X	13172.90	12641.97	12837.76	15355.20	20068.05
Y-Y	14903.95	16999.09	19342.29	16856.24	21846.64

Table 4 shows the variations of the global stiffness of the study models as a function of the stiffness irregularity in a specific floor.
In the x-x direction, cases 1, 2, 3 and 4, which have stiffness irregularity in a specific floor, have experienced a variation in stiffness greater than the regular structure. In case 1 there is a greater variation in its lateral stiffness compared to the other cases. In the y-y direction three different circumstances occur. In the first place, the variation of stiffness in case 2 and 4 of values 45.24% and 45.35% respectively, barely exceeds the regular case. On the other hand, we have case 3 that has a lower stiffness variation response compared to the regular case. And finally, the variation in stiffness of 48.70% that corresponds to case 1 is higher than the other cases.

Table 4. Variation of the overall stiffness of the structure - no P ∆ effect.

Direction	Case 1	Case 2	Case 3	Case 4	Case regular
X-X	56.30%	50.62%	53.06%	50.15%	42.06%
Y-Y	48.70%	45.24%	44.32%	45.20%	45.20%

Table 5 shows the variations of the global stiffness of the study models as a function of the stiffness irregularity in a specific floor and the consideration of the P-delta effect.
When the P-delta effect is considered in the case studies, a greater reduction in stiffness is generated. In the x-x direction, there is an increase in the variation of stiffness in cases 2, 3 and 4 caused by the P-delta effect. Regarding the analysis in the y-y direction, there is a similar response from the case studies. When considering the P-delta effect and the stiffness irregularity in the structures, there is an increase in the degradation of stiffness in cases 1, 2 and 4.

Table 5. Variation of the overall stiffness of the structure - with P ∆ effect.

Direction	Case 1	Case 2	Case 3	Case 4
X-X	54.40%	56.88%	57.13%	50.53%
Y-Y	59.85%	50.03%	43.24%	52.77%

4. Conclusions

According to Eurocode 8, the 20-story building should take into consideration the P-Delta effect, because the stability index is greater than 0.1, but when considering the effect for cases of stiffness-type irregularity, its most critical variation percentages are 16.50%, 11.0%, 14.00%, of drifts, shear force, and moments per floor respectively; which are considerable. In a structure of 20 floors or more, a seismic performance analysis must be carried out considering the P-Delta effect obligatorily; because the design variation when incorporating this effect plus the irregularity in stiffness is considerable. When the stiffness irregularity occurs on floors 3 and 5; they are more considerable, so while the floors are affected, they are of a lower level; the greater the P delta effect.
The presence of the stiffness irregularity in tall buildings leads to a degradation of the stiffness and resistance of the structure. This degradation of stiffness and strength is greater in a structure with stiffness irregularity than a regular structure. In the non-linear static analysis procedure, in the cases that consider stiffness irregularity, variations in stiffness of up to 56.30% have been obtained. Therefore, the stiffness irregularity directly influences the degradation of the stiffness of the structure.

In structures that consider the stiffness irregularity and the P-delta effect, a greater number of plastic hinges are formed in structural elements, consequently the rigidity of the structural elements decreases and the overall rigidity of the structure is degraded.

With the incorporation of the stiffness irregularity and the consideration of the P-delta effect in tall structures, variations in stiffness of up to 59.85% have been obtained, that is, with the P delta effect there is a greater degradation of the overall stiffness of the structure.

5. References

[1] Pattar, A. N., & Muranal, S. M. 2017 *P-delta effect on multi-storey buildings. i-manager's Journal on Structural Engineering* 6(3), pp 1-11.

[2] Wei, B., Xu, Y., & Li, J. 2012 *Treatment of P-Δ effects in displacement-based seismic design for SDOF systems*. Journal of Bridge Engineering 17(3), pp 509-518.

[3] Fard Mousavi, S. M., & Sensoy, S. 2019 *Direct estimation of the P-delta effect through the stability-coefficient-response-spectra by introducing the first-storey-single-degree-of-freedom system*. Bulletin of Earthquake Engineering 17, pp 3495-3516.

[4] Black, E. F. 2011 *Use of stability coefficients for evaluating the P–Δ effect in regular steel moment resisting frames*. Engineering Structures 33(4), pp 1205-1216.

[5] Adam, C., & Jäger, C. 2012 *Simplified collapse capacity assessment of earthquake excited regular frame structures vulnerable to P-delta*. Engineering Structures 44, pp 159-173.

[6] Adam, C., Ibarra, L. F., & Krawinkler, H. 2004 *Evaluation of P-delta effects in non-deteriorating MDOF structures from equivalent SDOF systems*. In: 13th World Conference on Earthquake Engineering, pp 1-15. Vancouver.

[7] Neuss, C. F., & Maison, B. F. 1984 *Analysis for P-delta effects in seismic response of buildings*. Computers & Structures 19(3), pp 369-380.

[8] Pourali, N., Khosravi, H., & Dehestani, M. 2018 *An investigation of P-delta effect in conventional seismic design and direct displacement-based design using elasto-plastic SDOF systems*. Bulletin of Earthquake Engineering 17, pp 313-336.

[9] Zhao, W., Hu, X., & Long, Z. 2016 *Seismic ductility reduction of flexural-type structures with vertical irregularities*. The Open Civil Engineering Journal 10, pp 1-11.

[10] Shehu, R., Angjeliu, G., & Bilgin, H. 2019 *A simple approach for the design of ductile earthquake-resisting frame structures counting for P-delta effect*. buildings 9(10), pp 1-16.

AUTHOR INDEX